FLORA

OF

TROPICAL EAST AFRICA

PREPARED AT THE ROYAL BOTANIC GARDENS/KEW
WITH ASSISTANCE FROM THE EAST AFRICAN HERBARIUM

EDITOR

R.M.POLHILL, B.A., Ph.D., F.L.S.

ORCHIDACEAE (Part 3)

BY

P. CRIBB, M.A., Ph.D.

CRC Press
Taylor & Francis Group
Boca Raton London New York

CRC Press is an imprint of the
Taylor & Francis Group, an **informa** business
A BALKEMA BOOK

CRC Press
Taylor & Francis Group
6000 Broken Sound Parkway NW, Suite 300
Boca Raton, FL 33487-2742

ISBN-13: 978-9-061-91353-5 (hbk)

Publisher's Note

The publisher has gone to great lengths to ensure the quality of this reprint but points out that some imperfections in the original copies may be apparent.

Visit the Taylor & Francis Web site at
http://www.taylorandfrancis.com

and the CRC Press Web site at
http://www.crcpress.com

FAMILIES OF VASCULAR PLANTS REPRESENTED IN THE FLORA OF TROPICAL EAST AFRICA

Foreword and preface (£0.90) Glossary (£1.80) Index of Collecting Localities (£15.00)

PTERIDOPHYTA

Aspidiaceae
Aspleniaceae
Blechnaceae
Cyatheaceae
Davalliaceae
Equisetaceae
Gleicheniaceae
Hymenophyllaceae
Isoetaceae
Lycopodiaceae
Marattiaceae
Marsileaceae
Ophioglossaceae
Osmundaceae
Parkeriaceae
Polypodiaceae
Psilotaceae
Pteridaceae
Salviniaceae
Schizaeaceae
Selaginellaceae
Vittariaceae

GYMNOSPERMAE (£1.80)

Cupressaceae
Cycadaceae
Podocarpaceae

ANGIOSPERMAE

Acanthaceae
Agavaceae
Aizoaceae (£1.80)
Alangiaceae (£0.90)
Alismataceae (£1.80)
Alliaceae
Aloeaceae — see Asphodelaceae
Amaranthaceae (£15.00)
Amaryllidaceae (£4.15)
Ampelidaceae — see Vitaceae
Anacardiaceae (£6.90)
Ancistrocladaceae (£1.50)
Annonaceae (£5.75)
Anthericaceae
Apocynaceae
Aponogetonaceae (£3.00)
Aquifoliaceae (£0.90)
Araceae (£8.25)
Araliaceae (£1.80)
Aristolochiaceae (£2.25)
Asclepiadaceae
Asparagaceae
Asphodelaceae
Avicenniaceae

Balanitaceae
Balanophoraceae
Balsaminaceae (£9.50)
Basellaceae (£0.90)
Begoniaceae
Berberidaceae (£0.90)
Bignoniaceae
Bischofiaceae — see Euphorbiaceae
Bixaceae (£0.90)
Bombacaceae (£3.00)
Boraginaceae
Brexiaceae (£0.90)
Buddlejaceae — see Loganiaceae
Burmanniaceae (£2.25)
Burseraceae
Butomaceae (£0.90)
Buxaceae (£0.90)

Cabombaceae (£0.90)
Cactaceae (£0.90)
Caesalpiniaceae — Leguminosae in part
Callitrichaceae
Campanulaceae (£1.80)
Canellaceae (£0.90)
Cannabaceae (£0.90)
Cannaceae
Capparidaceae (£3.95)
Caprifoliaceae (£0.90)
Caricaceae (£0.90)
Caryophyllaceae (£1.80)
Casuarinaceae (£1.65)
Cecropiaceae — see Moraceae
Celastraceae
Ceratophyllaceae (£1.65)
Chailletiaceae — see Dichapetalaceae
Chenopodiaceae (£1.80)
Chrysobalanaceae — see Rosaceae
Cochlospermaceae (£0.90)
Colchicaceae
Combretaceae (£4.50)
Commelinaceae
Compositae
Connaraceae (£1.80)
Convolvulaceae (£7.30)
Cornaceae (£0.90)
Costaceae — see Zingiberaceae
Crassulaceae (£9.00)
Cruciferae (£9.00)
Cucurbitaceae (£7.00)
Cyanastraceae — see Tecophilaeaceae
Cyclocheilaceae (£1.45)
Cyperaceae
Cytinaceae — see Rafflesiaceae

Dichapetalaceae (£3.00)
Dilleniaceae (£0.90)
Dioscoreaceae (£1.80)
Dipsacaceae (£1.80)
Dipterocarpaceae (£3.00)
Dracaenaceae
Droseraceae (£0.90)

Ebenaceae
Elatinaceae (£0.90)
Ericaceae
Eriocaulaceae
Eriospermaceae

Erythroxylaceae (£2.25)
Escalloniaceae (£0.90)
Euphorbiaceae
  Part 1 (£25.50)
  Part 2 (£18.00)

Fabaceae — see Leguminosae
Ficoidaceae — see Aizoaceae
Flacourtiaceae (£3.00)
Flagellariaceae (£0.90)
Fumariaceae (£0.90)

Gentianaceae
Geraniaceae (£1.80)
Gesneriaceae
Goodeniaceae (£0.90)
Gramineae (£53.60)
  Part 1 (£7.90)
  Part 2 (£12.30)
  Part 3 (£44.10)
Guttiferae (£1.80)

Haemodoraceae in part — see Agavaceae
Haloragaceae (£1.80)
Hamamelidaceae (£0.90)
Hernandiaceae (£2.25)
Hippocrateaceae — Celastraceae in part
Hugoniaceae — Linaceae in part
Hyacinthaceae
Hydnoraceae
Hydrocaryaceae — see Trapaceae
Hydrocharitaceae
Hydrophyllaceae (£1.50)
Hydrostachyaceae (£1.65)
Hymenocardiaceae — see Euphorbiaceae
Hypericaceae (£1.80)
Hypoxidaceae

Icacinaceae (£1.80)
Ilicaceae — see Aquifoliaceae
Illecebraceae — see Caryophyllaceae in part
Iridaceae
Irvingiaceae — see Ixonanthaceae in part
Ixonanthaceae (£2.25)

Juncaceae (£1.80)
Juncaginaceae (£0.90)

Labiatae
Lauraceae
Lecythidaceae (£0.90)
Leguminosae (£52.20)
  Part 1, Mimosoideae (£7.80)
  Part 2, Caesalpinioideae (£10.35)
  Part 3 }
  Part 4 } Papilionoideae (£47.70)
Lemnaceae (£1.80)
Lentibulariaceae (£1.80)
Linaceae (£1.80)
Lobeliaceae (£6.75)
Loganiaceae (£2.15)
Loranthaceae
Lythraceae

Malpighiaceae (£1.80)
Malvaceae
Marantaceae (£1.80)
Melastomataceae (£4.20)
Meliaceae
Melianthaceae (£0.90)
Menispermaceae (£1.80)
Menyanthaceae
Mimosaceae — Leguminosae in part
Molluginaceae — see Aizoaceae
Monimiaceae (£0.90)
Montiniaceae (£0.90)
Moraceae
Moringaceae (£2.25)
Musaceae
Myricaceae
Myristicaceae
Myrothamnaceae (£1.45)
Myrsinaceae (£3.45)
Myrtaceae

Najadaceae (£3.00)
Nectaropetalaceae — Erythroxylaceae in part
Nesogenaceae (£1.20)
Nyctaginaceae
Nymphaeaceae (£3.00)

Ochnaceae
Octoknemaceae — Olacaceae in part
Oenotheraceae — see Onagraceae
Olacaceae (£1.80)
Oleaceae (£1.80)
Oliniaceae (£0.90)
Onagraceae (£1.80)
Opiliaceae (£0.90)
Orchidaceae
  Part 1, Orchideae (£10.35)
  Part 2, Neottieae, Epidendreae (£18.00)
  Part 3, Epidendreae, Vandeae (£19.50)
Orobanchaceae (£0.90)
Oxalidaceae (£1.80)

Palmae (£6.90)
Pandaceae — see Euphorbiaceae
Pandanaceae
Papaveraceae (£0.90)
Papilionaceae — Leguminosae in part
Passifloraceae (£3.15)
Pedaliaceae (£1.80)
Periplocaceae — Asclepiadaceae in part
Phytolaccaceae (£0.90)
Piperaceae
Pittosporaceae (£1.80)
Plantaginaceae (£0.90)
Plumbaginaceae (£1.80)
Podostemaceae
Polygalaceae
Polygonaceae (£1.80)
Pontederiaceae (£0.90)
Portulacaceae
Potaliaceae — Loganiaceae in part
Potamogetonaceae
Primulaceae (£1.80)
Proteaceae
Ptaeroxylaceae

Rafflesiaceae
Ranunculaceae (£1.80)
Resedaceae (£0.90)
Rhamnaceae (£1.80)
Rhizophoraceae (£1.80)
Rosaceae (£2.80)
Rubiaceae
  Part 1 (£18.00)
  Part 2 (£25.50)
  Part 3
Ruppiaceae — see Potamogetonaceae
Rutaceae (£6.30)

FLORA OF TROPICAL EAST AFRICA

ORCHIDACEAE (Part 3)

P. J. CRIBB*

III. EPIDENDREAE (continued)

A key to the tribe *Epidendreae* is given in Part 2 of this account published in 1984. However, as this part will undoubtedly be used on occasion by those without immediate access to earlier parts of the Flora, a key to the genera of the *Epidendreae* included in this part is provided here.

KEY TO THE GENERA OF THE EPIDENDREAE IN PART 3

1. Plants epiphytic or lithophytic; some of the roots aerial, ascending around the bases of the pseudobulbs, tapering to fine points . . . . 43. **Graphorkis**
   Plants terrestrial 2
2. Lip 4-lobed, the lobes subequal; spur subglobose to cylindric . . . . 44. **Oeceoclades**
   Lip 3-lobed or entire; spur absent, saccate, clavate or cylindrical 3
3. Spur absent; lip deeply 3-lobed at the base with the side lobes almost free to the base; column very short, biauriculate, lacking a foot . . . . 46. **Pteroglossaspis**
   Spur present, saccate, clavate or cylindrical, or rarely absent; lip 3-lobed in the middle or apical half with the side lobes united to the rest of the lip for most of their length; column usually slender, elongate, lacking auricles, with a short to long foot . . . . 45. **Eulophia**

43. **GRAPHORKIS****

Thou. in Nouv. Bull. Sci. Soc. Philom. Paris 1: 318 (1809)

Epiphytic, sympodial herbs. Pseudobulbs clustered, conical-ovoid to cylindrical-fusiform, with several nodes, partially covered by fibrous persistent leaf-bases; roots of two types, those attached to substrate and those that are aerial, which are narrow, clustered and vertical. Leaves narrowly elliptic, acute to acuminate, plicate, petiolate. Inflorescence proteranthous, erect, paniculate, many-flowered; lowermost bracts conspicuous, uppermost bracts inconspicuous. Flowers resupinate, small to medium-sized, yellow, marked brown or purple. Sepals and petals free, spreading. Sepals spathulate to oblong-elliptic, subacute. Petals narrowly elliptic. Lip 3-lobed, spurred at base; disc with 2 keels; mid-lobe crenulate to bifid; spur often bent forward. Column with hirsute basal auricles; rostellum elongate; pollinia 2, waxy, sessile, attached to a solitary viscidium.

Five species in Madagascar and the Mascarene Is., and a single species in tropical Africa.

* With contributions by J.J. Wood, J. Stewart & S. Robbins

**By J. J. Wood

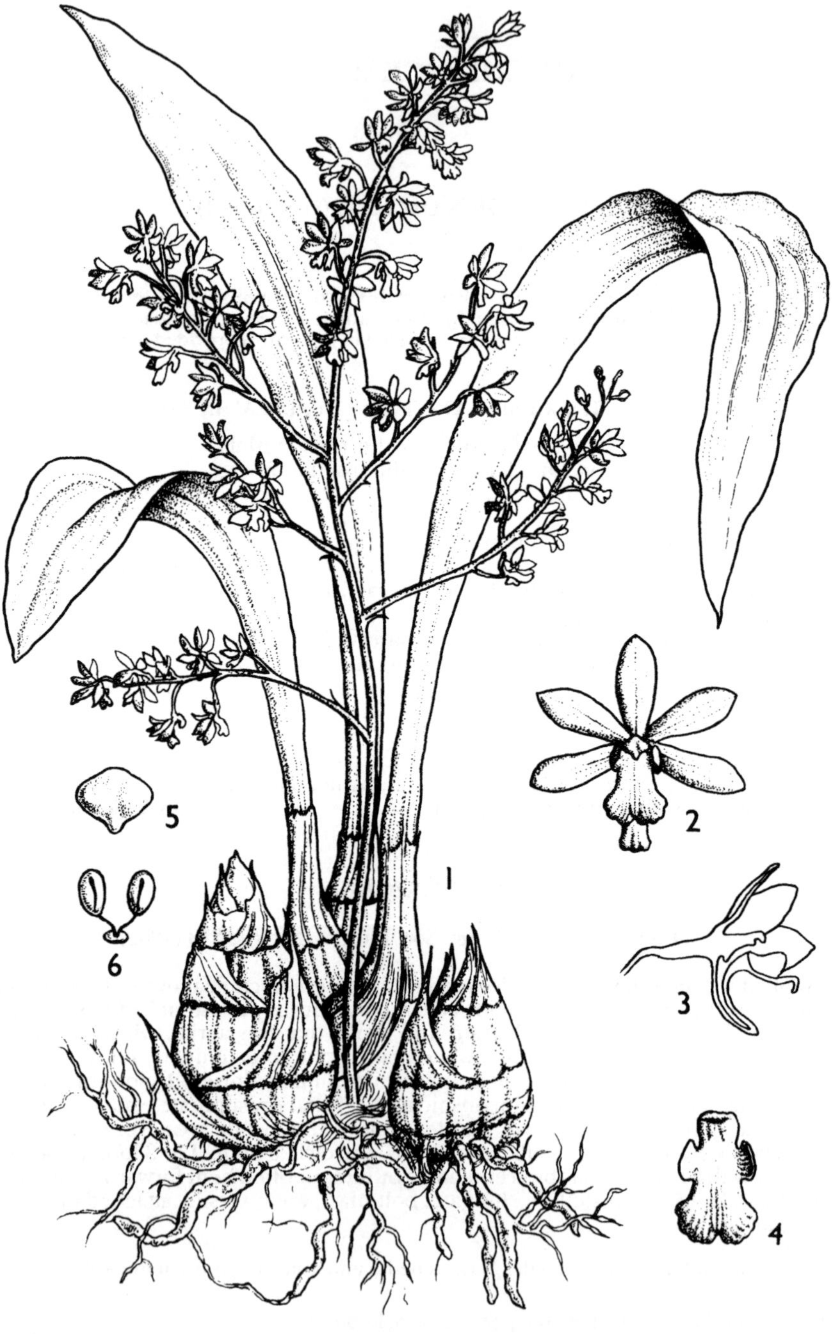

FIG. 89. *GRAPHORKIS LURIDA*—**1**, habit, × ⅔; **2**, flower, × 3⅔; **3**, longitudinal section of flower, × 2⅔; **4**, lip, × 9⅓; **5**, anther, × 14; **6**, pollinarium, × 14; 1, from *Brown* 361, 2–6, from *Eggeling* 5183. Drawn by Cherry Ann Lavrih.

G. lurida (*Sw.*) *O. Kuntze*, Rev. Gen. Pl. 2: 662 (1891); Summerh. in K.B. 8: 161 (1953); Piers, Orch. E. Afr., ed. 2: 191 (1968); Summerh. in F.W.T.A., ed. 2, 3: 251 (1968); Williamson, Orch. SC. Afr.: 150 (1977). Type: Sierra Leone, without exact locality, *Afzelius* (UPS, holo.)

An erect epiphytic herb. Pseudobulbs cylindrical-fusiform or conical-ovoid, yellowish, with 3–5 nodes, 3–9 cm. long, 1–3 cm. broad, 4–6-leaved. Leaves narrowly elliptic, acute to acuminate, plicate, articulated at a persistent sheathing base, thin-textured; median and some secondary nerves prominent and hard, partially remaining as sharp, rigid thorns after lamina is shed; sheaths blackish brown along margins and at base, becoming scarious and fibrous, 2–3(–5) cm. long; lamina, including petiole, 20–40 cm. long, 1.5–3.5(–4) cm. broad. Inflorescence erect, paniculate, 15–50 cm. long, appearing from base of pseudobulbs; branches spreading, 3–22 cm. long, many-flowered; sterile bracts 3–4, remote, ovate-elliptic, obtuse to acute, amplexicaul, 1.5–4 cm. long; bracts subtending branches ovate-elliptic or narrowly elliptic, acute, 0.8–2.5 cm. long; fertile bracts triangular-ovate, acute, 0.1–0.3 cm. long. Flowers with sepals purple or brown on the outside, pale green inside, petals pale green or cream flushed purple or brown, lip with pale greenish yellow side lobes and deep yellow mid-lobe. Pedicel with ovary slender, 10–15 mm. long. Sepals spathulate-oblong, obtuse, 5–6.5 mm. long, 1.5–2 mm. broad. Petals elliptic, obtuse, 4–6 mm. long, 2.5–3 mm. broad. Lip 3-lobed, 5–6 mm. long, 3 mm. broad; mid-lobe retuse to bifid; side lobes oblong, obtuse, erect; disc with 2 fleshy keels; spur cylindrical, obtuse, sharply bent forward, 3–4 mm. long. Column erect, oblong, biauriculate at base, 3 mm. long; anther ovate. Pl. 1. Fig. 89.

UGANDA. Bunyoro District: Busingiro, 1944, *Eggeling* 456!; Busoga District: between Kenya border and Jinja, 1968, *Piers*!; Mengo District: Entebbe, Nov. 1905, *E. Brown* 361!
TANZANIA. Ufipa District: near Lake Tanganyika, Feb. 1976, *C. Owens* in *Cribb*!
DISTR. **U** 2–4; **T** 4; Senegal, Guinea-Bissau, Guinée, Sierra Leone, Liberia, Ivory Coast, Ghana, Togo, Nigeria, Bioko [Fernando Po], Cameroon, Equatorial Guinea, Gabon, Zaire, Burundi
HAB. In forest, often epiphytic on palms, particularly *Hyphaene*; 1050–1300 m.

SYN. *Limodorum luridum* Sw. in Schrad., Neues Journ. Bot. 1: 87 (1805)
Eulophia lurida (Sw.) Lindl., Gen. Sp. Orch. Pl.: 182 (1833); Rolfe in F.T.A. 7: 55 (1897)
Eulophiopsis lurida (Sw.) Schltr., Die Orchideen: 348 (1914); Summerh. in F.W.T.A. 2: 446 (1936); Moreau in Journ. E. Afr. Nat. Hist. Soc. 17: 27 (1943); Piers, Orch. E. Afr: 66 (1959)

44. OECEOCLADES*

Lindl. in Bot. Reg. 18, sub t. 1522 (1832)

Aeceoclades Duchartre in Orbigny, Dict. 9: 170 (1849)
Eulophidium Pfitz., Entw. Natur. Anordn. Orch.: 87–89 (1887)

Terrestrial, rarely epiphytic herbs. Pseudobulbs close together, usually ovoid to fusiform, ± approximate, usually heteroblastic (with only one internode elongated, the remaining basal ones very short), apex 1–3-leaved, up to 15 cm. long and 3 cm. broad, but often narrower. Leaves usually with duplicate vernation, coriaceous, conduplicate, often variegated, usually petiolate, the petiole articulate some distance above the base and sometimes above the middle, the line of articulation consisting of a number of irregular blunt or acute teeth or occasionally ± regular. Inflorescences arising from the base of the pseudobulb, often exceeding the leaves, simply racemose or frequently paniculate; bracts inconspicuous, rarely with a basal extrafloral nectary. Flowers resupinate, rather small, thin in texture. Sepals and petals free, variously spreading, similar, the petals usually slightly shorter and broader. Lip decurved, spurred, 3- or apparently 4-lobed; side lobes erect; mid-lobe usually lobulate or emarginate; disc either with 2 approximate, quadrate or triangular calli at the spur entrance or with 3 variously thickened, parallel ridges which together with the lateral nerves are sparsely but distinctly papillose or hirsute. Column erect, short, oblique at the base or with a short foot; anther cucullate or cristate; pollinia 2, ovoid or pyriform, on a short or rudimentary stipe; viscidium large; stigmata confluent; rostellum short.

* By J.J. Wood

A genus of 31 species native to the tropics and subtropics of Africa, Madagascar, the Mascarene Is., Seychelles, SE. Asia, Australia, the Pacific Is., the West Indies, the Bahamas and South America.

Lindley established the genus *Oeceoclades* in 1832 and it is lectotypified by *Angraecum maculatum* Lindl. In 1887 Pfitzer established *Eulophidium* which he also based on *Angraecum maculatum* Lindl. while fully aware of Lindley's suggestion that the name *Oeceoclades* be applied only to *O. maculata*. Since both genera, *Oceoclades* Lindl. and *Eulophidium* Pfitz., are based on the same type, *Angraecum maculatum* Lindl., *Oeceoclades* was reinstated by Garay & Taylor (1976) because of the rules of priority.

Leaf or leaves plicate . . . . 7. *O. pulchra*
Leaf or leaves coriaceous:
  Pseudobulbs always 1-leaved; leaves petiolate, never broad and entirely green; disc and calli of lip entirely glabrous:
    Leaves greyish green with darker green banding or mottling, oblong-lanceolate or oblong-elliptic, usually 15–28 cm. long, petiole 1–8 cm. long, regularly articulated; pseudobulbs 2–4 cm. long; lip with 2 divergent basal keels, spur oblong-clavate; roots smooth . . . . 1. *O. maculata*
    Leaves entirely green, linear-lanceolate to broadly lanceolate, usually 30–45 cm. long, petiole 5.5–10 (–20) cm. long, irregularly articulated; pseudobulbs 5–7.5 cm. long; lip with 2 parallel basal keels, spur cylindrical; roots verruculose . . . . 2. *O. zanzibarica*
  Pseudobulbs usually 2- or 3-leaved; leaves sessile or petiolate, sometimes broad and entirely green; disc and calli of lip hirsute, papillose or glabrous:
    Leaves linear-ligulate to lanceolate, sessile, 1–1.2(–1.9) cm. broad, greyish green, mottled bluish grey and whitish grey; lateral sepals distinctly spathulate, 14–20 mm. long; lateral lobes of lip somewhat saccate at the base, callus glabrous . . . . 3. *O. decaryana*
    Leaves narrowly elliptic to ovate-elliptic, petiolate, more than 2.5 cm. broad, mottled or entirely green; lateral sepals not distinctly spathulate; lateral lobes of lip not saccate at the base, disc and callus papillose or hirsute:
      Leaves with pronounced pale green or white mottling; lip broader than long, calli divergent, spur strongly curved outwards towards underside of lip . . . . 4. *O. lubbersiana*
      Leaves entirely green; lip longer than broad, calli parallel, spur straight, descending:
        Plant (40–)50–80 cm. high; pseudobulbs caespitose; leaves (10–)13–18(–26) × (3–)4.5–6(–8) cm., petiole 6–7 cm. long; inflorescence 16–22(–30) cm. long, with extrafloral nectaries exuding viscid liquid; flowers large, greenish yellow with vinous nerves; lateral sepals 9.7–14 mm. long; lip 10–15 mm. long, spur cylindrical . . . . . 5. *O. saundersiana*
        Plant 12–21(–26) cm. high; pseudobulbs superposed; leaves (6–)8–11 × 2.5–4.5 cm., petiole 2–6 cm. long; inflorescence (6–)8–14 cm. long, without nectaries; flowers small, mostly white, with purple on lip; lateral sepals 7–9 mm. long; lip 7–8 mm. long, spur clavate . . . . 6. *O. ugandae*

1. **O. maculata** (*Lindl.*) *Lindl.*, Gen. Sp. Orch. Pl.: 237 (1833); Summerh. in B.J.B.B. 27: 397 (1957); Garay & Taylor in Bot. Mus. Leafl. Harv. Univ. 24: 266 (1976). Type: Brazil, without precise locality, cultivated *Loddiges* 34.10.16 (BM, holo.)

A terrestrial or epiphytic herb (16–)28–30(–40) cm. high, entirely glabrous; roots stout, verruculose. Pseudobulbs congested, ovoid, ovoid-oblong or stout and cylindrical, (2–)

2.5–4 cm. long, 1-leaved, concealed and exceeded by several pale brown, ovate-elliptic, acute, imbricate, evanescent cataphylls. Leaves oblong-oblanceolate or oblong-elliptic, acute, duplicate, particularly towards base; lamina (10–)15–28(–40) cm. long, 2.5–4.5(–6) cm. broad, with the median nerve prominent on lower surface, greyish green, with darker green transverse, irregular banding or mottling; petiole 1–8 cm., regularly articulate. Inflorescence simple or with 1 or 2 short branches, 5–12(–20) cm. long, 3–4 cm. broad, erect, lax, 6–12(–20)-flowered, sometimes branched at base; bracts lanceolate, acuminate, scarious, lowermost to 17 mm. long, uppermost to 4 mm. long. Flowers porrect; sepals and petals translucent greenish white or pinkish green, sometimes straw-coloured; lip greenish white or cream with rose-pink nerves on the side lobes and 2 rose-pink blotches at the base of the mid-lobe; calli white; spur greenish white, flushed rose-pink or brownish green; column cream; pedicel and ovary slender, 10–13(–15) mm. long. Dorsal sepal linear-ligulate or linear-elliptic, obtuse to acute, 3-nerved, incurved, (8–)10–12(–13) mm. long, 2–3.2 mm. broad; lateral sepals linear-ligulate or narrowly oblanceolate-oblong, falcate, obtuse to acute, 3-nerved, (6–)7.5–9(–11) mm. long, 2–3.3(–4.6) mm. broad. Petals oblong-elliptic, oblique, acute, 3-nerved, broader than sepals, (8.3–)10–11.2 mm. long, (2–)3.3–4.6 mm. broad. Lip projected forwards, articulated at the base, distinctly 3-lobed, subquadrate in outline, glabrous, 7–9 mm. long; mid-lobe broadly obovate or reniform, abruptly and shortly clawed at the base, lobulate, retuse or ± truncate, longer than side lobes, 3.5–4.2 mm. long, (6–)7–8(–10) mm. broad; side lobes semiobovate, rounded, not saccate at the base, erect, somewhat incurved, nerves not prominent, shorter than mid-lobe; disc with 2 approximate, obliquely semiobovate, divergent basal keels, glabrous; spur short, stout, oblong-clavate, rounded, straight, 4–5 mm. long. Column erect, oblong, oblique at base or with a short foot, 4–5 mm. long, 2 mm. broad; anther oblong, ± 2.2 mm. long.

UGANDA. Kigezi District: Impenetrable Forest, 1945, *C.G. van Someren*!; Busoga District: Igwe, 26 May 1951, *G.H. Wood* 265!

TANZANIA. Ulanga District: Mahenge, Mbangala, 20 Feb. 1932, *Schlieben* 1803!; Rungwe District: Mbaka Kilambo, 6 Mar. 1913, *Stolz* 1909!; Zanzibar I., Mvuleni, May 1933, *Vaughan* 2110!; Pemba I., Micheweni, 20 Apr. 1957, *Tremlett* 2!

DISTR. **U** 2, 3; **T** 6, 7; **Z**; **P**; Senegal, Guinea-Bissau, Sierra Leone, Liberia, Ghana, Togo, Nigeria, São Tomé, Gabon, Zaire, Congo-Brazzaville, Burundi, Sudan, Zambia, Zimbabwe, Angola, also U.S.A. (Florida), Bahamas, Dominican Republic, West Indies, Brazil, Guyana, Paraguay, Trinidad, Venezuela, Colombia, Peru, Bolivia, Argentina

HAB. Amongst leaf-mould in shade in rain-forest, rocky scrub near the sea, shady river banks or epiphytic on savannah trees; 0–1200 m.

SYN. *Angraecum maculatum* Lindl., Coll. Bot., t. 15 (May 1821)
Limodorum maculatum Lodd., Bot. Cab. 5, t. 496 (June 1821)
Geodorum pictum Link & Otto, Ic. Pl. Sel. 3: 35, t. 14 (July 1821). Lectotype chosen by Garay & Taylor (1976): plate published with protologue of a plant from Brazil, without precise locality, grown in Berlin (specimen at B†)
Aerobion maculatum (Lindl.) Sprengel, Syst. Veg. 3: 718 (1826)
Eulophia maculata (Lindl.) Reichb.f. in Walp., Ann. Bot. 6: 647 (1863)
Eulophidium maculatum (Lindl.) Pfitz., Entw. Nat. Anordn. Orch.: 88 (1887); Piers, Orch. E. Afr., ed. 2: 193 (1968); Summerh. in F.W.T.A., ed. 2, 3: 243 (1968); Williamson, Orch. SC. Afr.: 149 (1977)
Graphorchis maculata (Lindl.) O. Kuntze., Rev. Gen. Pl. 2: 662 (1891)
Eulophia ledienii N.E. Br. in K.B. 1889: 90 (1889); Rolfe in F.T.A. 7: 50 (1897). Type: Zaire, Lower Congo, *Ledien* (WRSL, holo.)
Eulophidium warneckeanum Kraenzl. in E.J. 33: 70 (1902). Type: Togo, near Lome, *Warnecke* 196 (B, holo.†)
E. ledienii (N.E.Br.) De Wild. in Ann. Mus. Congo, Bot., sér. 5, 1: 115 (1904); Robyns & Tournay, F.P.N.A. 3: 497 (1955); Piers, Orch. E. Afr., ed. 2: 193 (1968)
E. nyassanum Schltr. in E.J. 53: 593 (1915). Type: Tanzania, Rungwe District, near Mbaka Kilambo and in Mulinda Forest, *Stolz* 1909 (BM, K, iso.!)

2. **O. zanzibarica** (*Summerh.*) *Garay & Taylor* in Bot. Mus. Leafl. Harv. Univ. 24: 273 (1976). Type: Zanzibar, without precise locality, 1908, *Last* (K, holo.!)

A terrestrial herb 25–45(–70) cm. high, entirely glabrous. Roots smooth, fleshy, white, 3–5 mm. broad. Pseudobulbs cylindrical, 5–7.5 cm. long, 1.5–2 cm. broad, 1-leaved, enclosed when young by a few scarious, straw-coloured, ovate to lanceolate, acuminate, evanescent sheaths. Leaves linear-lanceolate to broadly lanceolate, acute or acuminate, conduplicate and cuneate towards base; lamina (7–)30–45 cm. long, 2–5(–6.5) cm. broad, green; petiole irregularly articulated, 5.5–10(–20) cm. long. Inflorescence erect, paniculate, elongated, 8–35(–40) cm. long, 4–6 cm. broad, many-flowered, lax or subdense; lowermost bracts narrowly elliptic, uppermost linear-subulate, acute, 1–15(–20) mm. long. Flowers with green sepals and petals; lip white or cream with purplish-red veins on the side lobes and 2 purple blotches at base of mid-lobe; pedicel and ovary slender, 1–1.8 cm. long. Dorsal sepal oblong-elliptic, mucronate, 3-nerved, 8–10.5 mm. long, 3–4.5 mm. broad, erect; lateral sepals oblique- or falcate-oblanceolate, mucronate, 4-nerved, 8–10.2 mm. long, 3–5 mm. broad, spreading. Petals oblong-elliptic, acute or mucronate, 3-nerved, 7.5–9.5 mm. long, 3–4.7 mm. broad. Lip not articulated at the base, distinctly 3-lobed, glabrous, 8–10.5 mm. long, 8–12.5 mm. broad; mid-lobe divided into 2 distinct, obliquely obovate or subquadrate, divaricate lobules, each 4–6 mm. long, with a mucro in the sinus between; side lobes oblong, rounded, not saccate at base, erect, shorter than mid-lobe; disc with 2 somewhat flattened, parallel basal keels and 2 distal slightly raised, papillose nerves; spur cylindrical, straight, obtuse, descending, 3–4 mm. long. Column porrect, oblong, 2.5 mm. long; foot short, 1–2 mm. long; anther quadrate, emarginate, glabrous, 1.5 mm. long. Pl. 1.

TANZANIA. Uzaramo District: near Dar es Salaam, Ubungo I., *Segerbäck* 1003! (colour slides only) & *Barrett* s.n.! (colour slides only); Zanzibar I., Chwaka, 25 Dec. 1930, *Vaughan* 1766!; Pemba I., Ras Kigomasha, 10 Dec. 1930, *Greenway* 2693!

DISTR. **T** 6; **Z**; **P**; not known elsewhere

HAB. In dense shade under mango trees, *Pandanus* forest over coral rocks, in water holes; near sea-level

SYN. *Eulophidium zanzibaricum* Summerh. in K.B. 1927: 417 (1927) & in B.J.B.B. 27: 403 (1957); Piers, Orch. E. Afr., ed. 2: 196 (1968)

3. **O. decaryana** (*H. Perr.*) *Garay & Taylor* in Bot. Mus. Leafl. Harv. Univ. 24: 262 (1976). Type: Madagascar, without proper locality, *Decary*, cultivated in Museum d'Histoire Naturelle in Paris, No. K 467 (P, holo.)

A terrestrial herb 50–60(–80) cm. high, entirely glabrous. Pseudobulbs ovoid, 5-angled, 4 cm. long, 2.3 cm. broad, (1–)2(–3)-leaved. Leaves linear-ligulate to lanceolate, acute, abbreviated towards the base, sessile, rigid, median nerve prominent on lower surface, (12–)15–23 cm. long, 1–1.2(–1.9) cm. broad, spreading, persisting for 2 or 3 seasons, greyish green, mottled bluish grey and whitish grey, sometimes suffused brown. Inflorescence simple, erect, 22–35(–60) cm. long, 3.5 cm. broad, laxly 12–20-flowered; bracts linear to linear-lanceolate, acute, 3-nerved, shorter than ovary, 6–15 mm. long. Flowers porrect; sepals and petals olive-green, with red or brownish purple nerves and markings; lip either bright yellow with red nerves or with the proximal lobes green with purple nerves and the distal lobes cream or pale green with purple nerves and a central yellow patch, calli lilac, spur green or orange; column yellowish green; pedicel and ovary slender, 12–25 mm. long. Dorsal sepal erect, oblanceolate to spathulate, apiculate, 3-nerved, 12–20 mm. long, 3–4 mm. broad; lateral sepals spathulate, apiculate, often somewhat falcate, deflexed, 14–20 mm. long, 2.7–4 mm. broad. Petals oblong-elliptic or obliquely ovate-elliptic, apiculate, 3-nerved, 7–10 mm. long, 4–7 mm. broad. Lip projecting forwards, articulated at the base, distinctly 3-lobed, rarely obscurely 4-lobed, 12–16 mm. long, 13–18 mm. broad; mid-lobe divided into 2 lobules, each broadly oblong or ± orbicular, somewhat deflexed, divaricate, margins sometimes gently undulate, nerves many, inconspicuous, vinous, longer than the side lobes; side lobes subquadrate or rounded, somewhat saccate at the base, erect, nerves 3–5, prominent, vinous; callus at the base of the lip, formed of 2 raised, horizontal, oblong, glabrous lobes which gradually extend towards the apical sinus as 2 minutely papillose, low, parallel keels, between which may be a third median keel; spur stout, cylindrical, often slightly dilated at the middle, weakly carinate, decurved, 4–6 mm. long. Column erect, ovoid-oblong, 4–5 mm. long, 3 mm. broad, with a short foot ± 2 mm. long; anther ± ovate, emarginate at apex, 2 mm. long and broad.

KENYA. Teita District: Sagala Hill, 22 June 1964, *Archer* 530!

DISTR. **K** 7; Mozambique, Zimbabwe, Madagascar

HAB. Rocky outcrop, in leaf-mould under bushes; 960 m.

SYN. *Eulophia decaryana* Perrier in Bull. Soc. Bot. Fr. 82: 154 (1935)
Lissochilus decaryanus (Perrier) Perrier in Humbert, Fl. Madag., Orch. 2: 32 (1941)
Eulophidium decaryanum (Perrier) Summerh. in B.J.B.B. 27: 395 (1957)

4. **O. lubbersiana** (*De Wild. & Laurent*) *Garay & Taylor* in Bot. Mus. Leafl. Harv. Univ. 24: 265 (1976). Type: Zaire, Sankuru, *Laurent* (BR, holo., K, iso.!)

A terrestrial herb 34–50 cm. high, glabrous except for the lip and column-foot. Pseudobulbs fusiform to ovoid, 2–4 cm. long, 1.5 cm. broad, 2-leaved. Leaves ovate-elliptic or elliptic, acute; lamina 6–10 cm. long, (2–)2.5–3.8 cm. broad, spreading, the median nerve prominent on lower surface, dark green with pronounced paler green or white mottling; petiole 3.5–8 cm. long, irregularly articulate. Inflorescence erect, simple, (8–) 12–14 cm. long, 3–4 cm. broad, lax or subdense, 8–18-flowered; lower bracts linear-lanceolate, acute; upper bracts triangular-ovate, acute, scarious, shorter than ovary, 2–6 mm. long. Flowers porrect or ± horizontal; sepals olive-green, whitish at base; petals olive-green or yellowish, with plum-purple spots and nerves; middle lobes of lip bright yellow, side lobes cream, with plum-purple nerves, disc white, spotted plum-purple, spur pale green; column green or white, with a plum-purple margin; anther yellow; pedicel and ovary slender, 7–12 mm. long. Dorsal sepal ligulate or narrowly oblanceolate, obtuse, 1-nerved, reflexed, 6–9 mm. long, 2–3.5 mm. broad; lateral sepals ligulate or ± subfalcate, often somewhat spathulate, obtuse, 1-nerved, spreading or deflexed, 7–9 mm. long, 2–2.5 mm. broad. Petals oblong-elliptic, apiculate, 3-nerved, 6.3–8 mm. long, 3.7–4 mm. broad. Lip projecting forwards, articulated at the base, distinctly 3-lobed, with hairs at the base and on the disc, otherwise glabrous, broader than long, 7.5–9 mm. long, 10–12.5 mm. broad; mid-lobe divided into 2 lobules, each flabellate, cuneate at the base, truncate, deflexed, divaricate, margins somewhat erose, nerves inconspicuous, longer than the side lobes; side lobes oblong, rounded, not saccate at the base, erect, nerves prominent; disc minutely papillose, hairy around the callus; basal callus formed of 2 low, rounded, papillose or hirsute, divergent keels, with longer hairs at their base; spur cylindrical, slightly attenuated towards apex, strongly curved outwards towards underside of lip, 2–3(–4) mm. long. Column erect, oblong, 4 mm. long, 2 mm. broad, with a short foot ± 1.5 mm. long; anther ovate, emarginate at apex, 1 mm. long and broad.

UGANDA. Bunyoro District: Budongo Forest, Apr. 1933, *Eggeling* 1295!; Toro District: Bwamba Forest, 20 Apr. 1971, *J. Stewart* 1356!
DISTR. U 2; Zaire
HAB. Forest, in dense shade; 1050 m.

SYN. *Eulophia lubbersiana* De Wild. & Laurent in Rev. Hort. Belg. 26: 4 (1900); Robyns & Tournay, F.P.N.A. 3: 503 (1955)
Eulophidium lubbersianum (De Wild. & Laurent) Summerh. in B.J.B.B. 27: 397 (1957)

5. **O. saundersiana** (*Reichb.f.*) *Garay & Taylor* in Bot. Mus. Leafl. Harv. Univ. 24: 270 (1976); P. Taylor in Bot. Mag. 181, t. 731 (1977). Type: Africa, West Coast, without precise locality, *Mann*, cult. Saunders (W, holo.)

A terrestrial herb (40–)50–80 cm. high, glabrous except for the callus on the lip. Roots numerous, smooth, fleshy, white, 3–4 mm. broad. Pseudobulbs caespitose, narrowly ovoid to fusiform, very slightly compressed, obscurely sulcate, erect or spreading, (4–)8–15 cm. long, the widest part up to 2.5 cm. broad, dark, slightly glossy green, 2- or occasionally 1-leaved, the base enclosed when young by 2 or 3 scarious, straw-coloured evanescent sheaths. Leaves narrowly elliptic or ovate-elliptic, sometimes oblique, acute or acuminate, margin gently undulate, obscurely nerved; lamina spreading, (10–)13–18(–26) cm. long, (3–)4.5–6(–8) cm. broad, slightly glossy dark green; petiole irregularly articulated, shorter than the lamina, 6–17 cm. long. Inflorescence erect, simple or with 1 or 2 short branches at the base, 16–22(–30) cm. long, 4–5 cm. broad, dense or lax, 15–28-flowered; bracts ovate-deltoid to narrowly ovate-elliptic, acute, 2–8 mm. long, with a basal extrafloral nectary exuding copious viscid sweet liquid. Flowers porrect at first, becoming somewhat deflexed with age; sepals and petals pale greenish yellow with 3–5 vinous nerves; lip pale yellow or greenish cream with vinous nerves on the lobes, green towards the spur-entrance; calli creamy white; spur reddish brown; column green, anther yellow; pedicel and ovary slender, 12–18 mm. long, reddish brown. Dorsal sepal narrowly oblong or oblong-obovate, obtuse or mucronate, 3-nerved, erect or somewhat reflexed, 8.5–10.5 (–14) mm. long, 3–5 mm. broad; lateral sepals narrowly oblong or oblong-obovate, obtuse or mucronate, 4-nerved, slightly concave distally, spreading, (9.7–)10.5–14 mm. long, 3–5 mm. broad. Petals ovate, obtuse, 5-nerved, curving forwards, (8–)9–13 mm. long, (4.5–)5–7 mm. broad. Lip bent forwards from and articulated at the base, distinctly 3-lobed, mostly glabrous, (10–)12–15 mm. long; mid-lobe divided into 2 lobules, each oblong to oblong-orbicular, usually divaricate, longer than side lobes, 8–12 mm. long, (7–)11–12(–14) mm. broad; side lobes ± orbicular, not saccate at the base, erect to spreading; disc proximally callose, papillose, sometimes with 3 or 4 low, longitudinal ridges distally; callus composed of 2 ± parallel fleshy, glabrous keels, shortly pubescent at the spur-entrance and at the apex; spur cylindrical, obtuse, descending, 4–6 mm. long. Column erect, oblong or ovoid, with a very short pubescent foot, 2.5–5 mm. long, foot 1 mm. long; anther ovate-oblong, papillose, ± 2 mm. long. Pl. 1. Fig. 90.

UGANDA. Bunyoro District: Budongo Forest, 12 July 1945, *A.S. Thomas* 4314!; Toro District: Bwamba, 28 Feb. 1939, *A.S. Thomas* 2787!; Masaka District: Malabigambo Forest, 3 km. SSW. of Katera, 3 Oct. 1953, *Drummond & Hemsley* 4587!
KENYA. Kwale District: Kwale Camp, *Forestry Dept.*, cult. Kew, Apr. 1908! & Shimba Hills, 22 Sept. 1962, *Archer* 411!; Kilifi District: Rabai, Nov. 1962, *Castelino* H347/62!
TANZANIA. Bukoba District: Minziro Forest, 27 Apr. 1960, *Procter* 1158!; Tanga District: Kange Gorge, 21 Mar. 1956, *Faulkner* 1830!; Ulanga District: Ujiji, 10 Sept. 1960, *Haerdi* 586/0!
DISTR. **U** 2, 4; **K** 7; **T** 1, 3, 6, 7; Sierra Leone, Liberia, Ivory Coast, Ghana, Nigeria, Cameroon, Gabon, Zaire, Zambia, Angola
HAB. Damp shady places in forest and thickets; sea-level–1200 m.

SYN. *Eulophia saundersiana* Reichb.f. in Bot. Zeit. 24: 378 (1866) & in Xenia Orch. 2: 174, t. 173 (1873); Rolfe in F.T.A. 7: 50 (1897); Robyns & Tournay, F.P.N.A. 3: 503 (1955)
Graphorchis saundersiana (Reichb.f.) O. Kuntze, Rev. Gen. Pl. 2: 662 (1891)
Lissochilus barombensis Kraenzl. in E.J. 17: 52 (1893). Type: Cameroon, Barombi, *Preuss* 546 (B, holo.†, K, iso.!)
Eulophia bieleri De Wild., Not. Pl. Util. Congo 1: 311 (1904). Type: Zaire, Mbandaka [Coquilhatville], 1903, *Bieler* in *Laurent* 206; cultivated at Eala, 1904 (BR, lecto.)
E. mildbraedii Kraenzl. in E.J. 43: 339 (1909). Type: Zaire, Semliki Plains, *Mildbraed* 2751 (B, holo.†, K, iso.!)
Eulophidium saundersianum (Reichb.f.) Summerh. in B.J.B.B. 27: 401 (1957); Piers, Orch. E. Afr., ed. 2: 193 (1968); Summerh. in F.W.T.A., ed. 2, 3: 243 (1968); U.K.W.F.: 782 (1974); Williamson, Orch. SC. Afr.: 148 (1977)

6. **O. ugandae** (*Rolfe*) *Garay & Taylor* in Bot. Mus. Leafl. Harv. Univ. 24: 273 (1976). Type: Uganda, Mengo District, Mabira Forest, *E. Brown* 443 (K, holo.!)

A terrestrial, occasionally epiphytic herb 12–21(–26) cm. high, glabrous except for the lip; roots smooth, fleshy, white, 3–5 mm. broad. Stem climbing. Pseudobulbs superposed, ± fusiform, attenuate towards apex, 7.5–11 cm. long, the widest part 5–8 mm. broad, 2–3-leaved, enclosed when young by several scarious, straw-coloured, tubular ovate-elliptic, acute, evanescent sheaths. Leaves narrowly elliptic or ovate-elliptic, acute; lamina (6–)8–11 cm. long, 2.5–4.5 cm. broad, green; petiole irregularly articulated, shorter than the lamina, 2–6 cm. long. Inflorescence erect, paniculate, compact, (6–)8–14 cm. long, 2–3 cm. broad, many-flowered, lax or subdense; lowermost bracts ovate-elliptic to narrowly elliptic; uppermost linear-subulate, acute, 4–7(–15) mm. long. Flowers small; sepals and petals white, sometimes flushed green; lip white, with a few purple nerves and a purple throat; column with a few purple flecks; pedicel and ovary 6–13 mm. long. Dorsal sepal oblong, ± acute, 3-nerved, 6.5–8.5 mm. long, 2 mm. broad; lateral sepals oblong or narrowly oblanceolate, acute, slightly carinate at apex, 3-nerved, 7–8(–9) mm. long, 2–2.5 mm. broad. Petals oblong-elliptic, mucronate, 3-nerved, 6–7.8 mm. long, 1.5–2.3 mm. broad. Lip not articulated at the base, distinctly 3-lobed, with very short hairs at the spur-entrance and along the nerves, otherwise glabrous, 7–8 mm. long, 4.3–6.2 mm. broad; mid-lobe truncate, emarginate or divided into 2 very short lobules, broadly obovate, somewhat undulate, 2.7–3 mm. long, 4 mm. broad; side lobes oblong, obtuse, not saccate at the base; disc with 2 rather obscure, low, parallel keels, otherwise smooth; spur short, clavate, much dilated and globose at the apex, descending, 3–4 mm. long. Column porrect or erect, oblong to clavate, with a very short foot, 3–5.5 mm. long, foot ± 1 mm. long; anther ovate-oblong, papillose at apex, 1.3 mm. long.

UGANDA. Kigezi District: Mitano Gorge, Feb. 1947, *Purseglove* 2336!; Mengo District: Kisubi, 3 Sept. 1946, *A.S. Thomas* 4517!; Mubende District: near Kiryanga, Oct. 1950, *Eggeling* 5999!
DISTR. **U** 2, 4; Ghana, Zaire and Ethiopia
HAB. Forest; 1170–1200 m.

SYN. *Eulophia ugandae* Rolfe in K.B. 1913: 339 (1913)
[*Eulophidium latifolium* sensu Summerh. in B.J.B.B. 27: 396 (1957) & in F.W.T.A., ed. 2, 3: 205 (1968), pro parte, *non* (Rolfe) Summerh. sensu stricto]

7. **O. pulchra** *Clements & Cribb*, Cat. Austr. Orch., ed. 2 (in press). Type: Réunion, *du Petit Thouars* (P, holo.)

A terrestrial herb 60–70 cm. tall. Perennating organs pseudobulbous, erect, approximate, cylindrical, 10–14 cm. long, 1–1.2 cm. diameter, green; rhizome 5–11 mm.

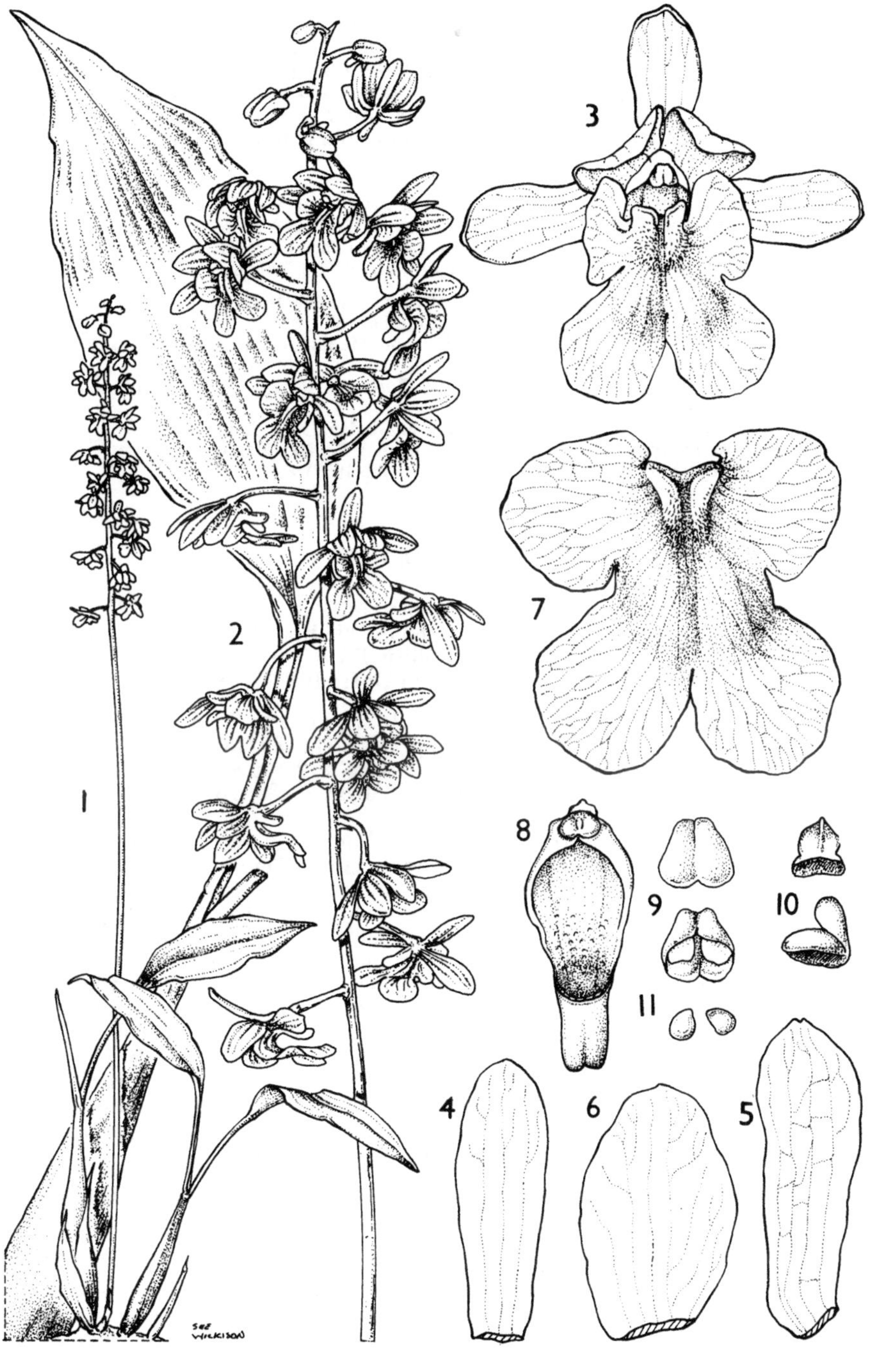

FIG. 90. *OECEOCLADES SAUNDERSIANA*—**1**, habit, × 1/6; **2**, habit, × 2/3; **3**, flower, × 2; **4**, dorsal sepal, × 3; **5**, lateral sepal × 3; **6**, petal, × 3; **7**, lip, × 3; **8**, column, ventral view, × 3; **9**, anther-cap, two views, × 6; **10**, stipes and viscidium, two views, × 17; **11**, pollinia, × 6. Drawn from Kew Spirit No. 18861-13 by Sue Wickison.

diameter, short; roots 3–4.5 mm. diameter, white. Leaves 2–4, erect, plicate, lanceolate or oblanceolate, acute, 30–80 cm. long, 1.2–9 cm. wide, longly petiolate. Inflorescence densely many-flowered; peduncle 4–5 mm. diameter, bearing 6 sheaths along length; rhachis 6–10.5 cm. long; bracts linear, acuminate, 10–26 mm. long. Flowers yellow or pale green marked with purple on side lobes of lip and with an orange callus; pedicel and ovary 13–21 mm. long. Dorsal sepal narrowly oblong-elliptic, apiculate, 12–15 mm. long; lateral sepals similar but falcate. Petals obovate-oblong, apiculate to subacute, 11–13 mm. long, 4.5–6 mm. wide. Lip 3-lobed but appearing 4-lobed when flattened, 7 mm. long, 18 mm. wide; side lobes large, rounded; mid-lobe transversely oblong, apiculate, strongly recurved; callus of 2 fleshy lobes at mouth of spur; spur globose, 3–4 mm. long. Column 5 mm. long, deflexed at apex. Fruit not known.

TANZANIA. Iringa District: Mwanihanga Forest Reserve, near Sanje, 7–8 June 1980, *Rodgers & Vollesen* 492!; Rungwe District: Chivanje Tea Estate, 6 Mar. 1978, *Leedal* 5015!
DISTR. T 7; Mozambique and Zimbabwe; Madagascar to Pacific Islands
HAB. In deep shade in riverine and montane rain-forest; 700–1400 m.

SYN. *Limodorum pulchrum* Thou., Orch. Iles Austr. Afr.: t. 43–44 (1822)
Eulophia pulchra (Thou.) Lindl., Gen. Sp. Orch. Pl.: 182 (1833)
Eulophia striata Rolfe in J.L.S. 29: 53 (1891). Type: Madagascar, near Fort Dauphin, *Scott Elliot* 2545 (K, holo.!)
E. silvatica Schltr. in E.J. 53: 586 (1915). Type: Tanzania, Rungwe District, Mulayala, *Stolz* 1910 (B, holo., K, record of holo.!)
Lissochilus pulcher (Thou.) Perr. in Humbert, Fl. Madag. Orch. 2: 40 (1941)
Eulophidium pulchrum (Thou.) Summerh. in B.J.B.B. 27: 400 (1957)
E. silvaticum (Schltr.) Summerh. in B.J.B.B. 27: 402 (1957)

45. **EULOPHIA**

Lindl. in Bot. Reg. 7, sub t. 573 (1821) & 8, t. 686 (1823) & Gen. Sp. Orch. Pl.: 180 (1833); Reichb.f. in Walp., Ann. Bot. Syst. 6: 644 (1863), Otia Bot. Hamb. 5: 66 (1878) & 6: 116 (1881); Hook.f. in Fl. Brit. Ind. 6: 1 (1891); Rolfe in F.T.A. 7: 47 (1897); Hall in S. Afr. Journ. Bot., Suppl. 5: 1–248 (1965); Seidenfaden in Opera Bot. 72: 28 (1983), *nom. conserv.*

Lissochilus Lindl. in Bot. Reg. 7, t. 573 (1821); Rolfe in F.T.A. 7: 70 (1897)
Cyrtopera Lindl., Gen. Sp. Orch. Pl.: 189 (1833)
Orthochilus A. Rich., Tent. Fl. Abyss. 2: 284, t. 82 (1850)
Hypodematium A. Rich., Tent. Fl. Abyss. 2: 286, t. 83 (1850)

Small, medium or large terrestrial or rarely lithophytic herbs; roots slender to stout, basal or adventitious, often with a well-defined velamen. Perennating organs stem-like, pseudobulbous or tuber-like, above the ground or more commonly underground, conical, cylindrical or irregular in shape, several-noded. Leaves usually present and green but in some species much reduced, scale-like and brown or buff; green leaves 1–many, thin-textured to fleshy or coriaceous, with or without prominent longitudinal veins, linear, lanceolate, ovate or elliptic, sheathing at the base; scale leaves when present sheathing. Inflorescences basal, laxly to subdensely many-flowered, usually racemose, rarely branching. Flowers small to large, sometimes showy and brightly coloured. Sepals and petals similar or with the petals much broader, free to base or with the lateral sepals fused at the base to the column-foot. Lip 3-lobed, spurred at the base, usually with a callus of ridges and/or papillae on upper surface. Column short to long, with or without a column-foot; anther-cap entire or 2-lobed at the apex; pollinia 2, subglobose; stipes solitary, triangular to oblong; viscidium oblong, elliptic or lunate.

A large genus of about 250 species widespread in tropical and southern Africa, Madagascar, the Mascarenes, tropical and subtropical Asia, SE. Asia, Australasia and the tropical Americas.

Eulophia is one of the largest genera of orchids in Africa and an extraordinarily diverse one. This, no doubt, reflects the wide range of habitats in which various species are to be found, from swamps and rain-forest to semi-deserts. Indeed, the delimitation of the genus itself is still the cause of some debate. The inclusion of *Lissochilus*, based on *L. speciosus*, in *Eulophia* has been a major source of controversy. Rolfe (1897) in the Flora of Tropical Africa accepted *Lissochilus* and distinguished it from *Eulophia* because of its "small spreading or reflexed sepals and broad often differently coloured sepals". Schlechter (1915) in his treatment of Stolz's orchid collections from Tanzania in Engler's Bot. Jahrb. was amongst the last to uphold *Lissochilus* stating that "it appears to me that (distinguishing) characters exist". However, he did not identify any distinguishing features. I have followed here Summerhayes (1968) in F.W.T.A. Orchidaceae, Hall (1965) in the Journal of South African Botany Suppl. 5 and other recent authors in including *Lissochilus* in *Eulophia*. However, in the absence of an explanatory argument I have not followed Geerinck (1987) who, in his Flore du Rwanda orchid account, has included the genera *Oeceoclades* and *Pteroglossaspis* in *Eulophia*.

Despite its morphological diversity, no satisfactory infrageneric treatment of *Eulophia* in Africa exists. In this treatment I have attempted to place closely related species together. Some groupings within the genus are quite distinct, for example species 24–37 and again species 55–67 in this account, and should be accorded taxonomic recognition at sectional rank or above. However, an infrageneric treatment needs to take into account species from outside the present study area and this work is as yet incomplete.

I have adopted a broad species concept for this treatment, in accordance with that adopted by Summerhayes (1968) and Hall (1965). Many of the species in *Eulophia* are widespread in Africa and there are few narrow endemics. Variations in plant stature, leaf size and flower size are found in several species, e.g. *E. speciosa*, *E. cucullata* and *E. streptopetala*. Some of this variation may be due to polyploidy which has been reported in several African species (Hall (1965) Poggio et al. in K.B. 41: 45–49 (1986)).

1. Plants saprophytic, lacking leaves entirely but with pale buff stems and bracts 2
   Plants autotrophic or only partly saprophytic, usually with green, greenish or purplish stems, leaves and bracts 3
2. Sepals and petals 15–25 mm. long; lip papillate all over, with a maroon mark at the base; bracts 28–50 mm. long . . . . . 38. *E. galeoloides*
   Sepals and petals 4.5–6 mm. long; lip with a 4-ridged basal callus but otherwide glabrous; bracts 3–4 mm. long . . . . . 39. *E. epiphanoides*
3. Inflorescence paniculate, with several spreading branches 4
   Inflorescence racemose, rarely with a single branch 5
4. Leaves coriaceous, thick, with serrated margins; sepals and petals 16–23 mm. long, recurved at the apex; lip 3-lobed . . . . . 15. *E. petersii*
   Leaves thin-textured, lacking serrated margins; sepals and petals 6–9 mm. long, not recurved at the apex; lip entire, spathulate . . . . . 18. *E. callichroma*
5. Leaves fleshy-coriaceous, with serrated margins 6
   Leaves thin-textured to somewhat fleshy, without serrated margins 7
6. Sepals and petals 19–25 mm. long; lip 18–20 mm. long; leaves 8 or more . . . . . 17. *E. taitensis*
   Sepals and petals 10 mm. or less long; lip 9 mm. or less long; leaves ± 4 . . . . . 16. *E. serrata*
7. Spur 15–25 mm. long, straight and parallel to the ovary, conical at the base, tapering to the apex; lip obscurely 3-lobed, with auriculate side lobes and a large purple or pink, flat, ecallose mid-lobe, 20–26 mm. long, 17–32 mm. wide . . . . . 2. *E. guineensis*
   Spur and lip not as above 8
8. Pseudobulbs 16–25 cm. long, 1–1.5 cm. in diameter, narrowly cylindrical; flowers green with a white lip marked with a transverse pink or purple band across the base of the mid-lobe; spur straight, clavate . . . . . 1. *E. euglossa*
   Pseudobulbs, if present, not as above 9
9. Sepals spathulate, rounded and somewhat concave at the apex, usually ± erect 10
   Sepals and petals not like that 13
10. Lip ± ecallose, pale lilac or white with the 5 central veins only slightly thickened; sepals white with a greenish apex; petals white . . . . . 7. *E. malangana*
    Lip with a 2–5-ridged callus, yellow, purple or rarely white; sepals yellow, brown or purple; petals yellow or rose-pink 11

11. Petals and lip yellow; petals yellow or brown; callus of 3 crenulate ridges raised in basal half and tapering to the apex . . . . 5. *E. angolensis*
    Petals and lip pink or rarely white; callus not as above 12
12. Lip 12–15 mm. long; callus of 2 subquadrate ridges in the basal half of the lip . . . . 10. *E. coeloglossa*
    Lip 18–21 mm. long; callus of 3 verrucose ridges at the base of the lip with 5 cristate ridges in front . . . . . 6. *E. caricifolia*
13. Callus with 2 prominent subquadrate raised flaps in basal part or middle of the lip 14
    Callus lacking such flaps 15
14. Sepals, petals and lip pink to rose-purple; lip 18–34 mm. long, 20–40 mm. wide when flattened, lacking papillae on the veins in the apical half . . . . 11. *E. cucullata*
    Sepals and petals olive-green or green; lip purple or rarely white, 14–19 mm. long, 8–12 mm. wide, papillate on veins in apical half . . . . 9. *E. alta*
15. Petals circular, broadly elliptic or obovate, as broad as long or almost so 16
    Petals linear, lanceolate, oblong or narrowly elliptic, much longer than broad 29
16. Lip more than 20 mm. long, adorned with 3–7 lamellate ridges; petals usually more than 20 mm. wide, pink to rose-purple 17
    Lip less than 16 mm. long, adorned with 1–7 low, rounded, rugulose ridges; petals less than 19 mm. wide, yellow to brown or rarely lilac, often veined with red 18
17. Lip much longer than broad when flattened, distinctly 3-lobed; callus of 3 ridges; petals 15–24 mm. wide, with slightly reflexed margins . . . . 3. *E. horsfallii*
    Lip as broad as long when flattened, obscurely 3-lobed; petals 25 mm. or more wide, concave . . . . . 4. *E. latilabris*
18. Flowers with a thin transverse lacerate-margined lamella near the base of the lip; lip lacking a spur at the base 19
    Flowers lacking such a transverse lamella at the base of the lip; spur usually present 20
19. Petals and lip yellowish; petals 5–6.5 mm. long . . . 56. *E. fridericii*
    Petals and lip lilac or pinkish; petals 9–11 mm. long 57. *E. gonychila*
20. Plants with prominent ovoid pseudobulbs above the ground; leaves plicate, usually more than 1 cm. wide 55. *E. streptopetala*
    Plants with underground tubers or rhizomes; leaves fleshy or plicate, usually less than 1 cm. wide 21
21. Flowers large; petals more than 11 mm. long; lip more than 12 mm. long 22
    Flowers small; petals 10 mm. or less long; lip 9 mm. or less long 24
22. Flowers bright yellow with a little red venation on the lip; base of lip obscurely conical, obtusely angled 58. *E. speciosa*
    Flowers not so coloured; petals reddish, brown or yellow veined with red; lip spurred at the base; spur 8–16 mm. long measured from the base of the ovary 23
23. Spur ± straight or decurved at the apex, cylindrical from a conical mouth, 13–16 mm. long from the base of the ovary . . . . 60. *E. orthoplectra*
    Spur upcurved at the apex, 8–9 mm. long from the base of the ovary . . . . 61. *E. schweinfurthii*
24. Spur very short or obscure, 1–2.5 mm. long 25
    Spur more than 3 mm. long 27
25. Lip obtuse at the base but lacking a true spur; lip mid-lobe reduplicate; petals and lip yellow . . . . 59. *E. tuberculata*
    Lip spurred at the base; lip mid-lobe flat or concave; petals and lip greenish or yellow, heavily marked with brown 26

26. Spur 1 mm. long or less; petals 5–5.5 mm. long . . . 64. *E. parvula*
Spur 1.5–2.5 mm. long; petals 6–8 mm. long . . . . 65. *E. pyrophila*
27. Callus on the lip 9-ridged; lip 10.5–11.5 mm. long 66. *E. stenoplectra*
Callus on the lip 3–5-ridged; lip less than 10.5 mm. long 28
28. Lip 6–10.5 mm. long; spur 3–5 mm. long; inflorescence produced before the leaves appear . . . . . . 62. *E. clitellifera*
Lip 4.5–5.5 mm. long; spur 5–6 mm. long; inflorescence produced with the young leaves . . . . 63. *E. leonensis*
29. Sepals strongly reflexed; petals elliptic, appressed to the column . . . . 8. *E. calantha*
Sepals and petals not as above 30
30. Side lobes of the lip spreading widely; sepals spreading widely; lip mid-lobe deflexed, broadly spreading 31
Side lobes and mid-lobe of the lip and the sepals not as above 33
31. Sepals lanceolate, greenish; lip and petals pale yellow; lip 23–29 mm. long; callus obscure; petals appressed to the column . . . . 12. *E. flavopurpurea*
Sepals oblanceolate, lilac; petals spreading, lilac; lip 12–22 mm. long, deep purple; callus of 5–9 ridges 32
32. Lip with 5–9 crenulate or papillate ridges on the mid-lobe 13. *E. cristata*
Lip with 5–9 smooth ridges on the mid-lobe . . . . 14. *E. livingstoniana*
33. Sepals, petals and lip with a similar base-colour: white, pink, purple, yellow, orange or green; sepals and petals similar in shape 34
Sepals and petals dissimilar in colour and often also in shape; lip of a different colour 48
34. Flowers bright orange 35
Flowers not orange-coloured 36
35. Sepals 4 mm. or more wide; lip broadly 3-lobed, 10 mm. or more wide; spur 1–2 mm. long . . . . 28. *E. aurantiaca*
Sepals linear, 3 mm. or less wide; lip narrowly 3-lobed, less than 9 mm. wide; spur absent . . . . 29. *E. walleri*
36. Flowers yellow or greenish, sometimes with markings of a different colour on the lip 37
Flowers white, pink, purple or brown, sometimes with markings of a different colour on the lip 43
37. Lip narrowly 3-lobed, with the side lobes almost as long as the narrow mid-lobe; lip with an orange mark on the lip . . . . 30. *E. nuttii*
Lip broadly 3-lobed, with the side lobes much shorter than the mid-lobe 38
38. Flowers green or yellowish green with brownish purple veining on the side lobes of the lip; spur strongly incurved-conical, up to 5 mm. long . . . . . . 33. *E. adenoglossa*
Flowers predominantly yellow or pale yellow; spur cylindrical 39
39. Sepals 14 mm. or less long; lip 11.5 mm. or less long; peduncle covered or almost covered by papery sheathing bracts 40
Sepals 17 mm. or more long; lip 12 mm. or more long; peduncle not entirely covered by sheathing bracts 41
40. Lip 8–11.5 mm. long, with orange or red papillae on the mid-lobe; sepals 9–14 mm. long; pedicel and ovary 12–21 mm. long . . . . 37. *E. odontoglossa*
Lip 6–7 mm. long, with yellow papillae on the mid-lobe; sepals 5–9.5 mm. long; pedicel and ovary 5–8 mm. long 35. *E. milnei*
41. Lip 12–14 mm. long; inflorescence secund; flowers pale yellow with a greenish mark on the lip, sometimes with a purplish border . . . . 24. *E. carsonii*
Lip more than 15.5 mm. long; inflorescence a dense head; flowers yellow or yellow with a purple or dark yellow mark on the lip 42

42. Flowers pale yellow to yellow, lacking a differently coloured mark on the lip; petals 15–26 mm. long; lip 15–24 mm. long . . . . 31. *E. subulata*
Flowers yellow with a maroon or orange mark on the lip; petals 28–32 mm. long; lip 23–26 mm. long . . . 32. *E. zeyheri*
43. Flowers white or pink with a dark maroon or purple spot on the lip 44
Flowers white, pink, purple or brown, but without a distinct darker spot on the lip 45
44. Flowers pure white with a dark maroon spot on the mid-lobe of the lip in front of 7 lines of yellow papillae . . . . 27. *E. thomsonii*
Flowers pale to darker pink with a purple spot on the lip; callus of 2 papillate ridges at the base and long papillae in front . . . . 25. *E. euantha*
45. Flowers dull purple or brown 46
Flowers not so coloured 47
46. Peduncle covered by papery sheathing bracts; petals 8–12 mm. long; lip 8–11.5 mm. long . . . . 37. *E. odontoglossa*
Peduncle not completely covered by sheathing bracts; petals 12–15 mm. long; lip 12 mm. or more long 26. *E. rara*
47. Inflorescence markedly secund; sepals and petals slender, 3–3.5 mm. wide; lip mid-lobe obovate, truncate 34. *E. ischna*
Inflorescence not secund; sepals and petals 6–10 mm. wide; lip mid-lobe subquadrate . . . . 36. *E. montis-elgonis*
48. Leaves 18–85 mm. wide, obviously plicate 49
Leaves grass-like, less than 10 mm. wide 50
49. Callus on the lip comprising 3 basal ridges with 5 rows of short papillae in front on the mid-lobe . . . . . 19. *E. stachyodes*
Callus on the lip comprising 2 basal ridges and a dense mass of long papillae in front on the mid-lobe 20. *E. elegans*
50. Plants with proximate cylindrical-conical pseudobulbs, usually partly or wholly above ground 51
Plants with underground, irregularly shaped tubers or fleshy rhizomes 60
51. Fruits erect 52
Fruits pendent 53
52. Spur straight, 8–10 mm. long; sepals 16–18 mm. long; lip 7–9 mm. wide; callus 3-ridged at base with papillae in front . . . . 40. *E. katangensis*
Spur incurved, 3–5 mm. long; sepals 11–14 mm. long; lip 9–12 mm. wide; callus of 3 basal ridges with 5–7 verrucose ridges in front . . . . 41. *E. rhodesiaca*
53. Lip purple or with a purple or lilac margin; spur deflexed, ± as long as the deflexed mid-lobe of the lip; leaves lanceolate, usually 5 mm. or more wide 54
Lip not as above; leaves 1–4 mm. wide 55
54. Petals 9–12 mm. long, 2–3 mm. wide; lip mid-lobe purple; spur 4–5 mm. long, somewhat sigmoid-decurved 49. *E. sabulosa*
Petals 8–8.5 mm. long, 1.5–2 mm. wide; lip green with a yellow centre and lilac margin to the mid-lobe; spur 3–4 mm. long, incurved . . . . 50. *E. seleensis*
55. Lip mid-lobe circular, about a third to half the length of the lip; spur straight, distinctly clavate . . . . 48. *E. penduliflora*
Lip mid-lobe ovate to oblong, about half the length of the lip; spur usually incurved and cylindrical, rarely clavate 56
56. Lip bright yellow . . . . 47. *E. mumbwaensis*
Lip white, green or pinkish 57
57. Lip 15–17 mm. long, 13 mm. wide when flattened; mid-lobe oblong-obovate, with an undulate margin; sepals green, ± shaded with pale brown; leaves 3–4 . . . 44. *E. tanganyikensis*
Lip less than 10 mm. long; mid-lobe ovate to lanceolate; sepals brown or maroon, 1–2 mm. wide; leaves 5–8 58

58. Sepals recurved; callus with short papillae on the three middle veins of the mid-lobe . . . . 51. *E. venulosa*
Sepals erect; callus with short papillae or crenulate ridges on the mid-lobe 59
59. Spur clavate, 3–4 mm. long; petals chocolate-brown; mid-lobe of the lip with 3–5 rows of crenulate ridges . . . . 45. *E. massokoensis*
Spur incurved, cylindrical, 4–5 mm. long; petals white; mid-lobe of the lip with 3 rows of papillae . . . . 46. *E. monotropis*
60. Fruits erect; flowers ± erect 61
Fruits pendent; flowers spreading to nutant 63
61. Flowers opening widely; sepals erect, 15–32 mm. long, usually about twice as long as the petals; spur 2 mm. long . . . . 52. *E. longisepala*
Flowers not opening widely; sepals not erect, 12–19 mm. long, only slightly longer than the petals; spur 2–6 mm. long 62
62. Sepals and petals greenish, flecked with red-brown; spur decurved, 5–6 mm. long; lip white with red veins on the side lobes, 9–9.5 mm. long; mid-lobe of the lip unspotted, with an undulate margin . . . . . . 53. *E. nyasae*
Sepals and petals deep purple or brown; spur ± straight, 4–5.5 mm. long; lip white with yellow side lobes, veined with purple, and a purple-spotted mid-lobe . . . 54. *E. subsaprophytica*
63. Petals yellow, 20–24 mm. long, 8–9 mm. wide; lip yellow with red-veined side lobes, 15–21 mm. long, 10–15 mm. wide . . . . 22. *E. ovalis*
Petals white, green or yellowish green, less than 12 mm. long; lip white, green, yellow or yellowish green, less than 15 mm. long, 10 mm. wide 64
64. Spur markedly dilated at the apex 65
Spur cylindrical, ± incurved 66
65. Spur tightly incurved, 3–4 mm. long; lip yellow, 4–6 mm. long, 1.5–2 mm. wide . . . . 42. *E. monile*
Spur slightly incurved, 2–3 mm. long; lip white with a pale green callus, 6–8.5 mm. long, 2.5–4 mm. wide . . 43. *E. norlindhii*
66. Sepals erect, 9–17 mm. long; lip 9–15 mm. long, green with a white or yellow mid-lobe; callus of 3 fleshy ridges at base of lip and papillae in front on the mid-lobe 21. *E. kyimbilae*
Sepals not erect, 10–12 mm. long; flower not opening widely; lip white flushed with dull purple, 7–10 mm. long; callus of 2 basal ridges with papillae on the mid-lobe . . . . 23. *E. clavicornis*

1. **E. euglossa** (*Reichb.f.*) *Reichb.f.* in Bot. Mag. 92, t. 5561 (1866); Summerh. in F.W.T.A., ed. 2, 3: 247 (1968). Type: ex Sierra Leone, cult. *van Hees* (W, holo.!)

A terrestrial herb 60–150 cm. tall. Perennating organs above ground, pseudobulbous, cylindrical-conical, swollen at base, 5–10-noded, 16–25 cm. long, 1–1.5 cm. diameter; roots basal, 3 mm. diameter, white. Leaves 5–10, plicate, ovate to lanceolate, acuminate, 16–42 cm. long, 2.5–5.5 cm. wide, petiolate, lowermost sheathing. Inflorescence laxly many-flowered; peduncle stout, 10–12 mm. diameter, bearing several bracts, lowermost sheathing, upper ones lanceolate; rhachis 10–38 cm. long; bracts spreading, lanceolate, acuminate, 15–36 mm. long. Flowers not opening widely, subnutant; sepals green with a brownish apex; petals similar; lip white with a pink or purple band across base of mid-lobe and a greenish spur; pedicel and ovary 17–21 mm. long. Dorsal sepal lanceolate, acuminate, 20–24 mm. long, 3 mm. wide; lateral sepals similar, 18–24 mm. long, 3 mm. wide. Petals lanceolate, acuminate, 15–18 mm. long, 3 mm. wide. Lip 3-lobed, 13–14 mm. long, 7.5–9 mm. wide; side lobes narrow, subacute; mid-lobe elliptic, obtuse to acute, 8 mm. long, 6.5 mm. wide, margins undulate; callus of 2–3 obscure ridges in basal half of mid-lobe; spur clavate, 5–7 mm. long, pointing backwards and slightly decurved. Column 5 mm. long, papillose at base; foot 1 mm. long. Fruit pendent, 23–30 mm. long.

UGANDA. Masaka District: Sese Is., Bugala I., cult. Entebbe Botanic Garden, 28 Feb. 1945, *Greenway & Thomas* 7208! & Serinya I., 22 Feb. 1933, *A.S. Thomas* 839!; Mengo District: Nakaziba Forest, June 1949, *Sangster* 1040!

DISTR. U 4; Sierra Leone, Liberia, Ivory Coast, Nigeria, Cameroon, Gabon and Ethiopia

HAB. In dense forest of *Uapaca guineensis, Harungana madagascariensis, Maesopsis eminii* etc. on red-brown sandy loam over ironstone; also in secondary forest; 1200–1300 m.

SYN. *Galeandra euglossa* Reichb.f. in Bot. Zeit. 10: 935 (1852)

2. **E. guineensis** *Lindl.* in Bot. Reg. 8, t. 686 (1823); Summerh. in F.W.T.A., ed. 2, 3: 246 (1968). Type: Sierra Leone, without exact locality, *G. Don* (K, holo.!)

A terrestrial or lithophytic herb (14–)30–65 cm. tall. Perennating organs above ground, pseudobulbous, conical, proximate, 3–3.5 cm. long, 1.5–2 cm. diameter; roots from base of pseudobulbs, 4–6 mm. diameter. Leaves 3–4, plicate, suberect, elliptic, acute, 10–35 cm. long, 3–9.5 cm. wide, shortly petiolate, articulated to conduplicate leaf-base. Inflorescence laxly 5–many-flowered, opening before or with leaves; peduncle 4–9 mm. diameter, bearing up to 7 papery sheaths in lower half; rhachis 7–29 cm. long; bracts suberect, papery, lanceolate, acuminate, 10–30 mm. long, 2–4 mm. wide. Flowers subnutant, showy; sepals and petals purplish brown; lip pinkish purple with a paler or white base and spur; pedicel and ovary 13–25 mm. long. Dorsal sepal reflexed, linear-lanceolate, acuminate, 16–26 mm. long, 3–4.5 mm. wide; lateral sepals similar. Petals linear-lanceolate, acuminate, 15–25 mm. long, 4–5 mm. wide. Lip obscurely 3-lobed at base, obovate, obtuse, 20–35 mm. long, 13–32 mm. wide; mid-lobe 20–26 mm. long, 17–32 mm. wide; callus absent or basal veins only slightly thickened; spur straight, tapering from a broad mouth, very slender at apex, 15–25 mm. long. Column 5–7 mm. long, 4–5 mm. broad; foot absent. Fruit pendent, 2–2.5 cm. long, with a persistent column. Pl. 2. Fig. 91.

UGANDA. Karamoja, May 1948, *Eggeling* 5795!; Bunyoro District: Kiryandongo, Mar. 1943, *Purseglove* 1328!; Busoga District: 9.5 km. N. of Busia, Buteba [Butiba], 4 Apr. 1951, *G.H. Wood* 262!

KENYA. Uasin Gishu District: Turbo, near Kipkarren R., May 1950, *Watney* in *Bally* B7780!; N. Kavirondo District: S. Elgon, above Kimilili, Mar. 1963, *Tweedie* 2586!

TANZANIA. Ufipa District: Rukwa, Kipangati, 4 Dec. 1950, *Bullock* 3513!; Morogoro, 24 Dec. 1934, *E.M. Bruce* 367!; Rungwe District: Bundali Mts., 55 km. from Kyela on Isoko Road, Ikikota, 12 Feb. 1976, *Cribb & Grey-Wilson* 10700!

DISTR. **U** 1–3; **K** 2–5; **T** 4, 6, 7; tropical Africa from Sierra Leone to Ethiopia and south to Malawi and Zambia, also in N. Yemen and Oman

HAB. In shade or semi-shade amongst rocks, in scrub and in woodland; 600–2000 m.

SYN. *E. quartiniana* A. Rich., Tent. Fl. Abyss. 2: 284, t. 81 (1850). Type: Ethiopia, Tigre, *Quartin Dillon* (P, holo.)

E. guineensis Lindl. var. *purpurata* Kotschy in Sitzb. Akad. Wissenschaft. Wien 51: 352 (1865). Type: not located

E. guineensis Lindl. var. *kibilana* Schltr. in E.J. 53: 587 (1915). Type: Tanzania, Rungwe District, Kiwira [Kibila] R., near Kimomo, *Stolz* 1875 (B, holo., G, iso.!)

3. **E. horsfallii** (*Batem.*) *Summerh.* in F.W.T.A. 2: 442, 444 (1936); Hall in Journ. S. Afr. Bot., Suppl. 5: 153 (1965); Summerh. in F.W.T.A., ed. 2, 3: 250 (1968); J. Stewart in U.K.W.F.: 780 (1974); Williamson in Orch. SC. Afr.: 173 (1977); J. Stewart et al., Wild Orch. S. Afr.: 243 (1982). Type: Nigeria, Old Calabar R., cult. *J.B. Horsfall*, coll. *Cheetham* (K, holo.!)

A large terrestrial herb 1.2–3 m. tall. Perennating organ subterranean, rhizomatous, fleshy, 4–6 or more cm. long, 2–3 cm. diameter; roots elongate, 4–5 mm. diameter, white. Leaves suberect, 3–5, erect, lanceolate to oblanceolate, acuminate, 30–140 cm. long, 1.6–15.5 cm. wide, long-petiolate. Inflorescence erect, laxly 5–50-flowered; peduncle stout, up to 3.5 cm. diameter at base, bearing 5 tubular sheaths along length; rhachis 16–37 cm. long; bracts elliptic or obovate, obtuse to acuminate, 19–35 mm. long, 10–18 mm. wide. Flowers large, fleshy; sepals dull purple or brown; petals rose-purple; lip side lobes green striped with dull purple, mid-lobe purple; callus yellow; column and ovary purple; pedicel and ovary 30–40 mm. long. Dorsal sepal reflexed, oblanceolate, acuminate, 17–26 mm. long, 7–8 mm. wide; lateral sepals oblique at base but similar, 17–27 mm. long, 7–8 mm. wide. Petals covering column, elliptic, obtuse, 19–35 mm. long, 15–24 mm. wide. Lip 3-lobed, 20–40 mm. long, 14–40 mm. wide; side lobes erect, semicircular; mid-lobe elliptic-oblong, obtuse, with undulate margins; callus in basal two-thirds of lip, tricristate, raised at base, undulate and lower in front; spur conical, 4–8 mm. long, slightly retrorse. Column clavate, 10 mm. long. Fruit pendent, ellipsoidal, 30–65 mm. long. Pl. 2. Fig. 92/1–10, p. 431.

FIG. 91. *EULOPHIA GUINEENSIS*—**1**, habit, × ½; **2**, inflorescence, × 1; **3**, column, × 3; **4**, anther-cap, from above, × 4; **5**, anther-cap, from below, × 4; **6**, pollinarium, × 4. Drawn by W.E. Trevithick. Reproduced from Flora of West Tropical Africa.

UGANDA. Bunyoro District: near Hoima, Baker's View road, km. 1, 26 May 1959, *Lind* 2436!; Mbale District: Elgon, 28 Oct. 1923, *Snowden* 816!; Mengo District: Buruli County, near Sezibwa R., Apr. 1931, *Sandford* 2032!

KENYA. W. Suk/Elgeyo District: Cherangani, 5 Apr. 1959, *Symes* 525!; S.Nyeri District: Mt. Kenya, Castle Forest, 21 June 1973, *Whitehead* 8!; Kwale District: Shimba Hills, near Kivukoni entrance to National Park, 23 Nov. 1971, *Bally & Smith* 14360!

TANZANIA. Bukoba District: near Bugene, Oct. 1931, *Haarer* 2298!; Lushoto District: E. Usambara Mts., Amani, Sangarawe, 10 Feb. 1937, *Greenway* 4893!; Rungwe Mt., 25 Oct. 1956, *Richards* 6796! & 18 Sept. 1919, *Geilinger* 2454!; Zanzibar I., Masingini ridge, 28 Nov. 1930, *Greenway* 2605!; Pemba I., Dec. 1908, *Barrand!*

DISTR. **U** 2–4; **K** 2–5, 7; **T** 1, 3, 4, 6–8; **Z**, **P**; throughout tropical Africa, south to South Africa (E. Transvaal and Natal)

HAB. By streams, rivers and lakes, in marshes, bogs and in swampy places in open and in forests; sea-level to 2500 m.

SYN. *Lissochilus roseus* Lindl. in Bot. Reg. 29, Misc.: 25 (1843) & 30, t. 12 (1844). Type: Sierra Leone, cult. *Rucker* (K, holo.!)

L. horsfallii Batem. in Hook., Bot. Mag. 91, t. 5486 (1865)

L. welwitschii Reichb.f. in Flora 48: 188 (1865). Type: Angola, Huilla, Monino, *Welwitsch* 710 (W, holo.!, K, iso.!)

L. porphyroglossus Reichb.f. in Otia Bot. Hamb. 1: 61 (1878). Type: Sudan, Niamniam, *Schweinfurth* 22 (W, holo.!)

L. sandersonii Reichb.f. in Otia Bot. Hamb. 1: 62 (1878). Type: South Africa, Natal, *Sanderson* (K, lecto. icon. del. Sanderson)

Eulophia elliotii Rendle in J.B. 33: 169 (1895). Type: Uganda, Ruwenzori, *Scott Elliot* 7924 (BM, holo., K, iso.!)

Lissochilus sceptrum Schltr. in E.J. 53: 589 (1915). Type: Tanzania, Rungwe District, Kyimbila, *Stolz* 452a (B, holo.)

L. eleogenus Schltr. in E.J. 53: 589 (1915). Type: Tanzania, Rungwe District, Bomalakitana, *Stolz* 452 (B, holo., K, iso.!)

NOTE. *E. horsfallii* is a very variable species in both stature and flower size but this variation seems to be more or less continuous and does not warrant taxonomic recognition as far as I can determine.

4. **E. latilabris** *Summerh.* in F.W.T.A. 2: 442, 444 (1936), in K.B. 1936: 223 (1936) & in F.W.T.A., ed. 2, 3: 250 (1968); J. Stewart in U.K.W.F.: 780 (1974); Williamson, Orch. SC. Afr.: 173 (1977). Type: Sudan, Niamniam, *Schweinfurth* 3776 (W, holo., K, iso.!)

A large terrestrial herb 75–120 cm. tall. Perennating organs underground, obliquely conical, 3–4 cm. tall, 2–3 cm. diameter, borne on a horizontal fleshy rhizome 1–1.5 cm. in diameter; roots 3.5 mm. diameter, white. Leaves 3–5, erect, lanceolate, acuminate, 30–70 cm. long, 3.8–8 cm. wide. Inflorescence produced before the leaves, densely many-flowered; peduncle stout, 1–1.8 cm. diameter at base, bearing 4–5 loose obtuse sheaths along length; rhachis 11–21 cm. long; bracts elliptic, acute or acuminate, 30–55 mm. long, 16–40 mm. wide, yellowish to greenish purple. Flowers large, fleshy, subnutant; sepals greenish purple; petals pinkish mauve; lip pale pinkish mauve with dark purple-veined side lobes and a bright yellow callus; pedicel and ovary 25–30 mm. long. Dorsal sepal reflexed, obovate, apiculate, 20–23 mm. long, 8–10 mm. wide; lateral sepal similar but oblique at base, 21–27 mm. long, 11–15 mm. wide. Petals porrect, almost circular, 24–27 mm. long, 28–30 mm. wide. Lip obscurely 3-lobed, 27–35 mm. long, 30–32 mm. wide; side lobes broadly rounded; mid-lobe transversely oblong to semicircular, 12 mm. long, 19 mm. wide; callus of 5 ridges, the 3 inner ones raised at base, undulate and lower towards apex; spur obliquely narrowly conical, pointing backwards, 5–8 mm. long. Column 12–14 mm. long. Fruit pendent, ellipsoidal, 35 mm. long. Pl. 2. Fig. 92/11–18.

UGANDA. W. Nile District: 2 km. NW. of Maracha Rest Camp, 21 July 1953, *Chandler* 14!; Acholi District: Gulu, Apr. 1938, *Eggeling* 3540!; Bunyoro District: 1.5 km. on Baker's View road, near Hoima, 26 May 1959, *Lind* 2437!; Mbale District: Elgon, 15 Apr. 1924, *Snowden* 876!

KENYA. N. Kavirondo District: near Malikisi [Malakisi], Apr. 1938, *Tweedie* 429!

TANZANIA. Buha District: Kasulu, 25 Jan. 1931, *Rounce* D11!; Songea District: ± 12 km. E. of Songea, Nonganonga stream, 27 Dec. 1955, *Milne-Redhead & Taylor* 7919!

DISTR. **U** 1–3; **K** 5; **T** 4, 8; West Africa across to the Sudan, Mozambique, Zambia, Angola, Malawi, Zimbabwe and Botswana

HAB. In marshes, bogs and seasonally wet grassland; 1300–1600 m.

SYN. *Lissochilus schweinfurthii* Reichb.f. in Otia Bot. Hamb. 1: 60 (1878), *non Eulophia schweinfurthii* Kraenzl. Type as for species

L. nyassae Rolfe in F.T.A. 7: 86 (1897). Type: Malawi, without exact locality, *Buchanan* 1066 (K, holo.!)

L. leopoldii Kraenzl. in Th. Dur. & Schinz, Mat. Fl. Congo 3: 53 (1899). Type: Zaire, near Lomani, *Laurent* (BR, holo.!)

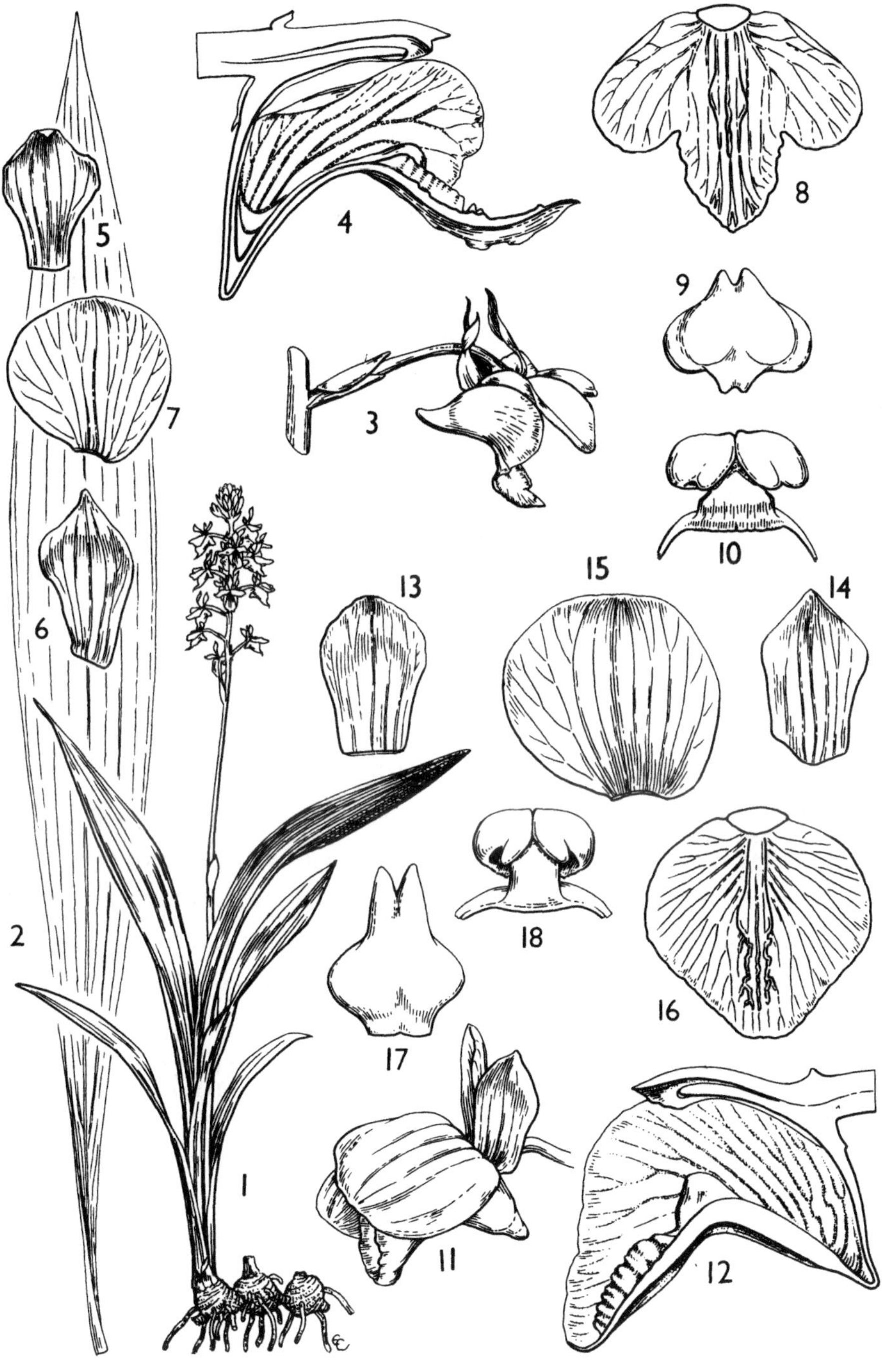

FIG. 92. *EULOPHIA HORSFALLII*—**1**, habit, × 1/15; **2**, leaf, × 2/9; **3**, flower, × 2/3; **4**, longitudinal section of lip and column, × 1½; **5**, dorsal sepal, × 1; **6**, lateral sepal, × 1; **7**, petal, × 1; **8**, lip, × 1; **9**, anther-cap, × 4; **10**, pollinarium, × 4. *E. LATILABRIS*, - **11**, flower, × 2/3; **12**, longitudinal section of lip and column, × 1½; **13**, dorsal sepal, × 1; **14**, lateral sepal, × 1; **15**, petal, × 1; **16**, lip, × 1, **17**, anther-cap, × 4; **18**, pollinarium, × 4. 1–10 from *Milne-Redhead & Taylor* 8344; 11–18, from *Milne-Redhead & Taylor* 7919. Drawn by Eleanor Catherine.

5. **E. angolensis** (*Reichb.f.*) *Summerh.* in K.B. 13: 76 (1958); Hall in Journ. S. Afr. Bot., Suppl. 5: 157 (1965); Piers, Orch. E. Afr., ed. 2: 157 (1968); Summerh. in F.W.T.A., ed. 2, 3: 249 (1968); J. Stewart in U.K.W.F.: 776 (1974); Williamson, Orch. SC. Afr.: 170 (1977); J. Stewart et al., Wild Orch. S. Afr.: 244 (1982). Type: Angola, Huilla, *Welwitsch* 734 (W, holo.!, G, K, iso.!)

A terrestrial herb 60–210 cm. tall. Perennating organs fleshy, subterranean, rhizomatous, stout, cylindric, many-noded, branching, 1–5 cm. diameter, with white, 2 mm. diameter roots emerging all over surface. Leaves 2–4, plicate, erect, linear-lanceolate, acuminate, 50–130 cm. long, 0.7–4.3 cm. wide. Inflorescence laxly many-flowered, rarely 1-branched; peduncle stout, 1–2 cm. diameter, bearing 5 acute to acuminate sheaths; rhachis 16–40 cm. long; bracts ovate, lanceolate or oblanceolate, 1–2.5 cm. long. Flowers spreading or subnutant; sepals purple-brown or yellow; petals and lip yellow; pedicel and ovary 14–21 mm. long. Dorsal sepal erect or reflexed, ligulate-spathulate, obtuse, concave towards apex, 16–27 mm. long, 5.3–7.5 mm. wide; lateral sepals similar but suberect or reflexed, 16–25 mm. long, 5.3–7.5 mm. wide. Petals parallel to and covering column, elliptic, rounded at apex, 16–25.5 mm. long, 9.5–13.8 mm. wide. Lip 3-lobed, saccate at base, 16–25 mm. long, 10–16 mm. wide; side lobes elliptic, rounded; mid-lobe oblong or obovate, obtuse or rounded, undulate on margin; callus of 3 tall crenulate ridges, flexuous in front. Column 10–11 mm. long; foot 3–4 mm. long; anther-cap 2-horned at apex. Fruit pendent, 3–4.5 cm. long. Pl. 2. Fig. 93/1–9.

UGANDA. Ankole District: Igara, Mitoma, Oct. 1938, *Purseglove* 433!; Masaka District: 3.5 km. W. of Katera, 4 Oct. 1953, *Drummond & Hemsley* 4617!; Mengo District: Mawakota, Mar. 1905, *E. Brown* 175!

KENYA. Kiambu, 12 June 1930, *Armstrong* 263!; Nairobi, 29 July 1913, *Battiscombe* 782!; N. Kavirondo District: 7 km. N. of Kakamega, Mar.–Apr. 1948, *Barney* 3!

TANZANIA. Ufipa District: Chapota, 11 Mar. 1957, *Richards* 8630!; Uzaramo District: Dar es Salaam, Pugu road, 8 Jan. 1939, *Vaughan* 2714!; Iringa District: Sao Hill, Dec. 1959, *Watermeyer* 217!; Songea District: 21 km. N. of Songea, R. Lumecha, 3 Jan. 1956, *Milne-Redhead & Taylor* 8122!

DISTR. **U** 1–4; **K** 3–7; **T** 1, 4, 6–8; throughout tropical and South Africa from the Gambia to Natal

HAB. In marshes, swamps and seasonally flooded grasslands often by lakes and rivers; sea-level to 2300 m.

SYN. *Cymbidium angolense* Reichb.f. in Flora 48: 188 (1865)
Lissochilus angolensis (Reichb.f.) Reichb.f. in Otia Bot. Hamb. 1: 64 (1878)
L. paludicolus Reichb.f. in Otia Bot. Hamb. 1: 63 (1878). Type: Sudan, Niamniam, *Schweinfurth* 3983 (W, holo.)
L. buchananii Reichb.f. in Otia Bot. Hamb. 1: 64 (1878). Types: South Africa, Natal, *Buchanan* (K, syn.!) & *Sanderson* (K, syn.!)
L. lindleyanus Reichb.f. in Otia Bot. Hamb. 1: 65 (1878). Type: Nigeria, Nupe, *Barter* 1486 (K, holo.!)
L. platypterus Reichb.f. in Flora 65: 533 (1882). Type: Angola, *Mechow* 415 (W, holo.)
Eulophia buchananii (Reichb.f.) Bolus in J.L.S. 25: 185 (1889)
Lissochilus antunesii Rolfe in Bol. Soc. Brot. 7: 236 (1890). Type: Angola, Huila, *Antunes* 19 (K, holo.!)
L. validus Rendle in J.B. 33: 197 (1895). Type: Burundi, without exact locality, *Scott-Elliot* 8347 (BM, holo.!)
L. validus Rendle var. *minor* Rendle in J.B. 33: 197 (1895). Type: Uganda, Kampala, *Scott-Elliot* 7305 (BM, holo.!)
L. latus Rolfe in F.T.A. 7: 79 (1897). Type: Zambia, Lake Mweru [Moero], *Carson* 28 (K, holo.!)
Eulophia lindleyana (Reichb.f.) Schltr., Westafr. Kautsch.-Exped.: 279 (1900)
Cyrtopera stolziana Kraenzl. in E.J. 33: 69 (1902). Type: Tanzania, Rungwe District, Kondeland, *Stolz* 13 (B, holo.)
Lissochilus eylesii Rendle in J.B. 43: 53 (1905). Type: Zimbabwe, Matopo Hills, *Eyles* 150 (BM, holo.!)
L. ugandae Rolfe in Bot. Mag. 131, t. 8044 (1905). Type: Uganda, Entebbe, *Mahon* (K, holo.!)
Eulophia stolziana (Kraenzl.) Engl. in V.E.3(2): 444 (1908), in obs.
Lissochilus mildbraedii Kraenzl. in E.J. 43: 338 (1909). Type: Tanzania, near Bukoba, Kiboroga, *Mildbraed* 277 (B, holo.)
L. stolzianus (Kraenzl.) Schltr. in E.J. 53: 588 (1915)
L. katentaniensis De Wild. in B.J.B.B. 6: 91, t. 14/6–12 (1919). Type: Zaire, Shaba, Biano [Katentania], *Homble* 765 (BR, holo.!)

NOTE. This species is variable in flower colour and the degree of development of the callus. However, this variation does not appear to warrant taxonomic recognition.

6. **E. caricifolia** (*Reichb.f.*) *Summerh.* in F.W.T.A. 2: 444 (1936), in K.B. 1936: 225 (1936) & in F.W.T.A., ed. 2, 3: 249 (1968). Type: Gabon, without exact locality, *Griffon de Bellay* (W, holo.!)

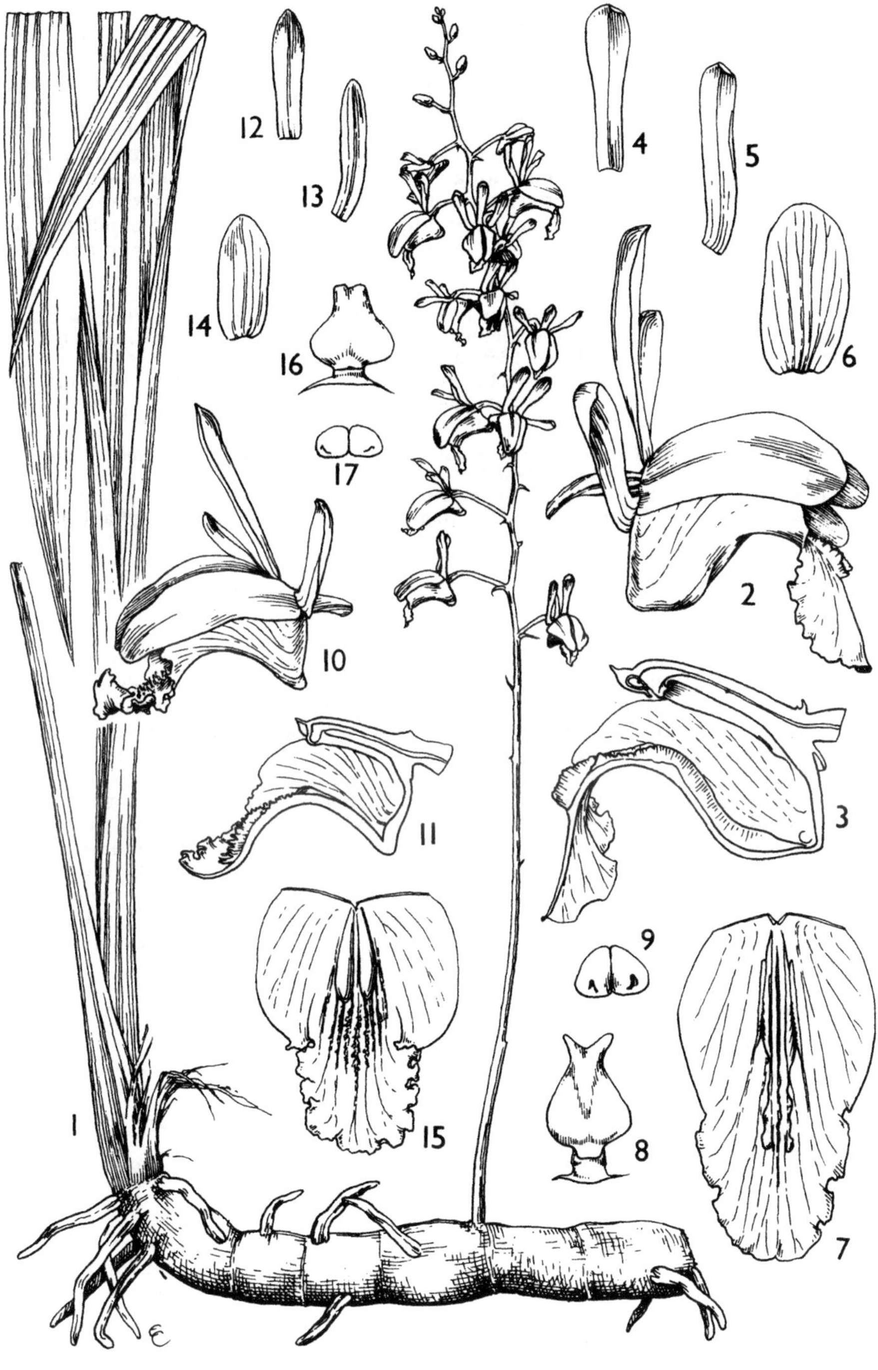

FIG. 93. *EULOPHIA ANGOLENSIS*—**1**, habit, × 1/3; **2**, flower, × 1½; **3**, longitudinal section of lip and column, × 1½; **4**, dorsal sepal, × 1; **5**, lateral sepal, × 1; **6**, petal, × 1; **7**, lip, × 1½; **8**, anther, × 4; **9**, pollinia, × 4; *E. CARICIFOLIA*—**10**, flower, × 1½; **11**, longitudinal section of lip and column, × 1½; **12**, dorsal sepal, × 1; **13**, lateral sepal, × 1; **14**, petal, × 1; **15**, lip, × 1½; **16**, anther, × 4; **17**, pollinia, × 4. 1, from photo.; 2–9, from *Drummond & Hemsley* 4614; 10–17, from *Foy* 6605. Drawn by Eleanor Catherine.

A terrestrial herb 50–130 cm. tall. Perennating organs subterranean, rhizomatous, cylindrical, fleshy, 15 or more cm. long, 6–8 mm. diameter; roots dense along rhizome, 2 mm. diameter. Leaves 4, erect, plicate, linear, acuminate, 60–140 cm. long, 6–14 mm. wide, long-petiolate. Inflorescence laxly few-flowered; peduncle 5 mm. diameter; rhachis 9–40 cm. long; bracts lanceolate-ovate, acute, 5–10 mm. long. Flowers spreading to subnutant; sepals grey-red to maroon; petals pink to white; lip pink veined red with a mauve mid-lobe and yellow callus; pedicel and ovary 21–24 mm. long. Dorsal sepal erect, spathulate, obtuse, 15–21 mm. long, 3–4 mm. wide; lateral sepals erect, spathulate, obtuse, 16–21 mm. long, 3–4 mm. wide. Petals porrect over column, elliptic, rounded, 14.5–18 mm. long, 9–10 mm. wide. Lip 3-lobed, 18–21 mm. long, 10–13.5 mm. wide; side lobes semi-elliptic; mid-lobe elliptic with undulate margin; spur saccate, 5 mm. long, truncate and recurved slightly at apex; callus of 3 basal verrucose ridges merging into 5 cristate-papillate ridges above. Column 8–9 mm. long; foot 5–5.5 mm. long; anther bilobed. Fruit pendent, 14–17 mm. long. Fig. 93/10–17.

UGANDA. Mengo District: Busiro County, ± 1 km. E. of Jungo Church, 30 Apr. 1969, *Lye* 2721! & Kampala, Kabaka's Lake, Apr. 1935, *Synge* 1949! & Kiagwe, Namanve, Mar. 1932, *Eggeling* 573!
TANZANIA. Bukoba, 18 Jan. 1959, *Lind* 2366! & Jan. 1932, *Haarer* 2434!
DISTR. **U** 2, 4; **T** 1; Guinée, Sierra Leone, Liberia, Ivory Coast, Ghana, Nigeria, Cameroon, Gabon, Central African Republic, Zaire, Burundi, Zambia and Angola
HAB. In wet grassland, marshes and swamps; 1200–1350 m.

SYN. *Lissochilus longifolius* Benth. in Hook., Niger Fl.: 530 (1849), *non Eulophia longifolia* Schltr. Type: Ghana, Grand Bassa, *Vogel* (K, holo.!)
L. caricifolius Reichb.f. in Linnaea 41: 74 (1877)
L. wilsonii Rolfe in F.T.A. 7: 79 (1897). Type: Uganda, Toro District, Busongora [Usongora], *Wilson* 23 (K, holo.!)

7. **E. malangana** (*Reichb.f.*) *Summerh.* in K.B. 11: 232 (1956); Williamson, Orch. SC. Afr.: 172 (1977). Type: Angola, Malange, *Mechow* 365 (W, holo.!, K, record of holo.!)

A terrestrial herb 60–100 cm. tall. Perennating organ underground, rhizomatous, 6 or more cm. long, 6–9 mm. diameter; roots 3.5 mm. diameter, white. Leaves 3–6, linear, erect, acuminate, 30–46 cm. long, 4–14 mm. wide, lowermost sheathing. Inflorescence laxly to densely many-flowered, produced before the leaves; peduncle 4 mm. diameter, bearing 3–4 well-spaced sheaths along length; rhachis 6–15 cm. long; bracts elliptic to lanceolate, acute, 6–10 mm. long. Flowers large, spreading; sepals white with greenish brown tips; petals and lip white, lip flushed with lilac in basal half; pedicel and ovary 10–20 mm. long. Dorsal sepal erect, spathulate, rounded at apex, 10–13 mm. long, 6–8.5 mm. wide; lateral sepals similar, 12–13 mm. long, 6–9.5 mm. wide. Petals porrect over column, elliptic-circular, rounded or shortly apiculate, 9–13 mm. long, 8–11 mm. wide, slightly concave. Lip obscurely 3-lobed, 14–21.5 mm. long, 8–12 mm. wide; side lobes narrowly oblong; mid-lobe large, oblong-subcircular, rounded in front, with undulate margins; callus obscure, with 5 central veins somewhat thickened; spur conical, recurved, 2–4 mm. long. Column 5–6 mm. long. Fruit pendent, 20–23 mm. long. Pl. 3.

TANZANIA. Iringa District: Sao Hill, Dec. 1959, *Watermeyer* 212! & 35 km. past Iringa on Mbeya road, 31 Jan. 1979, *Cribb, Grey-Wilson & Mwasumbi* 11226!
DISTR. **T** 7; Zaire, Zambia and Angola
HAB. In seasonally wet grassland and swamps; 1850–2100 m.

SYN. *Lissochilus malanganus* Reichb.f. in Flora 65: 533 (1882)
Eulophia warburgii Schltr. in Warb., Kunene-Sambesi-Exped.: 214 (1903). Type: Angola, Habungu, *Baum* 495 (B, holo., K, iso.!)
E. rigidifolia Kraenzl. in E.J. 51: 387 (1914). Type: Zambia, Kankasa, *Kassner* 2092 (B, holo., K, iso.!)

8. **E. calantha** *Schltr.* in Warb., Kunene-Sambesi-Exped.: 215 (1903); Piers, Orch. E. Afr., ed. 2: 167 (1968); Summerh. in F.W.T.A., ed. 2, 3: 249 (1968); Williamson, Orch. SC. Afr.; 170 (1977). Type: Angola, Longa, near Chijija, *Baum* 631 (B, holo., G, K, iso.!)

A terrestrial herb, 25–78 cm. tall. Perennating organs subterranean, tuberous, irregularly fusiform, obliquely conical or ellipsoidal, tuberous, 1.5–3 cm. long, 8–15 mm. diameter, borne 6–25 mm. apart on a slender rhizome; roots scattered along rhizome and tuber, 1–1.5 mm. diameter. Leaves 1–2, 3-veined, erect, 25–40 cm. long, 5 mm. wide, subtended by 2 sheaths. Inflorescence produced before the leaves, 3–7-flowered; peduncle 18–60 cm. long, slender, bearing 3–4 scattered acute sheaths along length;

rhachis 2.5–13 cm. long; bracts lanceolate, acuminate and bristle-like at the apex, 8–20 mm. long. Flower spreading; sepals white to pinkish flushed light mauve and green at the tips; petals white; lip white with a pink or rosy mauve margin and yellow callus; column white with greenish anther; pedicel and ovary 12–17 mm. long. Dorsal sepal strongly reflexed, oblong-lanceolate, acuminate, 1.3–1.6 cm. long, 3–4 mm. wide; lateral sepals strongly reflexed, oblong, acuminate, 1.3–1.6 cm. long, 3–4 mm. wide. Petals porrect, elliptic or elliptic-obovate, obtuse, 1.7–2 cm. long, 8–10 mm. wide. Lip 3-lobed, curved in side view, 1.4–2 cm. long, 9–14 mm. wide; side lobes obliquely oblong, rounded; mid-lobe transversely oblong, obtuse or emarginate; callus of 3–5 fleshy rugulose ridges slightly dilated towards apex of lip; spur conical-incurved, 5–7 mm. long. Column 7–8 mm. long; foot 1–2 mm. long. Fruit pendent, ± 20 mm. long. Fig. 94/1–8, p. 436.

UGANDA. Mengo District, Buruli, 5 Apr. 1963, *C. Leakey* 22!
TANZANIA. Buha District: Malagarasi R, NE. of Uvinza, 23 Dec. 1949, *Bullock* 81!; Songea District: Ulamboni valley, ± 11 km. W. of Songea, 1 Jan. 1956, *Milne-Redhead & Taylor* 7992!
DISTR. **U** 4; **T** 4, 8; Guinée to Zaire, Angola, Zambia and Malawi
HAB. In marshes, swamps and seasonally wet grassland; 950–1300 m.

SYN. *Lissochilus kapandensis* De Wild. in B.J.B.B. 6: 90 (1919). Type: Zaire, Shaba, Kapanda, *Homblé* 973 (BR, holo.!, K, record of holo.!)
Eulophia calantha Schltr. var. *kubangensis* Schltr. in Warb., Kunene-Sambesi-Exped.: 216 (1903). Type: Angola, Kubango, *Baum* 382 (K, record of holo.!)

9. **E. alta** (*L.*) *Fawcett & Rendle*, Fl. Jam. 1: 112 (1910); Summerh. in F.W.T.A., ed. 2, 3: 249 (1968); Williamson, Orch. SC. Afr.: 169 (1977). Type: based on Plumier's *Limodorum foliis nervosis lanceolatis*, etc. in Burm., Pl. Amer., t. 189 (1758)

A large terrestrial herb 1–2 m. tall with white roots 1.5 mm. diameter. Perennating organ a fleshy elongate subterranean rhizome ± 1 cm. diameter. Leaves 4, plicate, erect, lanceolate, acuminate, up to 120 cm. long, 5–7.5 cm. wide, long-petiolate, sheathing at base. Inflorescence densely many-flowered; peduncle 8–10 mm. diameter, green, bearing 3 sheaths along length; rhachis 16–35 cm. long; bracts reflexed, linear-aristate, acuminate, 10–30(–80) mm. long. Flower not opening widely; sepals olive-green or green; lip and petals red-purple (albino variants are rare); pedicel and ovary 15–23 mm. long. Dorsal sepal oblong-lanceolate, subacute, 18–20 mm. long, 4–5 mm. wide; lateral sepals oblong or ligulate, subacute to acute, 18–22 mm. long, 4–6.5 mm. wide. Petals oblong-ligulate, acute to obtuse, 15–19 mm. long, 3–5 mm. wide. Lip 3-lobed in apical half, saccate at base, 14–19 mm. long, 8–12 mm. wide; side lobes rounded; mid-lobe semicircular, 5–6 mm. long, 8 mm. wide, with an undulate margin; callus of 2 obliquely subquadrate flap-like ridges in middle of lip and with papillae on the central 5 viens of the mid-lobe. Column 7–10 mm. long; foot 4 mm. long. Fruit pendent, 30 mm. long.

UGANDA. Mengo District: Buruli, *Sanford* 2124! & Entebbe, edge of Lake Victoria, Dec. 1935–Jan. 1936, *Chandler* 1578! & Kiagwe, Namanve, May 1932, *Eggeling* 711!
DISTR. **U** 4; Guineé, Sierra Leone, Liberia, Ivory Coast, Ghana, Nigeria, Cameroon, Gabon, Central African Republic, Sudan, Ethiopia, Zaire, Angola, Zambia, Zimbabwe, also in tropical America
HAB. Terrestrial in papyrus and other swamps and in wet grassland; 1150–1300 m.

SYN. *Limodorum altum* L., Syst. Nat., ed. 12, 2: 594 (1767)
Dendrobium longifolium Kunth, Nov. Gen. Sp. 1: 360 (1815). Type: Colombia, Popayan Province, *Bonpland & Humboldt* (P, holo.)
Cyrtopodium woodfordii Sims in Bot. Mag. 43, t. 1814 (1816). Type: plate 1814 in Bot. Mag., based on Brazil, *Woodford*
Cyrtopera woodfordii (Sims) Lindl., Gen. Sp. Orch. Pl.: 189 (1833)
Eulophia woodfordii (Sims) Rolfe in F.T.A. 7: 68 (1897)
E. longifolia (Kunth) Schltr., Orchideen: 347 (1914)

10. **E. coeloglossa** *Schltr.* in Warb., Kunene-Sambesi-Exped.: 216 (1903); Williamson, Orch. SC. Afr.: 171 (1977); J. Stewart et al., Wild Orch. S. Afr.: 244 (1982). Type: Angola, Cubango, Cabindere, *Baum* 367 (B, holo., G, K, iso.!)

A terrestrial herb 30–65 cm. tall. Perennating organ subterranean, rhizomatous, elongate, 4–6 mm. diameter; roots 2 mm. diameter, white. Leaves 4–5, distichous, in a fan, linear, acute, 14–35 cm. long, 4–8 mm. wide. Inflorescence laxly 6–13-flowered; peduncle 2–5 mm. diameter, bearing 3 short acute sheaths; rhachis 6–19 cm. long; bracts lanceolate, acute to acuminate, 4–14 mm. long. Flowers subnutant; sepals glossy, golden-brown to purple-brown; petals pink or yellow; lip pink or mauve in front, rarely yellow;

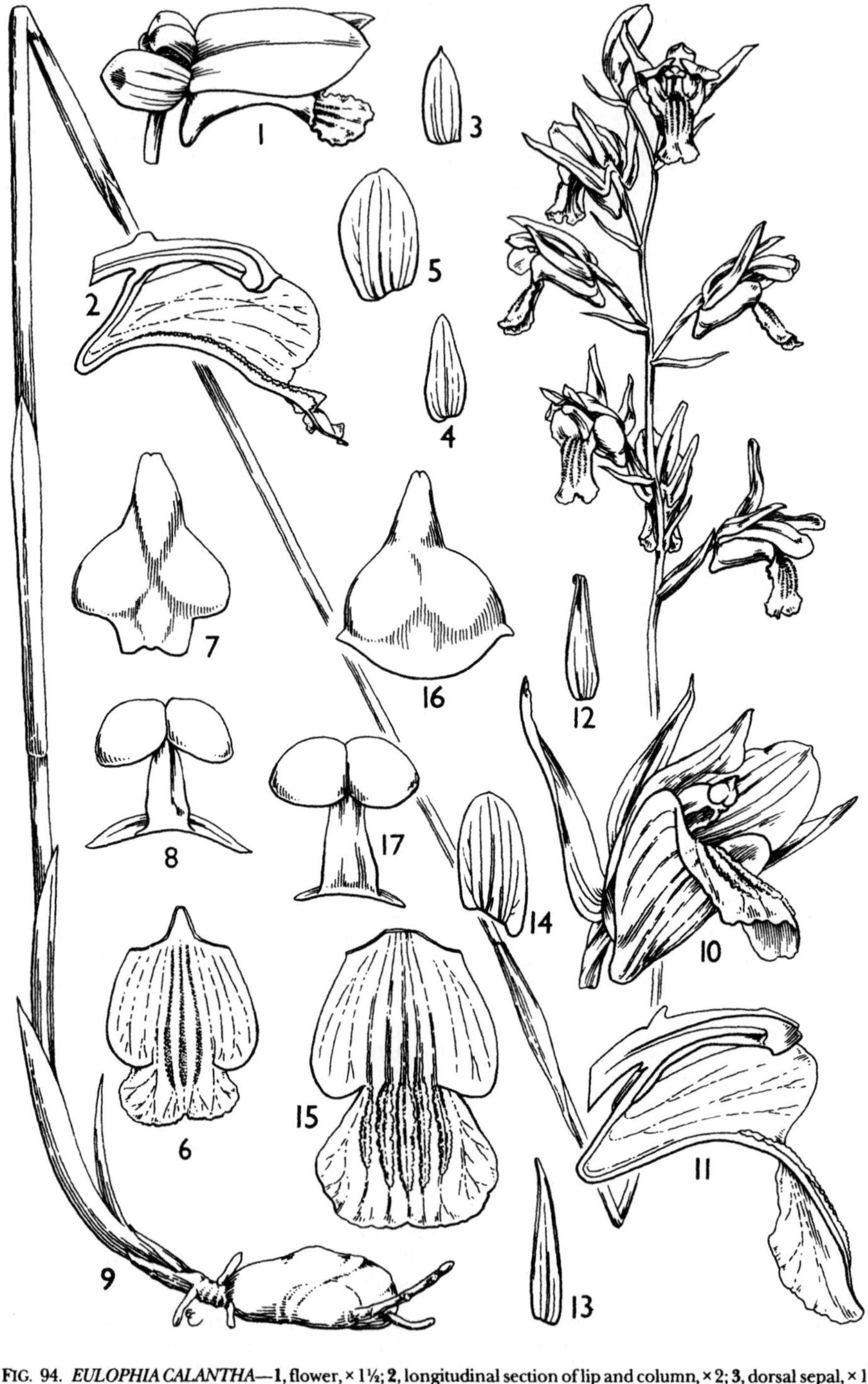

FIG. 94. *EULOPHIA CALANTHA*—**1**, flower, × 1½; **2**, longitudinal section of lip and column, × 2; **3**, dorsal sepal, × 1; **4**, lateral sepal, × 1; **5**, petal, × 1; **6**, lip, × 1½; **7**, anther-cap, × 8; **8**, pollinarium, × 8. *E. FLAVOPURPUREA*—**9**, habit, × ⅔; **10**, flower, × 1½; **11**, longitudinal section of lip and column, × 2; **12**, dorsal sepal, × 1; **13**, lateral sepal, × 1; **14**, petal, × 1; **15**, lip, × 1½; **16**, anther-cap, × 8; **17**, pollinarium, × 8. 1–8, from *Milne-Redhead & Taylor* 7992; 9–17, from *Milne- Redhead & Taylor* 8348. Drawn by Eleanor Catherine.

pedicel and ovary 10–20 mm. long. Dorsal sepal erect, concave, spathulate, rounded, 14–16 mm. long, 3.5–5.5 mm. wide; lateral sepals similar, 14.5–17 mm. long, 3–5.6 mm. wide. Petals porrect, concave, elliptic, obtuse or rounded, 12–16 mm. long, 7–11 mm. wide. Lip obscurely 3-lobed, deeply concave, 12–15 mm. long, 10 mm. wide, saccate at base; side lobes rounded; mid-lobe oblong-elliptic, obtuse; callus of 2 fleshy ridges in basal half of lip; spur conical, 6 mm. long. Column 8–10 mm. long; foot 1 mm. long. Fruit pendent, 18 mm. long, purple-brown. Pl. 2. Fig. 95/1–10, p. 438.

TANZANIA. Songea District: 12 km. E. of Songea, 28 Dec. 1955, *Milne-Redhead & Taylor* 7941!
DISTR. **T** 8; Zambia, Malawi, Angola and Zimbabwe
HAB. In marshes and swamps; 1000–1100 m.

11. **E. cucullata** (*Sw.*) *Steud.*, Nom. Bot., ed. 2: 205 (1840); Hall in Journ. S. Afr. Bot., Suppl. 5: 139 (1965); Summerh. in F.W.T.A., ed. 2, 3: 249 (1968); Piers, Orch. E. Afr., ed. 2: 169 (1968); J. Stewart in U.K.W.F.: 780 (1974); Williamson, Orch. SC. Afr.: 172 (1977); J. Stewart et al., Wild Orch. S. Afr.: 241 (1982). Type: West Africa, without exact locality, *Afzelius* (S, holo.)

A terrestrial herb 40–130 cm. tall. Perennating organs irregularly fusiform-conical potato-like tubers, 4–5.5 cm. long, 2–3 cm. diameter; roots emerging all over tuber and at base of new growth, 1–2.5 mm. diameter, white. Leaves 3–4, plicate, erect, linear, acuminate, 20–70 cm. long, 0.4–1.7 cm. wide; leaf-sheath purple-veined. Inflorescence produced before or with young leaves, laxly 3–8-flowered; peduncle stout, 3–12 mm. diameter, bearing 3 tubular acute sheaths in lower part; rhachis 6–25 cm. long; bracts spreading, linear, acuminate, 35–55 mm. long. Flowers showy, variable in size; sepals maroon, brown or ochre; petals rose-purple; lip rose-purple with a white to cream throat speckled and streaked with purple, side lobes greenish striped with brown; pedicel and ovary 8–25 mm. long. Dorsal sepal erect or reflexed, lanceolate, acuminate, 14–24 mm. long, 3–8 mm. wide; lateral sepals erect or reflexed, obliquely lanceolate, acuminate, 14–25 mm. long, 3–10 mm. wide. Petals parallel to column, elliptic or obovate, rounded, 15–24 mm. long, 10–21 mm. wide. Lip 3-lobed, saccate, 18–34 mm. long, 20–40 mm. wide; side lobes erect, rounded; mid-lobe occasionally clawed, transversely oblong, emarginate, 11–12 mm. long, 16–22 mm. wide; callus of 2 parallel quadrate ridges at base of mid-lobe. Column white, 13–15 mm. long; foot 1 mm. long. Fruit pendent, ± 20 mm. long. Pl. 2. Fig. 95/11–19, p. 438.

UGANDA. Acholi District: between Opit and Paimol, near Paimol, 9 May 1961, *Lind* 2984!; Toro District: Busongoro, June 1933, *Brasnett* 1352!; Teso District: Serere, Kamod swamp, 22 July 1934, *Synge* 758!; Mengo District: near Entebbe, Feb. 1932, *Eggeling* 4286!
KENYA. Turkana District: Lake Turkana [Rudolf], 12 Sept. 1899, *Wellby*!; Trans-Nzoia District: Kitale, Mar. 1935, *Thorold* 3200!; Kwale District: Shimba Forest, 26 Apr. 1962, *Lucas, Jeffrey & Kirrika* 246!
TANZANIA. Bukoba District: Kamachumu road, Oct. 1931, *Haarer* 2204!; Pangani District: Mkwaja, Mkaramo, 25 Jan. 1956, *R. Tanner* 2558!; Kondoa District: Salanka, 10 Jan. 1962, *Polhill & Paulo* 1117!; Mbeya District: 8 km. NE. of Tunduma, 10 Jan. 1975, *Brummitt & Polhill* 13683!
DISTR. **U** 1–4; **K** 1–3, 5, 7; **T** 1–8; **Z**; throughout tropical and southern Africa from Sierra Leone to South Africa (Natal), also in Madagascar
HAB. In grassland, wooded grassland, bushland and woodland; 200–2300 m.

SYN. *Limodorum cucullatum* Sw. in K. Svenska Vetensk.-Akad. Handl. 21: 243 (1800) & in Pers., Syn. Pl. 2: 521 (1807)
Lissochilus arenarius Lindl. in J.L.S. 6: 133 (1862). Type: Niger Exped., *Barter* 1488 (K, holo.!)
L. dilectus Reichb.f. in Otia Bot. Hamb. 1: 62 (1878). Type: Angola, Pungo Andongo, *Welwitsch* 886 (W, holo.!, K, iso.!)
Eulophia arenaria (Lindl.) Bolus in J.L.S. 25: 185 (1889)
E. dilecta (Reichb.f.) Schltr. in Westafr. Kautsch.-Exped.: 279 (1900)
E. monteiroi (Rolfe) Butzin in Willdenowia 7: 589 (1975)
Lissochilus monteiroi Rolfe in F.T.A. 7: 83 (1987). Type: Angola, Benguela, *Monteiro* (K, holo.!)

12. **E. flavopurpurea** (*Reichb.f.*) *Rolfe* in F.T.A. 7: 65 (1897); Summerh. in F.W.T.A., ed. 2, 3: 249 (1968); Williamson, Orch. SC. Afr.: 170 (1977). Type: Sudan, Niamniam, *Schweinfurth* 3546 (W, holo.!, K, record of holo.!)

A terrestrial herb 60–95 cm. tall. Perennating organs subterranean, tuberous, irregularly shaped, 3–4 cm. long, 1–2 cm. diameter, 2–3-noded; roots scattered, 1 mm. diameter. Leaves 3–4, linear, acuminate, 5–35 cm. long, 2–7.5 mm. wide, basal 2 sheathing. Inflorescence produced before leaves have developed, laxly 1–8-flowered; peduncle up to 75 cm. long, bearing 3 widely spaced acute bracts; rhachis 2–20 cm. long;

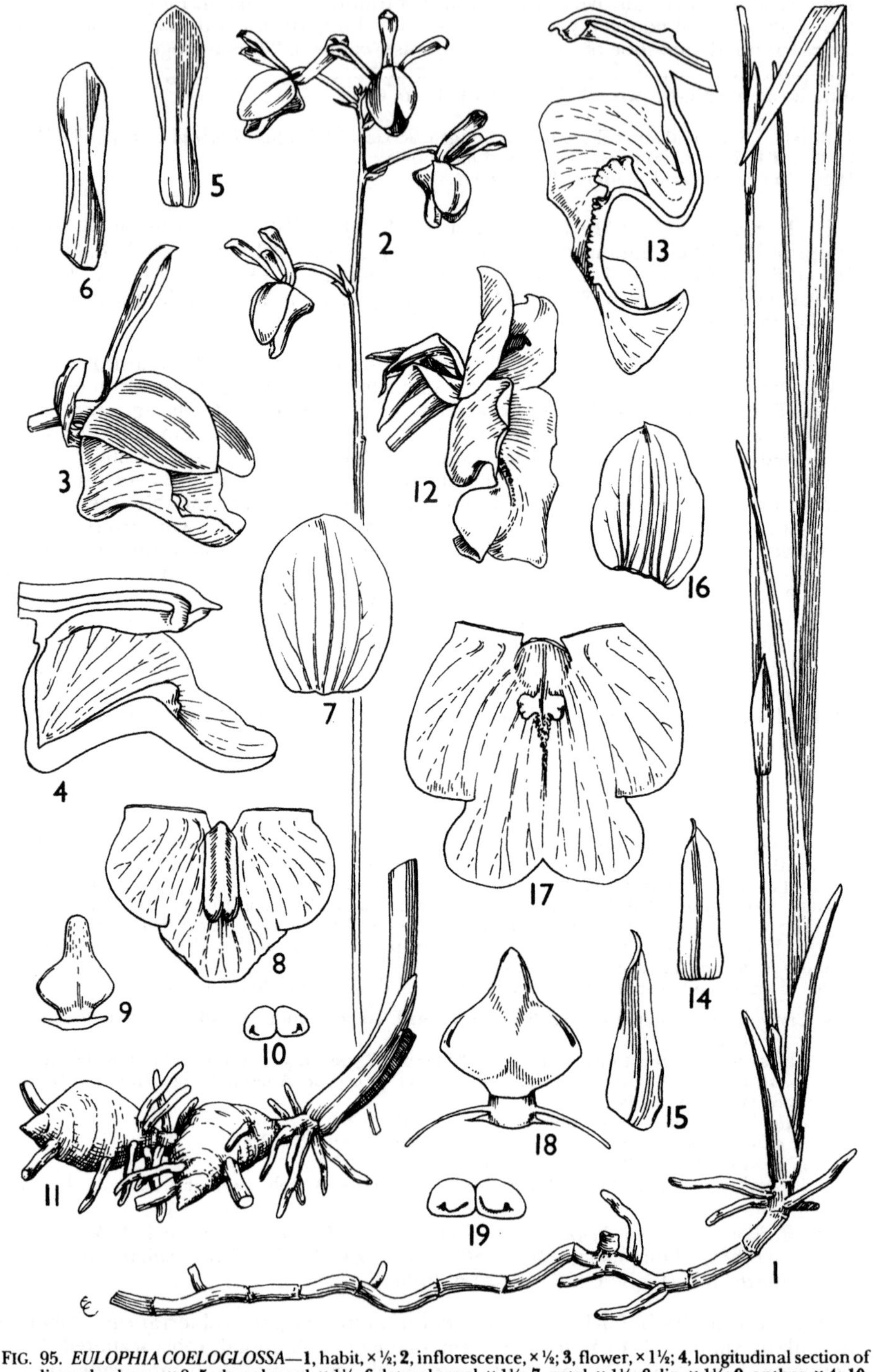

FIG. 95. *EULOPHIA COELOGLOSSA*—**1**, habit, × ½; **2**, inflorescence, × ½; **3**, flower, × 1½; **4**, longitudinal section of lip and column, × 2; **5**, dorsal sepal, × 1½; **6**, lateral sepal, × 1½; **7**, petal, × 1½; **8**, lip, × 1½; **9**, anther, × 4; **10**, pollinia, × 4. *E. CUCULLATA*—**11**, rhizome, × ½; **12**, flower, × 1; **13**, longitudinal section of lip and column, × 1½; **14**, dorsal sepal, × 1; **15**, lateral sepal, × 1; **16**, petal, × 1; **17**, lip, × 1; **18**, anther, × 4; **19**, pollinia, × 4. 1–10, from *Milne-Redhead & Taylor* 7941; 11–19, from *Milne-Redhead & Taylor* 7976B. Drawn by Eleanor Catherine.

bracts spreading, linear, acuminate, 15–30 mm. long. Flowers subnutant, not opening widely; ovary and sepals pale greenish, usually tinged dull purple; petals pale greenish yellow with a purplish midvein; lip yellowish green; pedicel and ovary 13–21 mm. long. Dorsal sepal spreading, lanceolate, acute, 2–3.4 cm. long, 4 mm. wide; lateral sepals similar but oblique at base, 2.4–3 cm. long, 4–7 mm. wide. Petals porrect, elliptic-ovate, obtuse, 2–2.3 cm. long, 7–9.5 mm. wide. Lip 3-lobed, 2.3–2.9 cm. long, 1–1.5 cm. wide; side lobes narrowly oblong, rounded in front; mid-lobe deflexed, flabellate-obovate, 1.5 cm. long, 1.2 cm. wide; callus of 7–11 low verrucose ridges on mid-lobe; spur conical, 4–7 mm. long. Column 12–13 mm. long; foot 1 mm. long. Fruit pendent, 24–28 mm. long. Fig. 94/9–17.

UGANDA. W. Nile District: Adumi, Apr. 1941, *Eggeling* 4286!; Bunyoro District: Kibangya, 16 May 1941, *A.S. Thomas* 3907!; Mengo District: Buruli County, Apr. 1963, *C. Leakey* 20!
TANZANIA. Mbeya District: 8 km. NE. of Tunduma, 10 Jan. 1975, *Brummitt & Polhill* 13685!; Songea District: Ulamboni valley, ± 11 km. W. of Songea, 7 Jan. 1956, *Milne-Redhead & Taylor* 7996B!
DISTR. **U** 1, 2, 4; **T** 7, 8; Ivory Coast, Ghana, Togo, Nigeria, Cameroon, Central African Republic, Zaire, Sudan, Zambia and Malawi
HAB. In wet grassland and swamps; 600–1650 m.

SYN. *Cyrtopera flavopurpurea* Reichb.f. in Otia Bot. Hamb. 1: 68 (1878)
Lissochilus millsonii Rolfe in F.T.A. 7: 79 (1897). Type: Nigeria, Yomba, *Millson* 86 (K, holo.!)
Eulophia tubifera Kraenzl. in E.J. 28: 169 (1900). Type: Cameroon, without exact locality, *Dusen* 258 (B, holo., K, record of holo.!)
Lissochilus lacteus Kraenzl. in E.J. 43: 399 (1909). Type: Togo, Agame Mts., *Baumann* 24 (B, holo., K, record of holo.!)
L. andersonii Rolfe in K.B. 1910: 159 (1910). Type: Ghana, Aburi, *Anderson* (K, holo.!)
L. johnsonii Rolfe in K.B. 1910: 160 (1910). Type: Ghana, Accra Plains, *Johnson* 854 (K, holo.!)
L. seretii De Wild. in Ann. Mus. Congo, Bot., sér. 5, 3: 180 (1910). Type: Zaire, between Vankerkhovenville and Faradje, *Seret* 529 (BR, holo.!, K, iso.!)
L. corbisieri De Wild. in B.J.B.B. 6: 80, t. 15/5–9 (1919). Types: Zaire, Welgelegen, *Homblé* 586 (BR, syn.!) & Kafubo Plain, *Hock* (BR, syn.!)

13. **E. cristata** (*Sw.*) *Steud.*, Nom. Bot., ed. 2, 1: 605 (1840); Summerh. in F.W.T.A., ed. 2, 3: 249 (1968). Type: Sierra Leone, without exact locality, *Afzelius* (S, holo.!)

A terrestrial herb 60–130 cm. tall. Perennating organs underground, tuberous, potato-like, cylindric-ellipsoidal, 3–9 cm. long, 1.3–4 cm. diameter, whitish, in chains; roots 2 mm. diameter, white, produced all over tubers. Leaves 4–6, erect, plicate, lanceolate, acuminate, 40–70 cm. long, 0.7–3.5 cm. wide, petiolate, basal 1–3 sheathing. Inflorescence appearing before leaves, laxly 10–30-flowered; peduncle 3–8 mm. diameter, with 3–5 sheaths along length; rhachis 13–44 cm. long; bracts aristate, acuminate, 10–30 mm. long. Flowers with lilac sepals and petals, a bright dark purple lip and greenish column; pedicel and ovary 15–27 mm. long. Dorsal sepal oblanceolate to narrowly elliptic, acute, 14–23 mm. long, 3–5.5 mm. wide; laterals similar but slightly larger and oblique at base. Petals elliptic, obtuse, 12.5–19 mm. long, 6.5–9.5 mm. wide. Lip 3-lobed, 12.5–22 mm. long, 10–13.5 mm. wide; side lobes erect, semicircular, recurved; mid-lobe deflexed, somewhat convex, elliptic, obtuse, crisped on margin; callus of 2 raised rounded ridges at base with 5–9 crenulate ridges in front on mid-lobe; spur slightly upcurved, conical, 2.5–5 mm. long. Column 7–7.5 mm. long; foot 1 mm. long. Fruit pendent, 30 mm. long. Fig. 96/1, p. 441.

UGANDA. Karamoja District: Opotipot, 8 Jan. 1958, *Beadle* 21!; Bunyoro District: Buhanika, Jan. 1943, *Purseglove* 1241!; Mengo District: Bugerere, Busana, Mar. 1932, *Eggeling* 529!
DISTR. **U** 1, 2, 4; Senegal and Gambia across to Sudan and Ethiopia
HAB. In grassland, often seasonally burnt and in *Commiphora-Albizia* bushland and woodland; 1100–1700 m.

SYN. *Limodorum cristatum* Sw. in Schrad., Neues Journ. Bot. 1: 86 (1805)
Lissochilus purpuratus Lindl. in J.L.S. 6: 133 (1862). Type: Niger Exped., Abbeokuta, *Barter* 3331 (K, holo.!)
L. heudelotii Reichb.f. in Otia Bot. Hamb. 1: 63 (1878). Type: Senegal, without exact locality, *Heudelot* (W, holo.!)
L. uliginosus Rolfe in K.B. 1913: 340 (1913). Type: Ghana, Chama, *Chipp* 192 (K, holo.!)

14. **E. livingstoniana** (*Reichb.f.*) *Summerh.* in K.B. 2: 132 (1948); J. Stewart in U.K.W.F.: 780 (1974); Williamson, Orch. SC. Afr.: 173 (1977); J. Stewart et al., Wild Orch. S. Afr.: 251 (1982). Types: Malawi, Manganja Hills, Margomero, *Meller* (W, syn.!, K, isosyn.!) & Manganja Hills, *Waller* (W, syn., K, isosyn.!)

A terrestrial herb 50–100 cm. tall. Perennating organs underground, tuberous, irregularly cylindrical, horizontal, branching occasionally, 5–9.5 cm. long, 1.1–2.5 cm. diameter; rhizome short, up to 1 cm. diameter; roots scattered along tuber, 1.5 mm. diameter, white. Leaves 3–6, linear, acute, 10–40 cm. long, 5–17 mm. wide, lowermost 1–2 sheathing. Inflorescence produced before leaves develop, laxly few–many-flowered; peduncle 37–75 cm. long, with 2–3 sheaths in lower half; rhachis 10–26 cm. long; bracts spreading, linear-lanceolate, acuminate, lowest 17–19 mm. long. Flowers with pale to darker lilac sepals and petals; lip purple with darker callus ridges and green or yellowish side lobes flushed and edged with pink; pedicel and ovary 22–30 mm. long. Dorsal sepal reflexed, lanceolate or oblanceolate, acute or obtuse, 15–20 mm. long, 4–5 mm. wide; lateral sepals similar but slightly wider. Petals erect, elliptic-obovate, obtuse, 14–20 mm. long, 8–9 mm. wide. Lip 3-lobed, 12–17 mm. long; side lobes erect, rounded; mid-lobe reduplicate, upcurved towards apex, elliptic, obtuse, 8–12 mm. long; callus of 5 low ridges to apex of mid-lobe; spur narrowly conical, 3–7 mm. long, slightly upcurved at apex. Column clavate, 7–9 mm. long, purple or pink. Fruit pendent, 18–20 mm. long. Pl. 3. Fig. 96/2–10.

UGANDA. W. Nile District: Paidha [Pyeda], Mar. 1934, *Tothill* 1580!; Kigezi District: Kinkizi [Kinkisi], Kirenzi [Karensi], Mar. 1946, *Purseglove* 1954!; Mubende, 22 Jan. 1931, *Hill* 42!

KENYA. Trans-Nzoia, Mar. 1932, *Tweedie* 28!; Uasin Gishu District: Kipkarren, Mar. 1932, *Brodhurst Hill* 699!; Kwale District: Shimba Hills, Lango ya Mwagandi [Longomwagandi], 29 Feb. 1968, *Magogo & Glover* 851!

TANZANIA. Biharamulo District: Kashasha, 10 Aug. 1960, *R. Tanner* 5070!; Mpanda District: Silkcub Highlands, 6 Dec. 1956, *Richards* 7167!; Kondoa District: Kikori, 5 Nov. 1929, *B.D. Burtt* 2761!; Zanzibar I., *Lyne* 70!

DISTR. **U** 1, 2, 4; **K** 3, 5, 7; **T** 1, 3–5, 7, 8: **Z**; Sudan, Zaire, Rwanda, Zambia, Mozambique, Malawi, Angola, Zimbabwe and Botswana, also in Madagascar and the Comoro Is.

HAB. In grassland or in deciduous woodland of *Brachystegia*, *Berlinia*, etc.; sea-level to 2200 m.

SYN. *Lissochilus livingstonianus* Reichb.f. in Otia Bot. Hamb. 2: 114 (1881)

L. malangensis Reichb.f. in Otia Bot. Hamb. 2: 114 (1881). Type: Angola, Malange, *Mechow* 261 (W, holo.!)

L. fallax Reichb.f. in Otia Bot. Hamb. 2: 115 (1881). Type: Kenya, Mombasa, *Hildebrandt* 1951 (W, holo.!)

L. rutenbergianus Kraenzl. in Abh. Naturw. Ver. Bremen 7: 257 (1882). Type: Madagascar, Mahazamba, *Rutenberg* (BREM, holo.)

L. affinis Rendle in J.B. 33: 193 (1895). Type: Burundi [Urundi], without exact locality, *Scott Elliot* 8173, in part (BM, holo., K, iso.!)

L. mediocris Rendle in J.B. 33: 193 (1895). Type: Kenya, Central Kavirondo District, Samia, *Scott Elliot* 7128 (BM, holo., K, iso.!)

L. cornigerus Rendle in J.B. 33: 194 (1895). Type: Uganda, E. side of Lake Albert Edward Nyanza, Kahimbe, *Scott Elliot* 8031 (BM, holo., K, iso.!)

L. gracilior Rendle in J.B. 33: 195 (1895). Type: Burundi [Urundi], without exact locality, *Scott Elliot* 8173, in part (BM, holo., K, iso.!)

L. gracilior Rendle var. *angusta* Rendle in J.B. 33: 195 (1895). Types: Uganda, E. of Lake Albert Edward Nyanza, *Scott Elliot* 8075 (BM, syn., K, isosyn.!) & Tanzania, Bukoba District, Karagwe, *Scott Elliot* 8209 (BM, syn., K, isosyn.!)

Eulophia jumelleana Schltr. in Ann. Mus. Col. Marseille, sér. 3, 1: 172 (1913). Type: Madagascar, Mahevarano, *Jumelle* 84 (P, holo.)

E. robusta Schltr. in Ann. Mus. Col. Marseille, sér. 3, 1: 172 (1913), in obs., nomen, *non* Rolfe (1910), based on Madagascan material

Lissochilus jumelleanus (Schltr.) Schltr. in Beih. Bot. Centr. 33: 421 (1915)

L. laggiarae Schltr. in F.R. 15: 332 (1918). Type: Madagascar, without exact locality, *Laggiara* (B, holo.)

L. faradiensis De Wild. in B.J.B.B. 6: 83 (1919), as "*faradjensis*". Type: Zaire, near Faradje, *Seret* 527 (BR, holo.!)

NOTE. This is a very variable species in both stature and flower size but the variation appears to be continuous and not to warrant taxonomic recognition.

15. **E. petersii** *Reichb.f.* in Flora 48: 186 (1865); Hall in Journ. S. Afr. Bot., Suppl. 5: 163 (1965); Piers, Orch. E. Afr., ed. 2: 179 (1968); J. Stewart in U.K.W.F.: 776 (1974); Williamson, Orch. SC. Afr.: 159 (1977); J. Stewart et al., Wild Orch. S. Afr.: 245 (1982). Type: Mozambique, without exact locality, *Peters* (W, holo.)

A large terrestrial herb 100–350 cm. tall. Perennating organs pseudobulbous, conical, 6–23 cm. tall, 1.4–4 cm. diameter, 4–6-noded, green turning yellow, covered with 5 papery sheaths when young; roots stout, 5 mm. diameter, white. Leaves 2–3(–4), erect or suberect, fleshy-leathery, somewhat conduplicate, linear-ligulate, acute, 14–80 cm. long, 1.4–6 cm.

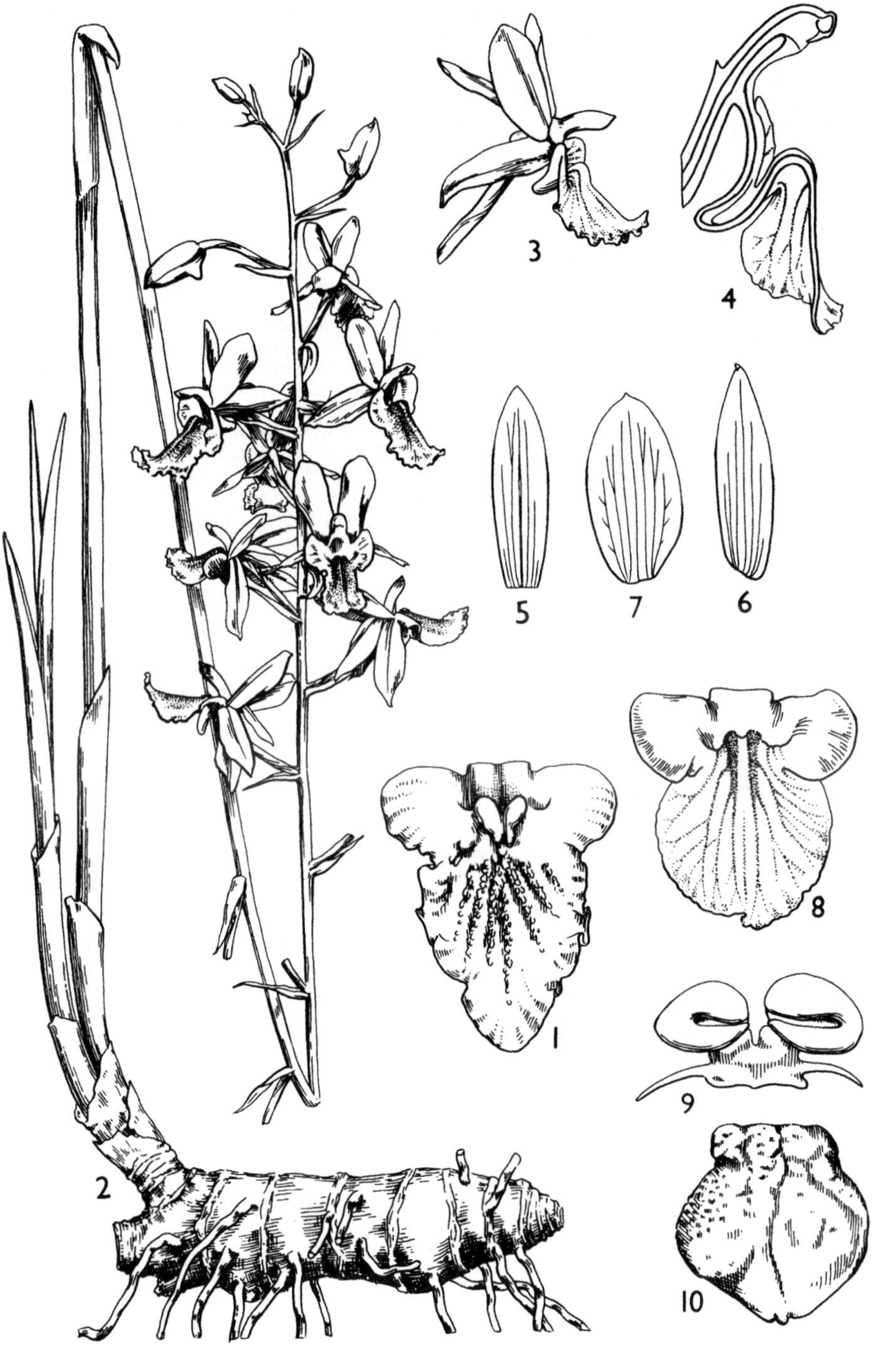

FIG. 96. *EULOPHIA CRISTATA*—**1**, lip, × 2. *E. LIVINGSTONIANA*—**2**, habit, × ⅔; **3**, flower, × 1; **4**, longitudinal section of lip and column, × 2; **5**, dorsal sepal, × 1½; **6**, lateral sepal, × 1½; **7**, petal, × 1½; **8**, lip, × 2; **9**, pollinarium, × 10; **10**, anther-cap, × 10. 1, from *Hoyle* 727; 2,3, from *Milne-Redhead & Taylor* 7805; 4–8, from *Robinson* 2472; 9,10, from *Piers* 80. Drawn by Eleanor Catherine.

wide, serrate on margins. Inflorescence 3–7-branched, laxly many-flowered; branches to 35 cm. long; peduncle stout, up to 90 cm. long, 1–1.4 cm. diameter, with 4 well-spaced bracts; bracts lanceolate, acuminate, 5–16 mm. long, 4–5 mm. wide. Flowers fleshy; sepals olive-green flushed or striped maroon; lip white, veined red or purple; pedicel and ovary 20–26 mm. long. Dorsal sepal erect, oblanceolate, acuminate or apiculate, 17–23 mm. long, 4–5 mm. wide, recurved at apex; lateral sepals similar. Petals oblanceolate or oblong, shortly apiculate, recurved at apex, 16–18 mm. long, 5–7 mm. wide. Lip 3-lobed, 14–20 mm. long, 8–15 mm. wide; side lobes elliptic; mid-lobe circular to subquadrate, emarginate or obtuse, 4 mm. long, 6 mm. wide; callus of 3 fleshy ridges, raised and erose on the mid-lobe; spur incurved, cylindric-subclavate, 4–6 mm. long. Column 8–9 mm. long; foot 1–2 mm. long. Fruit pendent, ellipsoidal, 40–46 mm. long. Fig. 97/1–12.

UGANDA. W. Nile District: Aringa, near summit of Midigo Hill, Apr. 1941, *Eggeling* 4289!; Bunyoro District: Kibiro [Kibero], flowered in Entebbe Botanic Garden, Nov. 1945, *A.S. Thomas* 3754!

KENYA. Northern Frontier Province: El Wak, 29 May 1952, *Gillett* 13392!; Turkana District: 1 km. NE. of Loiya, on Lodwar road, 8 Nov. 1977, *Carter & Stannard* 260!; Kwale District: Kinango to Mariakani road, 3.6 km. from Kinango, 29 Aug. 1953, *Drummond & Hemsley* 4083!

TANZANIA. Mwanza District: coast of Lake Victoria, Bwiro [Bwiru] Bay, June 1935, *B.D. Burtt* 5216!; Korogwe District: N. of Makuyuni [Makauni] on Korogwe to Mombo road, 9 Mar. 1956, *Greenway* 8876!; Uzaramo District: Dar es Salaam, 2 Oct. 1941, *Moreau* 223a!; Zanzibar I., Chukwani, 23 Apr. 1950, *R.O. Williams* 21!

DISTR. **U** 1, 2; **K** 1–4, 7; **T** 1–3, 5, 6; **Z**; Zaire, Ethiopia and Sudan south to South Africa, also in Arabia

HAB. By sea in sandy soil, in rocky places, in *Acacia-Commiphora*, *Grewia* and other thickets and bushland; sea-level to 1800 m.

SYN. *Galeandra petersii* Reichb.f. in Linnaea 20: 679 (1847)
Eulophia schimperiana A. Rich., Tent. Fl. Abyss. 2: 283 (1850). Type: Ethiopia, *Schimper* 1592 (P, holo., G, iso.!)
E. caffra Reichb.f. in Flora 48: 186 (1865). Type: South Africa, cult. *Warner* (W, holo.)
E. baginsensis Reichb.f. in Otia Bot. Hamb. 1: 66 (1878). Type: Sudan, Niamniam, *Schweinfurth* 24 (W, holo., K, iso.!)
E. longepedunculata Rendle in J.L.S. 30: 382 (1895). Type: Tanzania, between the coast and Uyui, *W.E. Taylor* (BM, holo.!)
E. coleae Rolfe in F.T.A. 7: 56 (1897). Type: Somalia, Golis range, Woob, *Cole* (K, holo.!)
E. phillipsiae Rolfe in F.T.A. 7: 56 (1897). Type: Somalia, Golis range, Woob, *Lort Phillips* (K, holo.!)
E. gumbariensis De Wild. in B.J.B.B. 6: 117 (1919). Type: Zaire, Gumbari, *Seret* 465 (BR, holo.!)

NOTE. *E. petersii* is a variable species particularly in its pseudobulb size and leaf width. These variations, however, do not seem to merit taxonomic recognition.

16. **E. serrata** *Cribb*, nom. nov. pro *E. stricta* Rolfe (1897), *non* Lindl. (1833). Type: Kenya, 40–60 km. from coast, *Johnston* (K, holo.!)

A terrestrial herb up to ± 65 cm. tall. Perennating organs not seen but said to be pseudobulbous, inconspicuous; roots thick, white. Leaves 3–4, fleshy, linear-lanceolate, acute, 14–35 cm. long. 1.5–2.2 cm. wide, with finely serrated margins. Inflorescence erect, unbranched, 40–65 cm. tall, laxly up to 20-flowered; peduncle 4 mm. diameter; rhachis up to 35 cm. long; bracts lanceolate, acuminate, 6–9 mm. long. Flowers fleshy, dull green with maroon streaks on sepals and petals; lip white with maroon streaks; pedicel and ovary 11–12 mm. long. Dorsal sepal oblong-lanceolate, obtuse, 10–11 mm. long, 4 mm. wide; lateral sepals similar. Petals porrect, ovate, subacute, 7–8 mm. long, 5 mm. wide. Lip 3-lobed, 9 mm. long and wide; side lobes erect, semicircular, rounded and erose in front; mid-lobe transversely elliptic-oblong, obtuse, with erose-undulate margins; callus of 3 ridges, rugulose and undulate towards apex on base of mid-lobe; spur 2–3 mm. long, slightly clavate. Column 4.5 mm. long; anther-cap bilobed. Fruit pendent, 23–25 mm. long.

KENYA. Kilifi District: near Ganze, fl. in cult. Feb. 1986, *S.A. Robertson* 4037B!

DISTR. **K** 7; not known elsewhere

HAB. In shallow sandy soil over rock in *Commiphora-Acacia* bushland; 150–200 m.

SYN. *E. stricta* Rolfe in Fl. Trop. Afr. 7: 51 (1897), *non* Lindl., Gen. Sp. Orch. Pl.: 185 (1833), *nom. illegit.*

17. **E. taitensis** *Cribb & Pfennig* in Die Orchidee 28: 214 (1977). Type: Kenya, Teita Hills, *Pfennig* 1113 (K, holo.!)

A terrestrial herb 60–110 cm. tall. Perennating organs erect, proximate, pseudobulbous, narrowly cylindric-conical to fusiform, 7–22 cm. long, 1.5–3 cm. diameter, green, covered

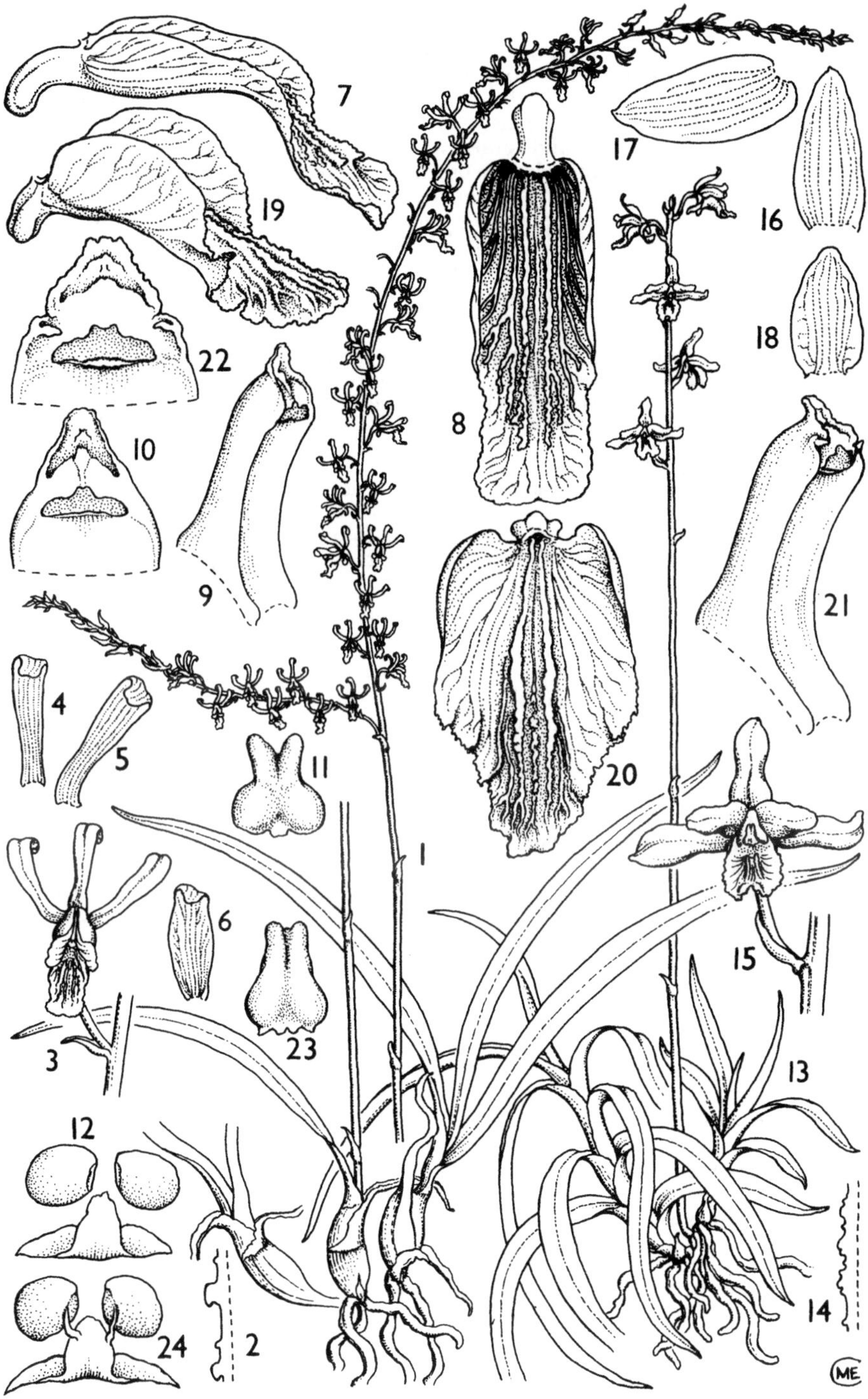

FIG. 97. *EULOPHIA PETERSII*—**1**, habit, much reduced; **2**, leaf-margin, × 6; **3**, flower, × ⅔; **4**, dorsal sepal, × 1; **5**, lateral sepal, × 1; **6**, petal, × 1; **7**, lip, side view, × 2; **8**, lip, front view, × 2; **9**, column, side view, × 3; **10**, column apex, front view, × 4; **11**, anther-cap, × 4; **12**, pollinarium, × 8. *E. TAITENSIS*—**13**, habit, much reduced; **14**, leaf-margin, × 6; **15**, flower × ⅔; **16**, dorsal sepal, × 1; **17**, lateral sepal, × 1; **18**, petal, × 1; **19**, lip, side view, × 2; **20**, lip, front view, × 2; **21**, column, side view, × 3; **22**, column apex, front view, × 4; **23**, anther-cap, × 3; **24**, pollinarium, × 5, 1,3–12, from *Pfennig* 1031; 2, from *Carter & Stannard* 260; 13, from *Meyer* 1432; 14–24, from *Pfennig* 1113. Drawn by Mrs M.E. Church.

when young by sheathing leaf-bases; roots stout, long, 6 mm. diameter, white. Leaves 6–14, fleshy, conduplicate, linear, acute or acuminate, 11–50 cm. long, 1.4–3 cm. wide, margins serrulate. Inflorescence laxly 5–15-flowered; peduncle bearing 4 short acute sheaths along length; rhachis 20–45 cm. long; bracts lanceolate, acuminate, 9–11 mm. long. Flowers fleshy; sepals and petals yellow-green flushed with brown at tips; lip white or cream with a yellow callus and veined with red; pedicel and ovary 15–27 mm. long. Dorsal sepal spreading, oblanceolate, acute or apiculate, 22–25 mm. long, 4–8 mm. wide; lateral sepals lanceolate or oblong-lanceolate, acute or apiculate, 22 mm. long, 6–7 mm. wide. Petals oblong-lanceolate, obtuse or acuminate, 19–20 mm. long, 4.5–9 mm. wide. Lip obscurely 3-lobed, 18–20 mm. long, 15–17 mm. wide; side lobes elliptic; mid-lobe ± quadrate; callus of 3 ridges, verrucose but not raised towards apex; spur 3 mm. long. Column 11 mm. long; foot 3 mm. long. Fruit pendent, 36–40 mm. long, 16 mm. diameter. Pl. 3. Fig. 97/13–24.

UGANDA. Karamoja District: Amudat, May 1948, *Eggeling* 5820! & Feb. 1957, *Tweedie* 1487!; Teso District: Kyere, Mar. 1938, *Chandler* 546!; Mengo District: Buruli, Apr. 1931, *Sanford* 2129!
KENYA. Kwale District: Mwawesa [Maweza], without date, *R.M. Graham* 388!
TANZANIA. Tanga District: Moa, *Boscawen* in *Bally* 8146!; Uzaramo District: 13 km. on Pugu road, 18 Sept. 1938, *Vaughan* 2483!
DISTR. **U** 1, 3, 4; **K** 7; **T** 3, 6; not known elsewhere
HAB. In thickets of *Acacia-Ormocarpum* on sandy soils, coastal bushland and in woodland; sea-level to 1300 m.

18. **E. callichroma** *Reichb. f.* in Otia Bot. Hamb. 2: 116 (1881). Type: Malawi, Manganga Hills, *Meller* (W, holo.!, K, iso.!)

A robust terrestrial or lithophytic herb up to 50 cm. tall. Perennating organ above ground, pseudobulbous, erect, conical, 3–10 cm. long, 1.4–2.1 cm. diameter; roots from base, fleshy, brittle, 3–5 mm. diameter, white. Leaves 7–11, distichous, recurving, thinly coriaceous, tapering from a broad sheathing base, linear above, acuminate, up to 40 cm. long, 6–8 mm. wide. Inflorescence arising from upper nodes of the pseudobulb, produced before the leaves, 1–4-branched, laxly many-flowered; peduncle stout, 16–21 cm. long, with 2–3 acute 15 mm. long sheaths along length; bracts linear, acuminate, 2–16 mm. long. Flower not opening widely; sepals and petals olive-brown to yellow-green; lip whitish green with brown lateral venation; pedicel and ovary 9–21 mm. long. Dorsal sepal linear-subspathulate, acute, 8–9 mm. long, 1 mm. wide; lateral sepals linear, acute, 8–9 mm. long, 1 mm. wide. Petals linear, acute, 6–7.5 mm. long, 1 mm. wide. Lip oblanceolate, obtuse or rounded, 7.9 mm. long, 1.5 mm. wide, apical margin undulate; callus of 5 ridges rugulose in apical half of lip, smooth and lower towards base; spur deflexed, cylindrical, 4–5.5 mm. long. Column 4–4.5 mm. long; foot 1 mm. long. Fruit pendent, 25 mm. long, 6 mm. diameter. Fig. 98.

TANZANIA. Songea District: ± 12 km. N of Songea, between Luhira and Lumecha Rivers, 4 May 1956 (sterile), *Milne-Redhead & Taylor* 9997!
DISTR. **T** 8; Zambia, Malawi, Mozambique and Zimbabwe
HAB. On sandy soil and rocks in *Brachystegia-Uapaca* woodland; 100–1100 m.

SYN. *Acrolophia paniculata* Cribb in K.B. 32: 173 (1977). Type: Zimbabwe, Harare [Salisbury], *Drummond* 4927 (K, holo.!, SRGH, iso.!)

19. **E. stachyodes** *Reichb.f.* in Otia Bot. Hamb. 1: 66 (1878); Summerh. in F.W.T.A., ed. 2, 3: 247 (1968); Piers, Orch. E. Afr., ed. 2: 185 (1968); J. Stewart in U.K.W.F.: 782 (1974); Williamson, Orch. SC. Afr.: 154 (1977). Type: Sudan, Niamniam, *Schweinfurth* 3554 (W, holo., K, iso.!)

A terrestrial herb 30–80 cm. tall. Perennating organs underground, tuberous, ellipsoidal or cylindric, 3–4.6 cm. long, 1 cm. diameter; roots 1.5–2 mm. diameter, emerging all over tuber, white. Leaves 4–5, suberect, plicate, lanceolate or oblanceolate, acuminate, 17–100 cm. long, 3–8.5 cm. wide, long-petiolate, basal 1–2 sheathing. Inflorescence produced with the leaves, densely many-flowered; peduncle 3–5 mm. diameter; rhachis 6–26 cm. long; bracts linear-lanceolate, acuminate, 10–55 mm. long. Flowers spreading; sepals dull greenish, bronze or greenish red; petals white, lip green with a purple callus; spur green flushed with purple; pedicel and ovary 15–16 mm. long. Dorsal sepal narrowly oblong, obtuse to acute, 18–23 mm. long, 2.5–4.5 mm. wide; lateral sepals similar. Petals elliptic-oblong, obtuse, 16–18 mm. long, 6.5–8 mm. wide. Lip 3-lobed, 14–15 mm. long, 9 mm. wide; side lobes broadly oblong, rounded in front; mid-lobe

FIG. 98. *EULOPHIA CALLICHROMA*—**1**, **2**, habit, × ⅔; **3**, flower, × 3; **4**, dorsal sepal, × 4; **5**, lateral sepal, × 4; **6**, petal, × 4; **7**, lip, × 4; **8**, column, side view, × 8; **9**, column, front view, × 8; **10**, anther-cap, × 14; **11**, pollinarium, × 14; **12**, capsule, × 1. 1, from *Milne-Redhead & Taylor* 9997; 2, from *Williamson & Gassner* 2315; 3–11, from *Greenway* 8082; 12, from *Grosvenor* 1352. Drawn by Mrs M.E. Church.

oblong, obtuse, 6 mm. long, 4.5 mm. wide, undulate on margin; callus of 3 ridges with 3–5 rows of short papillae on mid-lobe; spur incurved, conical-cylindrical, 4 mm. long. Column 6.5–7 mm. long. Fruit pendent, 20–25 mm. long. Pl. 3. Fig. 99/1–9.

UGANDA. Acholi District: Imatong Mts., Apr. 1938, *Eggeling* 3614!; Bunyoro District: Budongo, 18 Apr. 1932, *Hancock* T1100!; Masaka District: Kiwala, Apr. 1916, *Dummer* 2805!
KENYA. Nandi District: Kaimosi, 1 km. N. of Yala R. bridge, May 1944, *C.G. Rogers* 719! & Nandi, May 1944, *Carroll* H31!; N. Kavirondo District: Kakamega, E. of Mumias, Apr. 1934, *Dale* in *F.D.* 3269!
TANZANIA. Buha District: Kalinzi, Dec. 1956, *Procter* 614!; Ulanga District: ± 10 km. S. of Mahenge, Mzelezi, 19 Jan. 1979, *Cribb, Grey-Wilson & Mwasumbi* 11066! & 5 km. S. of Mahenge, above Isongo, 20 Jan. 1979, *Cribb, Grey-Wilson & Mwasumbi* 11099!
DISTR. **U** 1, 2, 4; **K** 3, 5; **T** 4, 6; Nigeria to Sudan and Ethiopia and south to Zimbabwe
HAB. In grassland, bushland and woodland; 950–2100 m.

SYN. *E. kirkii* Rolfe in F.T.A. 7: 66 (1897). Type: Mozambique, Moramballa, *Kirk* (K, holo.!)
E. lambii Rolfe in K.B. 1914: 212 (1914). Type: Nigeria, Bauchi Plateau, *Lamb* (K, holo.)
E. variopicta Chiov. in Ann. Bot. Roma 9: 135 (1911). Type: Ethiopia, Amhara, *Chiovenda* 692 (FT, holo.!)

20. **E. elegans** *Schltr.* in E.J. 53: 585 (1915). Types: Tanzania, Rungwe District, Bundali Mts., *Stolz* 2556 (B, syn., K, isosyn.!) & Rungwe Mt., *Stolz* 1980 (B, syn., K, Z, isosyn.!)

A terrestrial herb 60–125 cm. tall. Perennating organ subterranean, tuberous, ellipsoidal, 2–2.5 cm. long, 1–1.3 cm. diameter; roots 2 mm. diameter, white. Leaves 3–4, plicate, erect, lanceolate to oblanceolate, acute, 24–40 cm. long, 1.4–3.5 cm. wide, lowermost sheath-like. Inflorescence produced before or with leaves, 9–20-flowered; peduncle 8 mm. diameter, covered by 4–5 papery sheaths; rhachis 12–38 cm. long; bracts spreading, linear-lanceolate, acuminate, 9–60 mm. long. Flowers spreading; sepals and petals green with red-brown veins; lip white at apex, with purple papillae and green side lobes veined with purple; pedicel and ovary 16–30 mm. long. Dorsal sepal narrowly oblong, acute, 12–14 mm. long, 4 mm. wide, keeled on outside; lateral sepals similar but oblique at base, 15–20 mm. long, 4–5 mm. wide. Petals oblong-elliptic, obtuse, 11–15 mm. long, 4.5–5 mm. wide. Lip 3-lobed, 14–20 mm. long, 7–15 mm. wide; side lobes oblong-elliptic, rounded in front; mid-lobe decurved, oblong, obtuse; callus of 2 basal ridges, densely long-papillate in front on mid-lobe; spur short, conical, 2–2.5 mm. long. Column 7–9 mm. long; foot 3.5 mm. long. Fruit pendent, 20–25 mm. long. Fig. 99/10–18.

TANZANIA. Mbeya range, N. side, 16 Mar. 1960, *Kerfoot* 1725!; Rungwe District: Bundali Mts., 3 Mar. 1914, *Stolz* 2556! & Rungwe Mt., 18 Apr. 1913, *Stolz* 1980!
DISTR. **T** 7; Malawi
HAB. In wet grassland at edge of riparian forest; 1000–1650 m.

21. **E. kyimbilae** *Schltr.* in E.J. 53: 584 (1915). Type: Tanzania, Rungwe District, Kyimbila, Mbaka, *Stolz* 1123 (B, holo., K, record of holo.! & iso.!)

A terrestrial herb 35–85 cm. tall. Perennating organs underground, ovoid-conical, 1–2 cm. tall, 2–2.5 cm. diameter, in chains; roots 1.5 mm. diameter, white. Leaves 4–5, in a fan, linear-lanceolate, acute, 30–55 cm. long, 2–16 mm. wide, basal leaf sheathing. Inflorescence flowering with leaves, laxly few-flowered; peduncle 2–4.5 mm. diameter, almost covered by 4–5 sheathing bracts; rhachis 9–16 mm. long; bracts spreading, lanceolate, acuminate, 6–35 mm. long. Flowers not opening widely; sepals and petals green usually veined purple or brown; lip green at base, white or yellow above, sometimes purple-flushed; pedicel and ovary 15–21 mm. long. Dorsal sepal linear-lanceolate, acute, 9–12 mm. long, 2–3 mm. wide; lateral sepals similar but oblique at base, 9–17 mm. long, 2–5 mm. wide, keeled on outer side. Petals narrowly elliptic, acute, 8–12 mm. long, 3.5–5 mm. wide. Lip 3-lobed, 9–15 mm. long, 6–10 mm. wide; side lobes obliquely oblong, rounded in front; mid-lobe oblong-flabellate, rounded or truncate; callus of 3 fleshy ridges in basal half of lip, with several rows of papillae on mid-lobe; spur shortly cylindrical, incurved, 1.5–2.5 mm. long. Column 4–6 mm. long; foot 2 mm. long. Fruit pendent, 24–25 mm. long.

UGANDA. Mengo District: Kampala, 21 July 1935, *Hancock & Chandler* 1!
KENYA. Trans-Nzoia District: Moi's [Hoey's] Bridge, June 1936, *Mainwaring* K1! & near Elgon Club, Mar. 1938, *Tweedie* 439!
TANZANIA. Mbeya Range, 1961, *Kerfoot* in *E.A.H.* 4/61!; Rungwe District: Poroto Mts., below Isongole, 21 Feb. 1976, *Cribb & Grey-Wilson* 10824! & 6 Feb. 1979, *Cribb, Grey-Wilson & Mwasumbi* 11305!; Songea District: Matengo Hills, Lupembe Hill, 3 Mar. 1956, *Milne-Redhead & Taylor* 8965!

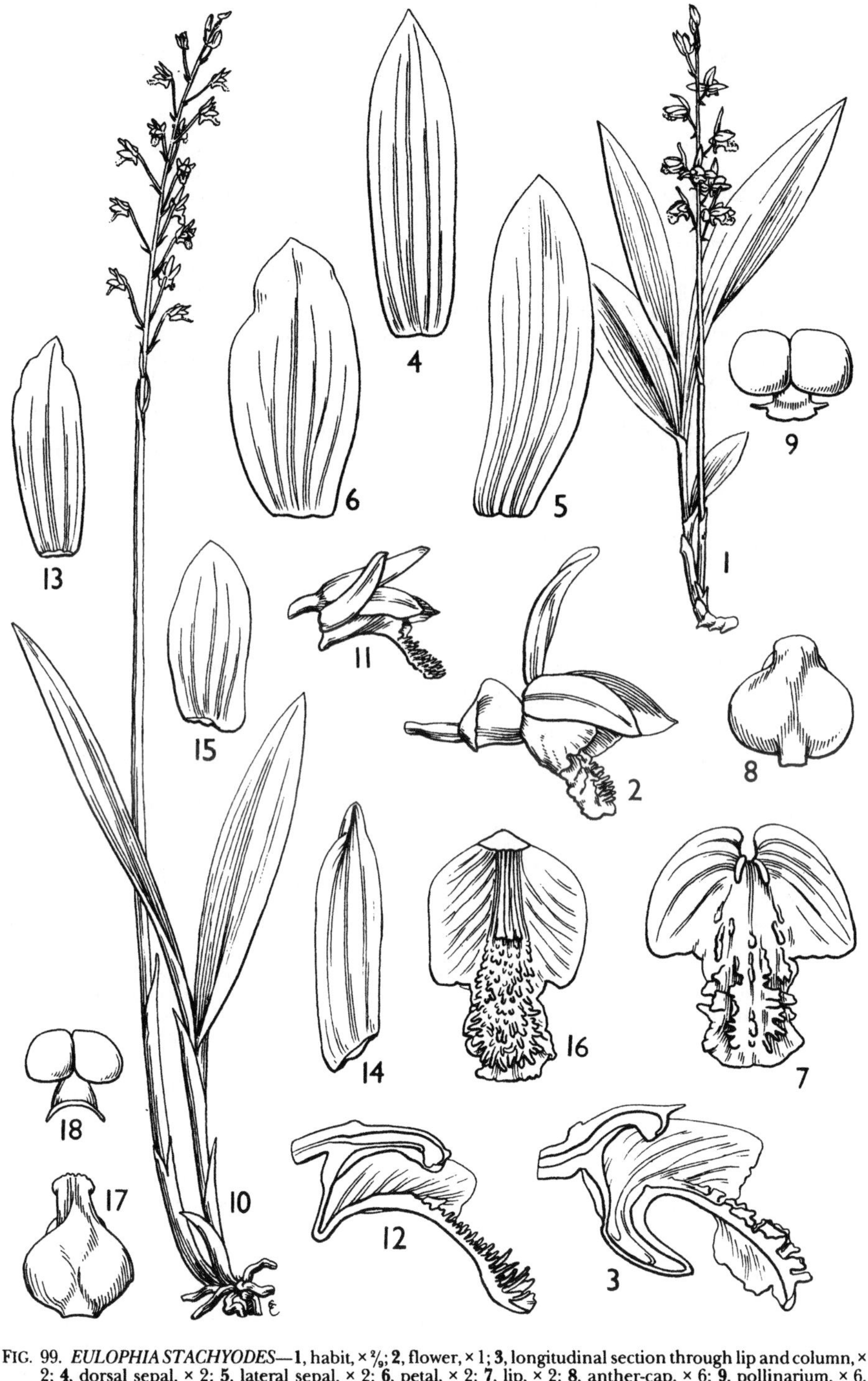

FIG. 99. *EULOPHIA STACHYODES*—**1**, habit, × $^2/_9$; **2**, flower, × 1; **3**, longitudinal section through lip and column, × 2; **4**, dorsal sepal, × 2; **5**, lateral sepal, × 2; **6**, petal, × 2; **7**, lip, × 2; **8**, anther-cap, × 6; **9**, pollinarium, × 6. *E. ELEGANS*—**10**, habit, × $^2/_9$; **11**, flower, × 1; **12**, longitudinal section through column and lip, × 2; **13**, dorsal sepal, × 2; **14**, lateral sepal, × 2; **15**, petal, × 2; **16**, lip, × 2; **17**, anther-cap, × 6; **18**, pollinarium, × 6. 1, from *Purseglove* 432; 2–9, from *Cribb & Grey-Wilson* 11066; 10, from *Stolz* 2556; 11–18, from *Robson* 1451. Drawn by Eleanor Catherine.

DISTR. **U** 4; **K** 3; **T** 7, 8; Zambia, Malawi and Zimbabwe
HAB. In seasonally wet grassland and on forest fringes; 1200–2150 m.

22. **E. ovalis** *Lindl.* in Comp. Bot. Mag. 2: 202 (1837); Hall in Journ. S. Afr. Bot., Suppl. 5: 87 (1965); J. Stewart et al., Wild Orch. S. Afr.: 235 (1982). Type: South Africa, Ado, *Drège* (K, holo.!)

A terrestrial herb 12–35 cm. tall. Perennating organ underground, tuberous, obliquely conical, 2 cm. tall, 2.3 cm. diameter, covered by fibrous remains of leaf-bases; roots 2 mm. diameter, white. Leaves 4–7, arranged in a fan, plicate, linear or linear-lanceolate, acute, 10–35 cm. long, 0.4–2 cm. wide. Inflorescence erect, produced with the leaves, laxly to subdensely few-flowered; peduncle bearing 4 whitish acuminate sheaths along length; rhachis 3–8 cm. long; bracts papery, lanceolate, acuminate, 10–27 mm. long. Flower with pale or brownish green sepals; petals yellow, often finely spotted pink within; lip yellow with purple-veined side lobes and a pinkish spur; pedicel and ovary 13–17 mm. long. Dorsal sepal oblong-lanceolate, acute or apiculate, 20–25 mm. long, 3–6 mm. wide; lateral sepals linear-lanceolate, acute, 20–22 mm. long, 3–6 mm. wide. Petals elliptic, acute to obtuse, 20–24 mm. long, 8–9 mm. wide, lying over column. Lip 3-lobed, 15–21 mm. long, 10–15 mm. wide; side lobes rounded in front; mid-lobe subcircular-elliptic, rounded or obtuse; callus of 2 basal ridges and papillae on 3 central veins in basal half of mid-lobe in front; spur incurved, cylindrical, 3–4 mm. long. Column 3–5 mm. long. Fruit pendent, 20–25 mm. long.

subsp. **bainesii** (*Rolfe*) *A. V. Hall* in Journ. S. Afr. Bot., Suppl. 5: 91 (1965); J. Stewart et al., Wild. Orch. S. Afr.: 235 (1982). Type: Tropical South Africa, *Baines* (K, holo.!)

The description given above refers to this subspecies. Fig. 100/1–11.

KENYA. Machakos/Masai District: Chyulu Hills, Jan. 1984, *Benn*!
TANZANIA. Ufipa District: Chito Mt., 6 Feb. 1971, *Sanane* 1524!; Iringa District: Iheme, 23 Feb. 1962, *Polhill & Paulo* 1587!; Njombe District: Kipengere Mts., 10 Jan. 1957, *Richards* 7648!
DISTR. **K** 4/6; **T** 4, 7; Zambia, Malawi, Zimbabwe and South Africa
HAB. In upland grassland; 1600–2600 m.

SYN. *E. bainesii* Rolfe in F.T.A. 7: 65 (1897)
E. monticola Schltr. in E.J. 53: 582 (1915). Type: Tanzania, Njombe District, near Madehani, *Stolz* 2449 (B, holo., K, iso.!)
E. transvaalensis Schltr. in Ann. Transv. Mus. 10: 237 (1924), *nom. illegit.*, *non* Rolfe (1910). Types: South Africa, Transvaal, *Korthals* (PRE, syn.) & *Jenkins* (B, syn.)
E. transvaalensis Schltr. var. *thorncroftii* Schltr. in Ann. Transv. Mus. 10: 237 (1924). Type: South Africa, Transvaal, *Thorncroft* (PRE, holo.)
E. pretoriensis H. Bolus in Fl. Pl. Afr. 13, sub t. 500 (1933). Types: South Africa, Transvaal, The Willows, *Stent* in *G.H.* 9949 (PRE, syn.) & Premier Mine, *F.A. Rogers* (BOL, syn.)
E. complanata Verdoorn in Fl. Pl. Afr. 27, sub t. 1056 (1949), nom. nov. pro *E. transvaalensis* Schltr.

NOTE. *E. ovalis* comprises two subspecies. Subsp. *bainesii* is the only one found in the Flora area. The typical subspecies, confined to the E. Cape, Natal, the Transvaal and Lesotho in southern Africa, is distinguished readily by its smaller differently coloured flowers with white petals speckled with lilac and a white lip. The spur, 2.5–5.6 mm. long, is also conical and not compressed.

23. **E. clavicornis** *Lindl.* in Comp. Bot. Mag. 2: 202 (1837); Hall in Journ. S. Afr. Bot., Suppl. 5: 75 (1965); J. Stewart et al., Wild Orch. S. Afr.: 234 (1982). Type: South Africa, Katberg, *Drège* (K, holo.!)

A terrestrial herb, 26–45 cm. tall. Perennating organs subterranean, tuberous, flattened-conical, in chains, 1.2–1.5 cm. long, 2 cm. diameter; roots 2 mm. diameter, white. Leaves 5–6 in a fan, plicate, linear, acuminate, 17–40 cm. long, 3–8 mm. wide, lowermost sheathing. Inflorescence produced with leaves, laxly few–many-flowered; peduncle 2–3 mm. diameter, with 4–5 papery acuminate sheathing bracts along length; rhachis 6–16 cm. long; bracts spreading, lanceolate, acuminate, 9–18 mm. long. Flowers rather small, often self-pollinating; sepals brown or green flushed with brown; petals white flushed with pink; lip white or yellowish flushed dull red-purple on edges; callus yellowish or greenish; pedicel and ovary 10–14 mm. long. Dorsal sepal oblong-elliptic, apiculate, 10–11 mm. long, 3–4 mm. wide, keeled on outer side; lateral sepals similar but oblique, 11–12 mm. long, 4–4.5 mm. wide. Petals elliptic, obtuse or apiculate, 8.5–10 mm. long, 3.5–5 mm. wide. Lip 3-lobed, 7.5–10 mm. long, 5–9 mm. wide; side lobes rounded in front;

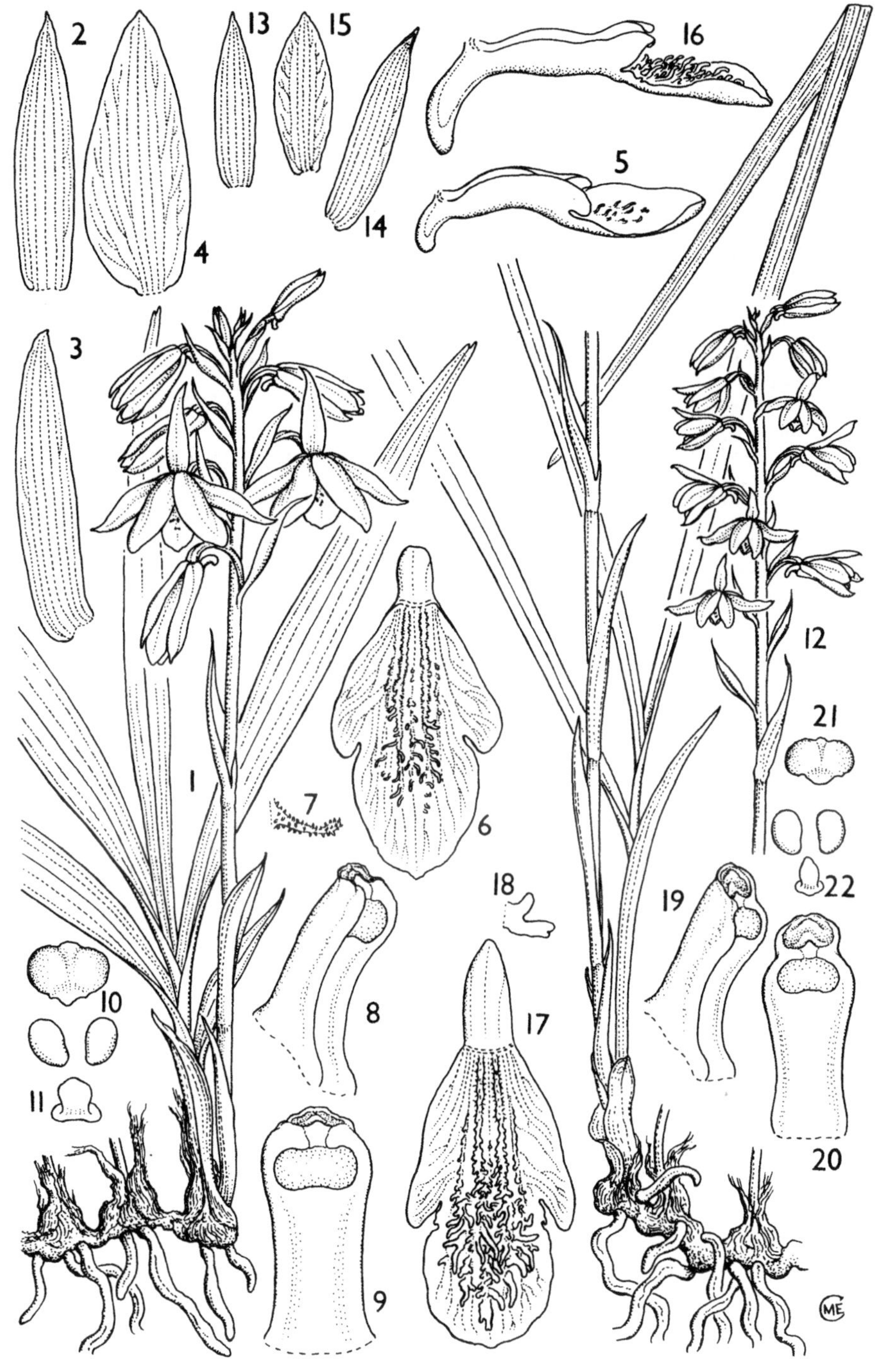

FIG. 100. *EULOPHIA OVALIS* subsp. *BAINESII*—**1**, habit, × ⅔; **2**, dorsal sepal, × 2; **3**, lateral sepal, × 2; **4**, petal, × 2; **5**, lip, side view, × 2; **6**, lip, spread out, × 2; **7**, detail of papilla showing small papillae, × 12; **8**, column, side view, × 6; **9**, column, front view, × 6; **10**, anther-cap, × 6; **11**, pollinarium, × 8. *E. CLAVICORNIS* var. *NUTANS*—**12**, habit, × ⅔; **13**, dorsal sepal, × 2; **14**, lateral sepal, × 2; **15**, petal, × 2; **16**, lip, side view, × 4; **17**, lip, spread out, × 4; **18**, papilla, × 12; **19**, column, side view, × 6; **20**, column, front view, × 6; **21**, anther-cap, × 6; **22**, pollinarium, × 8. 1–11, from *Richards* 7648; 12, from *Leedal* 370; 13–22, from *Leedal* 5306. Drawn by Mrs M.E. Church.

mid-lobe subcircular-oblong, obtuse, 3 mm. long, 4.5 mm. wide; callus of 2 ridges below and 4 lines of papillae on mid-lobe, all papillate; spur incurved, cylindrical-clavate, 2–3 mm. long. Column 2–3 mm. long; foot 0.5 mm. long. Fruit pendent, 18–20 mm. long.

var. **nutans** (*Sond.*) *A.V. Hall* in Journ. S. Afr. Bot., Suppl. 5: 77, 82 (1965); Williamson, Orch. SC. Afr.: 159 (1977); J. Stewart et al., Wild. Orch. S. Afr.: 234 (1982). Types: South Africa, Uitenhage, *Ecklon & Zeyher* (W, syn.!) & Katriviersberg, *Ecklon & Zeyher* (W, syn.!)

The description given above relates to this variety. Fig. 100/12–22, p. 449.

KENYA. N. Nyeri District: Nanyuki, Sept. 1943, *Copley* 119!; Kisumu-Londiani District: Kedowa, June–July 1933, *R.M. Graham* in *F.D.* 3223! & Londiani, 18 May 1957, *Smart* 622! & 27 July 1951, *D. Davis* 181! Machakos/Masai District: Chyulu Hills, 16 Dec. 1947, *J. Williams* in *Bally* 6286!
TANZANIA. Moshi District: Ngare Nairobi, 26 Feb. 1976, *Beesley* 297!; Rungwe District: Bundali Mts., Ibaba to Isoko road, 14 Feb. 1976, *Cribb & Grey-Wilson* 10717!; Njombe District: Kipengere Mts., tributary of R. Rumakali, 14 Jan. 1957, *Richards* 7752!
DISTR. **K** 4–6; **T** 2, 7, ?8; Ethiopia, Malawi, Zambia, Zimbabwe and South Africa, also in Madagascar and Yemen
HAB. In grassland and bushland rarely in cypress plantations in S. Tanzania; 1650–2450 m.

SYN. *E. nutans* Sond. in Linnaea 19: 73 (1846)
E. vaginata Ridl. in J.L.S. 21: 467 (1885). Type: Madagascar, Ankaratra, *Hildebrandt* 3864 (BM, holo., K, iso.!)
E. crinita Rolfe in F.T.A. 7: 58 (1897). Type: Malawi, Shire Highlands, *Buchanan* (K, holo.!)
E. ukingensis Schltr. in E.J. 53: 579 (1915). Type: Tanzania, Njombe District, Ukinga, *Stolz* 2495 (B, holo., K, BR, iso.!)

NOTE. Three varieties of *E. clavicornis* have been recognised by Hall (1965). Of these only var. *nutans* is found in the Flora area. The typical variety and var. *inaequalis* (Schltr.) A.V. Hall are both confined to South Africa and are readily distinguished by their flowering before the production of leaves and by their flower colour. In var. *clavicornis* the petals and lip are lilac-coloured, while in var. *inaequalis* the petals and lip are bright yellow.

24. **E. carsonii** *Rolfe* in F.T.A. 7: 64 (1897); Williamson, Orch. SC. Afr.: 154 (1977). Type: Zambia, Fwambo, *Carson* (K, holo.!)

A terrestrial herb 22–30 cm. tall. Perennating organs subterranean, tuberous, irregularly ellipsoidal, somewhat flattened, 2–3 cm. long, 1.2–2 cm. diameter, in chains; roots from base of the new growth, 1–1.5 mm. diameter. Leaves 1–3, erect, plicate, grass-like, linear, acuminate, 15–31 cm. long, 1–6 mm. wide. Inflorescence produced as leaves develop, densely 2–10-flowered; peduncle slender, 1.5–2 mm. diameter, with 2 spaced sheaths; rhachis 2–8 cm. long; bracts spreading or reflexed, lanceolate, acuminate, 8–22 mm. long. Flowers secund, subnutant, pale lemon yellow with greenish disc on lip sometimes edged with purple; pedicel and ovary 8–12 mm. long. Dorsal sepal lanceolate or oblong-lanceolate, acute or acuminate, 1.7–2 cm. long, 3–5 mm. wide; lateral sepals similar, 1.7–2.2 cm. long, 3–6 mm. wide. Petals oblong-elliptic, acute or obtuse, 1.4–1.6 cm. long, 4–5 mm. wide. Lip 3-lobed, 1.2–1.4 cm. long, 6.5–11 mm. wide; side lobes obtuse or rounded; mid-lobe much longer, oblong or oblong-elliptic, obtuse or apiculate, erose; callus of 2 papillose ridges in basal half with long papillae on mid-lobe surface; spur cylindrical, slightly incurved, 2–2.5 mm. long. Column 3–4.5 mm. long; foot 3 mm. long. Fruit pendent, 11–14 mm. long.

TANZANIA. Ufipa District: Chapota, 3 Dec. 1949, *Bullock* 2001!; "Tanganyika Plateau", *Carson*!
DISTR. **T** 4, ?7; S. Zaire, Zambia and Malawi
HAB. In *Brachystegia* woodland and occasionally in grassland; 1150–2100 m.

25. **E. euantha** *Schltr.* in E.J. 53: 586 (1915); Williamson, Orch. SC. Afr.: 154 (1977). Type: Tanzania, Mbeya/Chunya District, Usafwa, *Stolz* 197 (B, holo., K, record or holo.!)

A terrestrial herb 20–40 cm. tall. Perennating organs subterranean, tuberous, irregularly spherical, potato-like, 2–2.5 cm. long, 1.6–2 cm. diameter, in strings; roots 1–1.5 mm. diameter. Leaves 1–2, erect, grass-like, linear, acuminate, 12–32 cm. long, 1.5–6 mm. wide. Inflorescence densely 4–10-flowered; peduncle 1–1.5 mm. diameter, with 3 short sheaths in basal half; rhachis 2–8 cm. long; bracts spreading, lanceolate, acuminate, 10–25 mm. long. Flowers not opening widely, spreading but becoming nutant, pink or whitish pink with a mauve mark on disc of lip; pedicel and ovary 9–11 mm. long. Dorsal sepal oblong, acute or apiculate, 21–27 mm. long, 7 mm. wide; lateral sepals similar but oblique at base, 19–25 mm. long, 5–9 mm. wide, keeled on outside. Petals elliptic or oblong-elliptic, obtuse or apiculate, 18–23 mm. long, 7–9 mm. wide. Lip not recurved, 3-lobed, 16–24 mm. long, 10–19 mm. wide; side lobes small, rounded; mid-lobe much larger, transversely oblong to circular, obtuse; callus of 2 short papillate ridges to base of mid-lobe, sparsely long-papillose on mid-lobe; spur cylindrical, incurved, 2–3 mm. long. Column 4–5 mm. long; foot 1 mm. long. Fruit pendent, ± 25 mm. long. Pl. 4. Fig. 101/1, 2, p. 452.

TANZANIA. Ufipa District: new Sumbawanga road, 4 Dec. 1960, *Richards* 13648!; Chunya District: N. Lupa Forest Reserve, 95 km. N. of Mbeya on Itigi road, 17 Nov. 1962, *Boaler* 722!; Songea District: ± 13 km. E. of Songea, 22 Dec. 1955, *Milne-Redhead & Taylor* 7873!
DISTR. T 4, 7, 8; Burundi, Zaire, Zambia, Mozambique, Malawi and Zimbabwe
HAB. In deciduous woodland, on rocky outcrops and at edges of wet grassland; 1000–1650 m.

SYN. *E. speciosa* Rolfe in F.T.A. 7: 63 (1897), *nom. illegit.*, *non* (Lindl.) Bolus (1889). Type: Tanzania, between Lakes Tanganyika and Nyasa, *Johnston* (K, holo.!)

26. **E. rara** *Schltr.* in E.J. 53: 582 (1915). Type: Tanzania, Rungwe District, Mulinda, *Stolz* 1815 (B, holo., K, record of holo.!)

A terrestrial herb 20–36 cm. tall. Perennating organs underground, tuberous, ovoid-spherical or fusiform, 15–30 mm. long, 13–24 mm. diameter, drying orange; roots fine, 1–1.5 mm. diameter, from base of new growth. Leaves 4–5, erect, linear, acute, 20–30 cm. long, 4–5 mm. wide, basal 3 sheathing. Inflorescence laxly few-flowered; peduncle 2 mm. diameter, with 2–3 tubular subacute sheaths in basal half; rhachis 2–7 cm. long; bracts reflexed, linear-lanceolate, acuminate, 10–20 mm. long. Flowers subnutant, not opening widely; sepals and petals dull purple or maroon with paler bases to the segments; spur greenish to creamy purple; pedicel and ovary 12–14 mm. long. Dorsal sepal lanceolate, acuminate, 14–16.5 mm. long, 3–3.5 mm. wide; lateral sepals similar. Petals lanceolate, acute, 12–15 mm. long, 4 mm. wide. Lip 3-lobed, 12 mm. long, 8–12 mm. wide; side lobes erect, rounded; mid-lobe obovate or elliptic, obtuse, with erose margins; spur 1.5–2 mm. long, cylindric-clavate. Column 3–4 mm. long; foot 1 mm. long. Fruit pendent, 12 mm. long.

TANZANIA. Ufipa District: above Msanzi, 13 Dec. 1958, *Richards* 10340! & 10341!; Mbeya District: Mbozi circle, Mchembo Estate, 11 Jan. 1961, *Richards* 13842!
DISTR. T 4, 7; Zambia, Malawi, Mozambique and Zimbabwe
HAB. In deciduous woodland and grassland; 800–1800 m.

27. **E. thomsonii** *Rolfe* in F.T.A. 7: 66 (1897): Williamson, Orch. SC. Afr.: 154 (1977). Type: Tanzania, between Lakes Tanganyika and Malawi [Nyassa], *Thomson* (K, holo.!)

A terrestrial herb 22–40 cm. tall. Perennating organs underground, tuberous, irregularly fusiform to oblong-ellipsoidal, 2.5–4 cm. long, 1.5–2.5 cm. diameter, 3–5-noded; rhizome short, 6 mm. diameter; roots slender, 1.5 mm. diameter, mainly from rhizome. Leaf solitary, erect, grass-like, linear, acuminate, 24–32 cm. long, 6–8 mm. wide, subtended by 2 sheaths. Inflorescence produced before or with leaf, densely many-flowered in a dense subsperical head 5–7 cm. across; peduncle bearing 2–3 sheaths in lower half; rhachis 1.5–4 cm. long; bracts linear-lanceolate, acuminate, 1.5–2.5 cm. long. Flowers not opening widely, white with a deep purple spot on the creamy white lip; pedicel and ovary 10–14 mm. long. Dorsal sepal lanceolate, acute, 19–20 mm. long, 7–8 mm. wide; lateral sepals similar but 21–25 mm. long, 7–10 mm. wide. Petals narrowly elliptic-lanceolate, acute, 18–24 mm. long, 5–6 mm. wide. Lip 3-lobed, straight with a slightly recurved mid-lobe, 16.5–20 mm. long, 11–14 mm. wide; side lobes erect, narrowly elliptic, rounded at apex; mid-lobe elliptic-subcircular, rounded at apex; callus of 7 low papillose ridges with papillae at apex on base of mid-lobe; spur conical, recurved, 1–3 mm. long. Column 5–7 mm. long; foot 1 mm. long. Fruit pendent, 15–20 mm. long. Pl. 4. Fig. 101/3–13, p. 452.

TANZANIA. Ufipa District: Malonje, 20 Nov. 1949, *Bullock* 1870!; Iringa District: Sao Hill, Dec. 1959, *Watermeyer* 190!; Songea District: Matengo Hills, Litembo rest camp, 14 Oct. 1956, *Semsei* 2602!
DISTR. T 4, 7, 8; Zambia and Malawi
HAB. In montane grassland and burnt open woodland of *Protea* and *Erythrina;* 1800–2550 m.

28. **E. aurantiaca** *Rolfe* in F.T.A. 7: 67 (1897). Types: Tanzania, 'Tanganyika Plateau', *Carson* (K, syn.!) & between Lakes Tanganyika and Malawi [Nyasa], *Johnston* (K, syn.!)

A terrestrial herb 30–60(–75) cm. tall. Perennating organs subterranean, irregularly ellipsoidal-conical, tuberous, successively produced annually in a row, 3–4 cm. long, 2.5–3

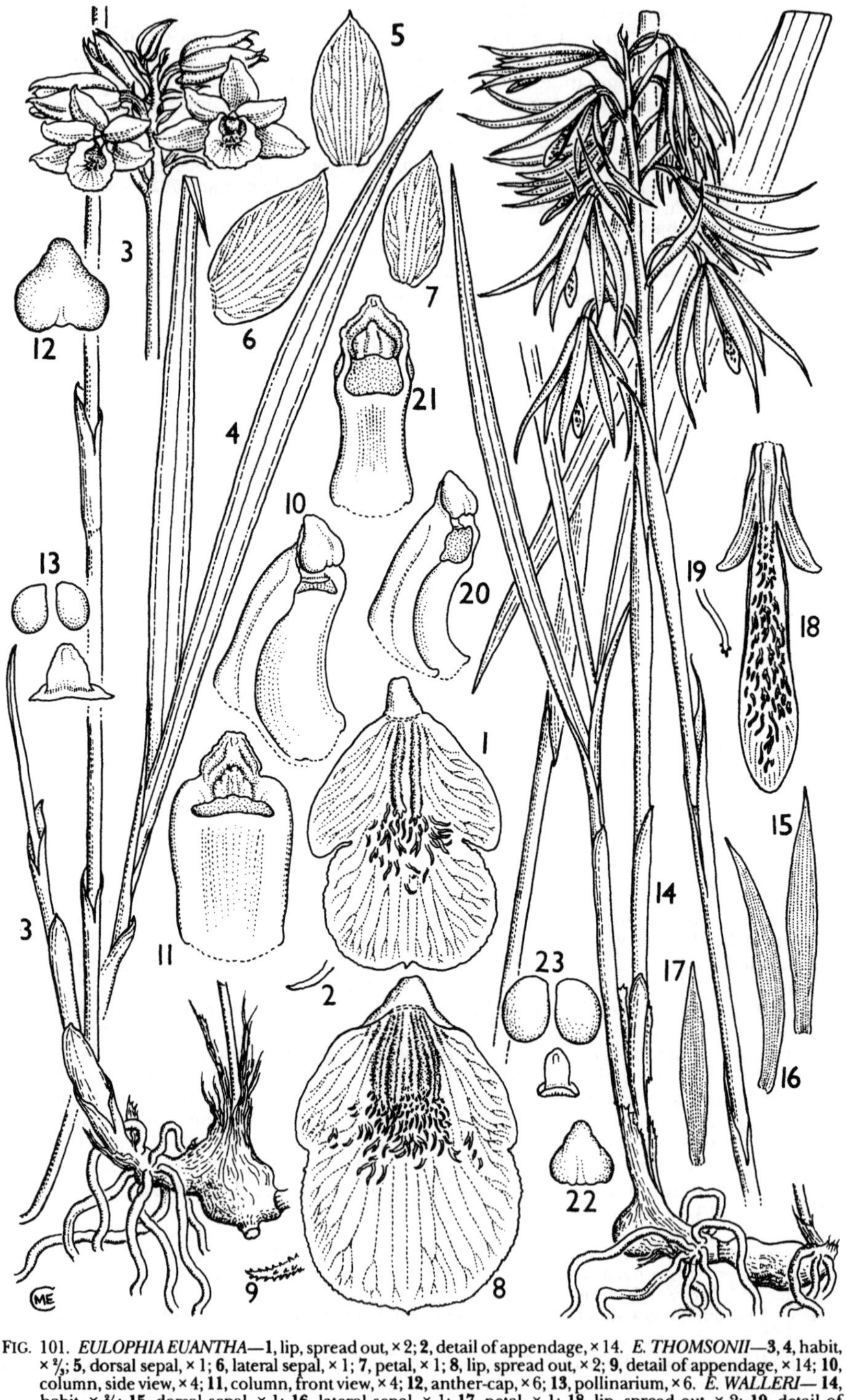

FIG. 101. *EULOPHIA EUANTHA*—**1**, lip, spread out, × 2; **2**, detail of appendage, × 14. *E. THOMSONII*—**3**, **4**, habit, × ⅔; **5**, dorsal sepal, × 1; **6**, lateral sepal, × 1; **7**, petal, × 1; **8**, lip, spread out, × 2; **9**, detail of appendage, × 14; **10**, column, side view, × 4; **11**, column, front view, × 4; **12**, anther-cap, × 6; **13**, pollinarium, × 6. *E. WALLERI*— **14**, habit, × ⅔; **15**, dorsal sepal, × 1; **16**, lateral sepal, × 1; **17**, petal, × 1; **18**, lip, spread out, × 2; **19**, detail of appendage, × 14; **20**, column, side view, × 6; **21**, column, front view, × 6; **22**, anther-cap, × 12; **23** pollinarium, × 12. 1,2, from *Richards* 13648; 3,5–13, from *Robson* 372; 4, from *Michelmore* 1067; 14–23, from *Robson* 1511. Drawn by Mrs M.E. Church.

cm. diameter, drying yellow; roots from base of new growth, 1 mm. diameter, white. Leaves 2–3, plicate, erect, linear, acuminate, 20–65 cm. long, 4–14 mm. wide, long-petiolate. Inflorescence laxly 8–20-flowered, produced before or with young leaves; peduncle almost covered by 2–3 sheathing bracts, upper one linear-lanceolate; rhachis 6–20 cm. long, bracts linear-lanceolate, acuminate, 12–21 mm. long. Flowers orange or orange-red, spreading to subnutant, not opening widely, slightly fragrant; pedicel and ovary 12–20 mm. long. Dorsal sepal linear-lanceolate, acute or acuminate, 3.1–4 cm. long, 6–7 mm. wide; lateral sepals similar but oblique at base, 2.8–4.2 cm. long, 5–7 mm. wide. Petals linear-lanceolate, acute or acuminate, 2.2–2.7 cm. long, 6–7 mm. wide. Lip 3-lobed, porrect, 1.6–2 cm. long, 1–1.2 cm. wide; side lobes rounded; mid-lobe much longer, emarginate, 8 mm. long, 9 mm. wide; callus of 2 papillose ridges in basal half with central area of mid-lobe densely long-papillate; spur very shortly conical, 1–2 mm. long. Column 6–8 mm. long; foot 3–4 mm. long. Fruit pendent, 23–25 mm. long, with persistent perianth.

TANZANIA. Ngara District: Bushubi, Rulenge, Goyagoya, 5 Oct. 1960, *R. Tanner* 5263!; Kigoma District: 152 km. from Mpanda on Uvinza road, 23 Nov. 1962, *Verdcourt* 3431!; Ufipa District: Chapota, 3 Dec. 1954, *Richards*!
DISTR. **T** 1, 4, 7; Burundi, Zaire and Zambia
HAB. In woodland of *Brachystegia*, *Combretum*, *Dalbergia*, *Vitex*, *Strychnos*, *Pterocarpus*, etc. and in grassland and bushland; 1550–2300 m.

SYN. *E. leopoldii* De Wild. & Th. Dur. in B.S.B.B. 38: 215 (1900). Type: Zaire, near Dembo, *Gillet* (BR, holo.)

29. **E. walleri** (*Reichb.f.*) *Kraenzl.* in P.O.A. C: 157 (1895); Williamson, Orch. SC. Afr.: 157 (1977). Type: Malawi, Manganja, *Waller* (W, holo., K, iso.!)

A terrestrial herb up to 1 m. tall. Perennating organs underground, tuberous, irregularly cylindrical-conical, 2.5–4 cm. long, 1.5–2 cm. diameter, proximate in chains; roots 1 mm. diameter, from base of new growth, greyish. Leaves 2–3, linear, acuminate, grass-like, 30–55 cm. long, 3–19(–26) mm. wide. Inflorescence densely 8–20-flowered; peduncle covered by 6–7 sheathing acute bracts; rhachis 6–16 cm. long; bracts setose, up to 40 mm. long. Flowers subnutant, bright orange; petals cream at base; lip greenish at base with side lobes striped with purple and a yellow callus; pedicel and ovary 6–10 mm. long. Dorsal sepal linear, acuminate or acute, 3–4.5 cm. long, 3–5 mm. wide; lateral sepals similar but longer, up to 5.4 cm. long. Petals linear, acute, 2.8–4 cm. long, 4–5 mm. wide. Lip straight, 3-lobed, 2–2.7 cm. long, 7–9 mm. wide, lacking a spur; side lobes oblong-elliptic, very short, recurved at apex; mid-lobe spathulate-elliptic, 15–16 mm. long; callus of 2 ridges at base with many long papillae on mid-lobe in front; spur absent. Column 4–5 mm. long, green with a yellow anther. Fruit pendent, 15–20 mm. long. Pl. 4. Fig. 101/14–23.

TANZANIA. Shinyanga, Nov. 1938, *Koritschoner* 2197!; Lindi District: Selous Game Reserve, 17 km. SW. of Kingupira, 5 Mar. 1976, *Vollesen* in *M.R.C.* 3342!
DISTR. **T** 1, 8; S. Zaire, Zambia, Mozambique, Malawi, Zimbabwe and Botswana
HAB. In seasonally wet grassland and in *Brachystegia* woodland; 175–1600 m.

SYN. *Cyrtopera walleri* Reichb.f. in Otia Bot. Hamb. 2: 117 (1881)
Eulophia angustiflora Kraenzl. in E.J. 43: 396 (1909). Type: Mozambique, Lukungui, *Tiesler* 47 (B, holo., K, record of holo.!)
E. piscicelliana Buscal. & Schltr. in Buscal. & Muschl. in E.J. 49: 463 (1913). Type: Zambia, between Buana Mukuba and Sekontui, *Duchess of Aosta* 440 (FT, holo.)

30. **E. nuttii** *Rolfe* in F.T.A. 7: 63 (1897); Williamson, Orch. SC. Afr.: 157 (1977). Type: Zambia, Fwambo, *Nutt* (K, holo.!)

A terrestrial herb 17–45 cm. tall. Perennating organs underground, tuberous, conical, ± 1 cm. tall, 1–1.2 cm. diameter; roots slender, 1 mm. diameter. Leaves 3, erect, linear, acute, 20–45 cm. long, 2–3 mm. wide, overtopping inflorescence. Inflorescence produced with leaves, densely several-flowered, 17–28 cm. tall; peduncle covered by papery acuminate sheaths; rhachis 8–16 mm. long; bracts linear, acuminate, 8–16 mm. long. Flowers lemon-yellow with a deeper orange-yellow lip; pedicel and ovary 8–10 mm. long. Dorsal sepal elliptic-oblong, acuminate, 18 mm. long, 5 mm. wide; lateral sepals similar, 20–22 mm. long, 7–8 mm. wide. Petals elliptic-oblong, acuminate, 12 mm. long, 4 mm. wide. Lip 3-lobed, 8–12 mm. long, 5.5–8 mm. wide; side lobes almost as long as mid-lobe, rounded in front; mid-lobe ovate-elliptic, obtuse; callus of 2 ridges, shortly papillate at apex on base of mid-lobe; spur cylindric, slightly incurved, 3–3.5 mm. long. Column 3–4 mm. long; foot 1–2 mm. long. Fruit not known.

TANZANIA. Dodoma District: Kazikazi, 14 Jan. 1933, *B.D. Burtt* 4623!; Rungwe District: Ulambya, Itumba, 28 Feb. 1970, *Leedal* 631!
DISTR. **T** 5, 7; Zambia
HAB. In wet grassland; 1300–1600 m.

31. **E. subulata** *Rendle* in J.B. 33: 167 (1895); Piers, Orch. E. Afr., ed. 2: 186 (1968); J. Stewart in U.K.W.F.: 782 (1974). Type: Uganda, Katonga R., *Scott Elliot* 7417 (BM, holo., K, iso.!)

A terrestrial herb 20–75 cm. tall. Perennating organs underground, tuberous, irregularly conical to ellipsoidal, 25–32 mm. long, 15–30 mm. diameter; roots fine, 1.5 mm. diameter, white. Leaves 1–2, erect, linear, acuminate, 15–40 cm. long, 2–15 mm. wide. Inflorescence produced with young leaves, clavate; peduncle 2.5–3.5 mm. diameter, bearing 4–5 papery but not overlapping sheaths along length; rhachis 2–11 mm. long; bracts reflexed, aristate, acuminate, 11–15 mm. long. Flowers densely clustered, pale yellow or yellow; pedicel and ovary 10–13 mm. long. Dorsal sepal lanceolate, acute to acuminate, 19–21 mm. long, 4–7 mm. wide; lateral sepals obliquely oblong-lanceolate, acute, 21–37 mm. long, 8–11 mm. wide. Petals oblong, acute or acuminate, 17–26 mm. long, 3.5–8 mm. wide. Lip 3-lobed, 15.5–24 mm. long, 10.5–13.5 mm. wide; side lobes rounded in front; mid-lobe circular to elliptic, obtuse or rounded; callus of 2 ridges with a few papillae in front on base of mid-lobe; spur pendent, subclavate-cylindrical, 2.5–4 mm. long. Column 2–3 mm. long; foot 2 mm. long. Fruit pendent, 16–20 mm. long.

UGANDA. Ankole District: near Mbarara, Feb. 1949, *J.G. Williams* in *Bally* 6598! & Nyabushozi, 3 Oct. 1970, *Katende* 610!; Mengo District: 74 km. from Kiwoko to Tweyanze, 23 Apr. 1956, *Langdale-Brown* 2077!
KENYA. Nairobi District: French Mission, Oct. 1931, *Napier* 1394!; Masai District: Lolgorien, Sept. 1933, *Napier* 5363!
TANZANIA. N. Mara District: Serengeti National Park, Tarime, Lamia area, 1.6 km. W. of Gongora marsh, 28 Feb. 1972, *Myles Turner* in *E.A.H.* 15169!
DISTR. **U** 2, 4; **K** 4, 6; **T** 1; Rwanda and Zaire
HAB. In seasonally wet grassland and swamps; 1100–1700 m.

32. **E. zeyheri** *Hook.f.* in Bot. Mag. 119, sub t. 7330 (1893); Summerh. in F.W.T.A., ed. 2, 3: 247 (1968); Piers, Orch. E. Afr., ed. 2: 188 (1968); J. Stewart in U.K.W.F.: 780 (1974). Types: South Africa, Transvaal, *Zeyher* (W, syn.) & Angola, Huilla, *Welwitsch* 718 (W, syn., K, P, isosyn.!)

A terrestrial herb 40–70 cm. tall. Perennating organs underground, tuberous, proximate, potato-like, fusiform-oblong, 5–10 cm. long, 2–2.5 cm. diameter, 4–5 mm. apart on a 4–5 mm. diameter rhizome; roots dense at ends of tubers, 1.5 mm. diameter, white. Leaves, 2–3, erect, lanceolate, acute, 30–50 cm. long, 1.3–4 cm. wide, subtended by 2 basal sheaths. Inflorescence produced before or with young leaves, apparently subumbellate, densely many-flowered 6–10 cm. across; peduncle stout, covered by 3–4 slightly inflated sheaths; rhachis 2–5 cm. long; bracts suberect to reflexed, subulate, acuminate, 25–40 mm. long. Flowers lemon yellow with a yellow, maroon or rich brown mark on the lip; pedicel and ovary 14–19 mm. long. Dorsal sepal elliptic-lanceolate, acuminate, 28–36 mm. long, 12–16 mm. wide; lateral sepals similar but larger, to 40 mm. long. Petals elliptic-lanceolate, acuminate, 28–32 mm. long, 12–14 mm. wide. Lip 3-lobed, 23–26 mm. long, 12–20 mm. wide; side lobes small, rounded in front; mid-lobe subcircular, obtuse to emarginate; callus of 2 ridges with long papillae in front on basal half of mid-lobe; spur conical, 2–4 mm. long. Column 6–7 mm. long; foot 3 mm. long. Fruit pendent, 18 mm. long. Pl. 4.

UGANDA. W. Nile District: Lendu, Zeu [Zeio], Apr. 1940, *Eggeling* 3898!; Ankole District: Bunyaruguru, Sept. 1938, *Purseglove* 371!; Kigezi District: Kyakabande, 20 Dec. 1933, *A.S. Thomas* 1084!
KENYA. Trans-Nzoia/N. Kavirondo District: Elgon, *E. & C. Lugard* 634!; N. Kavirondo District: 6.5 km. N. of Kakamega, without date, *Barney* 10!; Teita District: Mbololo, *Gardner* in *F.D.* 3018!
TANZANIA. Mwanza District: Rubondo I., 28 Oct. 1956, *Gane* 85!; Lushoto District: E. Usambara Mts., Monga, 5 Mar. 1941, *Greenway* 6128!; Buha District: 10 km. from Manyovu on Kasulu road, *Verdcourt* 3422!; Lindi District: Rondo Plateau, near Nahoro, 12 Dec. 1955, *Milne-Redhead & Taylor* 7618!

DISTR. **U** 1, 2; **K** 3, 5, 7; **T** 1–4, 6–8; Nigeria, Sudan, Zaire, Rwanda, Angola, Zambia, Malawi, Mozambique, Zimbabwe and South Africa
HAB. In grassland and wooded grassland; 1000–2100 m.

SYN. *E. bicolor* Reichb.f. & Sond. in Flora 48: 186 (1865), *non* Dalzell (1851), *nom. illegit.* Type as for species
Ornithochilus renschianus Reichb.f. in Flora 65: 532 (1882). Type: Angola, Malange, *Mechow* 299 (W, holo.)
O. mechowii Reichb.f. in Flora 65: 532 (1882). Type: Angola, Malange, *Mechow* 300 (W, holo.)
Eulophia woodii Schltr. in E.J. 20, Beibl. 50: 5 (1895). Type: South Africa, Natal, *J.M. Wood* 3577 (B, holo., NH, iso.)
E. renschianus (Reichb.f.) Th. Dur. & Schinz, Consp. Fl. Afr. 5: 25 (1895)
E. praestans Rendle in J.B. 33: 166 (1895). Type: Malawi, Shire Highlands, *Scott Elliot* 8616, in part (B, holo., K, iso.!)
E. dichroma Rolfe in F.T.A. 7: 62 (1897), *nom. nov. pro. E. welwitschii* Hook. f.
E. lujae De Wild. in Pl. Nov. Herb. Hort. Then., t. 17 (1904). Type: Mozambique, Morrumbala, *Luja* 391 (BR, holo.)
E. granducalis Kraenzl. in E.J. 43: 34 (1909). Type: Zaire, Lake Kivu, Wau. I., *Mildbraed* 1174 (B, holo., K, record of holo.!)

33. **E. adenoglossa** (*Lindl.*) *Reichb.f.* in Otia Bot. Hamb. 1: 66 (1878); Summerh. in F.W.T.A., ed. 2, 3: 247 (1968); J. Stewart in U.K.W.F.: 782 (1974); Williamson, Orch. SC. Afr.: 156 (1977). Type: Nigeria, Nupe, *Barter* (K, holo.!)

A terrestrial herb 45–75 cm. tall with fine roots 1–1.5 mm. diameter. Perennating structures underground, tuberous, ovoid or ellipsoid, 3–5 cm. long, 2–4 cm. diameter. Leaves 2–3, plicate, erect, linear, acuminate, 18–40 cm. long, 5–8 mm. wide. Inflorescence produced with the young leaves, up to 75 cm. tall, laxly many-flowered; peduncle 6 mm. diameter; rhachis up to 20 cm. long; bracts aristate or narrowly lanceolate, acuminate, 10–38 mm. long, 1–3 mm. wide. Flower subnutant; sepals and petals green or yellow-green; lip green or creamy green with brown-purple venation on the side lobes; column green with purplish tip; pedicel and ovary 12–16 mm. long. Dorsal sepal lanceolate, subacute, 15–26 mm. long, 4–6 mm. wide; lateral sepals obliquely lanceolate, acute, 17–28 mm. long, 5–7 mm. wide. Petals narrowly oblong, obtuse, 15–21 mm. long, 4.5–5.5 mm. wide. Lip 3-lobed, 20–21 mm. long, 10–11 mm. wide; side lobes rounded; mid-lobe subquadrate, emarginate or truncate; callus of 2 ridges terminating in papillae on mid-lobe; spur incurved, narrowly conical, 2–5 mm. long. Column 8–9 mm. long; foot 2 mm. long. Fruit pendent, 22 mm. long.

KENYA. N. Kavirondo District: near Malakisi, Apr. 1938, *Tweedie* 433!
TANZANIA. Rungwe District: Masoko, 9 Jan. 1913, *Stolz* 1808!
DISTR. **K** 5; **T** 7; Ghana, Nigeria, Zambia, Malawi, Mozambique, Angola, Zimbabwe and South Africa
HAB. In grassland, rocky ground and in *Brachystegia* woodland; 700–1600 m.

SYN. *Cymbidium adenoglossum* Lindl. in J.L.S. 6: 134 (1862)
Eulophia nigricans Schltr. in E.J. 20, Beibl. 50: 5 (1895). Type: South Africa, Natal, *J.M. Wood* 335 (B, holo.!, BM, K, iso.!)
E. aristata Rendle in J.B. 33: 169 (1895). Type: Malawi, Shire Highlands, *Scott Elliot* 8616 (BM, holo.!, K, iso.!)
E. lanceolata Rolfe in F.T.A. 7: 60 (1897). Type: Zambia, Fwambo, *Carson* 34 (K, holo.!)
E. baumiana Kraenzl. in Warb., Kunene-Sambesi-Exped.: 213 (1903). Type: Angola, Ungombekike, *Baum* 505 (B, holo.!, K, record of holo.!)

34. **E. ischna** *Summerh.* in K.B. 12: 125 (1957); Williamson, Orch. SC. Afr.: 159 (1977). Type: Angola, Cubango, *Baum* 341 (B, holo.!, K, Z, iso.!)

A terrestrial herb 27–70(–90) cm. tall. Perennating organs subterranean, tuberous, irregularly conical, flattened, 10–25 mm. long, 7–10 mm. high, consecutive on a short, 5–7 mm. long rhizome. Leaves erect, 2–3, linear, acuminate, up to 60 cm. long, 1.5–3 mm. wide, lowermost 1–2 sheathing. Inflorescence secund, laxly few–13-flowered; peduncle almost covered by 4–5 papery sheaths; rhachis 2–12(–17) cm. long; bracts lanceolate-elliptic, acuminate, 7–18 mm. long. Flowers subnutant; sepals and petals white or pale pink, often tinged with purple at base; lip brownish red with a yellow base and wine-red papillae; column purple or white flushed purple. Dorsal sepal linear-lanceolate, acuminate, 10–15(–19) mm. long, 3 mm. wide; lateral sepals similar but slightly falcate, 12–17(–20) mm. long, 3–3.5 mm. wide. Petals linear-lanceolate, acuminate, 10–16 mm. long, 3 mm. wide. Lip 3-lobed, 8–13 mm. long, 4–7 mm. wide; side lobes oblong, curved;

mid-lobe obovate, obtuse; callus of 2 raised short ridges between side lobes and several lines of papillae on mid-lobe; spur shortly conical, 1 mm. long, almost hidden by bases of lateral sepals. Column 4–5.5 mm. long. Fruit pendent, 14–15 mm. long.

TANZANIA. Iringa District: ± 12 km. W. of Sao Hill on Mbeya road, 9 Jan. 1975, *Brummitt & Polhill* 13649! & 31 Jan. 1979, *Cribb, Grey-Wilson & Mwasumbi* 11242!; Njombe District: Mwakete to Njombe road, 17 Jan. 1957, *Richards* 7849!; Songea District: 11 km. W. of Songea, Ulamboni valley, 31 Dec. 1955, *Milne-Redhead & Taylor* 7980!
DISTR. **T** 7, 8; Zaire, Zambia, Malawi and Angola
HAB. In marshes, swamps and seasonally wet grassland; (550–)960–2100 m.

SYN. *E. macra* Schltr. in Warb., Kunene-Sambesi-Exped.: 217 (1903), *non* Ridley (1886), *nom. illegit.* Type as for species

35. **E. milnei** *Reichb.f.* in Otia Bot. Hamb. 2: 116 (1881); Hall in Journ. S. Afr. Bot., Suppl. 5: 207 (1965); Summerh. in F.W.T.A., ed. 2, 3: 247 (1968); Williamson, Orch. SC. Afr.: 160 (1977); J. Stewart et al., Wild Orch. S. Afr.: 250 (1982). Type: Nimbo R., Benito Ground, *Milne* (W, holo.!)

A terrestrial herb 25–52 cm. tall. Perennating organs underground, tuberous, ovoid, in strings, 10–13 mm. tall, 8–10 mm. diameter, white; roots slender, 0.75–1 mm. diameter. Leaves 2–6, linear, acuminate, 10–30 cm. long, 1–2 mm. wide. Inflorescence produced before leaves, densely many-flowered; peduncle 1.5 mm. diameter, almost covered by 5 papery sheaths; rhachis 5–10 mm. long; bracts linear, acuminate, 10 mm. long. Flowers small but rather variable in size, yellow sometimes with reddish tinged side lobes to the lip; pedicel and ovary 5–8 mm. long. Dorsal sepal oblong-lanceolate, acute or obtuse, 5.5–8.5 mm. long, 2.3–3 mm. wide; lateral sepals similar, 5.5–9.5 mm. long, 2–3 mm. wide. Petals oblanceolate, subacute or acute, 5.7–8 mm. long, 2.4–3 mm. wide. Lip 3-lobed, 6–7 mm. long, 4–4.5 mm. wide; side lobes rounded in front; mid-lobe oblong-obovate, obtuse or emarginate; callus of 2–3 basal ridges with 3–5 lines of papillae in front on mid-lobe; spur cylindric, 2.5–3.5 mm. long. Column 2–2.7 mm. long. Fruit pendent, 20 mm. long. Fig. 102.

UGANDA. Lango District: Lira, Apr. 1938, *Eggeling* 3586!; Teso District: Serere, Apr. 1932, *Chandler* 589!; Mengo District: Buruli County, near Sezibwa [Sezibura] R., Mar. 1931, *Sandford* 2013!
KENYA. N. Kavirondo District: Broderick Falls, Apr. 1939, *Tweedie* 449!
TANZANIA. Iringa District: near Ifunda, Nov. 1928, *Haarer* 1615!; Njombe District: Mwakete to Njombe road, 17 Jan. 1957, *Richards* 7859!; Songea District: by Kimarampaka stream, ± 12 km. W. of Songea, 30 Dec. 1955, *Milne-Redhead & Taylor* 7963!
DISTR. **U** 1–4; **K** 5; **T** 7, 8; Sierra Leone to Sudan and south to Angola, Zambia, Malawi, Mozambique and South Africa
HAB. In seasonally wet grassland, marshes and swamps; 960–2000 m.

SYN. *E. lutea* Lindl. in J.L.S. 6: 132 (1862). Type: Nigeria, Nupe, *Barter* 1480 (K, holo.!)
E. gracillima Ridl. in J.B. 24: 292 (1886). Type: Nigeria, Calabar, *Robb* (BM, holo.!, K, record of holo.!)
E. lujaana Kraenzl. in B.J.B.B. 38: 217 (1900). Type: Zaire, Pool de Maleba [Stanley Pool], *Luja* 99 (BR, holo., K, record of holo.!)
E. coralloriziformis Schltr. in E.J. 20, Beibl. 50: 9 (1900). Types: South Africa, Umlomati Valley, *Galpin* 1221 (B, syn., K, isosyn.!), same locality, *Culver* 77 (B, syn.) & Transvaal, near Middelburg, *Schlechter* (B, syn., BOL, isosyn.!)
E. dictyostegioides Kraenzl. in E.J. 28: 168 (1900). Type: Angola, Huila, *Antunes* 26 (B, holo.)
E. warneckeana Kraenzl. in E.J. 33: 67 (1902). Type: Togo, Lome, *Warnecke* 328 (B, holo., K, record of holo. & iso.!)
E. pusilla Rolfe in K.B. 1914: 212 (1914). Type: Ghana, Affram Plains, *Burbridge* 245 (K, holo.!)
E. microdactyla Kraenzl. in E.J. 51: 389 (1914). Type: Cameroon, Garua, *Ledermann* 3385 (B, holo., K, record of holo.!)

NOTE. This species is very variable in flower size. More field work is needed to see if the variation has a regional basis and warrants taxonomic recognition.

36. **E. montis-elgonis** *Summerh.* in K.B. 1932: 509 (1932); Piers, Orch. E. Afr. ed. 2: 175 (1968); J. Stewart in U.K.W.F.: 780 (1974). Type: Kenya, Elgon, *Lugard* 663 (K, holo.!)

A terrestrial herb 30–70(–105) cm. tall. Perennating organs underground, tuberous, irregularly conical-ovoid, in a chain, 2–3 cm. tall, 2 cm. diameter; roots fine, 1 mm. diameter, white. Leaves 2–3, erect, linear, acuminate, plicate, 30–60 cm. long, 2.5–3 cm. wide. Inflorescence densely many-flowered; peduncle 15–25 cm. long, 5–8 mm. diameter, almost covered by 3–6 acuminate papery sheaths; rhachis 4–12 cm. long; bracts linear-

FIG. 102. *EULOPHIA MILNEI*—**A**, habit; **B**, leaf; **C**, dorsal sepal; **D**, lateral sepal; **E**, petal; **F**, lip and column, side view; **G**, lip with spur removed, front view; **H**, column. Drawn by Miss Stella Ross-Craig. Reproduced from Flora of West Tropical Africa.

aristate, acuminate, 16–33 mm. long. Flowers subnutant, pinkish mauve or white with a purple lip; pedicel and ovary 20–25 mm. long. Dorsal sepal ovate, acute or acuminate, 15–18 mm. long, 7–8.5 mm. wide; lateral sepals obliquely ovate-elliptic, acute or obtuse, 17.5–18 mm. long, 8–10 mm. wide. Petals elliptic, obtuse or apiculate, 13 mm. long, 6 mm. wide. Lip 3-lobed, 13–14.5 mm. long, 10–11 mm. wide; side lobes porrect, rounded in front; mid-lobe subquadrate; callus of 2 ridges with 5 rows of papillae in front; spur 3–3.5 mm. long. Column 6 mm. long; foot 2 mm. long. Fruit not known.

UGANDA. Karamoja District: Mt. Kadam [Debasien], 30 May 1939, *A.S. Thomas* 2941!
KENYA. Trans-Nzoia District: Kitale, June 1961, *Barnley* in *E.A.H.* 12364!; Kisumu-Londiani District: Mt. Songhor, July 1932, *Harvey* 199!
DISTR. **U** 1; **K** 3, 5; Sudan
HAB. Rocky hillsides and upland grassland with *Hagenia*; 2100–2850 m.

37. **E. odontoglossa** *Reichb.f.* in Linnaea 19: 373 (1847); Hall in Journ. S. Afr. Bot., Suppl. 5: 197 (1965); Williamson, Orch. SC. Afr.: 158 (1977); J. Stewart et al., Wild Orch. S. Afr.: 248 (1982). Type: South Africa, Natal, *Gueinzius* (W, holo.)

A terrestrial herb 60–100 cm. tall. Perennating organs underground, tuberous, irregularly fusiform-conical or subglobose, 2–4 cm. tall, 1.8–2.5 cm. diameter, white with brown nodes; roots fine, from base of new growth, 1 mm. diameter. Leaves 5–6, erect, plicate, oblanceolate, acuminate, 40–70 cm. long, 1–2.1 cm. wide, basal 3 sheathing. Inflorescence densely many-flowered; peduncle covered by 7 papery pale brown sheaths; rhachis 4–13 cm. long; bracts linear-aristate, 8–22 mm. long. Flowers yellow with yellow, orange or red papillae on lip, less commonly brown or crimson; pedicel and ovary 12–21 mm. long. Dorsal sepal ovate-elliptic, obtuse, 9.3–12 mm. long, 4–5.8 mm. wide; lateral sepals obliquely ovate, acute, 9–14 mm. long, 3.8–5.2 mm. wide. Petals obliquely elliptic, subacute to obtuse, 8–12 mm. long, 3.3–5 mm. wide. Lip 3-lobed, 8–11.5 mm. long, 4–8 mm. wide; side lobes porrect, acute to rounded at apex; mid-lobe oblong, subquadrate or obovate, obtuse or truncate; callus of 2 basal ridges with long papillae over basal three-quarters of mid-lobe; spur shortly conical-cylindric, 1–3 mm. long. Column 4–4.7 mm. long; foot 3–4.3 mm. long. Fruit pendent, 18 mm. long. Pl. 4.

UGANDA. W. Nile District: hill E. of Zeu Hill, 2 Sept. 1953, *Chancellor* 234!; Kigezi District: Maziba to Rwanda Border, Dec. 1944, *Purseglove* 1615!; Mengo District: Mawokota, Mar. 1905, *E. Brown* 144!
KENYA. Trans-Nzoia District: S. Elgon, Aug. 1943, *Tweedie* 620!; Kisumu-Londiani District: near Songhor [Sangor], July 1932, *Harvey* 198!
TANZANIA. Ngara District: Bushubi, Muronzi, 20 Dec. 1960, *R. Tanner* 5641!; Lushoto District: E. Usambara Mts., Kijango, Apr. 1935, *Greenway* 4041!; Iringa District: Dabaga Highlands, Kilolo, 10 Feb. 1962, *Polhill & Paulo* 1427!
DISTR. **U** 1, 2, 4; **K** 3, 5; **T** 1–4, 6–8; throughout tropical Africa from Guinée and Sierra Leone to Ethiopia and south to South Africa (Transvaal and Natal)
HAB. In grassland and bushland, and in rocky areas; 300–2350 m.

SYN. *Cyrtopera shupangae* Reichb.f. in Otia Bot. Hamb. 2: 116 (1881). Type: Mozambique, Shupanga, *Kirk* (K, holo.!)
C. papillosa Rolfe in K.B. 1893: 336 (1893). Type: South Africa, Natal, cult. *Lawrence* (K, holo.!)
C. holstiana Kraenzl. in Gilg in P.O.A. C: 157 (1895). Type: Tanzania, Usambara Mts, Simbili, *Holst* 2363 (B, holo., K, iso.!)
Eulophia shupangae (Reichb.f.) Kraenzl. in Gilg in P.O.A. C: 157 (1895)
E. missionis Rendle in J.B. 33: 168 (1895). Type: Malawi, Mt. Mulanje, *Scott Elliot* 8618 (BM, holo., K, iso.!)
E. chrysantha Schltr. in E.J. 20, Beibl. 50: 2 (1895). Type: South Africa, Natal, *Sanderson* 829 (B, syn.) & Inanda, *J.M. Wood* 785 (B, syn.)
E. graciliscapa Schltr. in E.J. 24: 418 (1897). Type: Malawi, without exact locality, *J.M. Wood* 5196 (B, holo., K, record of holo.!)
E. aurea Kraenzl. in Bull. Herb. Boiss. 5: 635 (1897). Type: Mozambique, Maputo, Delagoa Bay, *Junod* 187 (G, holo.)
E. johnstonii Rolfe in F.T.A. 7: 66 (1897). Type: Tanzania, Kilimanjaro, *Johnston* (K, holo.!)
E. panganiensis Kraenzl. in E.J. 28: 169 (1900). Type: Tanzania, Lushoto District, Kwashemshi, *Buchwald* 673 (B, holo., K, record of holo.!)
E. durbanensis Rolfe in K.B. 1917: 83 (1917). Type: South Africa, Natal, *J.M. Wood* 11775 (K, holo.!)
E. propinqua Hutch. in K.B. 1921: 401 (1921). Type: Nigeria, between Hepham and Ropp, *Lely* 353 (K, holo.!)

NOTE. Two colour variants of this species are found in the southern part of the Flora area. The typical yellow-flowered form usually with red papillae on the lip is a very common plant in grass-land, but a form with brownish or reddish flowers is sometimes found, particularly in areas **T**

4 and 7. Further investigation is needed to see if the latter form warrants formal taxonomic recognition.

38. **E. galeoloides** *Kraenzl.* in E.J. 24: 508 (1898); Summerh. in F.W.T.A., ed. 2, 3: 247 (1968); J. Stewart in U.K.W.F.: 780 (1974). Type: Tanzania, Lushoto District, Amani, *Braun* 670 (B, holo., K, iso.! & record of holo.!)

A terrestrial saprophyte 6–100 cm. tall. Perennating organ subterranean, rhizomatous, fleshy, 4–10 cm. or more long, 4–16 mm. diameter; roots absent. Leaves absent. Inflorescence lacking chlorophyll, densely to laxly 2-many-flowered; peduncle pale buff-coloured, bearing 2–4 lanceolate acuminate bracts above and 2–3 tubular sheathing ones below; rhachis up to 27 cm. long; bracts linear-lanceolate, acuminate, 28–50 mm. long, 6–9 mm. broad. Flowers white to pale buff-coloured with 2 dark purple marks at base of lip and red or purple hairs on side lobes; column cream; pedicel and ovary 25–35 mm. long. Dorsal sepal oblong-lanceolate, acute, 15–21 mm. long, 4–7 mm. wide; lateral sepals similar but oblique at base, 15–25 mm. long, 2–5 mm. wide. Petals oblong-spathulate, obtuse, 20–25 mm. long, 4–7 mm. wide. Lip 3-lobed, 20–21 mm. long, 12–14 mm. wide, papillate all over; side lobes rounded in front; mid-lobe elliptic, obtuse or rounded; callus of 3–4 fleshy ridges in basal third of lip; spur conical, 4–5 mm. long, acute, upcurved at tip. Column 9–10 mm. long; foot 1–2 mm. long. Fruit pendent, 15–25 mm. long. Pl. 4. Fig. 103/1–10, p. 460.

KENYA. N. Kavirondo District: Kakamega Forest, Mar. 1939, *Tweedie* 446!
TANZANIA. Arusha District: Mt. Meru, 13 Dec. 1970, *Richards* 26526!; Lushoto District: Amani, Muheza, 26 Nov. 1955, *R. Tanner* 2500!; Morogoro District: Uluguru Mts., Mgeta R., gorge below Hululu Falls, 15 Mar. 1953, *Drummond & Hemsley* 1601!; Njombe District: Milo, without date, *Archbold* 2625!
DISTR. **K** 5; **T** 2, 3, 6, 7; Ghana, Nigeria, Sudan and Zaire
HAB. In shade in evergreen forest, rarely in coffee plantations; 900–2300 m.

SYN. *E. goetzeana* Kraenzl. in E.J. 28: 372 (1900). Type: Tanzania, S. Uluguru Mts., *Goetze* 252 (B, holo., K, record of holo.!)
E. cyrtosioides Schltr., Westafr. Kautsch.-Exped.: 279 (1900). Type: Zaire, Lukulela, *Schlechter* 12643 (B, holo., K, record of holo.!)

39. **E. epiphanoides** *Schltr.* in E.J. 53: 575 (1915). Type: Tanzania, Rungwe District, Mulinda, *Stolz* 1437 (B, holo., K, record of holo.!)

A terrestrial saprophyte 9–25 cm. tall. Perennating organ subterranean, rhizomatous, fleshy, branching, 4 cm. or more long, 6 mm. diameter, almost covered by loose papery sheaths; roots absent. Leaves absent. Inflorescence slender, laxly 4–15-flowered; peduncle wiry, 1–3 mm. diameter; rhachis 2–10 cm. long; bracts lanceolate, acute, 3–4 mm. long. Flowers pinkish cream with brown veins; pedicel and ovary 4 mm. long, clavate, glabrous. Dorsal sepal oblanceolate, acute, 4.5–6 mm. long, 1 mm. wide; lateral sepals similar but oblique at base. Petals oblanceolate, acute, 4–5 mm. long, 1 mm. wide. Lip obscurely 3-lobed, 4–6 mm. long, 2–2.5 mm. wide; side lobes obscure; mid-lobe elliptic, obtuse; callus of 4 rounded ridges in basal half of lip; spur shortly conical, incurved, 1.75–2.5 mm. long. Column 2.5–3 mm. long; foot very short. Fruit pendent, ellipsoidal, 6–7 mm. long. Fig. 103/11–19, p. 460.

TANZANIA. Rungwe District: Mulinda, July 1912, *Stolz* 1437!
DISTR. **T** 7; known only from the type collection
HAB. Growing in red sandy soil in forest; 700–900 m.

40. **E. katangensis** (*De Wild.*) *De Wild.* in B.J.B.B. 6: 121 (1919); Williamson, Orch. SC. Afr.: 165 (1977). Type: Zaire, Shaba [Katanga], *Verdick* (BR, holo.!)

A terrestrial herb 11–44 cm. tall. Perennating organs underground, pseudobulbous, ovoid-conical, proximate in a row, 2–4 cm. tall, 8–35 mm. diameter, 3-noded; roots 1–1.5 mm. diameter, from base of pseudobulbs, white. Leaves 3, erect, linear, acute, 20–25 cm. long, 3–3.5 mm. wide. Inflorescence produced before leaves, 2–8-flowered; peduncle wiry, 1 mm. diameter, with 3 sheaths towards base; rhachis 4–20 cm. long; bracts lanceolate, acuminate, 2–6 mm. long. Flowers erect; sepals and petals whitish pink to greenish yellow, shaded with purple at apices of petals; lip mid-lobe pinkish mauve; pedicel and ovary 10–16 mm. long. Dorsal sepal linear, acute to acuminate, 16–18 mm. long, 2 mm. wide; lateral sepals similar. Petals linear-lanceolate, acute to apiculate, 11–12 mm. long, 2–2.5 mm. wide. Lip 3-lobed, 8–15 mm. long, 7–9 mm. wide; side lobes rounded;

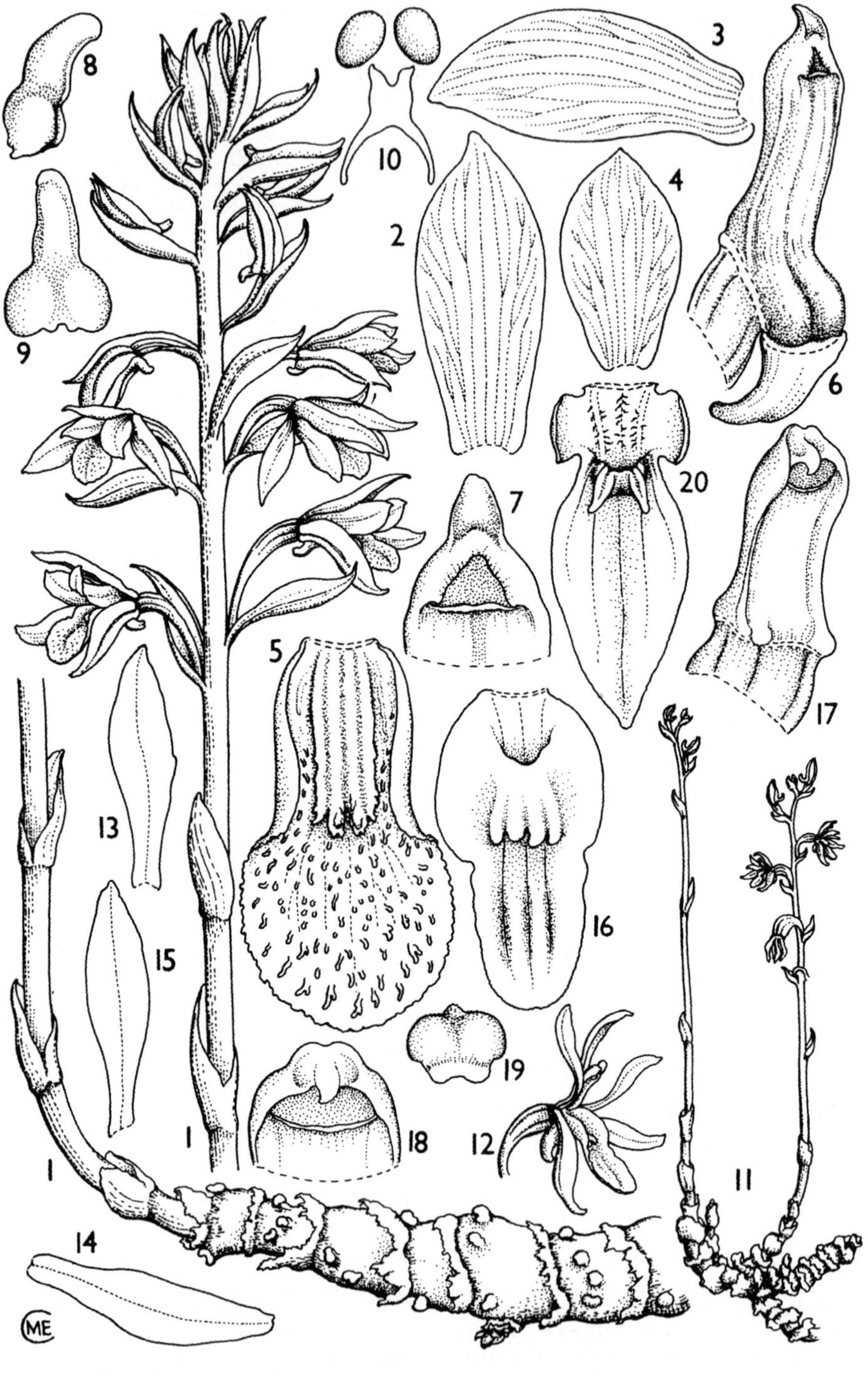

FIG. 103. *EULOPHIA GALEOLOIDES*—**1**, habit, × ⅔; **2**, dorsal sepal, × 2; **3**, lateral sepal, × 2; **4**, petal, × 2; **5**, lip, front view, × 3; **6**, column, × 4; **7**, column apex, front view, × 6; **8**, anther-cap, side view, ×8; **9**, anther-cap, front view, × 8; **10**, pollinarium, × 8. *E. EPIPHANOIDES*—**11**, habit, × ⅔; **12**, flower, × 3; **13**, dorsal sepal, × 6; **14**, lateral sepal, × 6; **15**, petal, × 6; **16**, lip, front view, × 8; **17**, column, × 8; **18**, column apex, front view, × 10; **19**, anther-cap, front view, × 12. *E.* sp. near *EPIPHANOIDES* (from Zambia)—**20**, lip, front view, × 6. 1–10, from *Drummond & Hemsley* 1601; 11–19, from *Stolz* 1437, 20, from *Richards* 10034. Drawn by Mrs M.E. Church.

mid-lobe subcircular, obtuse; callus of 3 ridges, papillose in front on mid-lobe; spur straight, at acute angle to ovary, cylindric from conical base, 8–10 mm. long. Column 6 mm. long. Fruit erect, 17–18 mm. long. Fig. 104/1–10, p. 462.

TANZANIA. Chunya District: Ukimbu, Masangati, 20 Oct. 1980, *Leedal* 6040!; Iringa District: Sao Hill, Oct. 1959, *Watermeyer* 187!
DISTR. **T** 7; Zaire, Zambia, Malawi and Zimbabwe
HAB. In bare and hardbaked soil on hilltops; 1500–2000 m.

SYN. *Lissochilus katangensis* De Wild. in Ann. Mus. Congo, Bot., sér. 4, 1: 22 (1902)
Eulophia paucisquammata De Wild. in B.J.B.B., 6: 125, t. 22/1–6 (1919). Type: Zaire, Shaba, Kisanfu, *Hock* (BR, holo.!)

41. **E. rhodesiaca** *Schltr.* in R.E. Fries, Wiss. Ergebn. Schwed. Rhod.-Kongo-Exped. 1911–12, 1: 248 (1916); Williamson, Orch. SC. Afr.: 166 (1977). Type: Zimbabwe, near R. Suera, *R.E. Fries* 585a (UPS, holo., K, record of holo.!)

A terrestrial herb 20–30 cm. tall. Perennating organs pseudobulbous, in closely spaced chains, conical, 2–3 cm. tall, 7–10 mm. diameter, 3–4-noded; roots 1.5 mm. diameter. Leaves 2–4, grass-like, 10–30 cm. long, 3–4.5 mm. wide. Inflorescence 2–5-flowered, produced before leaves; peduncle slender, wiry, 1 mm. diameter; rhachis 3.5–7 cm. long; bracts lanceolate, acuminate, 3–5 mm. long. Flowers erect; sepals buff or brown; petals pale yellow-green; lip mauve-pink with a maroon spur; pedicel and ovary 11–17 mm. long. Dorsal sepal linear-oblanceolate, acute, 11–13 mm. long, 2.5–3 mm. wide; lateral sepals similar but slightly longer. Petals oblong-elliptic, acute, 8.5–10 mm. long, 3.5–4 mm. wide. Lip 3-lobed in apical half, broadest across basal half, 8–10 mm. long, 9–12 mm. wide; side lobes elliptic; mid-lobe semicircular, obtuse, erose on margin; callus of 3 basal ridges with 3–7 verrucose ridges on mid-lobe in front; spur incurved, cylindrical, 12–13 mm. long. Column 6 mm. long; foot 3 mm. long. Fruit erect, narrowly ellipsoidal, 16–18 mm. long.

TANZANIA. Rungwe District: Ulambya, E. of Mlale, 21 Aug. 1977, *Leedal* 4602!
DISTR. **T** 7; Angola, Zambia and Zimbabwe
HAB. In deciduous woodland or on rocks; 1400–1700 m.

42. **E. monile** *Reichb.f.* in Flora 50: 105 (1867); Summerh. in F.W.T.A., ed. 2, 3: 247 (1968). Type: Angola, Huilla, *Welwitsch* 709 (W, holo.!)

A terrestrial herb 20–40 cm. tall. Perennating organs underground, tuberous, conical or obliquely cylindrical-conical, approximate, 2.5–5 cm. tall, 1–2 cm. diameter, several-noded, drying dull orange; roots wiry, 0.5–1 mm. diameter. Leaves 5–7, vestigial, linear, acuminate, 3–10 cm. long, 1–2 mm. wide, green or brown, basal 2–3 sheathing. Inflorescence 6–13-flowered; peduncle arising from upper node of tuber, with a short sheathing acuminate sterile bract in the middle; rhachis 6–19 cm. long; bracts lanceolate, acuminate, 3–6 mm. long. Flowers not opening widely, possibly self-pollinating; sepals yellow-brown suffused with purple within; petals and lip yellow-green with purple margins and marginal veins and a yellow callus and spur; pedicel and ovary 7–12 mm. long. Dorsal sepal linear, acuminate, 7–11 mm. long, 1 mm. wide, recurved towards apex; lateral sepals similar. Petals porrect, linear, acute, 7–10 mm. long, 1 mm. wide. Lip 3-lobed, 4–6 mm. long, 1.5–2 mm. wide; side lobes rounded in front; mid-lobe elliptic; callus of 3 ridges, papillate in front on mid-lobe; spur incurved, clavate, 3–4 mm. long. Column 2–2.5 mm. long. Fruit pendent, cylindric, 12–15 mm. long.

TANZANIA. Rungwe District: Chivanjee, 10 Feb. 1976, *Cribb & Grey-Wilson* 10677a!; Tunduru District: 112 km. W. of Tunduru, 21 Dec. 1955, *Milne-Redhead & Taylor* 7848! & 7848a!
DISTR. **T** 7, 8; tropical Africa from Sierra Leone to Angola, Zambia, Malawi and Zimbabwe
HAB. In loamy and sandy soil in *Brachystegia* woodland; 850–1150 m.

SYN. *E. brevipetala* Rolfe in F.T.A. 7: 53 (1897). Type: Sierra Leone, above Falaba, *Scott Elliot* 5224 (K, holo.!)
E. concinna Schltr. in E.J. 53: 581 (1915). Type: Tanzania, Rungwe District, Bomalakitana, *Stolz* 1052 (B, holo., K, record of holo. & iso.!)
E. exilis Schltr in E.J. 53: 581 (1915). Type: Tanzania, Rungwe District, Kiwira [Kibila], *Stolz* 1785 (B, holo., K, iso.!)
E. anisotepala Summerh. in K.B. 1927: 415 (1927). Type: Zambia, Mumbwa, *Macaulay* 968 (K, holo.!)

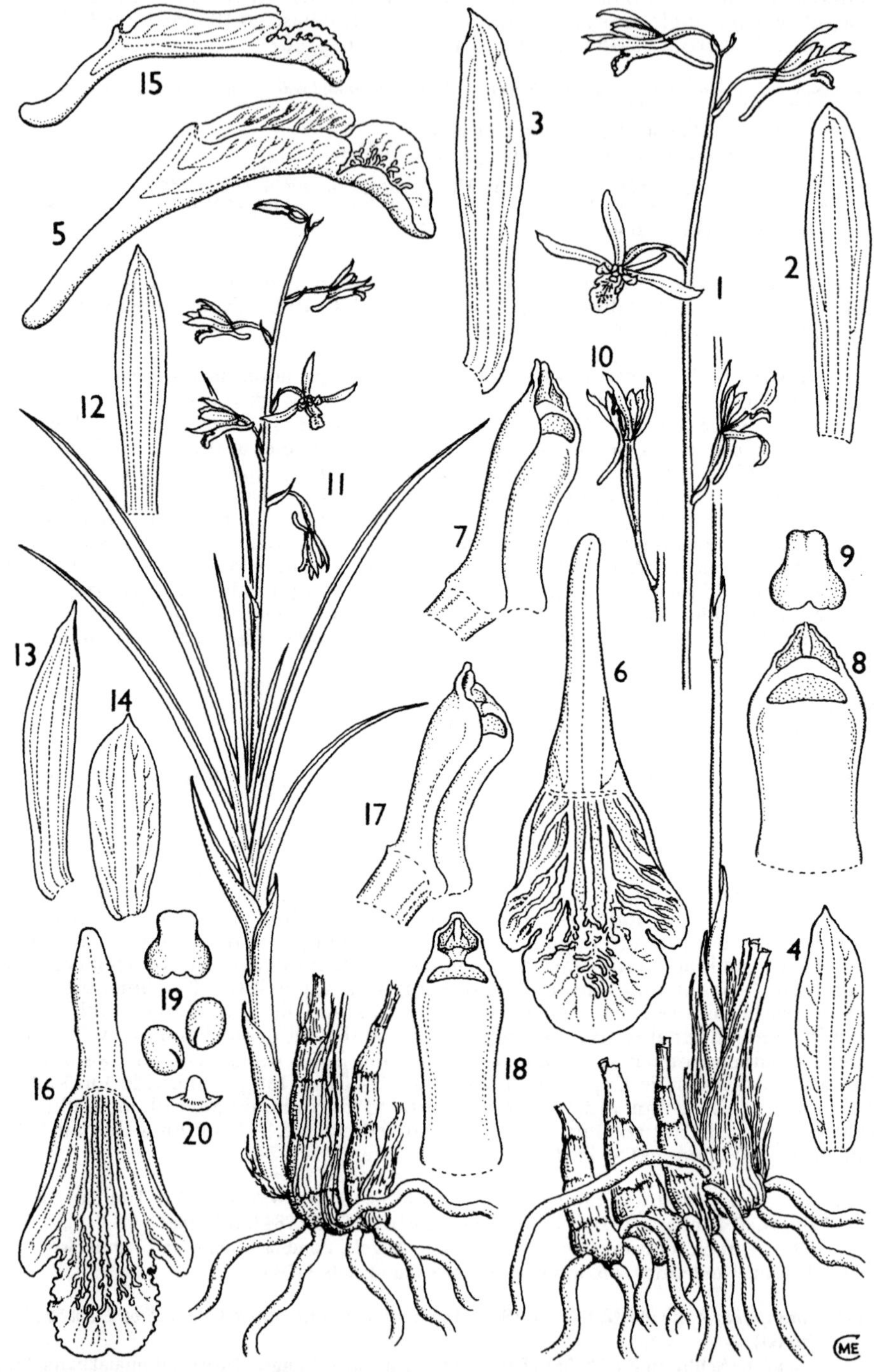

FIG. 104. *EULOPHIA KATANGENSIS*—**1**, habit, × ⅔; **2**, dorsal sepal, × 4; **3**, lateral sepal, × 4; **4**, petal, × 4; **5**, lip, side view, × 4; **6**, lip, spread out, × 5; **7**, column, side view, × 8; **8**, column, front view, × 8; **9**, anther-cap, × 8; **10**, capsule, × ⅔. *E. PENDULIFLORA*—**11**, habit, × ⅔; **12**, dorsal sepal, × 4; **13**, lateral sepal, × 4; **14**, petal, × 4; **15**, lip, side view, × 4; **16**, lip, spread out, × 5; **17**, column, side view, × 8; **18**, column, front view, × 8; **19**, anther-cap, × 8; **20**, pollinarium, × 18. 1–9, from *Schaijes* 1989; 10, from *Leedal* 4601; 11–20, from *Milne-Redhead & Taylor* 7832. Drawn by Mrs M.E. Church.

43. **E. norlindhii** *Summerh.* in Bot. Notis. 1937: 197 (1937); Williamson, Orch. SC. Afr.: 160 (1977). Type: Zimbabwe, Inyanga, *Fries, Norlindh & Weimarck* 3267a (K, holo.!)

A terrestrial herb 10–35 cm. tall. Perennating organs underground, tuberous, elliptic or cylindrical, 3.3–4 cm. long, 1.3–2 cm. diameter, whitish or purplish, ± 5-noded; rhizome up to 12 cm. long, 4–5 mm. diameter; roots 1 mm. diameter, white. Leaves 2–7, vestigial, linear, acuminate, up to 10 cm. long, 3 mm. wide. Inflorescences 1–3, 7–20-flowered; peduncle with a short sheath in the middle; rhachis 3.5–15 cm. long; bracts lanceolate, acuminate, 4–7 mm. long. Flowers possibly self-pollinating, not opening widely, spreading to subnutant; sepals yellow-green to brown; petals yellow-green; lip white with a pale green callus and side lobes and a brown spur; pedicel and ovary 7–9 mm. long. Dorsal sepal linear, acute, 7–10 mm. long, 1 mm. wide; lateral sepals similar. Petals linear, acute, 6–9 mm. long, 1 mm. wide. Lip 3-lobed, 6–8.5 mm. long, 2.5–4 mm. wide; side lobes narrowly oblong, obtuse; mid-lobe obovate, obtuse; callus of 3 ridges with 3 rows of papillae in front on mid-lobe; spur sometimes slightly incurved, clavate, 2–2.5 mm. long. Column 2.5–3 mm. long; foot obscure. Fruit pendent, 15 mm. long.

TANZANIA. Mbeya District: Chimala escarpment, 31 Dec. 1970, *Leedal* 558! & 17 Feb. 1976, *Cribb & Grey-Wilson* 10802!; Rungwe District: Chivanjee, 15 Dec. 1980, *Leedal* 6219!; Tunduru District: ± 112 km. W. of Tunduru, 21 Dec. 1955, *Milne-Redhead & Taylor* 7859!
DISTR. **T** 7, 8; S. Zaire, Zambia, Malawi, Mozambique and Zimbabwe
HAB. In *Brachystegia* woodland and on river banks in black loam; 870–1900 m.

NOTE. Peloric forms with lip-like petals are recorded from neighbouring Malawi.

44. **E. tanganyikensis** *Rolfe* in F.T.A. 7: 59 (1897). Type: Zambia, Fwambo, *Nutt* (K, holo.!)

A terrestrial herb 40–60 cm. tall. Perennating organs pseudobulbous, conical, 8–15 mm. long, 10–20 mm. diameter, on a rhizome 3.5–4 mm. diameter; roots slender 1.5 mm. diameter. Leaves 3–4, plicate, erect, linear, acute, up to 40 cm. long, 4 mm. wide. Inflorescence laxly few–10-flowered; peduncle 3–4 mm. diameter, bearing 3–4 sheaths in lower two-thirds; rhachis 9–12 cm. long; bracts linear-lanceolate, acuminate, 5–20 mm. long. Flowers not opening widely; sepals green shaded light brown; petals white; lip white faintly veined with purple and with white or red papillae; pedicel and ovary 13–21 mm. long. Dorsal sepal linear, acute, 13–17 mm. long, 3 mm. wide; lateral sepals similar. Petals parallel to column, oblong, acute, 13–15 mm. long, 5 mm. wide. Lip 3-lobed, 15–17 mm. long, 13 mm. wide; side lobes erect, elliptic, rounded in front; mid-lobe oblong-obovate, obtuse or emarginate, with an undulate margin; callus of 3 short fleshy ridges with 5–7 rows of papillae in front on to middle of mid-lobe; spur parallel to ovary but pointing down at apex, cylindrical, 3.5–5 mm. long. Column 5–6 mm. long; foot 3 mm. long. Fruit pendent, ± 24 mm. long.

TANZANIA. Ufipa District: 19 km. from Kawimbe, 29 Jan. 1957, *Richards* 8052!
DISTR. **T** 4; Zambia, Zaire, Malawi and Zimbabwe
HAB. In swamps and wet grassland; 1100–1700 m.

SYN. *E. chlorotica* Kraenzl. in E.J. 51: 389 (1914). Type: Zaire, Kantu, *Kassner* 2370 (B, holo., K, iso.!)

45. **E. massokoensis** *Schltr.* in E.J. 53: 580 (1915). Type: Tanzania, Rungwe District, Masoko [Massoko], *Stolz* 1831 (B, holo., K, record of holo. & iso.!)

A terrestrial herb 20–35 cm. tall. Perennating organs conical to ovoid, proximate, 1.5–4 cm. tall, 1–1.5 cm. diameter; roots 1 mm. diameter, white. Leaves 6–7, vestigial, in a fan, up to 16 cm. long, 2 mm. wide. Inflorescence laxly 5–10-flowered; peduncle slender, wiry, 1 mm. diameter, bearing a short sheath in the middle; rhachis 5–11 cm. long; bracts linear-lanceolate, acuminate, 3–7 mm. long. Flowers chocolate-brown with a white or cream lip, brown spur and greenish column; pedicel and ovary 7–11 mm. long. Dorsal sepal erect, linear, acuminate, 11–15 mm. long, 1 mm. wide; lateral sepals similar. Petals porrect, narrowly oblong, acute or acuminate, 8–11 mm. long, 1.5–2 mm. wide, recurved at apex. Lip 3-lobed, 10–13 mm. long, 5.5–7 mm. wide; side lobes narrowly oblong, rounded in front; mid-lobe elliptic, obtuse; callus of 3 fleshy ridges on disc and 3–5 lines of flap-like papillae on mid-lobe; spur clavate, little incurved, 3–4 mm. long. Column 4 mm. long; foot obscure. Fruit pendent, 16–18 mm. long.

TANZANIA. Ufipa District: Chapota, 3 Dec. 1949, *Bullock* 1978!; Rungwe District: Masoko, 9 Jan. 1913, *Stolz* 1831!; Songea District: ± 16 km. W. of Songea, 2 Jan. 1956, *Milne-Redhead & Taylor* 8113!

DISTR. T 4, 7, 8; Zaire, Zambia
HAB. In *Brachystegia-Uapaca* woodland in shade; 700–2500 m.

46. **E. monotropis** *Schltr.* in R.E. Fries, Wiss. Ergebn. Schwed. Rhod.-Kongo-Exped. 1911–12, 1: 248 (1916); Williamson, Orch. SC. Afr.: 165 (1977). Type: Zambia, Kalungwishi, *R.E. Fries* 631a (B, holo., K, record of holo.!)

A terrestrial herb 30–43 cm. tall. Perennating organs pseudobulbous, proximate, conical, 2–4 cm. tall, 1–1.5 cm. diameter; roots 1.5–2 mm. diameter, white. Leaves 7–8, erect, linear, acuminate, 5–20 cm. long, 1–1.5 mm. wide, basal 3 sheathing. Inflorescence 3–10-flowered, produced before the leaves; peduncle with a short acute sheath in the middle; rhachis 6–15 cm. long; bracts lanceolate, acuminate, 4–7(–11) mm. long. Flowers with brown or greenish sepals, pale pinkish white petals and a pinkish lip with red veins, white papillae and an olivaceous spur; pedicel and ovary 6–11 mm. long. Dorsal sepal linear-lanceolate, acute, 9–16 mm. long, 1 mm. wide; lateral sepals similar, 1.5 mm. wide. Petals elliptic, obtuse to apiculate, 8–13 mm. long, 2 mm. wide. Lip 3-lobed, 9 mm. long, 7 mm. wide; side lobes obliquely oblong, rounded; mid-lobe oblong-elliptic, obtuse or apiculate; callus of 3 ridges on disc with 3 rows of papillae on mid-lobe; spur incurved, conical, 4–7 mm. long. Column 4 mm. long; foot short. Fruit not known.

TANZANIA. Kilwa District: 1.6 km. S. of Njenje, 25 Nov. 1917, *McLoughlin* 32!
DISTR. T 8; S. Zaire, Zambia and Zimbabwe
HAB. Under trees in sandy soil in *Brachystegia* woodland; altitude not given

47. **E. mumbwaensis** *Summerh.* in K.B. 1927: 416 (1927); Williamson, Orch. SC. Afr.: 165 (1977). Type: Zambia, Mumbwa, *Macaulay* 859 (K, holo.!)

A terrestrial herb 28–75 cm. tall. Perennating organs pseudobulbous, proximate, conical, 2–4 cm. tall, 1–1.2 cm. diameter, drying dull orange, 3–4-noded; roots 1–1.5 mm. diameter, white. Leaves 4–8, vestigial, linear, acuminate, up to 9 cm. long, 2 mm. wide, green, lowermost 3 sheathing. Inflorescence laxly 5–9-flowered; peduncle with 1–2 distant sheaths along length; rhachis 15–27 cm. long; bracts lanceolate, acuminate, 5–9 mm. long. Flowers subnutant; sepals maroon-brown; petals pale brown striped with purple; lip yellow with purple veins on side lobes and a reddish spur; pedicel and ovary 7–14 mm. long, purple. Dorsal sepal spreading, linear, acuminate, 16–18 mm. long, 1 mm. wide; lateral sepals similar. Petals porrect, oblong-elliptic, obtuse or slightly apiculate, 8–10 mm. long, 2–3 mm. wide. Lip 3-lobed, 10–14 mm. long, 6–8 mm. wide; side lobes narrowly elliptic, rounded in front; mid-lobe elliptic, obtuse; callus of 3 fleshy ridges on disc and long papillate in front on mid-lobe; spur incurved, conical-subclavate, 4–5 mm. long. Column 3–4 mm. long; foot 1 mm. long. Fruit pendent, 22 mm. long.

TANZANIA. Mpanda District: Nyonga to Mpanda track, 8 km. from Uruwira, 30 Oct. 1959, *Richards* 11657!
DISTR. T 4; S. Zaire, Zambia, Angola and Zimbabwe
HAB. Dry rocky places in *Brachystegia* and other woodlands; 1200-1600 m.

48. **E. penduliflora** *Kraenzl.* in E.J. 30: 288 (1901); Williamson, Orch. SC. Afr.: 167 (1977). Type: Tanzania, Mbeya District, Mbozi Hill, *Goetze* 1383 (B, holo., K, record of holo.!)

A terrestrial herb 12–25(–40) cm. tall. Perennating organs pseudobulbous, closely clustered, erect, conical-cylindric, 2.5–4 cm. long, 6–8 mm. wide, 3–4-noded; roots 1.5 mm. diameter. Leaves 5–9, in a fan, not plicate, linear, acuminate, 5–30 cm. long, 1–2 mm. wide, lowermost 3–4 sheathing. Inflorescences 1–2, produced with developing leaves, laxly 4–10-flowered; peduncle wiry, 1 mm. diameter; rhachis 2–8 cm. long; bracts narrowly lanceolate, acuminate, 7–9 mm. long. Flowers with maroon or purple-brown sepals; petals green or yellow with a brown central vein; lip white or yellow or green sometimes with pale mauve on mid-lobe; pedicel and ovary 7–9 mm. long. Dorsal sepal reflexed, linear-oblanceolate, acute, 8.5–11 mm. long, 1.8–2.4 mm. wide; lateral sepals similar. Petals oblong-oblanceolate, acute or apiculate, 7–9 mm. long, 2–3.5 mm. wide. Lip 3-lobed, 6.5–8.6 mm. long, 3.4–5.7 mm. wide; side lobes obtuse to rounded at apex; mid-lobe circular, obtuse or emarginate; callus of 3 ridges with 5–7 lines of papillae on mid-lobe in front; spur clavate, parallel to or at an acute angle to ovary, 5–6 mm. long, slightly upcurved at apex. Column 3.5–4 mm. long; foot short. Fruit cylindrical, pendent, 16–18 mm. long. Fig. 104/11-20, p. 462.

TANZANIA. Morogoro District: Uluguru Mts., Mlali to Mgeta, 30 Dec. 1970, *Wingfield* in *D.S.M.* 2285!; Mbeya District: Unyiha, Hezya, 17 Nov. 1970, *Leedal* 504!; Tunduru District: W. of Puchapucha, 2.5 km. E. of R. Mawese, 19 Dec. 1955, *Milne-Redhead & Taylor* 7832!
DISTR. T 6–8; Nigeria, Cameroon, Zaire, Malawi and Zimbabwe
HAB. In grassland and *Brachystegia* woodland; 900–1400 m.

SYN. *E. lindiana* Kraenzl. in E.J. 51: 388 (1914). Type: Tanzania, Lindi District, Tendaguru, *Janensch & Hennig* 24 (B, holo., K, record of holo.!)
E. nana Schltr. in E.J. 53: 579 (1915). Type: Tanzania, Rungwe District, Bomalakitana, *Stolz* 1748 (B, holo., K, iso. & record of holo.!)
E. angustilabellata De Wild. in B.J.B.B. 6: 109 (1919). Type: Zaire, Kisanfu, *Hock* (BR, holo.!)

NOTE. This species is rather variable in lip shape and colour and petal colour. The variation seems to be continuous and not to warrant taxonomic recognition but further information is needed to clarify the nature of the variation.

49. **E. sabulosa** *Schltr.* in E.J. 53: 578 (1915); Williamson, Orch. SC. Afr.: 166 (1977). Type: Tanzania, Rungwe District, near Lehnin, *Stolz* 2575 (B, holo., K, iso.!)

A terrestrial herb 35–70 cm. tall. Perennating organs pseudobulbous, approximate, conical, in chains, 1.7–3.5 cm. tall, 1.5–2.3 cm. diameter; roots 1–1.5 mm. diameter. Leaves 7–8, in a fan, linear, acute or acuminate, 4–21 cm. long, 3–11 mm. wide, basal 3 sheathing. Inflorescences 1–2, laxly few–many-flowered; peduncle slender, 1–2 mm. diameter; rhachis 10–29 cm. long; bracts linear, acuminate, 4–10 mm. long. Flowers with green or brownish green sepals; petals green at base, white at apex; lip white with a purple mid-lobe, cream callus and purple-striped side lobes; pedicel and ovary 12–18 mm. long. Dorsal sepal fleshy, erect, linear-oblong, acuminate, 12–18 mm. long, 1 mm. wide; lateral sepals similar. Petals porrect, oblong-narrowly elliptic, acute or apiculate, 8–12 mm. long, 2–3 mm. wide, recurved at tip. Lip 3-lobed, 8–11 mm. long, 6–7.5 mm. wide; side lobes oblong, rounded in front; mid-lobe ovate, acute, recurved; callus of 2 ridges with several lines of long papillae in front on mid-lobe; spur cylindric, slightly sigmoid, 4–5 mm. long. Column 4–5 mm. long; foot 1 mm. long. Fruit pendent, ± 20 mm. long. Fig. 105/1–9, p. 466.

TANZANIA. Mbeya District: Chivanjee, 11 Feb. 1976, *Cribb & Grey-Wilson* 10696!; Iringa District: Sao Hill, Ipogoro to M'kawa track, 12 Dec. 1961, *Richards* 15580!; Tunduru District: near Tunduru, 20 Dec. 1917, *McLoughlin* 38!
DISTR. T 6–8; Zambia, Malawi and Zimbabwe
HAB. In *Brachystegia-Uapaca* woodland in sandy soil; 250–1350 m.

50. **E. seleensis** (*De Wild.*) *Butzin* in Willdenowia 7: 589 (1975). Type: Zaire, Pool de Maleba [Stanley Pool], *Gillet* 1449 (BR, holo.!)

A terrestrial herb 28–50 cm. tall. Perennating organs partially underground, proximate, pseudobulbous, conical, 1.8–4.2 cm. tall, 0.6–2.4 cm. diameter, drying ochre-yellow; roots 2.5–3.5 mm. diameter, white. Leaves 5–9, in a fan, linear, acute or acuminate, 12–19 cm. long, 2–8 mm. wide. Inflorescences 1–2, laxly 5–9-flowered; peduncle wiry, with a short sheathing bract in basal half; rhachis 7–19 cm. long; bracts linear-lanceolate, acuminate, 4–9 mm. long. Flowers erect; sepals and petals yellowish green, sometimes red or purplish mottled; lip green with a yellow centre, purple veins and a lilac edge to the mid-lobe; pedicel and ovary 13–22 mm. long. Dorsal sepal erect, linear-oblanceolate, acute, 12–16 mm. long, 1 mm. wide; lateral sepals similar. Petals narrowly elliptic, acute, 8–8.5 mm. long, 1.5–2 mm. wide. Lip 3-lobed, 7–10 mm. long, 5–6 mm. wide; side lobes elliptic-oblong, rounded in front; mid-lobe reflexed, subcircular to obovate; callus of 3 ridges with several papillate lines in front on mid-lobe; spur incurved, cylindrical, 3–4 mm. long. Column 4–5 mm. long; foot 1 mm. long. Fruit pendent, 20–23 mm. long. Fig. 105/10–18, p. 466.

TANZANIA. Ufipa District: Kawa [Kara] R., 11 Nov. 1961, *Richards* 15354! & Sumbawanga, Mbala [Abercorn] to Kasanga road, 24 Nov. 1959, *Richards* 11832!; Rungwe District: Ulambya, Mlale, 24 Oct. 1976, *Leedal* 3907!; Songea District: Songea to Liwale road, Ngorogo stream, 5 Nov. 1917, *McLoughlin* 24!; Tunduru District: 112 km. W. of Tunduru, 21 Dec. 1955, *Milne-Redhead & Taylor* 7845!
DISTR. T 4, 7, 8; S. Zaire, Zambia, Angola, Malawi and Zimbabwe
HAB. In *Brachystegia* woodland; 850–1600 m.

SYN. *Lissochilus seleensis* De Wild. in Not. Pl. Util. Congo: 131 (1903)
Eulophia tenuiscapa Schltr. in R.E. Fries, Wiss. Ergebn. Schwed. Rhod.-Kongo-Exped. 1911–12, 1: 249 (1916). Type: Zambia, near Kalambo, *R.E. Fries* 1343 (B, holo.)
E. recurvata Williamson in Pl. Syst. Evol. 134: 70 (1980). Type: Zimbabwe, Barotse Province, *Drummond & Cookson* 6251 (SRGH, holo., K, iso.!)

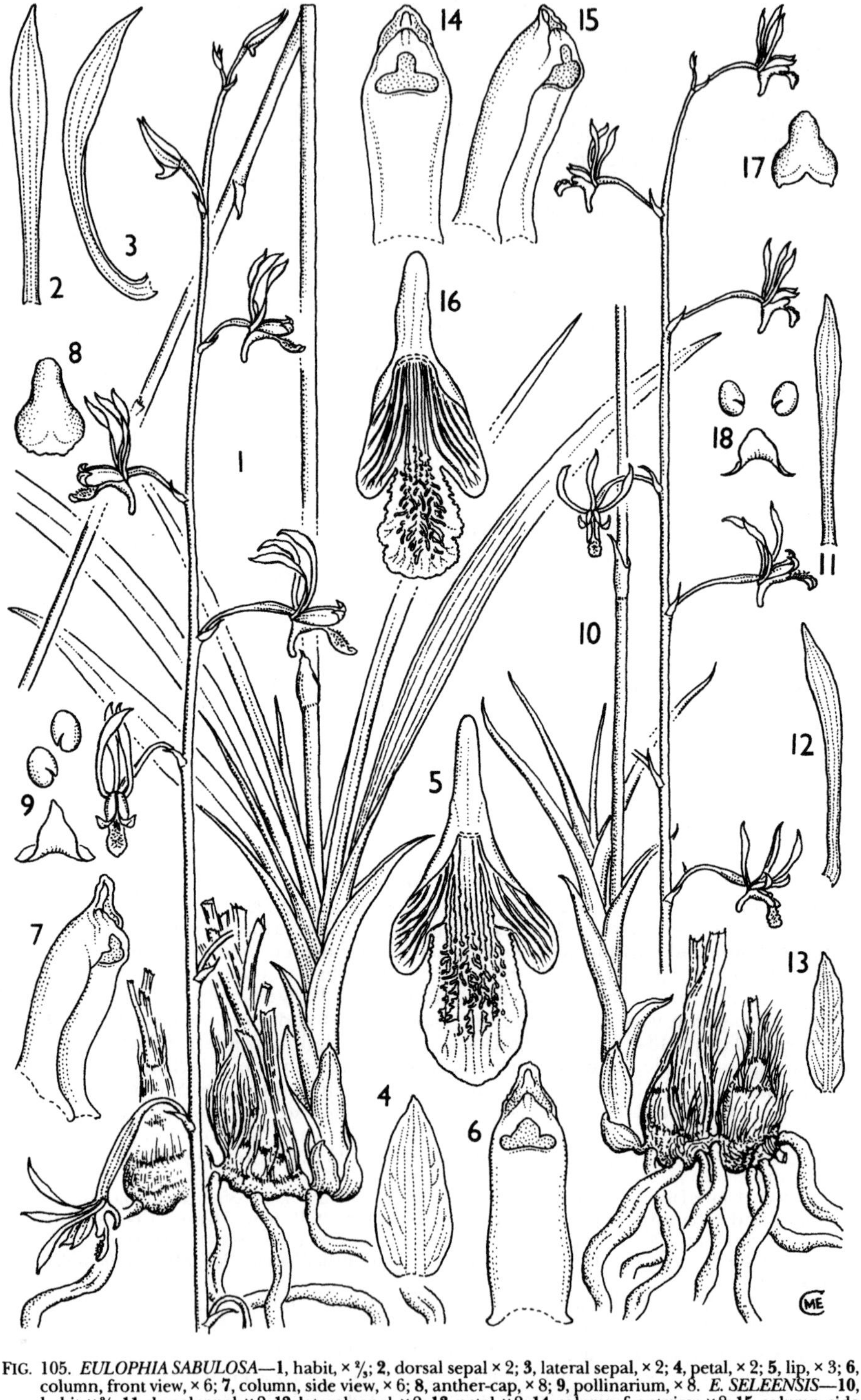

FIG. 105. *EULOPHIA SABULOSA*—**1**, habit, × ⅔; **2**, dorsal sepal × 2; **3**, lateral sepal, × 2; **4**, petal, × 2; **5**, lip, × 3; **6**, column, front view, × 6; **7**, column, side view, × 6; **8**, anther-cap, × 8; **9**, pollinarium, × 8. *E. SELEENSIS*—**10**, habit, × ⅔; **11**, dorsal sepal, × 2; **12**, lateral sepal, × 2; **13**, petal, × 2; **14**, column, front view, × 8; **15**, column, side view, × 8; **16**, lip, × 3; **17**, anther-cap, × 8; **18**, pollinarium, × 8. 1–9, from *Robson* 888; 10, from *la Croix* 510; 11–16, from *Schaijes* 2566; 17, 18, from *Schaijes* 2424. Drawn by Mrs M.E. Church.

51. **E. venulosa** *Reichb. f.* in Otia Bot. Hamb. 2: 116 (1881). Type: Malawi, Manganja Hills, *Meller* 72 (W, holo., K, iso.!)

A terrestrial herb 15–75 cm. tall. Perennating organs pseudobulbous, erect, ovoid-conical, 1–6 cm. long, 1–2 cm. diameter, 3-4-noded; roots 1–1.5 mm. diameter. Leaves 5–7, erect, linear, acuminate, 5–20 cm. long, 1–1.5 mm. wide. Inflorescences 1–2, produced before or with the leaves, laxly 9–20-flowered; peduncle with a short central sheath, red-brown; rhachis 10–28 cm. long; bracts lanceolate, acuminate, 2–9 mm. long. Flowers with red-brown sepals and petals and a white lip with purple-red veins; pedicel and ovary 8–12 mm. long. Dorsal sepal recurved, linear, acute, 15–17 mm. long, 2 mm. wide; lateral sepals similar. Petals narrowly elliptic-lanceolate, acute, 9–10 mm. long, 2.5 mm. wide. Lip 3-lobed, broadest across base, 11–17 mm. long, 5–8 mm. wide; side lobes narrowly oblong, rounded in front; mid-lobe elliptic or oblanceolate, acute; callus of 3 fleshy ridges with lines of long papillae in front on mid-lobe; spur sigmoid, cylindrical, often incurved at apex, 3–5.5 mm. long. Column 3–5 mm. long; foot 1–2 mm. long. Fruit pendent, 16–18 mm. long. Fig. 106/1–9, p. 468.

TANZANIA. Rungwe District: Chivanjee, 15 Dec. 1980, *Leedal* 6128!; Songea District: ± 16 km. W. of Songea, 2 Jan. 1956, *Milne-Redhead & Taylor* 8113; Tunduru District: 112 km. W. of Tunduru, 21 Dec. 1955, *Milne-Redhead & Taylor* 7846!

DISTR. T 7, 8; Zaire, Zambia, Angola, Malawi, Mozambique and Zimbabwe

HAB. In *Brachystegia* woodland; 700–1200 m.

SYN. *E. humilis* Rendle in J.B. 33: 165 (1895). Type: Malawi, Shire Highlands, *Scott Elliot* 8484 (BM, holo., K, iso.!)

E. paradoxa Kraenzl. in E.J. 33: 68 (1902). Type: Tanzania, near Songea, *Busse* 700 (B, holo., K, record of holo. & iso.!)

E. encyclioidea Schltr. in E.J. 53: 577 (1915). Type: Tanzania, Rungwe District, Masoko, *Stolz* 1051 (B, holo., K, record of holo. & iso.!)

52. **E. longisepala** *Rendle* in Trans. Linn. Soc., Bot., ser. 2, 4: 43 (1894); Hall in Journ. S. Afr. Bot., Suppl. 5: 135 (1965); Williamson, Orch. SC. Afr.: 164 (1977); J. Stewart et al., Wild Orch. S. Afr.: 241 (1982). Type: Malawi, Mulanje, *White* (BM, holo.!, K, iso.!)

A terrestrial semisaprophytic herb 10–75 cm. tall. Perennating organ underground, rhizomatous, branching, often vertical, several cm. long, 6–7 mm. diameter; roots 1 mm. diameter, white. Leaves 2–7, vestigial, linear, acuminate, 2–13 cm. long, 1.5–3 mm. wide, purplish. Inflorescences 1–3, produced before leaves or with young leaves; peduncle slender, wiry, 1–1.5 mm. diameter, red-green to purple, bearing 2 basal sheaths and 1–2 lanceolate, acuminate, sterile bracts above; rhachis 8–25 cm. long; bracts lanceolate, acuminate, 3–9 mm. long. Flowers spreading; sepals deep claret; petals similar or greenish red; lip mid- and side lobes claret-veined, callus greenish yellow; pedicel and ovary 11–16 mm. long. Dorsal sepal erect, linear, acute or acuminate, 15–32 mm. long, 1 mm. wide; lateral sepals similar. Petals porrect, narrowly elliptic, acute, 11–16 mm. long, 2.5–3 mm. wide. Lip 3-lobed, 10–14 mm. long, 6–10 mm. wide; side lobes oblong, rounded in front; mid-lobe transversely elliptic, obtuse to emarginate; callus of 3 ridges, papillate in 3–5 lines on mid-lobe; spur slightly incurved, shortly cylindrical, 2 mm. long. Column 5 mm. long. Fruit pendent, 25–27 mm. long. Fig. 106/10–19, p. 468.

TANZANIA. Rungwe District: Chivanjee, 10 Feb. 1976, *Cribb, Grey-Wilson & Mwasumbi* 10677! & 30 Dec. 1978, *Leedal* 5270!; Songea District: just E. of R. Mtandazi, 21 Jan. 1956, *Milne-Redhead & Taylor* 8541!; Tunduru District: ± 112 km. W. of Tunduru, 21 Dec. 1955, *Milne-Redhead & Taylor* 7860!

DISTR. T 7, 8; Zaire, Mozambique, Malawi, Zambia, Angola, Zimbabwe and South Africa (Natal)

HAB. In *Brachystegia* woodland, often in burnt areas; 150–1100 m.

SYN. *E. antennata* Schltr. in E.J. 26: 334 (1899). Types: Mozambique, Maputo, Delagoa Bay, *Schlechter* (B, syn.) & near Massinga, *Schlechter* (B, syn., Z, isosyn.!)

E. stolzii Schltr. in E.J. 53: 577 (1915). Type: Tanzania, Rungwe District, Masukulu [Mwasukulu], *Stolz* 739 (B, holo., K, record of holo. & iso.!, Z, iso.!)

E. triceras Schltr. in E.J. 53: 578 (1915). Type: Tanzania, Rungwe District, Masoko and Mulinda, *Stolz* 1806 (B, holo., K, record of holo. & iso.!, Z, iso.!)

E. bangweolensis Schltr. in R.E. Fries, Wiss. Ergebn. Schwed. Rhod.-Kongo-Exped. 1911–12, 1: 247 (1916). Type: Zambia, between Maumba and Kali, *Fries* 631 (B, holo., K, record of holo.!)

NOTE. A variable species particularly in its flower size, the length of the erect petals and the size of the callus on the lip. The variation appears, however, to lack discontinuities but further investigation is undoubtedly needed to assess the variation patterns in this species.

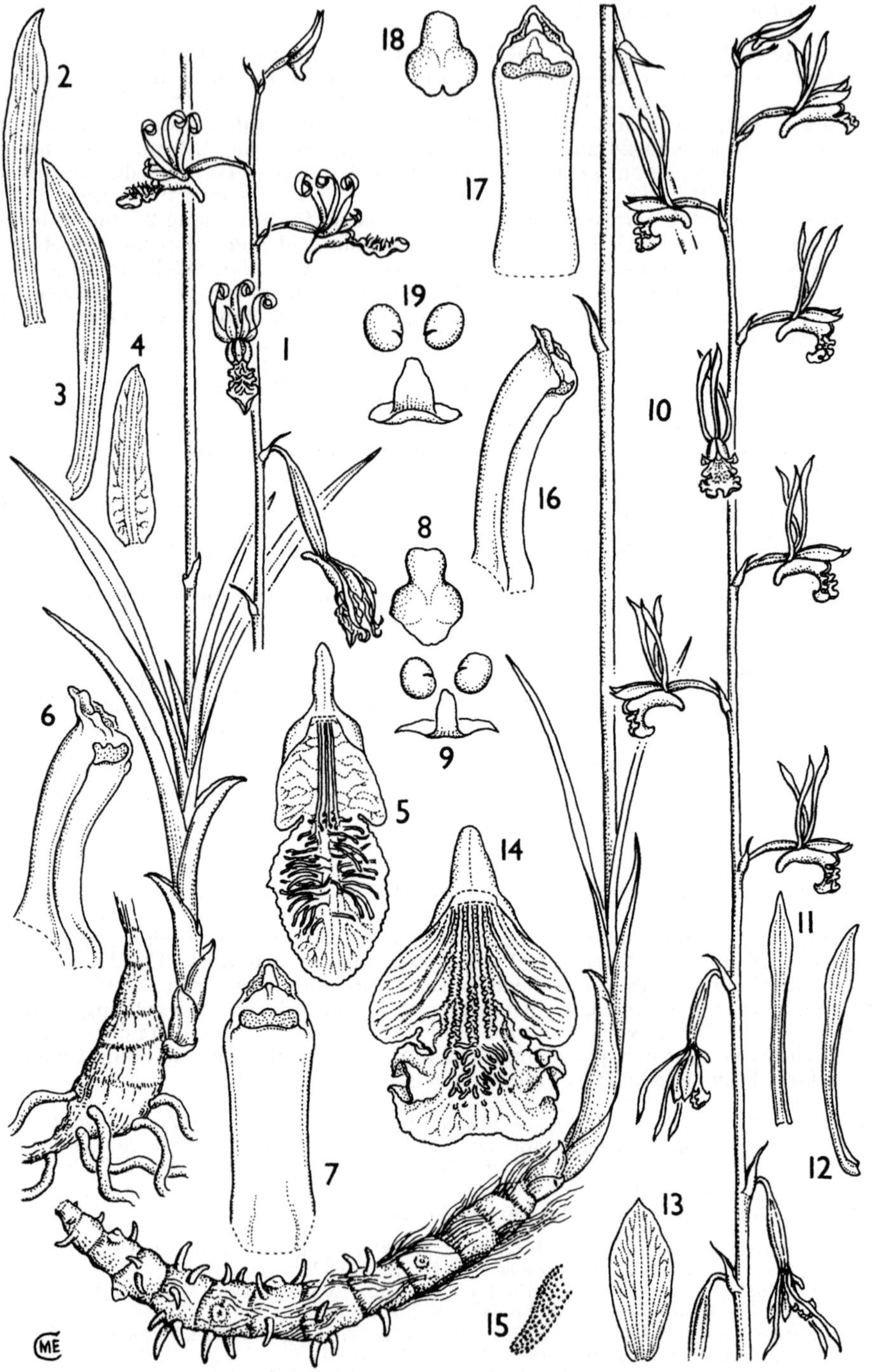

FIG. 106. *EULOPHIA VENULOSA*—**1**, habit, × 2/3; **2**, dorsal sepal, × 2; **3**, lateral sepal, × 2; **4**, petal, × 2; **5**, lip, × 2; **6**, column, side view, × 6; **7**, column, front view, × 6; **8**, anther-cap, × 6; **9**, pollinarium, × 8. *E. LONGISEPALA*—**10**, habit, × 2/3; **11**, dorsal sepal, × 2; **12**, lateral sepal, × 2; **13**, petal, × 2; **14**, lip, × 3; **15**, papilla, × 14; **16**, column, side view, × 6; **17**, column, front view, × 6; **18**, anther-cap, × 6; **19**, pollinarium, × 8. 1–9, from *Milne-Redhead & Taylor* 7846; 10–19, from *Milne-Redhead & Taylor* 8541. Drawn by Mrs M.E. Church.

53. **E. nyasae** *Rendle* in Trans. Linn. Soc., Bot., ser. 2, 4: 44 (1894). Type: Malawi, Mt. Mulanje, *Whyte* (BM, holo., K, iso.!)

A terrestrial subsaprophytic herb 25–40 cm. tall, drying black. Perennating organ underground, fleshy, rhizomatous, irregularly cylindrical, branching, 5–7 cm. long, 1.5–2.5 cm. diameter; roots emerging along rhizome, white. Leaves 2–4, vestigial, up to 6 cm. long, brown. Inflorescences 1–2, laxly many-flowered; peduncle with 4 imbricate lanceolate basal sheaths and another in the middle; rhachis 9–22 cm. long; bracts linear-setose, 9–18 mm. long, 2 mm. wide. Flowers not opening widely; sepals green speckled or flushed red-brown or entirely brown; petals green lined with brown; lip white with red veins; pedicel and ovary 11–12 mm. long. Dorsal sepal narrowly oblong, acute, 12–14 mm. long, 2–2.5 mm. wide; laterals similar, with a dorsal keel on midvein. Petals narrowly elliptic, acute, 10–12 mm. long, 3–4 mm. wide. Lip 3-lobed, 11–13 mm. long, 9–9.5 mm. wide; side lobes rounded and erose at apex; mid-lobe obovate-oblong, undulate; callus of 3 ridges with 3–5 ridges of long papillae on mid-lobe; spur straight or deflexed, clavate, up to 6 mm. long. Column 5 mm. long; foot 1–2 mm. long. Fruit ellipsoidal, pendent, 28–30 mm. long.

TANZANIA. Tabora District: NW. of Tabora, Panda area, Gombagulu, 29 Nov. 1949, *Shabani* 68!; Iringa District: Ipogoro to M'kawa track, 12 Dec. 1961, *Richards* 15565!; Tunduru District: 112 km. W. of Tunduru, 21 Dec. 1955, *Milne-Redhead & Taylor* 7862!
DISTR. **T** 4, 7, 8; S. Zaire, Zambia, Malawi, Angola and Zimbabwe
HAB. In *Brachystegia* woodland; 870–1680 m.

SYN. *E. florulenta* Kraenzl. in E.J. 33: 66 (1904). Type: Tanzania, near Songea, *Busse* 691 (B, holo., K, record of holo. & iso.!)
E. limodoroides Kraenzl. in E.J. 51: 387 (1914). Type: Zambia, near Chibenga stream, *Kassner* 2080 (B, holo., K, iso.!)
E. bletilloides Schltr. in E.J. 53: 576 (1915). Type: Tanzania, Rungwe District, Mulinda, *Stolz* 1800 (B, holo., K, iso.!)
E. ecalcarata Williamson in Journ. S. Afr. Bot. 46: 336 (1980). Type: Zambia, 17 km. E. of Lusaka, *Williamson* 216 (SRGH, holo., K, iso.!)
E. farcta Williamson in Journ. S. Afr. Bot. 46: 337 (1980). Type: Zambia, 80 km. W. of Solwezi, *Williamson & Simon* 1764 (SRGH, holo., K, iso.!)

NOTE. This is a very variable species in both flower size and lip ornamentation. Field-work might reveal patterns that might suggest the taxonomic recognition of some variants. Peloric forms in which the lip is petaloid or partly so are not uncommon, e.g. *Leedal* 341 and 881 from **T** 7. Williamson has described the peloric form as *E. ecalcarata* and the partially peloric form as *E. farcta* based on Zambian collections.

54. **E. subsaprophytica** *Schltr.* in E.J. 53: 576 (1915); Williamson, Orch. SC. Afr.: 162 (1977). Type: Tanzania, Rungwe District, Mulinda, *Stolz* 1807 (B, holo., K, record of holo.!)

A terrestrial saprophytic herb 50–80 cm. tall. Perennating organ rhizomatous, horizontal or vertical, underground, up to 15 cm. long, 8–14 mm. diameter, many-noded; roots vestigial, 1 mm. diameter. Leaves reduced to 2–3 linear, acuminate bracts at base of inflorescence, fleshy, 4–9 cm. long, purplish. Inflorescence laxly many-flowered; peduncle bearing 1–2 short tubular sheaths along length; rhachis 20–45 cm. long, tapering to apex; bracts lanceolate, acuminate, 4–6(–12) mm. long. Flowers spreading, becoming pendent; sepals deep purple or brown; petals brown or greenish purple; lip white with yellow side lobes, purple veins and purple flecks on the mid-lobe, spur yellowish suffused with purple; callus yellow at base, white above; pedicel and ovary 9–12 mm. long. Dorsal sepal recurved, linear, acute, 13–19 mm. long, 1.5 mm. wide; lateral sepals similar. Petals porrect, parallel to column, linear, acute, 12–14 mm. long, 2 mm. wide. Lip 3-lobed, 11–16 mm. long, 7–8 mm. wide; side lobes narrowly oblong, rounded in front; mid-lobe elliptic, obtuse, half length of lip; callus of 3 ridges with short papillae in lines in front on mid-lobe; spur straight, clavate, 4–5.5 mm. long. Column 6 mm. long; foot 0.5 mm. long. Fruit pendent, ellipsoidal, 12 mm. long.

TANZANIA. Iringa District: Mt. Jamulinga, 11 Dec. 1916, *McLoughlin* 29!; Rungwe District: Mulinda, 9 Jan. 1913, *Stolz* 1807!; Tunduru District: 112 km. W. of Tunduru, 11 km. E. of Songea boundary, 21 Dec. 1955, *Milne-Redhead & Taylor* 7861!
DISTR. **T** 7, 8; Mozambique, Malawi, Zambia and Zimbabwe
HAB. In *Brachystegia* woodland; 870–1450 m.

55. **E. streptopetala** *Lindl.* in Bot. Reg. 12, t. 1002 (1826); Hall in Journ. S. Afr. Bot., Suppl. 5: 123 (1965); J. Stewart et al., Wild Orch. S. Afr.: 239 (1982). Type: cult. *Colvill* (K, holo.!)

A terrestrial herb 50–150 cm. tall. Perennating organs pseudobulbous, clustered, conical to cylindric-fusiform, 2.5–10 cm. tall, 1.7–2.5 cm. diameter, covered in first year by leaf-bases; roots 4–6 mm. diameter, white. Leaves 4–9, plicate, in a fan, suberect, lanceolate, acuminate, 40–65 cm. long, 0.4–8 cm. wide. Inflorescence laxly many-flowered; peduncle 9 mm. diameter, with 5–11 sheaths along length; rhachis 15–35 cm. long; bracts lanceolate or ovate, acuminate, 7–20 mm. long, 5–8 mm. wide. Flowers spreading; sepals green or dull yellow veined and mottled with brown; petals yellow; lip darker yellow with reddish purple side lobes and a pale purple spur; column white; pedicel and ovary 15–22 mm. long. Dorsal sepal oblanceolate to elliptic, acute or obtuse, 8–18 mm. long, 3–9.5 mm. wide; lateral sepals similar but longer. Petals elliptic, circular or obovate, obtuse, 8–21 mm. long, 6–17 mm. wide. Lip difficult to flatten, 3-lobed, 8–12 mm. long, 7.5–11 mm. wide; side lobes erect, rounded; mid-lobe conduplicate, obovate or elliptic, 7.5–12 mm. long; callus of 3–5 low fleshy rugulose ridges on mid-lobe; spur conical, tapering to apex, 2–3.5 mm. long, pointing backwards. Column 5–8 mm. long. Fruit pendent, oblong-ellipsoidal, 35–50 mm. long. Pl. 3.

var. **streptopetala**

Leaves broad, usually more than 2 cm. wide when fully developed. Sepals 15 mm. or more long. Petals usually 15 mm. or more long and 10 mm. or more wide.

UGANDA. Toro District: Ruwenzori Mts., Kivata, 1894, *Scott Elliot* 7871!

KENYA. Uasin Gishu, Apr. 1932, *Harvey* 146!; Londiani, 10 May 1953, *Davis* 82!; Masai District: Ngong Hills, 2 Apr. 1939, *Bally* 8818!

TANZANIA. Masai District: E. Serengeti, Ngamuriali, 21 Nov. 1962, *Newbould* 6292!; Kondoa District: Great North Road, between Bereko and Kolo, 20 km. N. of Kondoa, 10 Jan. 1962, *Polhill & Paulo* 1121!; Songea District: Matengo Hills, Lupembe Hill, 10 Jan. 1956, *Milne-Redhead & Taylor* 8170!

DISTR. **U** 2; **K** 3–6; **T** 1–8; throughout tropical and southern Africa, also in SW. Arabia

HAB. In grassland and on forest margins; 1200–2550 m.

SYN. *Lissochilus streptopetalus* (Lindl.) Lindl., Gen. Sp. Orch. Pl.: 191 (1833)

L. krebsii Reichb.f. in Linnaea 20: 685 (1847). Type: South Africa, Cape of Good Hope, *Krebs* (W, holo.)

L. rueppelii Reichb.f., in Linnaea 20: 685 (1847), in obs. & 22: 865 (1849). Type: Ethiopia, between Halei and Temben, *Rueppel* (FR, holo.)

Hypodematium abyssinicum A. Rich., Tent. Fl. Abyss. 2: 286, t. 83 (1851). Type: Ethiopia, Ouodgerate, *Petit* (P, holo.)

Lissochilus paivaeanus Reichb.f. in Flora 48: 187 (1865). Type: Angola, Huilla, *Welwitsch* 731 (W, holo.)

L. grantii Reichb.f. in Trans. Linn. Soc., Bot., 29: 153 (1875). Type: Tanzania, Mpwapwa/Kilosa Districts, Rubeho Mts., *Grant* (BM, holo., K, iso.!)

L. oliverianus Reichb.f. in Trans. Linn. Soc., Bot., 29: 154 (1875). Type: Tanzania, Bukoba District, Karagwe, *Grant* 463 (BM, holo.)

L. oatesii Rolfe in Oates, Matabeleland 2: 409 (1889). Type: Zimbabwe, Matabeleland, *Oates* (K, holo.!)

Eulophia krebsii (Reichb.f.) Bolus in J.L.S. 25: 185 (1889)

Lissochilus graefei Kraenzl. in Gard. Chron., ser. 3, 11: 749 (1892). Type: cult. *Graef* (B, holo.)

L. graniticus Schweinf. in Bull. Herb. Boiss., 2, App. 2: 89, 112 (1894). Types: Eritrea, Ginda valley, near Filogobai, *Schweinfurth* 517 (G, syn.) & Accrur, 2205 (G, syn.)

L. stuhlmannii Kraenzl. in E.J. 19: 249 (1894). Type: Tanzania, Bukoba, *Stuhlmann* 3220 (B, holo.)

L. abyssinicus (A. Rich.) Th. Dur. & Schinz, Consp. Fl. Afr. 5: 27 (1895)

L. papilionaceus Rendle in J.B. 33: 170 (1895). Type: Zambia, Stevenson road, *Scott Elliot* 8268 (BM, holo., K, iso.!)

L. ruwenzoriensis Rendle in J.B. 33: 170 (1895). Type: Uganda, Toro District, Ruwenzori Mts., Kivata, *Scott Elliot* 7871 (BM, holo., K, iso.!)

L. shirensis Rendle, in J.B. 33: 171 (1895). Type: Malawi, Sotchi, *Scott Elliot* 8613 (BM, holo., K, iso.!)

L. carsonii Rolfe in F.T.A. 7: 90 (1897). Type: Zambia, Fwambo, *Carson* 123 (K, holo.!)

L. erythraeae Rolfe in F.T.A. 7: 91 (1897). Types: Ethiopia, Eritrea, *Schweinfurth & Riva* 1833 (K, syn.!) & *Schweinfurth* 168 (K, syn.!)

L. kirkii Rolfe in F.T.A. 7: 92 (1897). Type: Zanzibar, *Kirk* (K, holo.!)

L. ukingensis Schltr. in E.J. 53: 591 (1915). Type: Tanzania, Njombe District, Ukinga Mts., *Stolz* 2278 (B, holo.)

Eulophia grantii (Reichb.f.) Summerh. in K.B. 8: 148 (1953)

E. rueppelii (Reichb.f.) Summerh. in K.B. 8: 156 (1953)

E. paivaeana (Reichb.f.) Summerh. subsp. *borealis* Summerh. in K.B. 8: 152 (1953). Type: Uganda, Mengo District, Kampala, *Eggeling* 801 (K, holo.!)

var. **stenophylla** (*Summerh.*) *Cribb*, comb. et stat. nov. Type: Kenya, N. Nyeri District, Ndoro, *von Hoehnel* 78 (B, holo., K, record of holo.!)

Differs from the typical variety in having long, slender leaves, 4–14(–18) mm. wide; very small flowers with the sepals up to 14 mm. long and 5 mm. wide; the petals elliptic, up to 12 mm. long, 8 mm. wide; and the lip 8–10 mm. long and without a noticeable callus of thickened veins on the lip mid-lobe.

UGANDA. Acholi District; Chua, without date, *Eggeling* 2405!
KENYA. Naivasha District: by Lake Naivasha, 14 Dec. 1954, *Verdcourt* 1165!; Nairobi District: Thompson's Estate, 19 Apr. 1933, *Napier* 5207!; Machakos District: Kilima Kiu, Sept. 1943, *Bally* 2707!; Teita District: Teita Hills, 3 km. N. of Wundanyi, 2 May 1975, *Friis & Hansen* 2685!
DISTR. **U** 1; **K** 3–7; not known elsewhere
HAB. In shade of bushes or in open in grassland or on rocky hillsides; 1400–2100 m.

SYN. *Lissochilus micranthus* Kraenzl. in E.J. 17: 53 (1893), *non Eulophia micrantha* Lindl. (1833). Type as for *E. streptopetala* Lindl. var. *stenophylla*
Eulophia stenophylla Summerh. in K.B. 4: 434 (1949), nom. nov. pro *Lissochilus micranthus* Kraenzl.

56. **E. fridericii** (*Reichb.f.*) *A.V. Hall* in Journ. S. Afr. Bot., Suppl. 5: 131 (1965); Williamson, Orch. SC. Afr.: 175 (1977); J. Stewart et al., Wild Orch. S. Afr.: 240 (1982). Type: Angola, Ambaca District, nr. Zamba, *Welwitsch* 678 (BM, holo.!)

A terrestrial herb 17–50 cm. tall. Perennating organs subterranean, pseudobulbous, conical, 15–40 mm. long, 10–17 mm. wide, approximate in a string, yellow when dry; roots 2.5–3 mm. diameter, white. Leaves 3–5, spreading, ovate-elliptic, acute, 7.5–14 cm. long, 2.3–5 cm. wide, lowermost 2 sheathing. Inflorescence laxly 9–30-flowered, produced before or as the leaves develop; peduncle 10–27 cm. long; rhachis 6.5–16 cm. long; bracts linear-lanceolate, acuminate, 4–12 mm. long. Flowers spreading to subnutant; sepals brown and green mottled; petals yellow veined purple; lip dull yellow tinged pink with greenish yellow purple-veined side lobes; pedicel and ovary 8–12 mm. long. Dorsal sepal spreading, obovate, retuse or obtuse, 5–10 mm. long, 4–5 mm. wide; lateral sepals similar but oblique at base, 5–9.5 mm. long, 4.5–5 mm. wide. Petals slightly concave, subcircular, obtuse, 5–6.5 mm. long, 5–8 mm. wide. Lip 3-lobed, erect, 9–12 mm. long, 5.5–11.5 mm. wide; side lobes triangular; mid-lobe reduplicate, elliptic-triangular; callus of 5 fleshy rugulose ridges separated by a transverse lacerate flap from saccate base of lip. Column 2.5 mm. long; foot absent. Fruit pendent, 12–21 mm. long. Fig. 107/1–8, p. 472.

TANZANIA. Ufipa District: Chala Mt., 10 Dec. 1956, *Richards* 7204!; Mbeya District: Chimala escarpement, 6 Dec. 1963, *Richards* 18594!; Songea District: ± 112 km. W. of Tunduru, ± 11 km. E. of Songea, 21 Dec. 1955, *Milne-Redhead & Taylor* 7853!
DISTR. **T** 4, 7, 8; Mozambique, Malawi, Zambia, Zimbabwe, Angola and South Africa
HAB. In *Brachystegia* woodland and occasionally in rocky grassland; 870–2110 m.

SYN. *Lissochilus fridericii* Reichb.f. in Flora 50: 113 (1867)
L. roseolabius Schltr. in E.J. 53: 590 (1915). Type: Tanzania, Rungwe District, near Kyimbila, *Stolz* 1790 (B, holo., K, iso.!)
L. latifolius Schltr. in R.E. Fries, Wiss. Ergebn. Schwed. Rhod.-Kongo-Exped. 1911–12, 1: 246, t. 17/2 (1916). Type: Zambia, near Kalambo, *Fries* 1341 (B, holo., K, record of holo.!)
Eulophia roseolabia (Schltr.) Butzin in Willdenowia 7: 589 (1975)

NOTE. This is very closely allied to the following species differing in flower size, and lip and petal colouration. Further study is needed to determine the relationship between them.

57. **E. gonychila** *Schltr.* in Warb., Kunene-Sambesi-Exped.: 217 (1903); Williamson, Orch. SC. Afr.: 175 (1977). Type: Angola, Nambali, near Kubango, *Baum* 261 (B, holo.)

A terrestial herb 15–40 cm. tall. Perennating organs subterranean, proximate, pseudobulbous, conical, 4-noded, 2–3 cm. long, 1.5–2 cm. diameter, drying dull yellow; roots thick, from base of pseudobulbs, 4 mm. diameter, white. Leaves 3–4, plicate, oblong to oblong-elliptic, obtuse, 8–14 cm. long, 4–5.5 cm. wide, petiolate, articulated to sheathing base. Inflorescences appearing before leaves, 6–14-flowered; peduncle 8–17 cm. long, bearing 3 tubular sheaths along lower part and a linear-lanceolate one above; rhachis 6.5–12.5 cm. long. Flowers spreading to subnutant, strongly scented; sepals brown; petals lilac-pink veined with brown; lip pale lilac; column lilac with yellow anther; pedicel and ovary 10–14 mm. long. Dorsal sepal erect, oblong-obovate, obtuse, 11–14 mm.long, 7–8 mm. wide, undulate on margins; lateral sepals similar but recurved, 11–14 mm. long, 6–8 mm. wide. Petals broadly ovate-subcircular, obtuse, 9–11 mm. long, 9–11 mm. wide. Lip

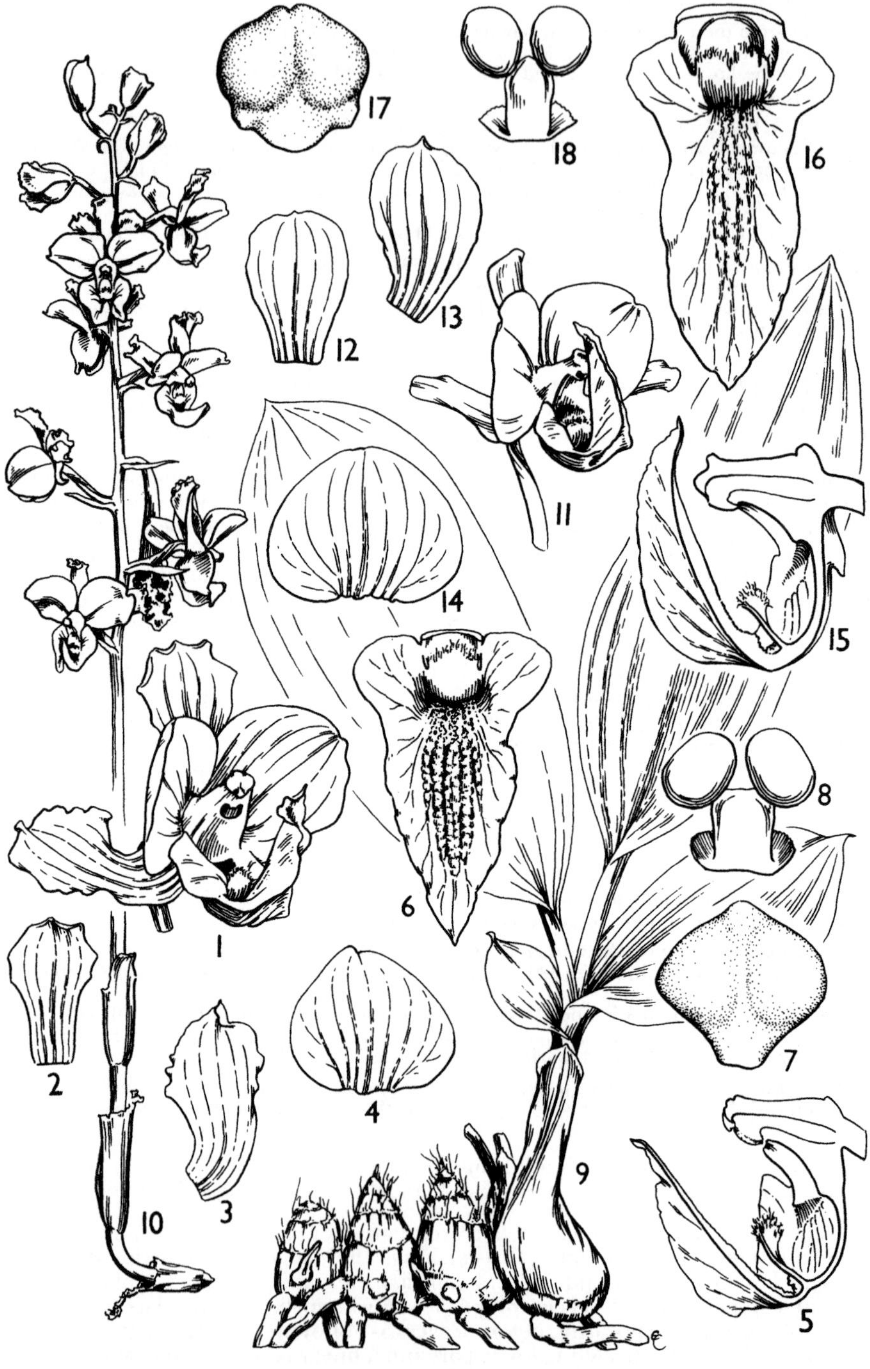

FIG. 107. *EULOPHIA FRIDERICII*—**1**, flower, × 2; **2**, dorsal sepal, × 2; **3**, lateral sepal × 2; **4**, petal, × 2; **5**, longitudinal section of column and lip, × 3; **6**, lip, × 3; **7**, anther-cap, × 11; **8**, pollinarium, × 18. *E. GONYCHILA*—**9**, habit, × ⅔; **10**, inflorescence, × ⅔; **11**, flower, × 2; **12**, dorsal sepal, × 2; **13**, lateral sepal, × 2; **14**, petal, × 2; **15**, longitudinal section of column and lip, × 3; **16**, lip, × 3; **17**, anther-cap, × 10; **18**, pollinarium, × 10. 1–8, from *la Croix* 263; 9, from *Morze* 4A; 10, from *Richards* 2052 and photo.; 11–18, from *Milne-Redhead & Taylor* 7853. Drawn by Eleanor Catherine.

obscurely 3-lobed, not easily flattened; side lobes erect, recurved, rounded, erose on margins; mid-lobe triangular, reduplicate, suberect, acute, 8–9 mm. long, 6–7 mm. wide; callus a lacerate transverse flap near base, with 1–3 slightly fleshy rugulose ridges in front; spur obscure. Column 3 mm. long, fleshy; foot absent. Fruit not known. Fig. 107/9–18

TANZANIA. Mpanda District: Nyonga-Mpanda road, 8 km. from Uruwira, 30 Oct. 1959, *Richards* 11660!; Rungwe District: Chivanjee, cult. Kew 13 July 1983, *Cribb, Grey-Wilson & Mwasumbi* 11506!
DISTR. **T** 4, 7; Burundi, Malawi and Zambia
HAB. In woodland; 1200–1500 m.

SYN. *Lissochilus brunneus* Kraenzl. in E.J. 48: 397 (1912). Type Tanzania, Rungwe District, Kyimbila, near Ulambya [Bulambya], *Stolz* 194 (B, holo.)
L. gonychilus (Schltr.) Schltr. in E.J. 53: 590 (1915)

58. **E. speciosa** (*Lindl.*) *Bolus* in J.L.S. 25: 184 (1889): Hall in Journ. S. Afr. Bot., Suppl. 5: 127 (1965); J. Stewart et al., Wild Orch. S. Afr.: 239 (1982). Type: South Africa, cult. *Griffin* (K, holo.!)

A terrestrial herb 60–120 cm. tall. Perennating organs underground, tuberous, irregularly conical, 4–6 cm. long, 2.5–4 cm. diameter, orange, separated by a stout rhizome up to 7 mm. diameter and 2 cm. long; roots 1.2–3 mm. diameter, white. Leaves fleshy, not plicate, 3–6, in a fan, 30–65 cm. long, 0.6–2.2 cm. wide, basal one sheathing. Inflorescence produced before the leaves, laxly many-flowered; peduncle stout, purplish, bearing 4–5 sheaths mostly near base, uppermost lanceolate; rhachis 11–29 cm. long; bracts lanceolate, acuminate, 4–14 mm. long. Flowers variable in size; sepals greenish; petals yellow; lip yellow with red veins on side lobes; column pale greenish; pedicel and ovary 12–19 mm. long. Dorsal sepal reflexed, elliptic, apiculate, 5–11 mm. long, 2–4 mm. wide; lateral sepals similar. Petals spreading, elliptic-ovate, obtuse, 11–18 mm. long, 8–15 mm. wide. Lip difficult to flatten, 3-lobed, 12–16 mm. long, 10–14 mm. wide; side lobes erect but reflexed above, obscurely rounded; mid-lobe convex, elliptic, subacute, 10–13 mm. long, 8–12 mm. wide; callus of 3–5 low fleshy rugulose ridges on mid-lobe; spur shortly conical, 1–3 mm. long. Column 4–6 mm. long. Fruit pendent, 3.5 cm. long. Pl. 3. Fig. 108/1–12, p. 475.

UGANDA. Acholi District: Gulu, 13 Apr. 1959, *Lind* 2416!; Bunyoro District: Pakanyi, Mar. 1943, *Purseglove* 1301!; Mengo District: 6.5–8 km. from Kampala, Oct. 1938, *Chandler* 2485!
KENYA. Machakos/Kitui District: Mtoni R., 17 Apr. 1902, *Kassner* 630!; Masai District: Lolgorien, Sept. 1933, *Napier* 5361!; Kilifi District: Sabaki R., Malindi to Jilore road, 26 Dec. 1954, *Verdcourt* 1191!
TANZANIA. Mwanza District: Nassa, Naruhandi, 22 Nov. 1951, *R. Tanner* 491!; Korogwe District: Kwalukonge, 11 Apr. 1969, *Faulkner* 4212!; Morogoro District: Uluguru Mts., between Morogoro and Morningside, 24 Jan. 1976, *Cribb & Grey-Wilson* 10407!; Songea District: ± 12 km. E. of Songea, at Nonganonga stream, 27 Dec. 1955, *Milne-Redhead & Taylor* 7729!; Zanzibar I., Tunguu, 24 Feb. 1933, *Vaughan* 2012!
DISTR. **U** 1, 2, 4; **K** 4, 6, 7; **T** 1–8; **Z**; **P**; SW. Arabia, Ethiopia and Sudan south to Zimbabwe and South Africa
HAB. In grassland, deciduous bushland and woodland; sea-level to 2000 m.

SYN. *Lissochilus speciosus* Lindl., Coll. Bot., t. 31 (1821)
L. calopterus Reichb.f. in Bonplandia 3: 218 (1855). Type: tropical Africa, *Peters* (W, holo., K, record of holo.!)
L. wakefieldii Reichb.f. & S. Moore in J.B. 16: 136 (1878). Type: Kenya, Nyika country, *Wakefield* (K, holo.!)
Eulophia dispersa N.E. Br. in K.B. 1892: 127 (1892). Type: Mozambique, Shupanga, *Kirk* (K, holo.)
Lissochilus brevisepalus Rendle in J.B. 33: 195 (1895). Types: Malawi, Soche, *Scott Elliot* 8526 (BM, syn., K, isosyn.!) & Ndurani, *Scott Elliot* 8511 (BM, syn.)
L. volkensii Rolfe in F.T.A. 7: 94 (1897). Type: Tanzania, Moshi District, Kahe, *Volkens* 2191 (K, holo.!)
L. rendlei Rolfe in F.T.A. 7: 96 (1897). Types: Zambia, Stevenson road, *Carson* in *Scott Elliot* 8266 (K, syn.!) & Fwambo, *Carson* 90 (K, syn.!)
L. homblei De Wild. in B.J.B.B. 6: 88 (1919). Types: Zaire, near Katentania, *Homblé* 768 (BR, syn.!) & Shaba [Katanga], without exact locality, *Verdick* 768 (BR, syn.!)
L. sapinii De Wild. in B.J.B.B. 6: 103 (1919). Type: Zaire, Bondo, *Sapin* (BR, holo.!)
Eulophia brevisepala (Rendle) Summerh. in K.B. 8: 147 (1953)

NOTE. This is a very widespread and variable orchid, particularly in its stature and flower size. However, the apparently continuous nature of the variation seen does not allow the recognition of infraspecific taxa.

59. **E. tuberculata** *Bolus* in J.L.S. 25: 184 (1889); Hall in Journ. S. Afr. Bot., Suppl. 5: 120 (1965); Williamson, Orch. SC. Afr.: 175 (1977); J. Stewart et al., Wild Orch. S. Afr.: 238 (1982). Type: South Africa, Zuureberg, *Drège* (K, holo.!)

A terrestrial herb 14–75 cm. tall. Perennating organs underground, tuberous, irregularly fusiform, 3–4.5 cm. long, 1 cm. diameter, borne 1 cm. apart on a 3 mm. diameter rhizome; roots 3 mm. diameter, white, all along rhizome. Leaf grass-like, linear, acuminate or acute, 20–25 cm. long, 6 mm. wide, subtended by 2 sheaths. Inflorescence produced before the leaf, laxly few–many-flowered; peduncle with 3–4 well-spaced sheaths in lower part; rhachis 4.5–10 cm. long; bracts lanceolate, acuminate, 3–5 mm. long. Flowers with purplish brown sepals, yellow petals and a yellow lip veined with red and with a yellow callus; pedicel and ovary 10 mm. long. Dorsal sepal oblong-obovate, apiculate, 6 mm. long, 3 mm. wide; lateral sepals similar but slightly larger. Petals subcircular, obtuse, 6–10 mm. long, 6–9 mm. wide. Lip 3-lobed, difficult to flatten, 8–11 mm. long; side lobes erect, oblong, obtuse; mid-lobe convex, elliptic; callus of 3 raised rugulose ridges on mid-lobe; spur obscurely conical, very short. Column 1–2 mm. long. Fruit pendent, 15–18 mm. long.

TANZANIA. Ufipa District: Malonje plateau, Mmemya [Mamya] Mt., 27 Oct. 1965, *Richards* 20635!; Iringa District: Kalumbara Mt., Mazombe, 25 Oct. 1982, *Leedal* 7194!

DISTR. **T** 4, 7; Zaire, Mozambique, Zambia, Malawi, Zimbabwe and South Africa

HAB. In burnt grassland, bare soil and woodland; 1450–2100 m.

SYN. *Lissochilus platypetalus* Lindl. in Hook., Comp. Bot. Mag. 2: 204 (1836), *nom. illegit.*, *non* Lindl. (1837). Type as for species.

L. verrucosus Rolfe in F.T.A. 7: 76 (1897). Type: Malawi, without exact locality, *Buchanan* 1067 (K, holo.!)

L. leendertziae Schltr. in Ann. Transv. Mus. 10: 239 (1924). Type: South Africa, Pretoria, *Leendertz* 281a (B, holo., K, PRE, iso.!)

60. **E. orthoplectra** (*Reichb.f.*) *Summerh.* in K.B. 1939: 499 (1939); Williamson, Orch. SC. Afr.: 176 (1977). Type: Sudan, Niamniam, *Schweinfurth* 3270 (W, holo.)

A terrestrial herb 60–100 cm. tall. Perennating organs underground, tuberous, irregularly ovoid-fusiform, flattened dorso-ventrally, 3.5–6 cm. long, 2–5 cm. diameter, up to 1 cm. apart on short rhizome; roots 3 mm. diameter, white. Leaves 2–4, fleshy, linear, acute, 18–50 cm. long, 0.7–2 cm. wide. Inflorescence laxly 6–20-flowered, produced before leaves; peduncle with 4–6 sheaths, 2–3 basal ones sheathing, upper lanceolate, spreading; rhachis 9–20 cm. long; bracts linear-lanceolate, acuminate, 8–12 cm. long. Flowers with red-brown sepals; petals yellow, red-brown within and red-veined; lip yellow with red-veined side lobes and a brown spur; pedicel and ovary 10–13 mm. long. Dorsal sepal reflexed, oblanceolate, apiculate, 8.5–12.5 mm. long, 3.8–6.3 mm. wide; lateral sepals similar. Petals elliptic-subcircular or elliptic-ovate, apiculate, 12–17.5 mm. long, 13–19 mm. wide. Lip 3-lobed, difficult to flatten, 11–15 mm. long; side lobes erect, rounded; mid-lobe convex, elliptic-obovate, obtuse or apiculate; callus of 5 fleshy rugulose ridges; spur deflexed or straight, conical-cylindrical with a broad mouth, 13–16 mm. long. Column 7–8 mm. long. Fruit pendent. Fig. 108/13–22.

UGANDA. Acholi District: near Kitgum Matidi, 4 Dec. 1957, *Langdale-Brown* 2422!; Toro District: Mwenge, near Buchuni Hill, Oct. 1950, *Osmaston* 771!; Mengo District: Bulwada [Balwada], Mar. 1932, *Eggeling* 595!

KENYA. Trans-Nzoia District: Moi's [Hoey's] Bridge, Jan. 1932, *Mainwaring* 1502!; Elgon, Oct.–Nov. 1930, *Lugard* 11!

TANZANIA. Arusha District: Nduruma, Arusha road, Dec. 1927, *Haarer* 934!; Mpanda District: near Lake Katabe, 13 km. from Sitalike, 6 Oct. 1970, *Richards & Arasululu* 26231!; Songea District: lower slopes of Litenga Hill, 19 Apr. 1956, *Milne-Redhead & Taylor* 9765!

DISTR. **U** 1–4; **K** 3, 5; **T** 1, 2, 4, 7, 8; Nigeria and Cameroon to Sudan and Ethiopia, Zaire, Rwanda, Malawi, Zambia, Mozambique and Zimbabwe

HAB. In grassland, wooded grassland, swamp and *Brachystegia* woodland; 1100–2100 m.

SYN. *Lissochilus orthoplectrus* Reichb.f. in Otia Bot. Hamb. 1: 63 (1878)

Eulophia bella N.E. Br. in Gard. Chron., ser. 3, 6: 210 (1889). Type: Malawi, Manganga Hill, *Meller*, cult. *Gumbleton* (K, holo.!)

Lissochilus milanjianus Rendle in Trans. Linn. Soc., Bot., ser. 2, 4: 46 (1894). Type: Malawi, Mulanji, *Whyte* (BM, holo.)

L. monticola Rendle in J.B. 33: 172 (1895). Type: Uganda, Ruwenzori, E. side, *Scott Elliot* 7552 (BM, holo., K, iso.!)

L. saccatus Rendle in J.B. 33: 172 (1895). Type: Kenya, Nandi, *Scott Elliot* 6988 (BM, holo., K, iso.!)

L. kraenzlinii Rolfe in F.T.A. 7: 97 (1897). Type: Tanzania, Kilimanjaro, *Volkens* 1617 (BM, holo.)

L. pulcher Schltr. in E.J. 53: 592 (1915). Type: Tanzania, Rungwe, *Stolz* 1381 (B, holo., K, iso.!)

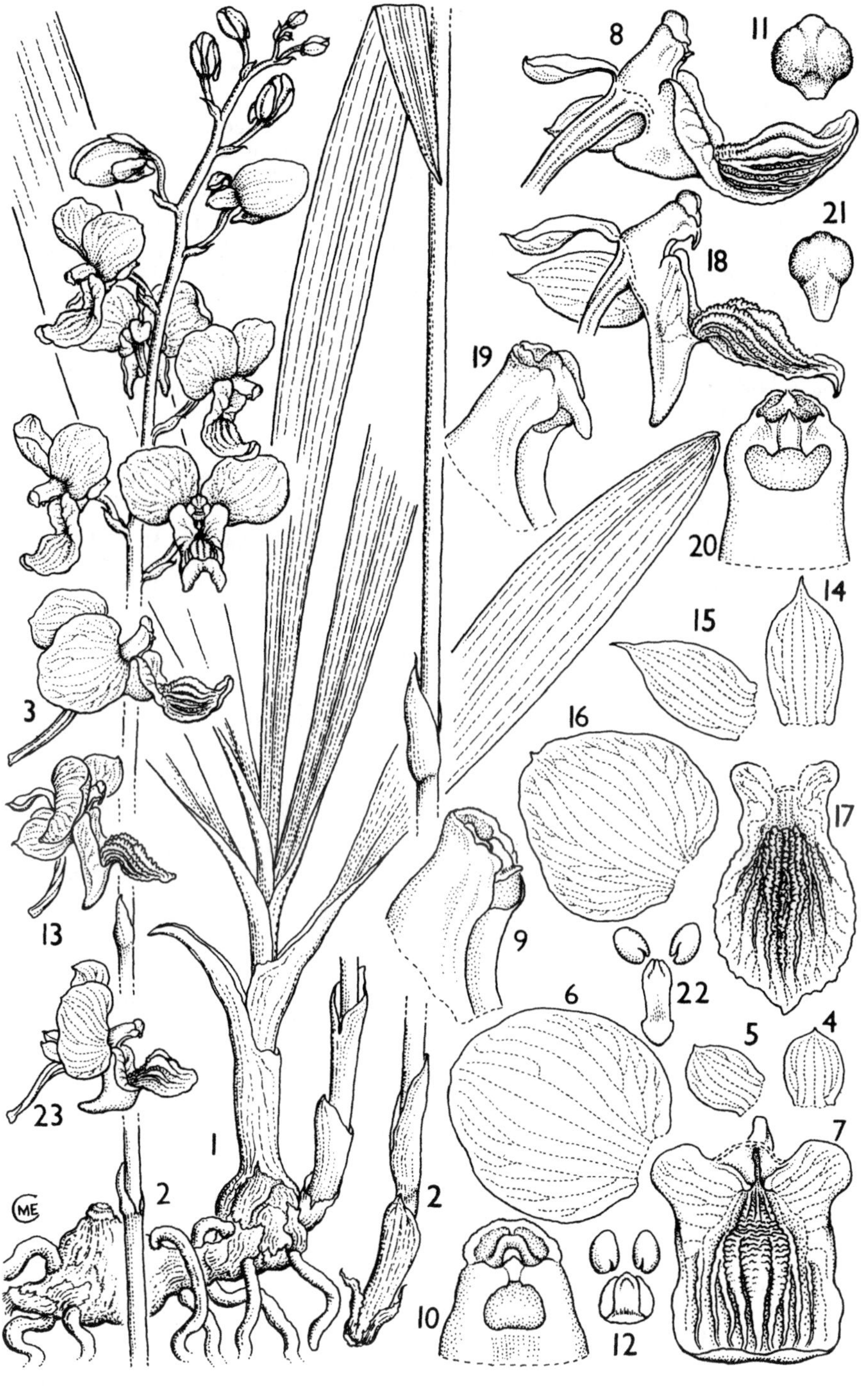

FIG. 108. *EULOPHIA SPECIOSA*—**1**, habit, × ⅔; **2**, inflorescence, × ⅔; **3**, flower, × 1; **4**, dorsal sepal, × 2; **5**, lateral sepal, × 2; **6**, petal, × 2; **7**, lip, × 2; **8**, column and lip, side view, × 2; **9**, column, side view, × 4; **10**, column with pollinarium removed, front view, × 4; **11**, anther-cap, × 4; **12**, pollinarium, × 4. *E. ORTHOPLECTRA*—**13**, flower, × 1; **14**, dorsal sepal, × 2; **15**, lateral sepal, × 2; **16**, petal, × 2; **17**, lip, × 2; **18**, column and lip, side view, × 2; **19**, column, side view, × 4; **20**, column, front view, with pollinarium removed, × 4; **21**, anther-cap, × 4; **22**, pollinarium, × 4. *E. SCHWEINFURTHII*—**23**, flower, × 1. 1, 3–12, from *Milne-Redhead & Taylor* 8516; 2, from *Milne-Redhead & Taylor* 7929; 13–22, from *Milne-Redhead & Taylor* 9765B; 23, from *Greenway & Kirrika* 11167. Drawn by Mrs M.E. Church.

61. **E. schweinfurthii** *Kraenzl.* in E.J. 17: 54 (1893); Hall in Journ. S. Afr. Bot., Suppl. 5: 114 (1965); Williamson, Orch. SC. Afr.: 175 (1977); J. Stewart et al., Wild Orch. S. Afr.: 239 (1982). Type: Sudan, Bongoland, *Schweinfurth* 2671 (B, holo., K, record of holo.!)

A terrestrial herb 45–110 cm. tall. Perennating organs underground, tuberous, in chains, irregularly cylindrical, somwhat dorso-ventrally flattened, 3.5–5 cm. long, 1.3–2.5 cm. diameter; roots 1.5–3 mm. diameter, white. Leaves 4–7, in a fan, slightly succulent, conduplicate when young, linear, acuminate, 21–38 cm. long, 2.5–10 mm. wide. Inflorescences 1–2, produced before the leaves; peduncle up to 7 mm. diameter, with 5–6 papery sheaths along length, lowermost 2 basal; rhachis 7–23 cm. long; bracts linear-lanceolate, acuminate, 4–15 mm. long. Flowers fleshy; sepals purplish; petals yellow with red veins; lip yellow with pale purplish side lobes and a purplish margin to the mid-lobe; pedicel and ovary 10–20 mm. long. Dorsal sepal reflexed, oblong, apiculate, 9–10.6 mm. long, 3.5–6 mm. wide; lateral sepals similar. Petals broadly ovate or subcircular, acute or obtuse, 11.5–16.5 mm. long, 9.5–17 mm. wide. Lip 3-lobed, 14–16 mm. long, 19–21 mm. wide; side lobes erect, rounded; mid-lobe convex, elliptic, 7–12 mm. long; callus of 3–5 verrucose ridges to apex on mid-lobe; spur upcurved-conical, 8–9 mm. long. Column 3–5 mm. long. Fruit pendent, 3–3.5 cm. long. Fig. 108/23, p. 475.

UGANDA. Kigezi District: Kisoro, 19 Apr. 1941, *A.S. Thomas* 3789!
KENYA. Machakos District: Makindu, 4 Apr. 1902, *Kassner* 834a!; Masai District: Nairobi to Arusha road, 3 Jan. 1963, *C. Leakey* 31!; Kwale District; just N. of Kwale, Cha Shimba [Pemba] R., 17 Mar. 1902, *Kassner* 359!
TANZANIA. Moshi, Dec. 1927, *Haarer* 917!; Ufipa District: Lake Kwela, near Mpui, 2 Aug. 1960, *Richards* 12920!; Masasi District: just W. of R. Bangwala, 15 Dec. 1955, *Milne-Redhead & Taylor* 7671!
DISTR. **U** 2; **K** 4, 6, 7; **T** 2, 4, 6–8; Sudan, Ethiopia, Zaire, Angola, Zambia, Malawi, Mozambique, Zimbabwe and Botswana
HAB. In grassland, thicket, bushland and woodland; 900–2100 m.

SYN. *Lissochilus aurantiacus* Reichb.f. in Flora 48: 187 (1865), *non Eulophia aurantiaca* Rolfe (1898). Type: Angola, Pungo Andongo, *Welwitsch* 691 (W, holo.)
L. holubii Rolfe in F.T.A. 7: 96 (1897). Type: N. Botswana, Leshumo Valley, *Holub* (K, holo.!)
L. smithii Rolfe in F.T.A. 7: 96 (1897), *non Eulophia smithii* Rolfe (1898). Type: Tanzania, Kilimanjaro, *C.S. Smith* (K, holo.!)
L. johnstonii Rolfe in F.T.A. 7: 97 (1897). Type: Kenya, Kilimanjaro Exped., 64–96 km. from coast, *Johnston* (K, holo.!)
L. pentheri Schltr. in E.J. 26: 339 (1899). Types: Zimbabwe, Matabeleland, *Penther* (B, syn.) & South Africa, Transvaal, *Schlechter* (B, syn.)
L. elegantulus Schltr. in R.E. Fries, Wiss. Ergebn. Schwed. Rhod.-Kongo-Exped. 1911–12, 1: (1916), *non Eulophia elegantula* Rolfe (1917). Type: Zambia, Mansa [Fort Rosebery], *Fries* 605 (B, holo.)
L. deschampsii De Wild. in B.J.B.B. 6: 81 (1919). Types: Zaire, near Moero, *Deschamps* (BR, syn.!) & between the Luapula and the Lufila, *Verdick* (BR, syn.!)
L. vermiculatus De Wild. in B.J.B.B. 6: 107 (1919). Type: Zaire, Shaba [Katanga], Lubumbashi [Elisabethville], *Hock* (BR, holo.!)
Eulophia involuta Summerh. in F.W.T.A. 2: 472 (1936), nom. nov. pro *Lissochilus smithii* Rolfe
E. compta Summerh. in K.B. 6: 472 (1952), nom. nov. pro *Lissochilus elegantulus* Schltr.
E. chrysops Summerh. in K.B. 13: 78 (1958), nom. nov. pro *Lissochilus aurantiacus* Reichb.f.

NOTE. This is a widespread and very variable species. Geerinck (1987), in the Flore du Rwanda 4, has treated it as a variety of *E. orthoplectra* but I think that the evidence for this is weak from the material I have studied. However, further work is certainly needed to clarify the considerable variation in flower size, lip shape and spur shape that is exhibited in *E. schweinfurthii.*

62. **E. clitellifera** (*Reichb.f.*) *Bolus* in J.L.S. 25: 184 (1889); Hall in Journ. S. Afr. Bot., Suppl. 5: 117 (1965); Williamson, Orch. SC. Afr.: 168 (1977); J. Stewart et al., Wild Orch. S. Afr.: 238 (1982). Type: South Africa, Natal, *Gueinzius* (W, holo.!)

A terrestrial herb 17–60 cm. tall. Perennating organs subterranean, pseudobulbous, conical, in chains, 2–3 cm. long, 2–2.5 cm. diameter; roots 2–4 mm. diameter, white. Leaves 5, in a fan, not plicate, falcate in side view, linear-lanceolate, acuminate, 20–70 cm. long, 7–9 mm. wide, lowermost 2–3 sheathing. Inflorescence produced before the leaves, laxly few–many-flowered; peduncle 2–3 mm. diameter, bearing 3–4 sheathing acute to acuminate bracts; rhachis 6–25 cm. long; bracts lanceolate, acuminate, 4–11 mm. long. Flowers spreading; sepals brown; petals and lip off-white to yellow with red to brown venation; pedicel and ovary 13–20 mm. long. Dorsal sepal spreading, oblong-elliptic, obtuse or apiculate, 5–6 mm. long, 2.6–4 mm. wide, concave; lateral sepals similar, 6–9 mm. long, 3–5.5 mm. wide. Petals spreading, ovate-elliptic to subcircular, obtuse, 7–10 mm. long, 7–8 mm. wide, concave. Lip 3-lobed, 6–10.5 mm. long, 7–10 mm. wide; side lobes as long as column, rounded or truncate at apex; mid-lobe subcircular or elliptic, obtuse or emarginate, 6 mm. long, 5.5 mm. wide; callus of 3–5 fleshy rugulose ridges on convex mid-lobe; spur conical, 3–5 mm. long. Column 1–2 mm. long, lacking a foot. Fruit pendent, 2–2.5 cm. long. Fig. 109/1–9, p. 478.

UGANDA. Ankole District: Ruizi R., 17 June 1950, *Jarrett* 313!
KENYA. Trans-Nzoia District: 19 km. from Kitale, Apr. 1967, *Tweedie* 3442!
TANZANIA. Musoma District: Tabora guard post, 20 Aug. 1962, *Greenway* in *E.A.H.* 12713!; Ufipa District: Chapota, 5 Dec. 1949, *Bullock* 2048!; Mbeya District: Kitulo Plateau, W. side of Ishinga Mt., 17 Aug. 1979, *Leedal* 5581!
DISTR. **U** 2; **K** 3; **T** 1, 4, 7; Ghana, Sudan, Zambia, Mozambique, Zimbabwe and South Africa
HAB. In grassland and open woodland, often seasonally burnt; 1000–2800 m.

SYN. *Lissochilus clitellifera* Reichb.f. in Linnaea 20: 687 (1847)
L. pulchellus Rendle in J.B. 33: 196 (1895). Type: Tanzania, Bukoba District, Kibwera, Burigi [Urigi] Lake, *Scott Elliot* 8126 (BM, holo.!, K, iso.!)
L. rehmannii Rolfe in Fl. Cap. 5(3): 55 (1912). Types: South Africa, Transvaal, *Rehmann* 4297 (K, syn.!) & *Tennant* 4040 (K, syn.!)
L. flexuosus Schltr. in Fries, Wiss. Ergebn. Schwed. Rhod.-Kongo-Exped. 1911–12, 1: 245, fig. 24 (1916), *non Eulophia flexuosa* Kraenzl. (1895). Type: Zambia, Luera R., *Fries* 591 (B, holo., K, record of holo.!)
Eulophia fractiflexa Summerh. in K.B. 8: 156 (1953), nom. nov. pro *L. flexuous* Schltr.

63. **E. leonensis** *Rolfe* in F.T.A. 7: 51 (1897); Summerh. in F.W.T.A., ed. 2, 3: 249 (1968). Type: Sierra Leone, Bafodeya, *Scott Elliot* 5536 (K, holo.!)

A terrestrial herb 30–60 cm. tall. Perennating organs underground, pseudobulbous, irregularly conical, 20–30 mm. tall, 20 mm. diameter, in chains, drying yellow-orange; roots 3 mm. diameter, white. Leaves 3–5, plicate, elliptic to lanceolate-elliptic, acute to acuminate, 8–28 cm. long, 3–6.5 cm. wide, articulated to a sheathing petiole. Inflorescence produced with young leaves or before, densely many-flowered; peduncle 2 mm. diameter, bearing 2–3 basal sheaths and 1–4 linear, acuminate, sterile bracts above; rhachis 6–17 cm. long; bracts linear, acuminate, 5–9 mm. long. Flowers spreading, becoming pendent; sepals greenish purple; petals and lip greenish yellow; pedicel and ovary 10–15 mm. long. Dorsal sepal oblanceolate, obtuse, 6.2–9 mm. long, 2.5–3 mm. wide; lateral sepals similar but oblique at base. Petals broadly ovate to elliptic, obtuse or rounded, 5–6 mm. long, 4–5.3 mm. wide. Lip 3-lobed, 4.5–5.5 mm. long, 6 mm. wide; side lobes rounded; mid-lobe oblong-elliptic; callus of 5 fleshy verrucose ridges on mid-lobe; spur at right-angles to ovary, cylindric, almost straight, 5–6 mm. long. Column 2–2.5 mm. long, united to side lobes of lip for half length. Fruit pendent, 18–20 mm. long.

UGANDA. W. Nile District: Ulepi, 1935, *Eggeling* 1958!; Bunyoro District: Busingiro Hill, Mar. 1938, *Eggeling* 3528!
DISTR. **U** 1, 2; Gambia to the Central African Republic
HAB. In short grass and on rocky hillsides; 1000–1500 m.

64. **E. parvula** (*Rendle*) *Summerh.* in K.B. 12: 125 (1957) & in F.W.T.A., ed. 2, 3: 247 (1968); J. Stewart in U.K.W.F.: 776 (1974); Williamson, Orch. SC. Afr.: 168 (1977). Type: Kenya, C. Kavirondo District, Samia, *Scott Elliot* 7081 (BM, holo., K, iso.!)

A terrestrial herb 22–60 cm. tall. Perennating organs underground, tuberous, obliquely irregular conical, in chains, 1.5–3.3 cm. long, 2.2–2.6 cm. diameter, yellow; roots 3–4 mm. in diameter. Leaves 7, not plicate, linear, acute, 10–30 cm. long, 8–12 mm. wide, lowest 3 sheathing. Inflorescence produced before leaves, laxly many-flowered; peduncle 2–3 mm. diameter, bearing 2–3 sheaths along length; rhachis 7.5–30 cm. long; bracts linear, acuminate, 4–8 mm. long. Flowers very small, yellow or greenish yellow spotted or marked with purple or brown; pedicel and ovary 15–17 mm. long. Dorsal sepal spreading, oblong-elliptic, obtuse, 4–6 mm. long, 2.2–2.5 mm. wide; lateral sepals similar. Petals subcircular to broadly ovate or elliptic, obtuse, 5–5.5 mm. long, 3.3–4.8 mm. wide. Lip 3-lobed, 5–5.8 mm. long, 6–6.2 mm. wide; side lobes elliptic, acute, fused in basal half to column; mid-lobe convex, circular, rounded or obtuse, 3–4 mm. long; callus of 3–5 fleshy rugulose ridges; spur conical, 1 mm. long. Column 1.5 mm. long. Fruit pendent, 16–21 mm. long. Fig. 109/10–18, p. 478.

UGANDA. Karamoja District: base of Mt. Moroto, without date, *Eggeling* 2614!; Toro District: Muhokya [Mohokya], 18 Feb. 1939, *A.S. Thomas* 2754!; Mengo District: Bugerere County, near R. Nile, Apr. 1930, *Hancock* 1553!

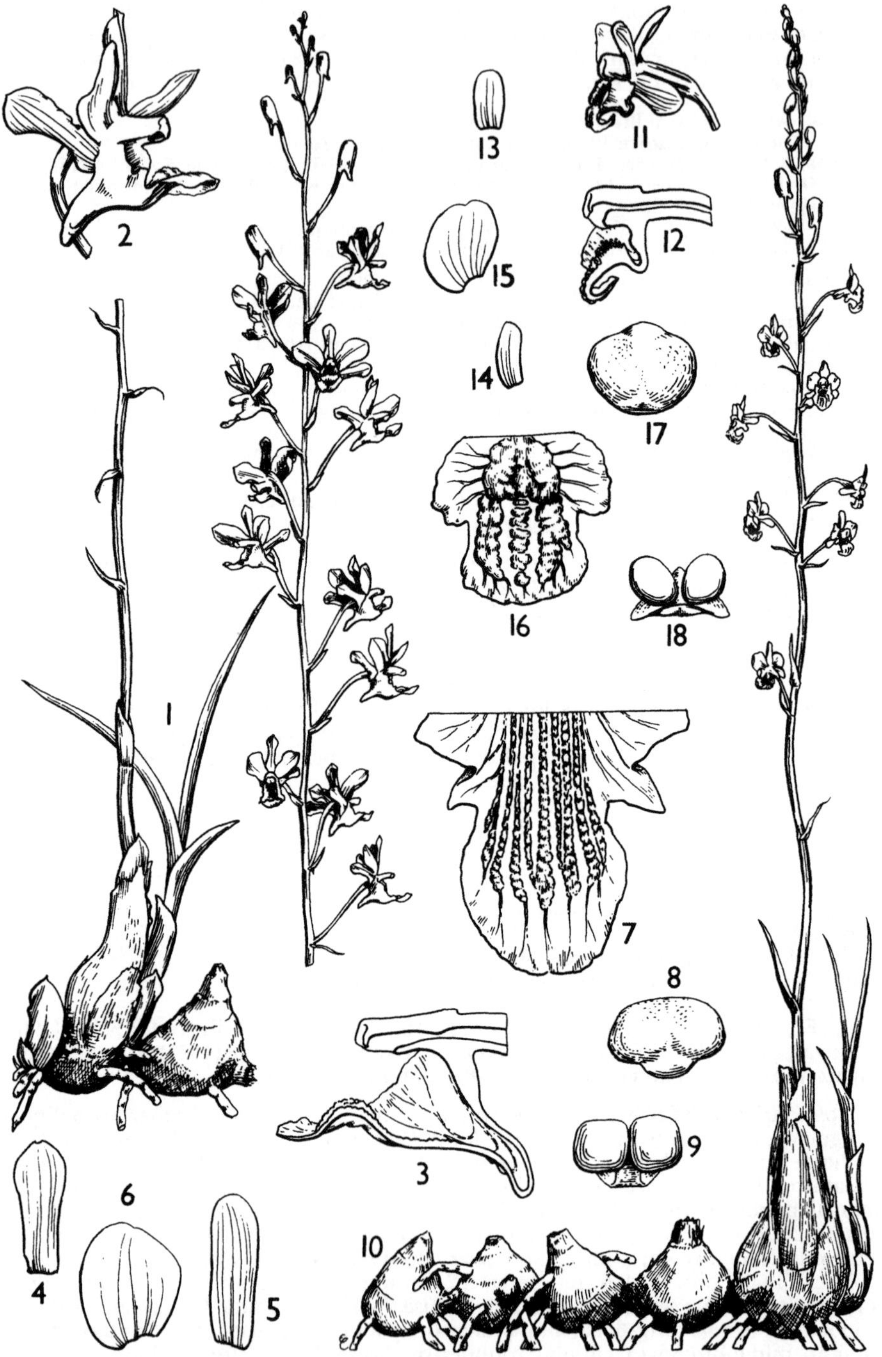

FIG. 109. *EULOPHIA CLITELLIFERA*—**1**, habit, $\times \frac{2}{3}$; **2**, flower, × 2; **3**, longitudinal section of lip and column, × 3; **4**, dorsal sepal, × 3; **5**, lateral sepal, × 3; **6**, petal, × 3; **7**, lip, × 5; **8**, anther-cap, × 8; **9**, pollinarium, × 8. *E. PARVULA*—**10**, habit, $\times \frac{2}{3}$; **11**, flower, × 2; **12**, longitudinal section of lip and column, × 3; **13**, dorsal sepal, × 3; **14**, lateral sepal, × 3; **15**, petal, × 3; **16**, lip, × 5; **17**, anther-cap, × 8; **18**, pollinarium, × 8. 1, from *Williamson* 2278 and *Bullock* 2048; 2–9, from *Bullock* 15296; 10, 11, from *Milne-Redhead* 825, *Milne-Redhead* 3019 and *Greatrex* 24841; 12–18, from *Greatrex* 24841. Drawn by Eleanor Catherine.

KENYA. C. Kavirondo District: Samia, Jan. 1894, *Scott Elliot* 7081!
TANZANIA. Kilwa District: Njenje to Paganyika road, near Mpurukasese [Mpurakasese], Nov. 1917, *McLoughlin* 25!
DISTR. **U** 1, 2, 4; **K** 5; **T** 8; Nigeria, Cameroon, Zaire, Rwanda, Burundi, Zambia, Malawi and Zimbabwe
HAB. In seasonally burnt grassland and dry bush country; 900–1600 m.

SYN. *Lissochilus parvulus* Rendle in J.B. 33: 196 (1895)
Eulophia tanganyikae Kraenzl. in B.S.B.B. 38: 216 (1900). Type: Zaire, Haut Marangu, *Debeerst* (BR, holo., K, part of holo.!)
E. ledermannii Kraenzl. in E.J. 48: 394 (1912). Type: Cameroon, *Ledermann* 2511 (B, holo., K, record of holo.!)

NOTE. The relationship of this and the following species needs investigating. They appear to differ in flower size and in details of lip ornamentation and spur length and shape.

65. **E. pyrophila** (*Reichb.f.*) *Summerh.* in K.B. 2: 132 (1948); J. Stewart in U.K.W.F.: 776 (1974); Williamson, Orch. SC. Afr.: 168 (1977). Type: Sudan, near Kuraggera, *Schweinfurth* 2795 (W, holo., K, iso.!)

A terrestrial herb 13–45 cm. tall. Perennating organs underground, tuberous, irregularly conical, 1–3 cm. long, 1–2.5 cm. diameter, in chains, yellow, 2–4-noded; roots 2–3 mm. diameter, white. Leaves 4–5, coriaceous, linear-lanceolate, acute, 3–14 cm. long, 2–2.5 mm. wide, lowermost sheathing. Inflorescence produced before the leaves, laxly many-flowered; rhachis 2–3 mm. diameter, with 3 short acute sheaths along length; bracts spreading, linear, acuminate, 3–9 mm. long. Flowers chocolate or dull brown with a cream or yellow lip striped brown and with a yellow callus; pedicel and ovary 9–20 mm. long, plum-coloured. Dorsal sepal oblong-elliptic, obtuse, 5–6.7 mm. long, 2.5–3.6 mm.wide; lateral sepals similar. Petals elliptic or oblong-elliptic, obtuse, 5.8–8 mm. long, 4.3–5.7 mm. wide. Lip 3-lobed, 5.7–7.5 mm. long, 5–6.8 mm. wide; side lobes triangular, subacute, united for half length to the column; mid-lobe convex, elliptic, margin undulate; callus of 5–9 fleshy rugulose ridges on to basal half of mid-lobe; spur conical, 1.5–2.5 mm. long. Column 2.5 mm. long. Fruit pendent, 17–20 mm. long.

UGANDA. Ankole District: W. Ankole, *Doggett*! & near Ndeizha [Ndeisha], Sept. 1941, *Eggeling* 4528!; Kigezi District: Kinkizi, Kasomi, Mar. 1946, *Purseglove* 1955!
KENYA. Trans-Nzoia District: near Kitale, Ley's Farm, Mar. 1966, *Tweedie* 3247!; Uasin Gishu District: Moiben, Sergoit Rock, 10 Mar. 1964, *Polhill* 2404!; Machakos/Masai District: Chyulu Hills, 22 Jan. 1947, *North* in *Bally* 5708!
TANZANIA. Bukoba District: Kabirizi, Oct. 1931, *Haarer* 2258!
DISTR. **U** 2; **K** 3–6; **T** 1; Ivory Coast to Ethiopia and south to Zimbabwe
HAB. In short, often burnt, grassland and amongst rocks; 900–2350 m.

SYN. *Lissochilus pyrophilus* Reichb.f. in Otia Bot. Hamb. 1: 65 (1878)
L. microceras Reichb.f. in Otia Bot. Hamb. 2: 115 (1881). Type: Malawi, Mt. Sochi, *Kirk* (W, holo., K, iso.!)
L. heteroglossus Reichb.f. in Otia Bot. Hamb. 2: 115 (1881). Type: Malawi, Upper Shire Valley, near Zemika, *Kirk* (W, holo., K, iso.!)
Eulophia sordida Kraenzl. in E.J. 33: 67 (1902). Type: Togo, Lome, *Warnecke* 95 (B, holo., K, record of holo.!)
Lissochilus maestus Merxm. in Trans. Rhod. Sci. Assoc. 43: 158 (1951). Type: Zimbabwe, Harare District, *Merxmüller* (SRGH, holo.)
Eulophia microceras (Reichb.f.) Summerh. in K.B. 8: 154 (1953)
E. maesta (Merxm.) Butzin in Willdenowia 7: 589 (1975)

66. **E. stenoplectra** *Summerh.* in K.B. 13: 81 (1958) & in F.W.T.A., ed. 2, 3: 249 (1968). Type: Sudan, between Mts. Konyi and Loka, *Myers* 8474! (K, holo.!)

A terrestrial herb 45–50 cm. tall. Perennating organs not seen. Leaves linear, acute, up to 85 cm. long, 12–16 mm. wide, not plicate. Inflorescence produced before leaves, laxly few-flowered; peduncle 2 mm. diameter, purple, with 3 sheathing and a lanceolate sterile bract along length; rhachis 10–20 cm. long; bracts lanceolate, acuminate, 4–6 mm. long. Flowers with greenish brown sepals and petals veined with purple; lip yellow with green margin and underside; pedicel and ovary 11–15 mm. long. Dorsal sepal spreading, oblong-subspathulate, obtuse, 6.5–7.5 mm. long, 2.5–2.8 mm. wide; lateral sepals similar. Petals elliptic-subcircular, obtuse, 7–8 mm. long, 4.5–6.5 mm. wide. Lip 3-lobed, difficult to flatten, 10.5–11.5 mm. long, 5–8 mm. wide; side lobes as long as column, rounded to subquadrate; mid-lobe elliptic-circular, rounded at apex, 3.5–5.3 mm. long, convex but with incurved margins; callus of 7–9 fleshy verrucose ridges; spur cylindric, pointing backwards at acute angle to ovary, 3.5–5.7 mm. long. Column 2.5 mm. long. Fruit pendent.

UGANDA. Teso District: Serere, Feb. 1933, *Chandler* 1085!
DISTR. U 3; Ghana and Sudan
HAB. In burnt grassland; 1100–1200 m.

46. PTEROGLOSSASPIS*

Reichb.f., Otia Bot. Hamb. 1: 67 (1878)

Terrestrial, sympodial herbs. Subterranean stem composed of a series of lobed fleshy tubers; aerial stem very short. Leaves 1–3, linear-lanceolate, long-petiolate, plicate, enclosed in sheaths towards base, arising with scape from the current year's growth. Scape erect, covered with subscarious, tubular sheaths. Inflorescence a several-flowered, short, dense, rarely elongated and lax, terminal raceme; bracts usually conspicuous, long, narrow, rarely broad, scarious, often exceeding the flowers. Flowers resupinate or non-resupinate, medium-sized. Sepals and petals similar, free, usually spreading, the petals usually shorter than the sepals. Lip sessile, 3-lobed, rarely entire, neither spurred, saccate nor clawed, flat, never concave, mid-lobe larger than side lobes, apex usually recurved, disc either entirely tuberculate or the nerves variously tuberculate, verruculose or lacerately keeled, median nerve of the side lobes often raised and ridge-like at the base. Column very short, rarely elongated, broad, curved and biauriculate at the base, the 2 auricles merging with the base of the lip, wings absent, foot absent or very rarely rudimentary, ± horizontal; anther terminal, ovate, obtuse, apiculate, unilocular; pollinia 2, subglobose, united on a short broad stipes; viscidium large, scale-like.

A genus of 7 species, 5 in tropical and subtropical Africa, with 1 in SE. U.S.A. and Cuba and 1 in Argentina and Brazil. Closely allied to *Eulophia*, particularly to *E. albo-brunnea* Kraenzl., *E. shupangae* (Reichb.f.) Kraenzl. and allied species, which are virtually identical in habit. *Pteroglossaspis* differs, however, in having a flat, unspurred lip and by the usually very short, biauriculate column.

Mid-lobe of lip with 3–7 tuberculate to lacerately-keeled nerves on the disc; nerves and ridges of both the disc and side lobes sometimes sparsely pilose . . . . 1. *P eustachya*
Mid-lobe of lip with nerves on the disc at the most only slightly verrucose, bearing a few irregular warts, never lacerately keeled; nerves and ridges of both the disc and side lobes papillose . . . . 2. *P. ruwenzoriensis*

1. **P. eustachya** *Reichb.f.*, Otia Bot. Hamb. 1: 67 (1878); Rolfe in F.T.A. 7: 99 (1897); J. Stewart in U.K.W.F.: 782 (1974). Type: Ethiopia, Tigre and Begemder, *Schimper* 1235** (W, holo., K, iso.!)

A terrestrial herb 22–65 cm. high, glabrous except for the lip. Leaves linear to linear-lanceolate, acute to acuminate, plicate; petiole 12–18 cm. long, enclosed in 1–3 greyish brown, tubular, obtuse to acute, subscarious sheaths below; lamina (26–)34–40 cm. long, 1.5–2 cm. broad. Inflorescence cylindrical or ovoid, many-flowered, dense; scape erect, bearing 4–6(–8) remote, greyish brown, tubular, obtuse to acute, subscarious sheaths; rhachis 2–8 cm. long; bracts pale brown, lanceolate or linear-lanceolate, often finely acuminate, scarious, with prominent venation, lowermost 3–8 cm. long, uppermost shorter. Flowers resupinate, variable in colour; sepals and petals white, dull yellow, suffused maroon or entirely purple-maroon; mid lobe of lip dark purple-maroon or crimson-lake, side lobes yellow or pale purple-red with darker purple-red veins; pedicel with ovary 1–2 cm. long. Sepals and petals obtuse and mucronate to acute, carinate at the base. Sepals oblong to oblong-elliptic; dorsal 10–11 mm. long, 3.5–4 mm. broad; laterals usually slightly oblique, 10 mm. long, 4–4.5 mm. broad. Petals narrowly oblong-elliptic, 9–11 mm. long, 3.5–4 mm. broad. Lip 3-lobed, 7–9 mm. long, 12–14 mm. broad (width measured across side lobes); mid-lobe broadly oblong-elliptic, oblong, obcordate or

* By J.J. Wood
** Mis-cited by Reichenbach as 1735.

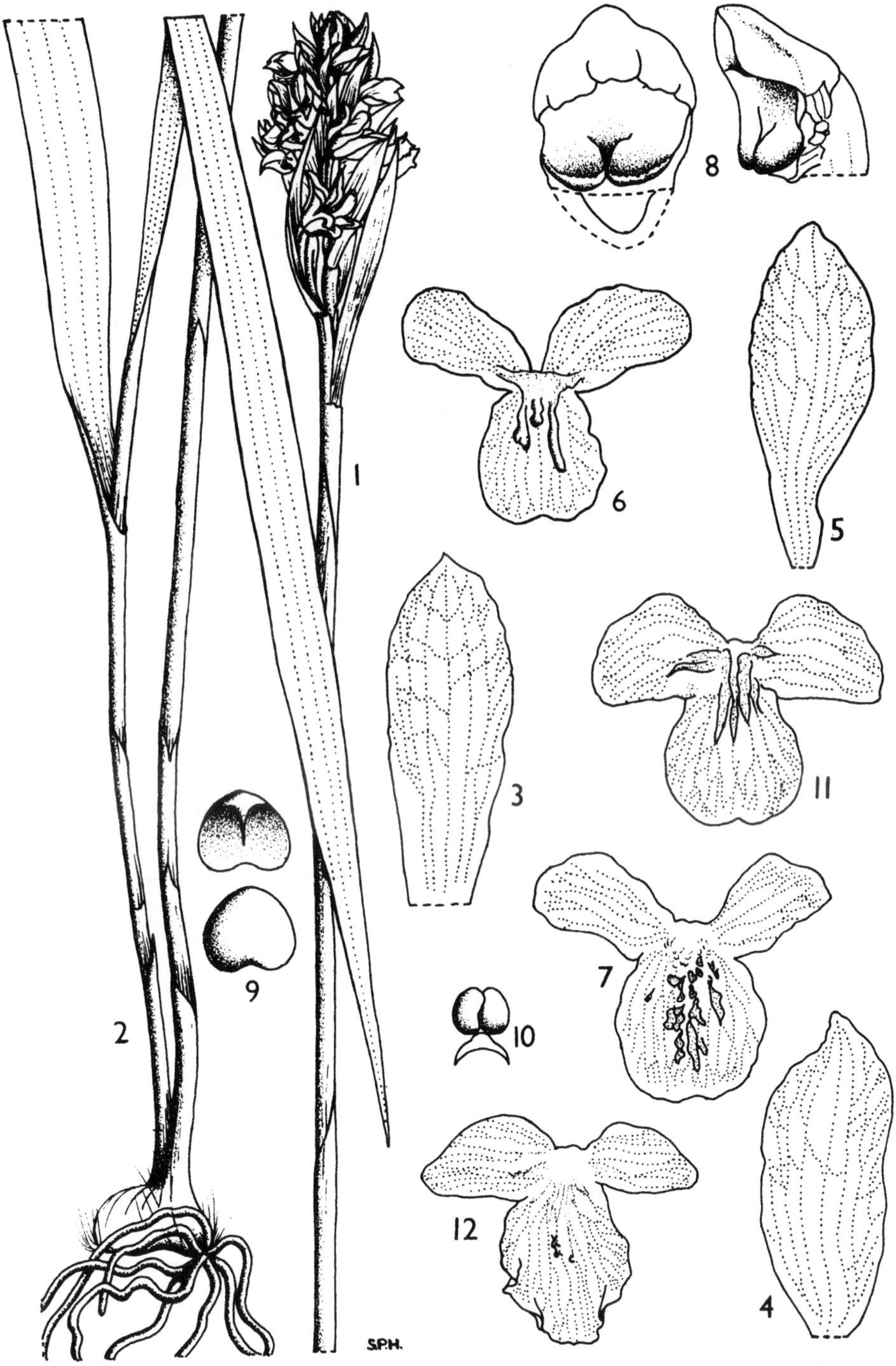

FIG. 110. *PTEROGLOSSASPIS EUSTACHYA*—**1**, inflorescence, × ⅔; **2**, habit, × ⅔; **3**, dorsal sepal, × 6; **4**, lateral sepal, × 6; **5**, petal, × 6; **6**, **7**, lip, × 4; **8**, column, × 6; **9**, anther-cap, × 6; **10**, pollinarium, × 6. *P. RUWENZORIENSIS*—**11**, **12**, lip, × 4. 1, 2, from *Bullock* 3685; 3–5, from *Gilbert & Thulin* 5906; 6, from *Bally* 6208; 7, from *Drake-Brockman* 209; 8, from *Schimper* 1935; 9, 10, from *Pollock* 17; 11, from *Copley* 97; 12, from *Leedal* 5026. Drawn by Susan Hillier.

quadrate-elliptic, obtuse or retuse, very rarely subentire, 6–8.5 mm. long, the disc bearing 3–7 tuberculate to lacerately-keeled veins, the veins and ridges of both disc and side lobes sometimes very sparsely pilose; side lobes broadly oblong, obtuse, often slightly curved, asymmetrical, divaricate, 5–6 mm. long, median vein usually raised and ridge-like at the base. Column straight above, 1.5–2(–3.5) mm. long, 1.5–2(–3) mm. broad; anther ovate, operculate, 1 mm. long, 1 mm. broad. Fig. 110/1–10.

UGANDA. W. Nile District: S. of Metu, Sept. 1953, *Chancellor* 243!; Ankole District: near Bugamba, Nov. 1940, *Eggeling* 4229!; Kigezi District: Kebisoni, Nov. 1946, *Purseglove* 2222!
KENYA. Uasin Gishu District: Kipkarren, Aug. 1931, *Brodhurst Hill* 306!; Kericho District: Sotik, May 1948, *Bally* 6208!
TANZANIA. Moshi District: Marangu, May 1893, *Volkens* 266!; Ufipa District: near Mmemya Mt., Feb. 1951, *Bullock* 3685!; Iringa District: Iheme, Feb. 1962, *Polhill & Paulo* 1609!
DISTR. **U** 1, 2; **K** 3, 5; **T** 2, 4, 7; Ethiopia, Mozambique and Zimbabwe
HAB. Wooded grassland, damp grassland, streamsides, short-grassed lava plains; 900–1900 m.

SYN. *P. engleriana* Kraenzl. in E.J. 19: 249 (1894); Rolfe in F.T.A. 7: 99 (1897); J. Stewart in U.K.W.F.: 782 (1974). Type: Tanzania, Moshi District, Marangu, *Volkens* 266 (B, holo., K, iso.!)

2. **P. ruwenzoriensis** (*Rendle*) *Rolfe* in F.T.A. 7: 100 (1897); Tweedie in Journ. E. Afr. Nat. Hist. Soc. 24: 17 (1964); Piers, Orch. E. Afr., ed. 2: 190 (1968); J. Stewart in U.K.W.F.: 783 (1974). Type: Uganda, Ruwenzori (eastern side), *Scott-Elliot* 7813, 7859 (BM, syn.!, K, isosyn.!) & 7551 (BM, syn.!)

A terrestrial herb 30–60(–93) cm. high, glabrous except for the lip. Leaves linear to linear-lanceolate, acute to acuminate, plicate; petiole 15–25(–38) cm. long, enclosed in 1–3 greyish brown, tubular, obtuse to acute, subscarious sheaths below; lamina 23–30(–40) cm. long, (1–)1.5–4 cm. broad. Inflorescence globular or ovoid, many-flowered, dense; scape erect, bearing 4–6(–8) remote, greyish brown tubular, obtuse to acute, subscarious sheaths; rhachis 3–6 cm. long; bracts pale brown, lanceolate or linear-lanceolate, often finely acuminate, scarious, with prominent venation, lowermost 3–8 cm. long, uppermost shorter. Flowers resupinate, variable in colour, white with a large dark purple-maroon central blotch on the lip and a few fine purple flecks at the base of the petals or entirely dark purple-maroon, sometimes flushed pale lilac or greenish on the outside, rarely mauve; pedicel with ovary 1–2 cm. long. Sepals and petals obtuse and mucronate to acute, carinate at the base. Sepals oblong-elliptic; dorsal subcuneate towards the base, 8–14 mm. long, 4–4.5 mm. broad; laterals usually slightly oblique, (9–)12–14 mm. long, 4.5 mm. broad. Petals narrowly oblong-elliptic, (10–)11–12(–14) mm. long, 3.5–4 mm. broad. Lip 3-lobed, (6.5–)8–12 mm. long, 8–12(–15) mm. broad (width measured across side lobes); mid-lobe oblong, oblong-elliptic or sometimes oblong-orbicular, obtuse and mucronate or retuse, rarely crenate, 6–7 mm. long, the disc bearing slightly verrucose veins, at most with only a few irregular warts, minutely papillose, the veins and ridges of both disc and side lobes never pilose; side lobes oblong, obtuse, often slightly curved, asymmetrical, usually enclosing the column, (4–)6–9 mm. long, median vein usually raised and ridge-like at the base. Column straight, 3.5–5 mm. long, 2.5–3 mm. broad; anther ovate, operculate, 1mm. long, 1.5 mm. broad. Fig. 110/11,12, p. 481.

UGANDA. W. Nile District: Chei, Sept. 1937, *Eggeling* 3404!; Toro District: Oruha Hill, May 1941, *Eggeling* 4333!; Mengo District: near Kampala, May 1929, *Liebenberg* 820!
KENYA. Nakuru District: Menengai Crater, Sept 1966, *Polhill* 199!; Kiambu District: Kabete, June 1961, *J.G. Williams* in *E.A.H.* 12318!; Londiani, July 1951, *Davis* 17!
TANZANIA. Biharamulo District: Rusumo Falls, Mar. 1960, *Tanner* 4783!; Ufipa District: Malonje Plateau, Feb. 1957, *Richards* 8428!; Iringa District: Sao Hill, *Watermeyer* 92!
DISTR. **U** 1, 2, 4; **K** 3–5; **T** 1, 4, 7; Zambia, Brazil and Argentina
HAB. Upland grassland, vlei grassland; 1300–2300 m.

SYN. *Eulophia ruwenzoriensis* Rendle in J.B. 33: 166 (1895)
Pteroglossaspis carsonii Rolfe in F.T.A. 7: 100 (1897), as "*carsoni*"; Williamson, Orch. SC. Afr.: 150 (1977). Types: Zambia, Lake Tanganyika, *Carson* & Fwambo, *Nutt* (both K, syn.!)
P. argentina Rolfe in K.B. 1906: 86 (1906). Type: Argentina, near Cordoba, *Stuckert* 6435 (K, holo.!)
Eulophia argentina (Rolfe) Schltr. in E.J. 53: 574 (1915), in obs.
Pteroglossaspis stricta Schltr. in E.J. 53: 574 (1915). Type: Tanzania, Rungwe District, Rutenganio [Rutuganjo], *Stolz* 1319 (B, holo., K, iso.!)

NOTE. This is an extremely interesting plant because of its curious disjunct distribution and is the sole representative of this genus occurring in South America. Other primarily African genera, such as *Eulophia* and *Oeceoclades*, each have one species native to America.

IV. VANDEAE

Lindl., Gen. Sp. Orchid. Pl.: 135 (1832); G.P. 3: 463 (1883); Senghas in Brieger et al. (eds.), Schlechter's Die Orchideen, ed. 3, 1: 973 (1986)

This tribe comprises three subtribes which are widespread in the tropics and subtropics of the Old World but have only a few representatives in the New World. All three subtribes are represented in East Africa as follows:—

1. Subtribe *Sarcanthinae* Benth. Four pollinia, often in 2 unequal pairs. *Acampe* (genus 47).
2. Subtribe *Angraecinae* Summerh. Two pollinia only; rostellum very short, deeply emarginate. *Angraecum* and *Jumellea* (genera 50, 51).
3. Subtribe *Aerangidinae* Summerh. Two pollinia only; rostellum long-beaked. Genera 52–74.

Dressler's (1981) classification, in his "The Orchids. Natural History and Classification", is the most widely used of the recent taxonomic treatments of the *Orchidaceae*. He divides the subfamily *Orchidoideae*, as understood here (see Summerhayes (1968) in F.T.E.A. Orchidaceae Part 1), into four subfamilies. Dressler's subfamily *Vandoideae* includes not only the tribe *Vandeae* Lindl. but also a further three tribes: *Polystachyeae* Pfitz., *Cymbidieae* Pfitz., and *Maxillarieae* Benth. The Polystachyeae includes the East African genera *Polystachya* and *Neobenthamia*, while *Eulophia*, *Oeceoclades*, *Graphorkis*, *Pteroglossaspis*, *Ansellia* and *Genyorchis* are included in the *Cymbidieae*. The third tribe *Maxillarieae* is confined to the New World tropics. However, in his most recent review of orchid classification, in "Classification of the Orchidaceae and their probable origin" in the journal Telopea, Dressler (1983) unites the *Vandoideae* and *Epidendroideae* in the subfamily *Epidendroideae*, and reinstates the tribe *Vandeae* in the sense used here. The classification of the Orchidaceae is currently the object of much research and speculation, but recent trends seem to indicate that the classification adopted by Summerhayes for this Flora will be supported in its broad outline if not in detail.

1. Plants without green leaves; roots greenish and photosynthetic but often appearing greyish white when dry	2
Plants with green leaves	3
2. Stems very short; roots densely clustered	53. **Microcoelia**
Stems elongate; roots emerging all along the stem	62. **Solenangis** (in part)
3. Lip very fleshy, erect; spur obscure, papillate within; pollinia 4, in 2 unequal pairs	47. *Acampe rigida*
Lip with a distinct short to long spur at the base; spur not papillate within; pollinia 2	4
4. Flowers borne in a ± globose head; rostellum distinctively reflexed in the basal half and parallel to the column and then abruptly recurved through 180° in the apical half	64. **Ancistrorhynchus**
Flowers either solitary or borne in a ± elongate raceme; rostellum not as above	5
5. Leaves bilaterally compressed, iridiform	6
Leaves dorsoventrally flattened or terete	9
6. Flowers green; stipes Y-shaped	73. *Ypsilopus viridiflorus*
Flowers white; stipes or stipites linear	7
7. Stem elongate; leaves distichously arranged along the stem	50. *Angraecum distichum*
Stem short; leaves arranged in a fan	8
8. Spur 10 mm. or more long, with a broad mouth and abruptly dilated apex; inflorescence-axis abbreviated, much shorter than the leaves	48. *Podangis dactyloceras*
Spur less than 5 mm. long, cylindrical to conical; inflorescence-axis elongate, as long as or longer than the leaves	56. **Bolusiella**
9. Rostellum very short, emarginate, not produced into a pendent or recurved flap or projection	9
Rostellum ± elongate, pendent or reflexed	11

10.	Lip ovate to cordate, with the base enfolding the column	50. **Angraecum**
	Lip lanceolate to narrowly subrhombic, with a narrow base not enfolding the column	51. **Jumellea**
11.	Flowers borne in whorls of 2–4 at each node of the inflorescence	12
	Flowers borne singly at the nodes of the inflorescence	13
12.	Lip obscurely 3-lobed, broader than long, 10 mm. or more long, fringed	54. *Diaphananthe fragrantissima*
	Lip simple, longer than broad, 4 mm. or less long, with an entire margin	57. **Chamaeangis**
13.	Lip geniculate towards the apex and abruptly dilated at the apex	14
	Lip not as above	15
14.	Lip 3-lobed, 10 mm. or less long and wide when flattened; rhachis markedly zig-zag; stem elongate	49. *Calyptrochilum christyanum*
	Lip entire, very large, 20–25 mm. wide when flattened, white with a green basal mark; rhachis ± straight; stem very short	63. *Eurychone rothschildiana*
15.	Sepals and petals united in the basal third; spur strongly recurved and pointing forwards at the apex; rostellum transversely hamate	52. *Rhaesteria eggelingii*
	Sepals and petals free to the base; spur and lip not as above	16
16.	Rostellum with 2 or 3 forwards-pointing acuminate lobes, almost as long as the column and often with these extending backwards as well	17
	Rostellum lobe or lobes pendent or reflexed	18
17.	Flower solitary; leaves ovate; rostellar lobes 2	68. *Distylodon comptum*
	Flowers in a raceme, longer than the linear leaves; rostellar lobes 3, trident-like in front	69. *Triceratorhynchus viridiflorus*
18.	Pollinia attached by separate stipites to 1 or 2 viscidia, rarely stipites coherent along length or part thereof	19
	Pollinia attached by a single stipe to the viscidium	29
19.	Flowers campanulate, white with a green anther; lip concave; spur clavate	55. **Margelliantha**
	Flowers not campanulate, white, yellow or green	20
20.	Pollinia with 2 stipites and viscidia	21
	Pollinia with 2 stipites attached to 1 viscidium	24
21.	Lip strongly 3-lobed in the middle; rostellum of 3 ± equal lobes	67. **Angraecopsis** (in part)
	Lip not as above; rostellum not as above	22
22.	Lip usually with a tooth-like callus at the base; rostellum smooth, with the mid-lobe a pendent peg-like projection, much exceeding the obscure side lobes	54. **Diaphananthe** (in part)
	Lip lacking a callus; rostellum of 3 equal lobes or with the outer lobes exceeding the mid-lobe	23
23.	Rostellar lobes papillate, with the outer lobes slender and exceeding the mid-lobe	66. **Mystacidium**
	Rostellar arms smooth, equal in length	67. **Angraecopsis** (in part)
24.	Flowers translucent pale yellow or pale green	25
	Flowers white, rarely marked with green on the lip	27
25.	Lip with a small tooth or swelling in the mouth of the spur	54. **Diaphananthe** (in part)
	Lip lacking a callus in the mouth of the spur	26

26. Inflorescences pendent, of many closely spaced pale green flowers; leaves very small, less than 5 cm. long . . . 67. **Angraecopsis** (in part)
    Inflorescences erect, laxly many-flowered; leaves usually more than 8 cm. long . . . 60. **Cribbia**
27. Lip with a dark green mark on the lip; inflorescences densely few–many-flowered . . . 65. **Sphyrarhynchus**
    Lip white; inflorescences laxly many-flowered . . . 28
28. Lip similar to the sepals and petals; rostellar side lobes much longer than the column; viscidium linear, hyaline or with a hard upper part and hyaline lower part . . . 61. **Cyrtorchis**
    Lip different in shape from the other segments; rostellar arms shorter than the column; viscidium cordate to suboblong, not linear nor hyaline nor bipartite . . . 59. **Rangaeris**
29. Lip cordate . . . 72. **Cardiochilos**
    Lip not cordate . . . 30
30. Lip 3-lobed in the middle or apical half . . . 31
    Lip entire, bilobed or only very obscurely 3-lobed . . . 32
31. Lip lacking auricles at the base; mouth of spur wide; viscidium very large compared to the very short stipes . . . 62. *Solenangis wakefieldii*
    Lip with basal auricles; mouth of the spur narrow; viscidium much shorter than the linear or Y-shaped stipes . . . 71. **Tridactyle** (in part)
32. Lip-lamina with a narrow basal claw and transversely oblong and bilobed at the apex . . . 74. **Nephrangis**
    Lip not as above . . . 33
33. Lip auriculate at the base . . . 34
    Lip not auriculate at the base . . . 35
34. Lip basal auricles fleshy and ridge-like; leaf usually lacking a ligulate extension at the apex of the leaf-sheath opposite the leaf . . . 71. **Tridactyle** (in part)
    Lip basal auricles not fleshy and ridge-like; leaf with a ligulate auricle at the apex of the leaf-sheath opposite the leaf . . . 70. **Eggelingia**
35. Leaves flat, ligulate, oblanceolate or obovate, dark green; flowers usually stellate; lip entire, lanceolate, pandurate or obovate; rostellum slender, entire; stipes entire . . . 58. **Aerangis**
    Leaves usually V-shaped in cross-section, linear; lip obscurely 3-lobed or subquadrate; rostellum bifid, usually dilated at the apex; stipes Y-shaped . . . 73. **Ypsilopus**

47. ACAMPE*

Lindl., Fol. Orch. 4, *Acampe*: 1 (1853), *nom. conserv.*

Epiphytic or lithophytic, monopodial herbs. Stem simple or branched, stout, leafy. Leaves distichous, conduplicate, apex mostly unequally bilobed, thickly coriaceous, unmarked. Inflorescence axillary, stout, racemose, corymbose or paniculate, shorter than the leaves; bracts inconspicuous, scale-like, persistent. Flowers non-resupinate, rather small, fleshy, fragile. Sepals and petals similar, flat or curved; lateral sepals adnate to the spur (when present); petals usually narrower and recurved at the apex. Lip positioned uppermost or pointing towards the apex of the inflorescence, saccate or with a short spur, adnate to the column, immobile, rigid, variously lobed, fleshy, muricate, papillose or dentate, often densely ciliate towards the base; spur, when present, lacking any internal tongue or median septum. Column short, fleshy, glabrous or papillose, with 2 narrow,

* By J.J. Wood

tooth-like stelidia; foot absent; anther ovate, apiculate; pollinia 4, globose, unequal, waxy, united in 2 pairs on a thick, subulate caudicle; viscidium small, oval; stigma transverse, concave; rostellum short, emarginate.

A genus of about 10 species distributed in SE. Asia, tropical and subtropical Africa, Madagascar and the islands of the western Indian Ocean.

A. pachyglossa *Reichb.f.*, Otia Bot. Hamb. 2: 76 (1881); Moreau & Moreau in Journ. E. Afr. Nat. Hist. Soc. 17: 25 (1943); Tweedie in Journ. E. Afr. Nat. Hist. Soc. 24: 54 (1964); Piers, Orch. E. Afr., ed. 2: 199 (1968); Williamson, Orch. SC. Afr.: 177 (1977). Type: Kenya, near Mombasa, *Hildebrandt* 1991 (W, holo.)

An erect or pendulous, epiphytic or lithophytic, glabrous (except lip) herb 18–30(–40) cm. high; roots long, branched, stout, 5–9 mm. broad, produced freely along stem. Stem stout, simple, rarely branched, leafy above, clothed in old leaf-sheaths below. Leaves ligulate, often arcuate, apex broad, retuse to unequally bilobed, with a mucro in the sinus, thick, rigid, coriaceous, conduplicate, articulated at a sheathing base, stiffly erect, (10–) 15–20(–30) cm. long, 1.5–2.5(–3.5) cm. broad. Inflorescence erect, racemose, corymbose or shortly paniculate, 3–8(–20) cm. long; branches (when present) 3–5 cm. long, few-many-flowered, often rather dense; bracts triangular-ovate, obtuse, margin membranous, those subtending a branch larger, amplexicaul, those subtending a flower smaller, appressed to the base of the pedicel and ovary, 2–4 mm. long. Flowers sweetly scented; sepals and petals greenish cream, greenish yellow or pale lemon-yellow with transverse crimson or reddish brown stripes and spots; lip white or pale pink, sparsely and similarly marked; column greenish cream, similarly marked; ovary subsessile, 7–10 mm. long. Sepals and petals thickly fleshy, strongly curved, forming a cup-like perianth. Sepals ovate-oblong or ovate-elliptic, obtuse, 10–14 mm. long, 5–7 mm. broad; lateral sepals slightly oblique, carinate, apex sometimes slightly papillose. Petals spathulate, often falcate, obtuse, 8–12 mm. long, 3–14 mm. broad. Lip thickly fleshy, 3-lobed, saccate at the base, 8–12 mm. long, 3–4 mm. broad (measured across mid-lobe), sac 2–4 mm. deep; mid-lobe ovate, flat or reflexed at apex, surface muricate, verruculose or rugose, papillose, disc with a small, erect, papillose, triangular basal keel, with usually 2 small, papillose protuberances either side, occasionally with a low median ridge, margin irregular; side lobes erect, truncate, 1.5–2 mm. high, densely ciliate inside. Column quadrate or rectangular, 1.5–3.5 mm. long, 3 mm. broad, usually with linear-subulate or triangular, glabrous or papillose stelidia; anther-cap ovate, with 2 small, obtuse dorsal ridges, 2 mm. long, 1 mm. broad; stigma broad. Pl. 1. Fig. 111.

KENYA. Kwale, Feb. 1942, *Copley* 25!; Mombasa District: Likoni, Aug. 1937, *Dale* in *F.D.* 3773!; Kilifi District: Malindi, Aug. 1944, *Tweedie* 634!

TANZANIA. Handeni District: 5 km. NE. of Gendagenda, 16 Nov. 1955, *Milne-Redhead & Taylor* 7317!; Tanga District: between Tanga and Mazumbai, Gombero [Gombopo], 6 Jan. 1976, *Cribb & Grey-Wilson* 10011!; Morogoro, 17 Dec. 1932, *Wallace* 580!; Zanzibar I., Sebleni, 24 Dec. 1929, *Vaughan* 1014!; Pemba I., Ras Mkumbuu, 22 Dec. 1930, *Greenway* 2767!

DISTR. **K** 7; **T** 3, 6, 8; **Z**; **P**; Zaire, Mozambique, Malawi, Zambia, Zimbabwe, Madagascar, Comoro Is., Seychelles, Aldabra, South Africa (E. Transvaal)

HAB. Epiphytic on trees in deciduous and coastal bushland, e.g. *Acacia, Combretum, Adansonia*, sometimes in association with *Ansellia gigantea* var. *nilotica*, rocks, sea cliffs, preferring semi-arid areas; sea-level–800 m.

SYN. *A. renschiana* Reichb.f., Otia Bot. Hamb. 2: 77 (1881); Perrier in Humbert, Fl. Madag., Orch. 2: 120 (1941). Type: Madagascar, Nosi-bé, *Hildebrandt* 3392 (W, holo., K, iso.!)

A. madagascariensis Kraenzl. in Gard. Chron., ser. 3, 10: 608 (1891). Type: Madagascar, cult. *Sander*, St. Albans (?B, holo.)

A. mombasensis Rendle in J.L.S. 30: 386 (1895); Moreau & Moreau in Journ. E. Afr. Nat. Hist. Soc. 17: 25 (1943); Piers, Orch. E. Afr., ed. 2: 198 (1968). Type: Kenya, Mombasa, *W.E. Taylor* (BM, holo.)

Saccolobium pachyglossum (Reichb.f.) Rolfe in F.T.A. 7: 132 (1897), *non* Lindl. (1858), *nom. illegit.*

S. mombasense (Rendle) Rolfe in F.T.A. 7: 132 (1897)

Acampe nyassana Schltr. in E.J. 53: 594 (1915); Moreau & Moreau in Journ. E. Afr. Nat. Hist. Soc. 17: 25 (1943). Type: Tanzania, Rungwe District, Kondeland, *Stolz* 766 (K, iso.!)

A. pachyglossa Reichb.f. subsp. *renschiana* (Reichb.f.) Senghas in Die Orchidee 15: 165 (1964); Piers, Orch. E. Afr., ed. 2: 200 (1968)

[*A. rigida* sensu Seidenf. in Bot. Mus. Leafl. Harv. Univ. 25: 60 (1977); Fosberg & Renvoize, Fl. Aldabra: 277 (1980); Seidenf. in Opera Bot. 95: 45 (1988), *non* (Smith) P.F. Hunt]

NOTE. *A. pachyglossa* appears related to and may possibly be conspecific with *A. praemorsa* (Roxb.) Blatter & McCann, a native of central and southern India and Sri Lanka. Detailed study of fresh material of both African and Asian plants is required.

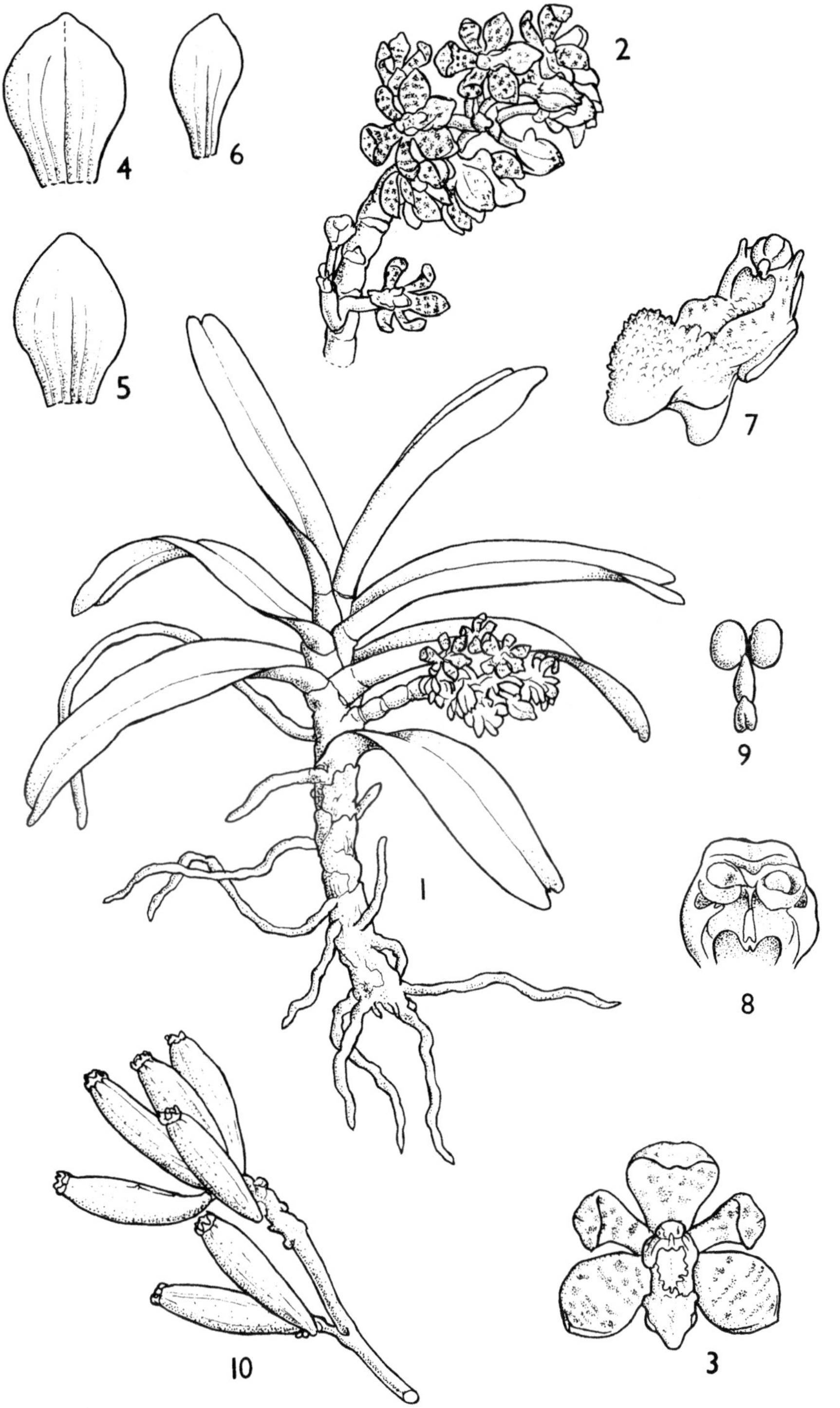

FIG. 111. *ACAMPE PACHYGLOSSA*—**1**, habit, × ⅓; **2**, inflorescence, × ⅔; **3**, flower, × 2; **4**, dorsal sepal, × 3; **5**, lateral sepal, × 3; **6**, petal, × 3; **7**, lip and column, × 2⅔; **8**, column, front view, × 6⅔; **9**, pollinarium, × 10; **10**, fruiting branch, × ⅔. Drawn from *Cribb & Grey-Wilson* 10011 by Cherry Ann Lavrih.

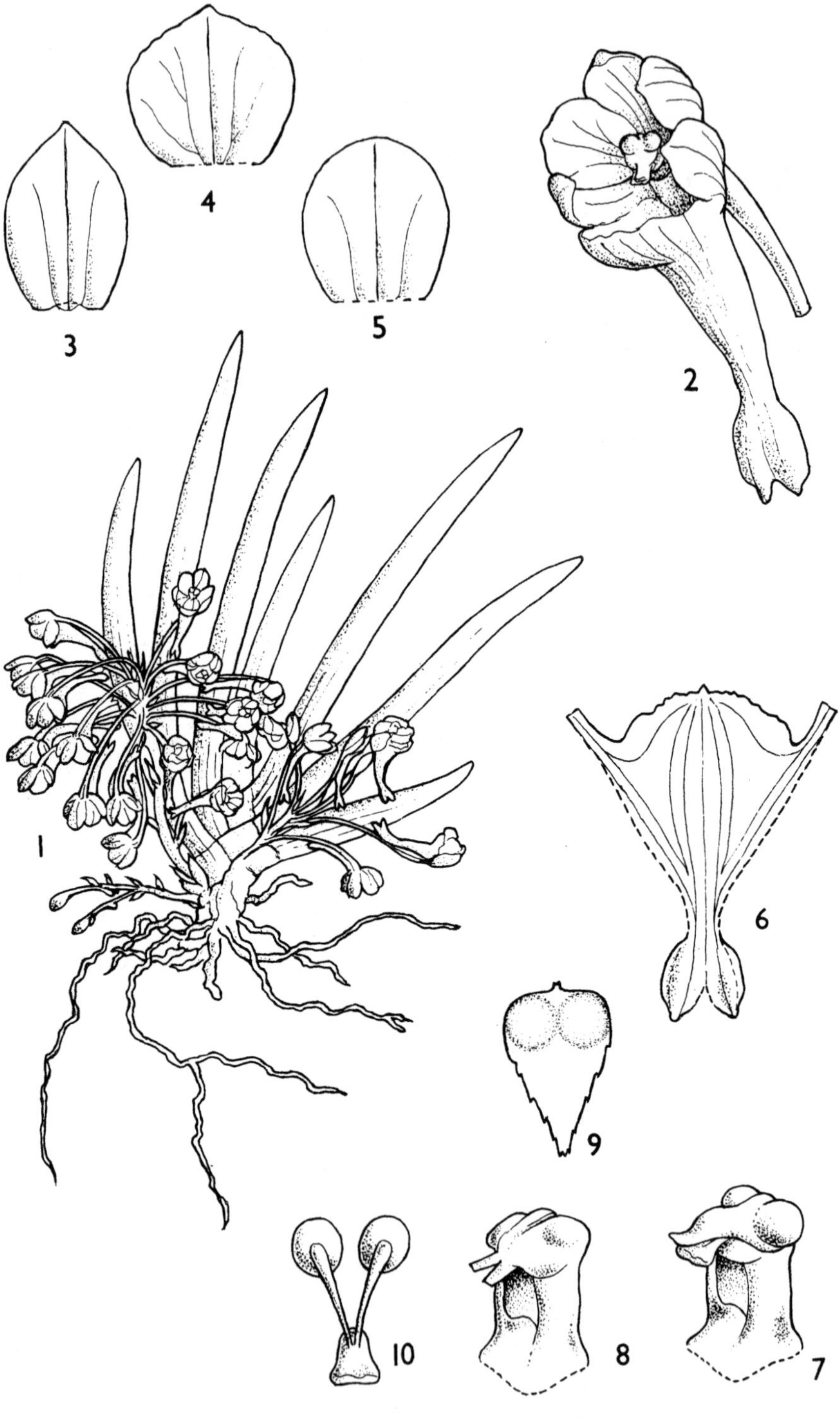

FIG. 112. *PODANGIS DACTYLOCERAS*—**1**, habit, × ⅔; **2**, flower, × 3; **3**, dorsal sepal, × 5⅓; **4**, lateral sepal, × 5⅓; **5**, petal, × 5⅓; **6**, lip, × 3; **7**, column with anther, × 8; **8**, column, without anther, × 8; **9**, anther-cap, × 12; **10**, pollinarium, × 12. 1, from *Lock* s.n. (unpublished illustration); 2–12, from *Meyer* K8. Drawn by Cherry Ann Lavrih.

48. PODANGIS*

Schltr. in Beih. Bot. Centr. 36: 82 (1918)

Epiphytic, entirely glabrous herb. Stem usually abbreviated, erect. Leaves iris-like, equitant, fleshy, with imbricate bases. Inflorescence an axillary open subumbellate raceme shorter than the leaves; bracts small, scarious. Flowers resupinate, narrow, medium-sized; ovary narrowly long-pedicellate. Sepals and petals free, spreading; petals slightly shorter. Lip orbicular, with a long, narrowly conical spur inflated and often bilobed at the apex. Column short, with a foot; rostellum narrow, bifid; anther terminal; pollinia 2, globose, each with a narrow stipes; viscidium solitary, relatively large.

A monotypic genus in tropical Africa.

P. dactyloceras (*Reichb.f.*) *Schltr.* in Beih. Bot. Centr. 36: 82 (1918); Moreau in Journ. E. Afr. Nat. Hist. Soc. 17: 24 (1943); Senghas in Die Orchidee 14: 68 (1963); Piers, Orch. E. Afr., ed. 2: 279 (1968); Summerh. in F.W.T.A., ed. 2, 3: 251 (1968). Type: Angola, Pungo Andongo, *Welwitsch* 677 (W, holo., K, iso.!)

Stem covered in roots and old leaf-sheaths below, leafy above, 1–8(–11) cm. long. Leaves 4–8, arranged in a fan, articulated at a sheathing base, linear-ligulate, often falcate, acute, fleshy, erect to spreading with both surfaces facing sideways, sheathing base 0.5–1 cm. long, lamina 4–10(–16) cm. long, 0.5–1.2 cm. broad. Inflorescences borne laterally in leaf-axil, few–20-flowered; peduncle and rhachis 1–4 cm. long; bracts pale brown, triangular-ovate, obtuse to acute, scarious, 2–5 mm. long. Flowers (including pedicels) glistening white, semi-transparent; ovary and anther-cap dark green. Pedicel with ovary 2–2.5 cm. long, ovary slightly ribbed. Sepals obtuse; dorsal sepal oblong, (3.5–)4–4.5 mm. long, 3 mm. broad; lateral sepals obliquely oblong-elliptic, carinate distally, 4–5 mm. long, 3 mm. broad. Petals transversely broadly obovate, obtuse, 4 mm. long, 5 mm. broad. Lip entire, free portion ± orbicular, 4–6 mm. long, 5 mm. broad, apex sometimes apiculate, margin often slightly erose; spur conical, broad at base, constricted in middle and inflated into 2 umbonate sacs or lobules at apex, 9–11 mm. long, 2–3 mm. broad at base. Column 1–2 mm. long, foot 2–2.5 mm. long; anther cucullate, apiculate; viscidium subquadrate. Pl. 1. Fig. 112.

UGANDA. Bunyoro District: Budongo Forest, Oct. 1935, *Eggeling* 2242! & Siba, Dec. 1942, *Eggeling* 5387!; Ankole District: Kalinzu Forest, 1977, *Meyer* K.8!
TANZANIA. Mwanza District: Uzinza, 24 km. S. of Geita gold mine, 13 Apr. 1937, *B.D. Burtt* 6497!; Morogoro District: Uluguru Mts., *Pócs* (pers. comm.)
DISTR. **U** 2; **T** 1, 6; Guinée, Sierra Leone, Ghana, Ivory Coast, Nigeria, Cameroon, Zaire, Angola, ? Madagascar
HAB. Forest; 750–1590 m.

SYN. *Listrostachys dactyloceras* Reichb.f. in Flora 48: 190 (1865); Th. Dur. & Schinz, Consp. Fl. Afr. 5: 49 (1895); Rolfe in F.T.A. 7: 168 (1897)
Angorchis dactyloceras (Reichb.f.) O. Kuntze, Rev. Gen. Pl. 2: 651 (1891)
Listrostachys forcipata Kraenzl. in E.J. 19: 254 (1894); Rolfe in F.T.A. 7: 168 (1897), in Orch. Rev. 10: 72 (1902), 16: 276 (1908) & 20: 305 (1912). Type: Cameroon, Buea, *Preuss* (B, holo.†)
Angraecum forcipatum (Kraenzl.) Engl., V.E. 2: 420 (1908)
Listrostachys saxicola Kraenzl. in E.J. 48: 399 (1912). Type: Cameroon, Tschape Pass, *Ledermann* 2783 (B, holo.†)

NOTE. In the Kew alcohol collection, there is a cultivated specimen (*Mason* 1891) reputedly from Madagascar, but its occurrence there seems unlikely.

49. CALYPTROCHILUM*

Kraenzl. in E.J. 22: 30 (1895)

Epiphytic, entirely glabrous monopodial herbs. Stems elongated, pendulous. Leaves fleshy, unequally bilobed at apex, alternate, twisted at base to lie in a parallel plane. Inflorescences abbreviated, lax or dense axillary racemes hidden beneath the leaves;

* By J.J. Wood

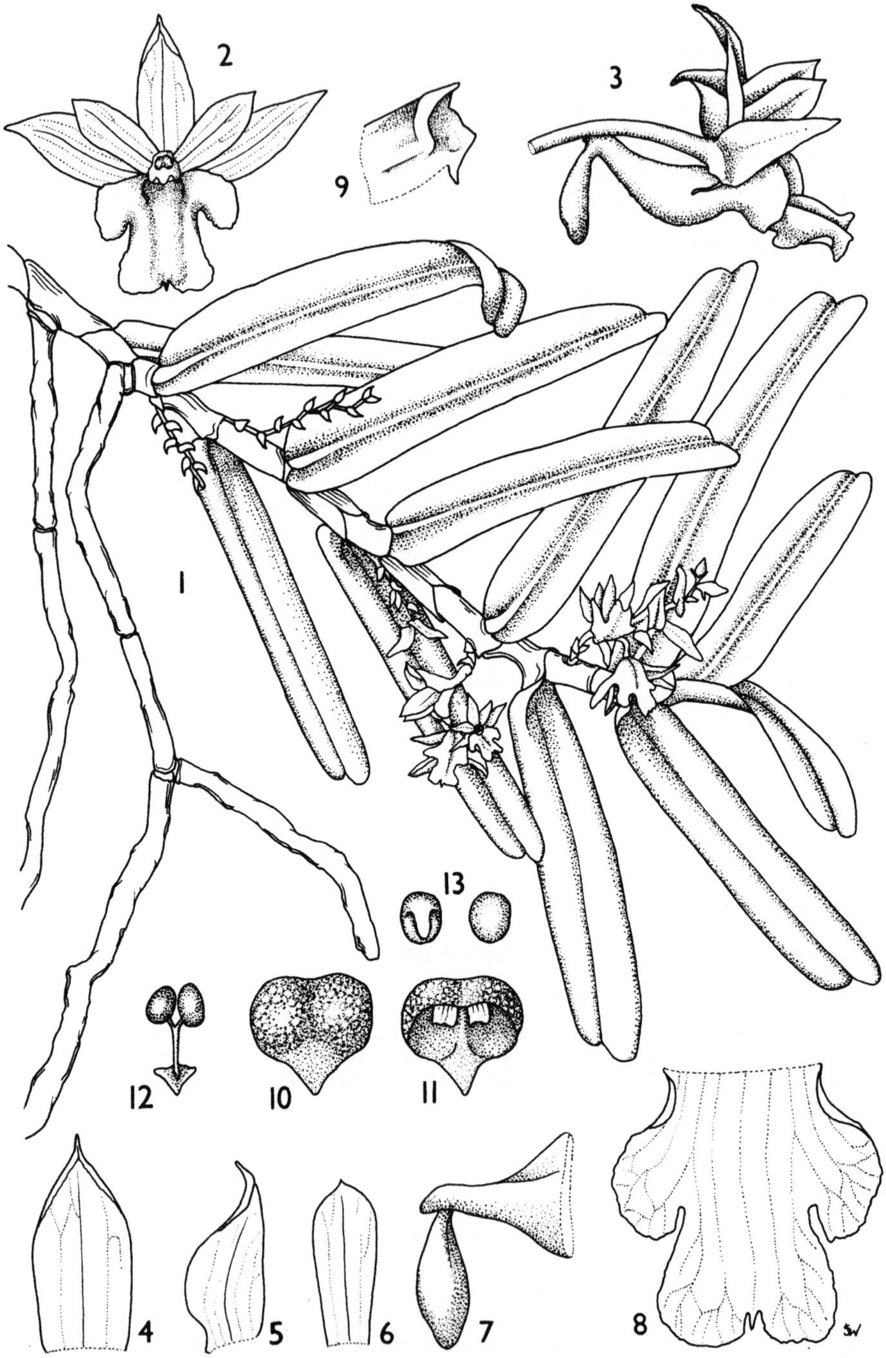

FIG. 113. *CALYPTROCHILUM CHRISTYANUM*—**1**, habit, × ⅔; **2**, flower, × 3; **3**, flower, side view, × 3; **4**, dorsal sepal, × 4; **5**, lateral sepal, × 4; **6**, petal, × 4; **7**, spur, side view, × 4; **8**, lip, × 4; **9**, column, × 8; **10**, anther-cap, front view, × 8; **11**, anther-cap, back view, × 8; **12**, pollinarium, × 6; **13**, pollinia, × 8. Drawn from *Pollock* in *Moreau* 601 by Sue Wickison.

bracts conspicuous or inconspicuous, scarious. Flowers resupinate, medium-sized. Tepals free, similar, apiculate, spreading, the petals slightly shorter. Lip 3-lobed, with a curved or geniculate spur which is inflated at the apex. Column short, with a small foot; rostellum prominent; anther apiculate; pollinia 2, globose, united to a long linear caudicle; viscidium solitary, large, triangular, grooved at the base to clasp rostellum.

A genus of two species in tropical Africa, extending as far south-east as Zimbabwe.

C. christyanum (*Reichb.f.*) *Summerh.* in F.W.T.A. 2: 450 (1936); Moreau in Journ. E. Afr. Nat. Hist. Soc. 17: 21 (1943); Summerh. in K.B. 2: 277 (1948), 13: 83 (1958) & in F.W.T.A., ed. 2, 3: 251 (1968); Piers, Orch. E. Afr., ed. 2: 249 (1968); Morris, Orch. Malawi: 79 (1970); Williamson, Orch. SC. Afr.: 182 (1977). Type: without proper locality, but believed to be W. Africa, *Christy* (W, holo.)

Stem simple, woody, rooting at nodes, leafy, 14–50 cm. long. Leaves ligulate, unequally bilobed, with sheathing base, coriaceous, fleshy, almost succulent, borne in one plane in 2 alternating rows, (4–)6–8(–13) cm. long, (0.8–)1.5–2(–2.5) cm. broad. Inflorescences much shorter than leaves, (3–)6–9(–12)-flowered; peduncle and rhachis fractiflex, 1.5–4 cm. long; bracts ovate, apiculate, scarious, 2–4 mm. long. Flowers scented or not, white or greenish white, fading to apricot; base of lip yellow or green; spur yellow at base or entirely green; pedicel with ovary 0.8–1 cm. long. Dorsal sepal ovate to ovate-elliptic, apiculate, 5–10 mm. long, 2.5–4 mm. broad; laterals similar but slightly oblique, apiculate. Petals oblong-elliptic, obtuse to apiculate, 5–8 mm. long, 2–3 mm. broad. Lip distinctly 3-lobed, 7–12 mm. long, 7–10 mm. broad across side lobes; side lobes short, rounded, erect to spreading, 2.5–5 mm. long, 3–5 mm. broad; mid-lobe oblong or ± rectangular, emarginate, 5–7 mm. long and broad, lobules entire or erose, with or without a mucro in the sinus; spur conical, strongly geniculate, broad below mouth, constricted in middle, inflated distally, 9–11 mm. long. Column 1–2 mm. long; anther ovate, apiculate, 2 mm. long and broad. Pl. 1. Fig. 113.

UGANDA. Bunyoro District: Budongo Forest, Dec. 1942, *Eggeling* 5185!; Mengo District: Entebbe Botanic Garden, 7 June 1946, *A.S. Thomas* 14434!; Mubende District: near Mubende, 1960, *Burgess* 9!

KENYA. Mombasa, 25 June 1879, *Kirk* 12!

TANZANIA. Rungwe District: Bomalakitana, 13 Dec. 1910, *Stolz* 469! & Masukulu, 1944, *Pollock* in *Moreau* 601!

DISTR. **U** 2, 4; **K** 7; **T** 7; Gambia, Mali, Guinea Bissau, Guinée, Sierra Leone, Liberia, Ivory Coast, Ghana, Nigeria, Cameroon, Central African Republic, Zaire, Sudan, Ethiopia (Eritrea), Mozambique, Malawi, Zambia, Zimbabwe, Angola

HAB. Forest, wooded grassland; 900–1200 m.

SYN. *Angraecum christyanum* Reichb.f. in Gard. Chron., n.s. 13: 806 (1880); Rolfe in F.T.A. 7: 142 (1897)
A. schoellerianum Kraenzl. in Schweinf. in Bull. Herb. Boiss. 2, App. 2: 112 (1894); Rolfe in F.T.A. 7: 144 (1897). Type: Ethiopia, Eritrea, *Schweinfurth* 169 (B, holo.)
A. malangeanum Kraenzl. in E.J. 22: 26 (1895). Type: Angola, Malange, *Mechow* 361 (B, holo.)
A. moloneyi Rolfe in F.T.A. 7: 145 (1897). Types: Sierra Leone, cultivated specimens, *Lord Scarborough* & Liberia, cultivated specimens, *Christy* & Nigeria, Lagos, *Moloney & Rowland* (K, syn.!)
A. mombasense Rolfe in F.T.A. 7: 145 (1897). Type: Kenya, Mombasa, *Kirk* 12 (K, holo.!)
A. zigzag De Wild., Not. Pl. Utiles Congo 1: 143 (1903). Type: Zaire, Basenga, *Gentil* 304 (BR, holo.)
A. pynaertii De Wild., Not. Pl. Utiles Congo 2: 160 (1906). Type: Zaire, Eala, *Pynaert* (BR, holo.)
A. ovalifolium De Wild., Not. Pl. Utiles Congo 2: 161 (1906). Type: Zaire, Eala, *Laurent* 686 (BR, holo.)
Rhaphidorhynchus moloneyi (Rolfe) Finet in Bull. Soc. Bot. Fr. 54, Mém. 9: 41 (1907)
Angraecum marsupio-calcaratum Kraenzl. in E.J. 51: 397 (1914). Type: Cameroon, Ngoni-Songolong, *Ledermann* 4305 (B, holo.)
Calyptrochilum mombasense (Rolfe) Schltr. in E.J. 53: 595 (1915)
C. orientale Schltr. in E.J. 53: 595 (1915). Type: Tanzania, Rungwe District, Bomalakitana, *Stolz* 469 (B, holo., K, iso.!)
C. schoellerianum (Kraenzl.) Schltr. in E.J. 53: 595 (1915)
Angraecum bokoyense De Wild. in B.J.B.B. 5: 182 (1916). Type: Zaire, Bokoyo, *Seret* 610 & Nala, *Boone* 109 (BR, syn.)
Calyptrochilum bokoyense (De Wild.) Schltr. in Beih. Bot. Centr. 36: 84 (1918)
C. malangeanum (Kraenzl.) Schltr. in Beih. Bot. Centr. 36: 84 (1918)
C. marsupio-calcaratum (Kraenzl.) Schltr. in Beih. Bot. Centr. 36: 84 (1918)
C. ovalifolium (De Wild.) Schltr. in Beih. Bot. Centr. 36: 85 (1918)
C. pynaertii (De Wild.) Schltr. in Beih. Bot. Centr. 36: 85 (1918)
C. zigzag (De Wild.) Schltr. in Beih. Bot. Centr. 36: 85 (1918)

50. ANGRAECUM

Bory, Voy. 1: 359, t. 19 (1804); Schltr. in Beih. Bot. Centr. 36: 157 (1918); Summerh. in K.B. 13: 260 (1958); Garay in K.B. 28: 496 (1973); Senghas in Schltr., Die Orchideen, ed. 3, 1: 987 (1986)

Angorkis Thou. in Nouv. Bull. Sci. Soc. Philom. Paris 1: 314 (1809), *nom. illegit.*
Aerobion Sprengel, Syst. Veg., ed. 16, 3: 679 (1826)
Macroplectrum Pfitz. in E. & P. Pf. 2,6 : 214 (1889)
Pectinaria (Benth.) Cordemoy in Rev. Gen. Bot. 11: 412 (1899)
Lepervenchea Cordemoy in Rev. Gen. Bot. 11: 415 (1899)
Ctenorchis K. Schum. in Just's Bot. Jahresb. 27(1): 467 (1901)
Monixus Finet in Bull. Soc. Bot. Fr. 54, Mém. 9: 15 (1907)

Dwarf to large epiphytic, lithophytic or rarely terrestrial herbs. Stems short to elongate, unbranched or branching, erect to pendent, covered by leaf-bases. Leaves thin-textured, fleshy or coriaceous, flattened or rarely iridiform, unequally bilobed at the apex, often twisted at the base and articulated to persistent leaf-bases. Inflorescences axillary, 1–many-flowered, racemose or rarely paniculate. Flowers fleshy, small to large, often stellate; ovary often twisted through 180°, sessile or pedicellate. Sepals and petals free, subsimilar. Lip concave, entire to 3-lobed, enveloping the column at the base, spurred at the base. Column fleshy, very short, lacking a foot; clinandrium rather shallow, cleft in front with a short tooth-like rostellum in the sinus; pollinia 2, globose, sulcate, attached to a common viscidium or each attached to a separate viscidium.

A genus of about 200 species from tropical Africa, Madagascar and the adjacent islands and Sri Lanka.

The genus *Angraecum* has had a complex history. It was used for many years as the genus to which most African and Madagascan monopodial orchids were assigned. Schlechter (1918), however, reduced it to more or less its present circumscription in his classic account of the taxonomy of this group of orchids. Since then Summerhayes (1958), Garay (1973) and Senghas (1986) have produced valuable accounts of *Angraecum.* Garay has produced an excellent infrageneric account of *Angraecum* which I have followed here. Nine sections are represented in East Africa:

1. Sect. *Dolabrifolia* (Pfitz.) Garay— species 1.
2. Sect. *Lemurangis* Garay— species 2.
3. Sect. *Nana* (Cordem.) Garay— species 3–5.
4. Sect. *Boryangraecum* Schltr.— species 6–8.
5. Sect. *Angraecoides* (Cordem.) Garay — species 9–11.
6. Sect. *Conchoglossum* Schltr— species 12–14, ?22.
7. Sect. *Arachnangraecum* Schltr— species 15–17.
8. Sect. *Afrangraecum* Summerh.— species 18–20.
9. Sect. *Angraecum* - species 21.

Leaves equitant, i.e. bilaterally flattened, fleshy, less than 1 cm. long:
- Stems elongate, over 10 cm. long; leaf-bases imbricate; inflorescences 1-flowered; sepals and petals 3–4 mm. long; spur 5–6 mm. long . . . . 1. *A. distichum*
- Stems short, less than 4 cm. long; leaf-bases not noticeably imbricate; inflorescences 2–4-flowered; sepals and petals 1.3–2.1 mm. long; spur 1.3–1.8 mm. long 2. *A. humile*

Leaves dorso-ventrally flattened or terete, sometimes twisted at the base to lie in one plane, thin-textured or coriaceous:
- Stem very short or almost absent, always less than 5(–6.5) cm. long:
  - Leaves terete, less than 1.5 mm. in diameter . . . . 8. *A. teres*
  - Leaves flattened, never terete, often twisted at the base:
    - Leaves over 1 cm. wide; plants over 10 cm. tall 6. *A. dives*
    - Leaves less than 5 mm. wide; plants dwarf, less than 7.5 cm. tall, usually much smaller:
      - Pedicel and ovary twisted through 180°, somewhat crozier-shaped . . . . 7. *A. sacciferum*
      - Pedicel and ovary not noticeably twisted or incurved:

Spur conical or saccate, less than 1.4 mm. long:
  Plant less than 2.5 cm. tall; leaves 0.5–2 cm. long, oblong or oblanceolate, rounded at the apex . . . . 3. *A. chamaeanthus*
  Plant 3.5–7 cm. tall; leaves linear, acute 5. *A. minus*
 Spur shortly cylindrical, 1.6–1.7 mm. long . . 4. *A. decipiens*
Stem elongate, mostly more than 10 cm. long and often much longer:
 Spur more than 3 cm. long, often much longer:
  Base of spur broadly conical, spur narrowed and somewhat geniculate in the middle; sepals and petals more than 4.3 cm. long:
   Spur 6–6.5 cm. long; sepals 4.3–5 cm. long; leaves elliptic, less than 10 cm. long . . . . . . 17. *A. spectabile*
   Spur 15–16.5 cm. long; sepals ± 7 cm. long; leaves narrowly elliptic, more than 12 cm. long 16. *A. infundibulare*
  Base of spur narrow, not broadly conical; spur straight, never geniculate:
 Spur less than 6 cm. long:
  Lip-lamina as broad as long or broader, apiculate, white; sepals and petals greenish or yellowish:
   Inflorescence short, 1(–2)-flowered; leaves 4–6 cm. long; flowers relatively thin-textured; sepals and petals slenderly acuminate . . . . 15. *A. conchiferum*
   Inflorescences mostly over 30 cm. long, many-flowered; leaves mostly more than 20 cm. long; flowers fleshy in texture; sepals and petals acute 21. *A. eburneum* subsp. *giryamae*
  Lip-lamina much longer than broad, acute; lip white marked with green at the base; sepals and petals white . . . . 20. *A reygaertii*
   Spur 15 cm. long or more . . . . 22. *A. keniae*
   Spur less than 2.6 cm. long:
    Spur less than 3 mm. long:
     Sepals and petals 6.7–8.5 mm. long; spur conical, slightly upcurved, 2.6 mm. long; lip cucullate, not recurved at the apex . . . . 12. *A. brevicornu*
     Sepals and petals 5–5.5 mm. long; spur subclavate, straight, 1.8 mm. long; lip with a somewhat recurved apex 14. *A. viride*
    Spur over 5 mm. long, often much longer:
     Leaves oblanceolate, more than 12 cm. long; pedicels emerging from the rhachis ± 5 mm. above the sheathing bracts . . . . . 18. *A. affine*
     Leaves oblong, elliptic or linear, less than 11 cm. long; pedicels emerging ± from the axils of the bracts:
      Inflorescences racemose, 4–8-flowered, up to 12 cm. or more long; lip abruptly slenderly apiculate at the apex; apicule 6–7 mm. long 19. *A. firthii*
      Inflorescences 1(–2)-flowered, less than 6 cm. long; lip tapering gradually to an acuminate apex:
       Inflorescences emerging in fascicles of 2–4 opposite the leaves:
        Leaves linear, apical lobes parallel or divergent, rounded; spur sigmoid, 16–20 mm. long 9. *A. chevalieri*
        Leaves falcate, linear-elliptic or oblong-elliptic, apical lobes convergent, acute; spur less than 10 mm. long . . . 11. *A. stolzii*
       Inflorescences solitary at each node opposite the leaves:

Plants scandent; stems erect, nodes mostly 1.5–2.5 cm. apart; flowers whitish; leaves straight . . . . 13. *A. erectum*
Plants epiphytic; stems pendent, nodes usually less than 1.2 mm. apart; flowers greenish to buff-orange; leaves falcate . . . . 10. *A. cultriforme*

1. **A. distichum** *Lindl.* in Bot. Reg. 21, t. 1781 (1836); Summerh. in K.B. 13: 261 (1958) & in F.W.T.A., ed. 2, 3: 254 (1968); Piers, Orch. E. Afr., ed. 2: 230 (1968). Type: Sierra Leone, cult. *Loddiges* (K, holo.!)

A dwarf epiphyte often forming considerable clumps. Stems arcuate, slender, seldom branching, 3–25 cm. long; roots fine, 0.5 mm. in diameter. Leaves distichous, bilaterally compressed, curved-elliptic, obtuse in side view, 5–11 mm. long, 3–5 mm. wide, glossy dark green, deciduous, articulated to overlapping leaf-bases. Inflorescences opposite leaves, 1-flowered, 8–9 mm. long; bracts elliptic, obtuse, 1–1.5 mm. long. Flowers white; pedicel and ovary 5–6.5 mm. long. Sepals elliptic or oblong-elliptic, obtuse, 3.6–4.5 mm. long, 1.5–2 mm. wide. Petals oblong or oblong-elliptic, obtuse, 3.2–4.2 mm. long, 1.2–1.3 mm. wide. Lip concave, obscurely 3-lobed, obovate, obtuse, 2.9–3.5 mm. long, 2.8–4 mm. wide; side lobes rounded; spur straight or slightly upcurved, tapering to an acute apex, 5.5–6 mm. long. Column 0.7 mm. long. Fruit 7–8 mm. long. Pl. 5.

UGANDA. Toro District: Fort Portal Forest Reserve, 19 July 1960, *Paulo* 585! & Toro, fl. cult. 26 Apr. 1921, *Lankester* in *Snowden* 738!
DISTR. **U** 2; Sierra Leone to Zaire
HAB. In upland rain-forest; 1500–1600 m.

2. **A. humile** *Summerh.* in K.B. 13: 269 (1958); Piers, Orch. E. Afr., ed. 2: 238 (1968); J. Stewart in U.K.W.F.: 785 (1974). Type: Kenya, N. Kavirondo District, Kakamega, *Carroll* in *Moreau* 413 (K, holo.!)

A dwarf epiphyte with elongate, rarely branching stems 1–4 mm. long; roots fine, 0.5 mm. in diameter. Leaves equitant, distichous, narrowly elliptic in side view, 6–7 mm. long, 2–3 mm. wide, light green. Inflorescences 2–4-flowered, 2–3 mm. long; peduncle very short, less than 1 mm. long; rhachis fractiflex; bracts ovate, subacute, 0.3–0.5 mm. long. Flowers greenish white; pedicel and ovary 3–4 mm. long. Sepals elliptic-oblong, obtuse, 1.4–1.7 mm. long, 0.7–0.8 mm. wide. Petals oblong-elliptic, obtuse, 1.3–1.4 mm. long, 0.7–0.8 mm. wide. Lip concave, ovate, acute, 1.6–1.7 mm. long, 1.3–1.5 mm. wide; spur clavate, with a broad mouth, parallel to the ovary, 1.4–1.8 mm. long. Column 0.3–0.4 mm. long; anther-cap pointed.

KENYA. Fort Hall District: E. Aberdare Mts., S. Mathioya, 12 Aug. 1952, *Archer* 33! & Aberdare Mts., 17 Oct. 1942, *C. van Someren* in *Moreau* 413A!; N. Kavirondo District: Kakamega, 30 Aug. 1943, *Carroll* in *Moreau* 413!
TANZANIA. Lushoto District: 1.6 km. E. of Magamba on Lushoto–Mkusi road, 3 June 1953, *Drummond & Hemsley* 2831!
DISTR. **K** 3–5; **T** 3; not known elsewhere
HAB. In riverine and dwarf montane forest, on mossy twigs; 1650–2500 m.

NOTE. The Tanzanian collection is similar to the Kenyan ones but differs in having fleshy narrower leaves, grooved on their upper surface, and larger flowers in which the lip is entire and the spur is not abruptly dilated at the apex.

3. **A. chamaeanthus** *Schltr.* in E.J. 53: 604 (1915); Summerh. in K.B. 13: 263 (1958); J. Stewart in U.K.W.F.: 783 (1974); Ball, S. Afr. Epiphytic Orch.: 46 (1978). Type: Tanzania, Rungwe District, near Kyimbila, *Stolz* 1168 (B, holo., K, iso.!)

A dwarf epiphyte with very short stems 1–4 mm. long; roots ± 1 mm. diameter. Leaves 2–4 in a fan, linear-ligulate, obtuse, 10–14 mm. long, 1.5–2.5 mm. wide. Inflorescences 4–10-flowered, 15–22 mm. long; peduncle 5 mm. long; bracts 0.5 mm. long. Flowers white; pedicel and ovary 1–2 mm. long. Sepals oblong-elliptic, rounded or obtuse, 1.3–1.4 mm. long, 0.7–0.8 mm. wide. Petals oblong or obovate, obtuse or rounded, 1.3–1.7 mm. long, 0.7–1.2 mm. wide. Lip very obscurely 3-lobed, broadly ovate or circular, obtuse, 1.2–1.5 mm. long, 0.9–1.3 mm. wide; spur conical, rounded at the apex, 0.8–1.2 mm. long. Column very fleshy, 0.3 mm. long; stipes entire, very short. Fruit ellipsoidal, 4 mm. long.

KENYA. Kiambu District: Kikuyu Escarpment Forest, 6 June 1961, *Smart* E.27!
TANZANIA. Iringa District: Mufindi, Lupembe Tea Estate, 16 Dec. 1970, *Jones, Paget-Wilkes & Pócs* 6318/AH!; Rungwe District: Mbaka, 13 Mar. 1912, *Stolz* 1168! & N. slope of Rungwe Mt., 8 Feb. 1961, *Richards* 14306!
DISTR. **K** 4; **T** 7; Malawi, Zimbabwe
HAB. In upland and dwarf montane forest; 1600–2400 m.

4. **A. decipiens** *Summerh.* in K.B. 20: 189 (1966); Piers, Orch. E. Afr., ed. 2: 230 (1968). Type: Kenya, Nairobi District, Langata Forest, *Archer* 30 (K, holo.!)

A dwarf epiphyte with a very short stem, less than 5 mm. long; roots terete, 0.5–1 mm. in diameter. Leaves twisted at the base to lie in one plane, arranged in a fan, distichous, linear-falcate, acuminate, 15–35 mm. long, 2.5–3.5 mm. wide, rather fleshy. Inflorescences arcuate, 20–45 mm. long, 5–15-flowered; pedicel very slender, 10–15 mm. long; bracts lanceolate, acute, ± 2 mm. long. Flowers pale to bright green; pedicel and ovary 2.5–4 mm. long; pedicel at an angle to the ovary. Dorsal sepal elliptic or obovate, subacute to obtuse, 1.7–2.3 mm. long, 0.8–1.1 mm. wide; lateral sepals obliquely elliptic-obovate, acute or subacute, 2–2.2 mm. long, 0.8–1 mm. wide. Petals elliptic, subacute or obtuse, 1.6–2 mm. long, 0.6–0.8 mm. wide. Lip entire, concave, ovate, obtuse, 1.5–2.1 mm. long, 1.2–1.4 mm. wide; spur ± straight, clavate, 1.4–1.7 mm. long, straight and ± parallel to the ovary. Column 0.3 mm. long; rostellum-lobes upcurved hooks. Fig. 114, p. 496.

KENYA. Nairobi District: Langata Forest, 20 Apr. 1964, *Archer* 30!
TANZANIA. Arusha District: Ngurdoto Crater, fl. cult. Nov. 1964, *J. & D. Stewart* 785!
DISTR. **K** 4; **T** 2; not known elsewhere
HAB. In upland forest in deep shade; 1650–2100 m.

NOTE. The generic placing of this species is uncertain. It has a rostellum in which the outer lobes are hook-like and upcurved. In this it is not unlike the equally diminutive *Angraecopsis parva* and it may well be better placed in that genus. However, the material currently available is not good enough to make such a transfer with confidence.

5. **A. minus** *Summerh.* in K.B. 13: 264 (1958); Williamson, Orch. SC. Afr.: 178 (1977); Ball, S. Afr. Epiphytic Orch.: 48 (1978). Type: Zimbabwe, Mutare [Umtali] District, *Wild* 4541 (K, holo.!, SRGH, iso.)

A dwarf epiphyte with a short stem, 2–10 mm. long; roots flattened, 0.5–1 mm. in diameter. Leaves distichous, in a fan, suberect to arcuate, linear, acute or subacute, 1.4–5.2 cm. long, 1–2 mm. wide, articulated at the base to a 6–9 mm. long leaf-sheath. Inflorescences suberect or erect, 3–5.2 cm. long, 3–14-flowered; peduncle slender, wiry, 1.4–2.1 cm. long; bracts broadly ovate, apiculate, sheathing at the base, 0.5–1 mm. long. Flowers white with a green spur; pedicel and ovary 1 mm. long. Dorsal sepal elliptic, obtuse or rounded, 1–1.1 mm. long, 0.7 mm. wide; lateral sepals obliquely oblong or obovate, obtuse, 1.3–1.5 mm. long, 0.5–0.6 mm. wide. Petals lanceolate, acute, 1.2 mm. long, 0.6–0.7 mm. wide. Lip fleshy, concave, broadly ovate, obtuse, 1.4 mm. long, 1.6 mm. wide; spur conical, slightly upcurved at the tip, 0.6 mm. long. Column 0.2–0.3 mm. long. Fruits elliptic, 2.5–3 mm. long.

TANZANIA. Lushoto District: E. Usambara Mts., Amani, 17 Aug. 1942, *Moreau* 122! & W. Usambara Mts., Mazumbai Forest Reserve, 8 Jan. 1976, *Cribb & Grey-Wilson* 10090!
DISTR. **T** 3; Zambia, Zimbabwe
HAB. In rain-forest and in plantations; 1000–1150 m.

6. **A. dives** *Rolfe* in F.T.A. 7: 142 (1897); Summerh. in K.B. 13: 265 (1958); Piers, Orch. E. Afr., ed. 2: 231 (1968). Type: Kenya, Mombasa, *Wakefield* (K, holo.!)

An erect epiphyte or lithophyte with a stem 1–4(–6.5) cm. long; roots 1.5–2 mm. in diameter. Leaves 2–6, coriaceous, linear, unequally roundly bilobed at the apex, 5–26 cm. long, 0.4–1.6 cm. wide, articulated obliquely to the leaf-bases. Inflorescences many, suberect to arching, 8–25-flowered, 3–25 cm. long; peduncle 1.5–6 cm. long, rather stout, bearing spaced sheaths along its length; bracts amplexicaul, ovate, acuminate, 2–3 mm. long. Flowers pale greenish yellow, fleshy, faintly scented; pedicel and ovary 4–5 mm. long. Sepals elliptic-lanceolate or lanceolate, acute or acuminate, recurved, 5 mm. long, 1.3–1.5 mm. wide. Petals lanceolate, acuminate, recurved, 5 mm. long, 1–1.2 mm. wide. Lip concave, ovate, acuminate, 4 mm. long, 2.5 mm. wide, slightly recurved; spur slightly upcurved beside ovary, subclavate, slightly dilated at the apex, 2.5–3 mm. long. Column 0.5 mm. long. Fruits 16–18 mm. long.

FIG. 114. *ANGRAECUM DECIPIENS*—**1**, habit, × 1; **2**, habit, × 2; **3**, flower, side view, × 6; **4**, flower, front view, × 6; **5**, dorsal sepal, × 10; **6**, lateral sepal, × 10; **7**, petal, × 10; **8**, lip and spur, × 10; **9**, column with anther-cap, × 10; **10**, column, anther-cap removed, × 10; **11**, anther-cap, back view, × 26; **12**, anther-cap, front view, × 26. Drawn from *Ombler* s.n. by Miss M.A. Grierson.

KENYA. Kwale District: Shimoni, 25 Oct. 1952, *Irwin* 92!; Kilifi District: 6.5 km. N. of Malindi, Sabaki, 2 Nov. 1961, *Polhill & Paulo* 691! & Gede Forest, 15 Oct. 1971, *Rathbun* in *E.A.H.* 14989!
TANZANIA. Tanga District: Sawa, 19 Aug. 1955, *Faulkner* 1701! & N. of Kwale, 13 July 1937, *Greenway* 4961!; Uzaramo District: near Dar es Salaam, Msasani, 10 Sept. 1939, *Vaughan* 2870!; Zanzibar I., N. of Mkokotoni, Oct. 1933, *Vaughan* 2170!; Pemba, Kohani I., 3 July 1957, *Potter* 3!
DISTR. **K** 7; **T** 3, 6; **Z**; **P**; Socotra
HAB. In coastal woodland, evergreen forest and scrub and on exposed coral rocks; sea-level–50 m.

NOTE. A decoction of the leaves is used to cure sores.

7. **A. sacciferum** *Lindl.* in Hook., Comp. Bot. Mag. 2: 205 (1837); Piers, Orch. E. Afr., ed. 2: 243 (1968); J. Stewart in U.K.W.F.: 783 (1974); Williamson, Orch. SC. Afr.: 178 (1977). Type: South Africa, Outniqualand, *Drège* (K, holo.!)

A dwarf epiphyte or rarely a lithophyte. Stems 0.5–2(–3) cm. long. Leaves few, distichous, coriaceous, linear to narrowly elliptic, acute at the unequally bilobed apex, 1.5–6 cm. long, 0.3–0.5 cm. wide, twisted at base to lie in one plane and articulated to overlapping leaf-bases. Inflorescences 1–several, 2–8 cm. long, 2–6-flowered; peduncle 2–2.5 cm. long, finely wiry; bracts elliptic, acute, 2–2.5 mm. long. Flowers green or yellow-green; pedicel and ovary strongly twisted, 2–3 mm. long. Sepals elliptic-oblong, obtuse, 2.7–3.7 mm. long, 1.7–2 mm. wide. Petals lanceolate, acute, 2.4–3.2 mm. long, 0.8–1.1 mm. wide. Lip deeply concave, broadly ovate, acute, 2.7–3.3 mm. long, 3.5 mm. wide; spur curved above the ovary, very shortly cylindrical, truncate, 1.5–2 mm. long. Column 0.8 mm. long. Fruit 7 mm. long.

UGANDA. Kigezi District: Rutenga, Apr. 1948, *Purseglove* 2620!
KENYA. Naivasha District: Kinangop, Apr. 1934, *Turner* in *C.M.* 6338!; N. Nyeri District: Nanyuki, 2 July 1943, *Moreau* 186A!; N. Kavirondo District: S. Elgon, Sosian R., June 1939, *Tweedie* 463!
TANZANIA. Lushoto District: W. Usambara Mts., Magamba Forest Reserve, 10 Jan. 1976, *Cribb & Grey-Wilson* 10140!; Morogoro District: Ukaguru Mts., road W. from Lake to Marangu, 1 Feb. 1976, *Cribb & Grey-Wilson* 10511!; Iringa District: Mufindi, Luisenga R., 16 Mar. 1962, *Polhill & Paulo* 1781!
DISTR. **U** 2; **K** 3–6; **T** 3, 6, 7; Cameroon, Zaire, Mozambique, Malawi, Zimbabwe and South Africa
HAB. In upland rain-forest and rarely on rocks by streams; 900–2200 m.

SYN. *A. parcum* Schltr. in E.J. 53: 604 (1915). Type: Tanzania, Rungwe District, Kyimbila, *Stolz* 640 (B, holo., K, iso.!)

8. **A. teres** *Summerh.* in K.B. 13: 265 (1958); Piers, Orch. E. Afr., ed. 2: 246 (1968). Type: Tanzania, Lushoto District, Amani, *Moreau* 49 (K, holo.!)

A small epiphyte with a very short stem, 0.5–1.5 cm. long, 0.3–0.5 mm. in diameter. Leaves straight or arcuate, 2–5 in a fan, terete, stiff, linear, apiculate, 4–15 mm. long, 0.1–0.2 mm. in diameter, dull green, articulated to overlapping leaf-bases. Inflorescences 4–17 cm. long, 2–9-flowered; peduncle wiry, slender, 2.5–8 cm. long; rhachis fractiflex; bracts ovate, acute, 1–1.5 mm. long. Flowers green with ochre tips to the segments and spur, nocturnally sweetly scented; pedicel and ovary 9–10 mm. long, ovary twisted through 180°. Dorsal sepal lanceolate, acuminate, 7.5–8 mm. long, 3.5–4 mm. wide; lateral sepals lanceolate, acuminate, 7–7.5 mm. long, 2.5–3 mm. wide. Petals lanceolate, acuminate, 8–8.5 mm. long, 1.5–2 mm. wide. Lip obscurely 3-lobed, ovate, obtuse, 5.5 mm. long, 3.5 mm. wide; spur ascending above the ovary, 2 cm. long. Column 0.8 mm. long.

TANZANIA. Lushoto District: E. Usambara Mts., Amani, 19 Mar. 1941, *Moreau* 49!
DISTR. **T** 3; known only from the type collection
HAB. Epiphytic on bole of *Cedrela odorata* and on small trees by the sea (*fide* Moreau); sea-level–600 m.

9. **A. chevalieri** *Summerh.* in K.B. 1936: 230 (1936) & in F.W.T.A., ed. 2, 3: 452 (1968). Type: Ivory Coast, Haut Sassandra, between Sogui and Koualé, *Chevalier* 21690 (P, holo.!, K, record of holo.!)

An epiphyte with 7–13 cm. long stems, 2 mm. in diameter; roots all along the stem, 2 mm. in diameter. Leaves distichous, in upper half of the stem, coriaceous, ligulate, unequally roundly bilobed at the apex, 28–52 cm. long, 4.5–5.5 mm. wide, twisted at the base to lie in one plane. Inflorescences 1–3 per node, up to 2.5 cm. long, 1–2-flowered; peduncle 6–7 mm. long; bracts narrowly elliptic, acute, 2–3 mm. long. Flowers parchment green; pedicel and ovary twisted just below the flower. Sepals lanceolate, acuminate,

8.2–12 mm. long, 1.6–2 mm. wide. Petals lanceolate, acuminate, 9–10.5 mm. long, 1–1.2 mm. wide. Lip concave, ovate, acuminate, recurved at apex, 7–10.5 mm. long, 2.1–3 mm. wide; spur filiform, rising above the ovary, ± straight to slightly sigmoid, slightly dilated at the apex, 16–20 mm. long. Column 1.5–2 mm. long; stipes short, bifid.

UGANDA. Bunyoro District: Budongo Forest, May 1940, *Eggeling* 3965!, Siba Block, Nov. 1943 (fruit) & May 1944 (flower), *Eggeling* 5464!; Kigezi District: Impenetrable Forest, Nyebeya, Oct. 1940, *Eggeling* 4146!
TANZANIA. Without exact locality, *Cuthbert!*
DISTR. **U** 2; **T** ?; also in Liberia, Ivory Coast, Guinée, Nigeria, Cameroon, Bioko, Gabon, Equatorial Guinea and Zaire
HAB. In rain-forest; 1100–1200 m.

10. **A. cultriforme** *Summerh.* in K.B. 13: 272 (1958); Williamson, Orch. SC. Afr.: 180 (1977); Ball, S. Afr. Epiphytic Orch.: 56 (1978). Type: Tanzania, Handeni District, Gendagenda, cult. H. Faulkner in *Milne-Redhead & Taylor* 7318 (K, holo.!)

A pendent epiphyte with an ancipitous stem, 4–15 cm. long, 3 mm. in diameter; roots all along stem. Leaves in 2 ranks, falcate, ligulate, obtuse at the unequally bilobed apex, 2.2–6 cm. long, 8–9 mm. wide, yellow- to bronze-green, twisted at the base to lie in one plane. Inflorescences 1 per node, 1-flowered, 0.5–3.5 cm. long; peduncle 5–25 mm. long; bracts lanceolate, acute, 2–3 mm. long. Flowers stellate, pale ochre yellow to dull brownish green; pedicel and ovary 6.5–10 mm. long. Sepals linear-lanceolate, acuminate, 10.2–19.3 mm. long, 2.2–3.8 mm. wide. Petals linear-lanceolate, acuminate, 10.3–14.5 mm. long, 2–2.8 mm. wide. Lip recurved, ovate, acuminate, 8.5–12.5 mm. long, 4.5–6 mm. wide; spur ± straight to slightly sigmoid, filiform, slightly dilated at the apex, 10.5–27 mm. long. Column 1.3–2.2 mm. long.

KENYA. Kwale District: Umba R., 8 km. N. of Tanzanian border, fl. cult. 7 July 1944, *Boscawen* in *Moreau* 724!
TANZANIA. Tanga District: Sigi Segoma, fl. cult. 6 June 1944, *Moreau* 842!; Handeni District: Kwa Mkono, 26 Mar. 1977, *Archbold* 2282!; Iringa District: Mufindi E., 31 Oct. 1947, *Greenway & Brenan* 8288!
DISTR. **K** 7; **T** 3, 7; Zambia, Mozambique
HAB. In *Brachystegia* woodland, dry evergreen forest and in evergreen bushland; sea-level–1800 m.

11. **A. stolzii** *Schltr.* in E.J. 53: 603 (1915); Summerh. in K.B. 13: 273 (1958); Williamson, Orch. SC. Afr.: 180 (1977). Type: Tanzania, Mt. Rungwe, *Stolz* 2508 (B, holo., K, iso.!)

An erect or pendent epiphyte 2.5–40 cm. long; roots 0.5-1.2 mm. in diameter. Leaves 2-ranked, all along the stem, linear, falcate, unequally acutely bilobed at the apex, 4.5–6.2 cm. long, 4–9 mm. wide, twisted at the base to lie in one plane, grass-green. Inflorescences 1–4 per node, 2–3 cm. long, 1-flowered; peduncle 9–17 mm. long; bracts ovate, acute or apiculate, 1.5–2 mm. long. Flowers cream or yellow-green turning orange with age; pedicel and ovary twisted, 4 mm. long. Sepals lanceolate, acuminate, 7.2–9 mm. long, 1.6–3 mm. wide. Petals lanceolate, acuminate, 7–9 mm. long, 1.2–1.6 mm. wide. Lip ovate, acuminate and recurved at the tip, 5–7.5 mm. long, 2.7–5 mm. wide; spur slightly sigmoid, filiform, 5.5–9 mm. long. Column 1.5–1.7 mm. long. Fruits 1.6–1.8 mm. long. Fig. 115.

TANZANIA. Lushoto District: E. Usambara Mts., Amani, 24 May 1943, *Moreau* 23!; Morogoro District: Kimboza [Kimbosa] Forest Reserve, 24 Jan. 1976, *Cribb & Grey-Wilson* 10405A!; Rungwe District: Poroto Mts., Isongole, Idweli, 6 Feb. 1979, *Cribb, Grey-Wilson & Mwasumbi* 11318!; Njombe District: Hagafilo R., 11 km. S. of Njombe, 10 July 1956, *Milne-Redhead & Taylor* 11033!
DISTR. **T** 3, 6, 7; Zaire, Malawi
HAB. In montane and riverine forest; 1700–2500 m.

NOTE. The **T** 3 and 6 collections cited are somewhat intermediate between *A. stolzii* and *A. cultriforme*. The relationship between these species needs further investigation.

12. **A. brevicornu** *Summerh.* in K.B. 16: 311 (1962). Type: Tanzania, Handeni District, Kideleko, *Nash* (K, holo.!)

A small epiphyte with a short bilaterally compressed stem, 5–7 cm. long, 1–2 mm. in diameter; roots 0.5-1 mm. in diameter. Leaves slightly leathery, alternate, falcate, ligulate, unequally acutely bilobed at the apex, 2.5–5 cm. long, 0.6–1.1 cm. wide, somewhat shiny, twisted at the base. Inflorescences 1 per node, 1-flowered, ± 1 cm. long; peduncle 4–8 mm. long; bracts lanceolate, acute, 1–2 mm. long. Flowers white with a faint pinkish or

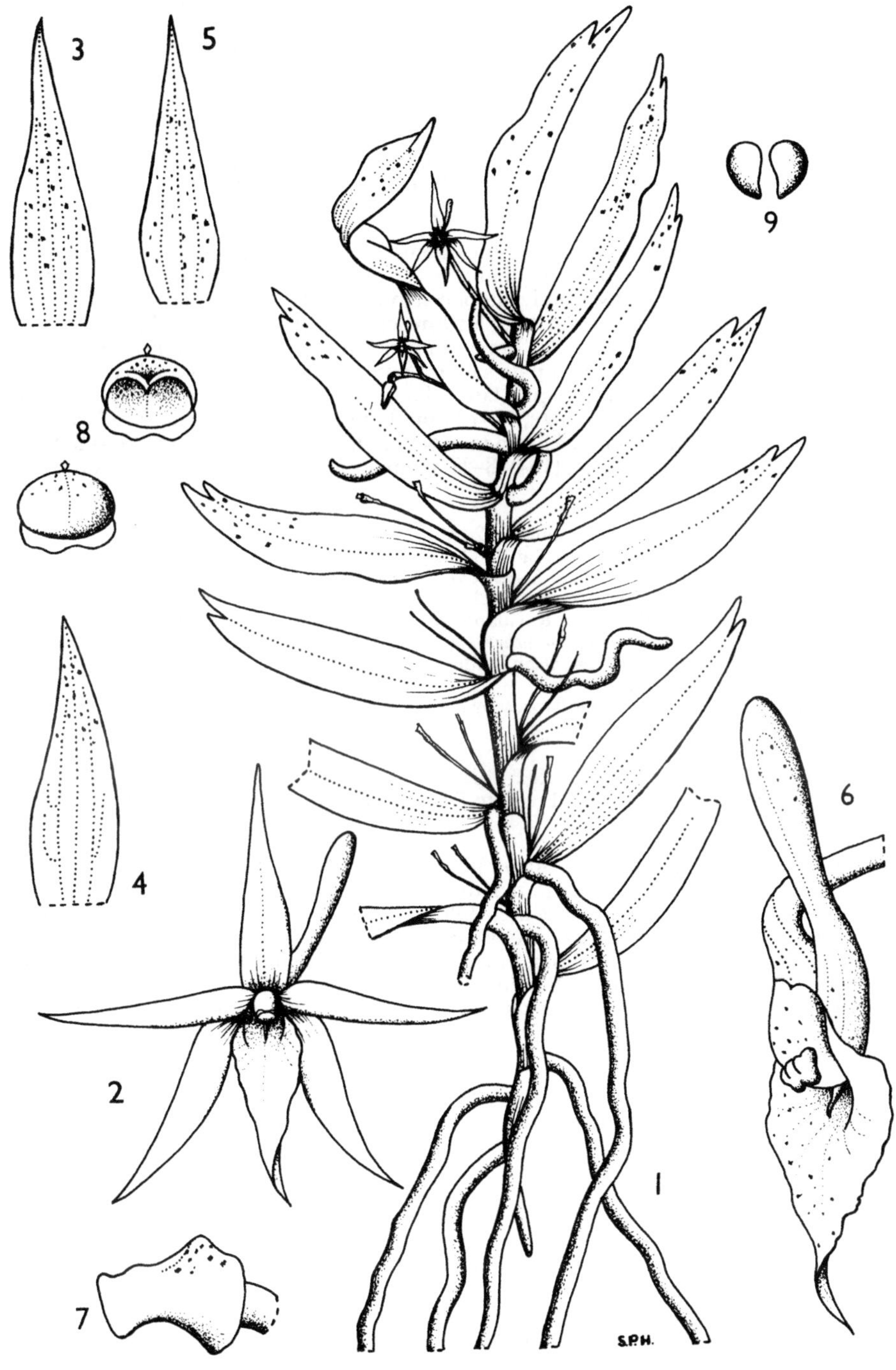

FIG. 115. *ANGRAECUM STOLZII*—**1**, habit, × 1; **2**, flower, × 4; **3**, dorsal sepal, × 6; **4**, lateral sepal, × 6; **5**, petal, × 6; **6**, lip with column, × 6; **7**, column, side view, × 12; **8**, anther-cap, × 12; **9**, pollinia, × 14. 1 from *Eggeling* 6102 and *Cribb & Grey-Wilson* 11318; 2–9, from *Cribb & Grey-Wilson* 11318. Drawn by Susan Hillier.

greenish pink tinge; pedicel and ovary curved, 7 mm. long. Sepals lanceolate, acuminate, 8–8.5 mm. long, 3–3.6 mm. wide. Petals elliptic-lanceolate, acute or acuminate, 6.5–7 mm. long, 2 mm. wide. Lip concave, ovate, acute, 3–4.5 mm. long, 3.5 mm. wide; spur conical-cylindrical, ± parallel to the ovary, 2.5–2.6 mm. long. Column 0.8–0.9 mm. long. Fig. 116.

TANZANIA. Handeni District: Kwa Mkono, fl. cult. 26 Mar. 1977, *Archbold* 2283! & Kideleko, U.C.M.A. Mission Station, 15 Mar. 1966, *Nash*!
DISTR. **T** 3; not known elsewhere
HAB. In lowland forest and on lemon trees; 500–600 m.

13. **A. erectum** *Summerh.* in K.B. 11: 232 (1956) & 13: 273 (1958); Piers, Orch. E. Afr., ed. 2: 231 (1968); J. Stewart in U.K.W.F.: 785 (1974); Williamson, Orch. SC. Afr.: 179 (1977). Type: Kenya, Nairobi, *Bally* in *C.M.* 7532 (K, holo.!, EA, iso.)

A scandent subepiphyte with ancipitous stems 20–40 cm. long and 3 mm. in diameter; roots all along stem, 1.5 mm. in diameter. Leaves coriaceous, distichous, 2-ranked, linear-lanceolate, unequally acutely bilobed at the apex, 2.8–6 cm. long, 0.5–1.3 cm. wide, articulated to leaf-sheaths 2–3 cm. long. Inflorescences 1 per node, 1–2-flowered, 2.7–4.5 cm. long; peduncle 0.8–2 cm. long; bracts amplexicaul, ovate, 2 mm. long. Flowers light yellow-green or whitish green to pale salmon-pink, scented at night; pedicel and ovary curved strongly, 4–7 mm. long. Sepals lanceolate, acuminate, 11–14 mm. long, 2.5–4 mm. wide. Petals lanceolate, acuminate, 10–12 mm. long, 1.8–2.8 mm. wide. Lip lanceolate, acuminate, 8.5–10.5 mm. long, 3.3–4.6 mm. wide, with a linear raised callus in the basal part; spur subsigmoid, filiform, (15–)18–28.5 mm. long. Column 1.5–1.7 mm. long. Pl. 5. Fig. 117, p. 502.

UGANDA. Karamoja District: Kokumongole, 28 May 1939, *A.S. Thomas* 2894!
KENYA. Laikipia District: Rumuruti, Mar. 1943, *Copley* 20A!; Kiambu District: 18 km. E. of Nairobi, Ruiru, Ruiru Falls, Nov. 1938, *C. van Someren* 51!; Masai District: Siyabei R. Gorge, Owor Oringenai Gulley, 14 July 1962, *Glover & Samuel* 3032!
TANZANIA. Musoma District: Mara R. Guard Post, 3 Oct. 1961, *Greenway* 10224!; Masai District: Olkeju Rongai R., 11 Nov. 1966, *Carmichael* 1301!; Mbulu District: headwaters of the Endabash R., fl. cult. 30 Nov. 1943, *Fuggles-Couchman* in *Moreau* 308A!
DISTR. **U** 1; **K** 3, 4, 6; **T** 1, 2; Zambia
HAB. In riverine forest and upland forest by river-beds and in gulleys, on vertical trunks in shade; 1300–2150 m.

14. **A. viride** *Kraenzl.* in E.J. 51: 395 (1914); Summerh. in K.B. 13: 273 (1958), *non* (Ridl.) Schltr. (1918). Type: Tanzania, Lushoto District, Amani, *Braun* in *Herb. Amani* 3209 (B, holo., K, iso.!)

An epiphyte with an elongate, bilaterally compressed stem up to 45 cm. long, 1.5–2 mm. in diameter; roots many in lower part of the stem, less than 1 mm. in diameter. Leaves in 2 ranks, linear-elliptic, unequally acutely bilobed at the apex, 2.7–5 cm. long, 0.4–0.6 cm. wide, twisted at the base to lie in one plane. Inflorescences 6–7 mm. long, 1–2-flowered; peduncle very short, 1–2 mm. long; bracts lanceolate, acuminate, 2–4 mm. long. Flowers greenish; pedicel and ovary 2–2.5 mm. long. Sepals lanceolate, acuminate, 4.5–5 mm. long, 0.8–1.6 mm. wide. Petals lanceolate, acuminate, 4 mm. long, 0.9–1 mm. wide. Lip concave, ovate, acuminate, 3.6–4 mm. long, 2.7–2.9 mm. wide; spur slightly sigmoid, clavate, 1.8–2 cm. long. Column 0.7–0.8 mm. long.

TANZANIA. Lushoto District: E. Usambara Mts., Amani, 7 Aug. 1944, *Moreau* 53!, 9 Apr. 1950, *Verdcourt* 144!, Kwamkoro, near Amani, 11 Sept. 1944, *Doughty* in *Moreau* 1008! & 15 Sept. 1944, *Doughty* in *Moreau* 880!
DISTR. **T** 3; not known elsewhere
HAB. In rain-forest; 850–1000 m.

SYN. *A. braunii* Schltr. in Beih. Bot. Centr. 36: 167 (1918), *nom. illegit.* Type: as for species.

15. **A. conchiferum** *Lindl.* in Hook., Comp. Bot. Mag. 2: 205 (1837); Bolus, Ic. Orch. Austr-Afr. 2, t. 4 (1911); Rolfe in Fl. Cap. 5(3): 70 (1912); Schltr. in Beih. Bot. Centr. 36: 170 (1918); Summerh. in K.B. 13: 275 (1958); Piers, Orch. E. Afr., ed. 2: 229 (1968); J. Stewart in U.K.W.F.: 785 (1974); Ball, S. Afr. Epiphytic Orch.: 58 (1978). Type: South Africa, Outniqualand, *Drège* (K, holo.!)

A straggling epiphyte with a branching, slightly compressed stem, 4–60 cm. long, ± 3 mm. in diameter; roots verrucose, 1–1.5 mm. in diameter. Leaves distichous, coriaceous,

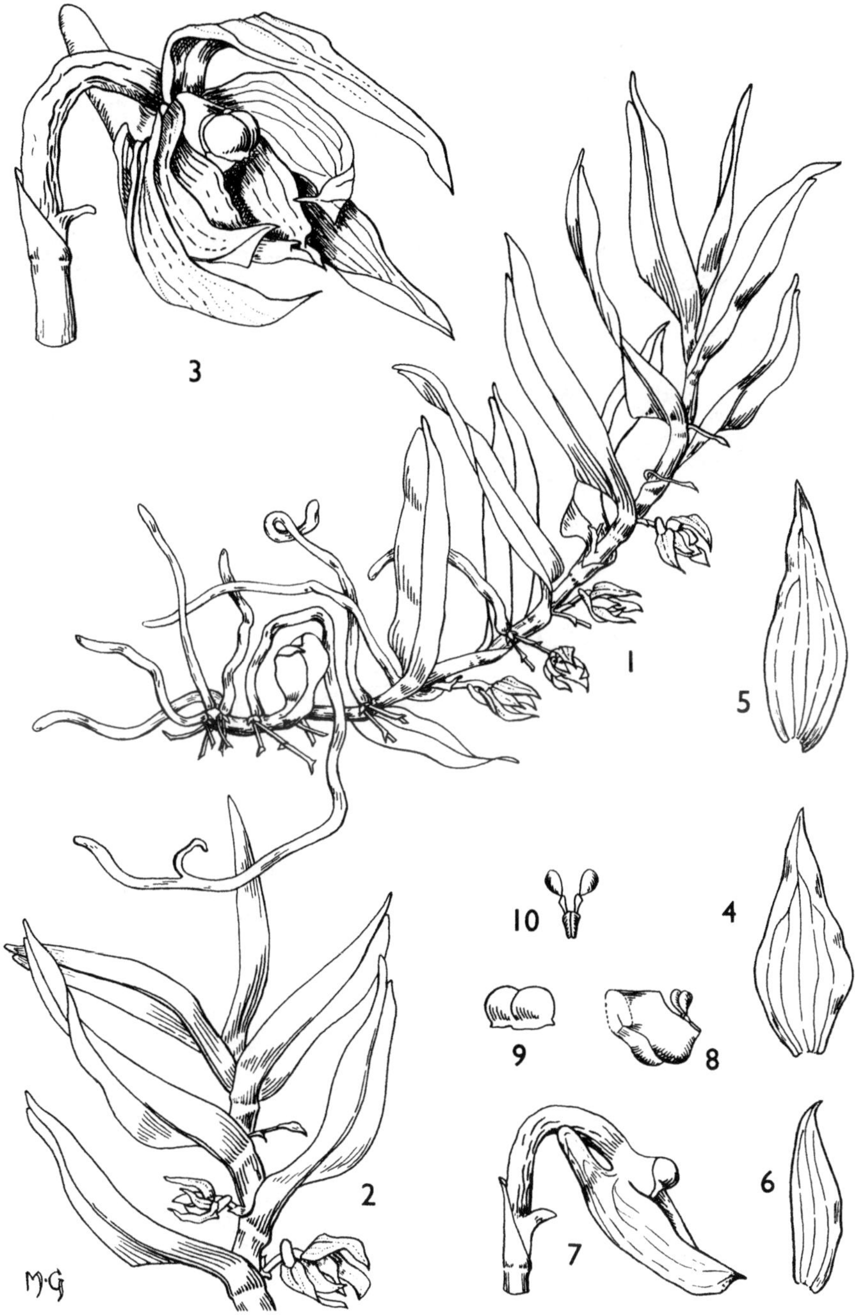

FIG. 116. *ANGRAECUM BREVICORNU*—**1**, flowering stem, × 1; **2**, apex of flowering stem, × 1½; **3**, flower, × 6; **4**, dorsal sepal, × 4; **5**, lateral sepal, × 4; **6**, petal, × 4; **7**, flower, tepals removed to show lip and column, × 4; **8**, apex of column, showing rostellum and pollinaria, × 7; **9**, anther, front view, × 7; **10**, pollinaria, × 7. Drawn from *Nash* s.n. by Miss M.A. Grierson.

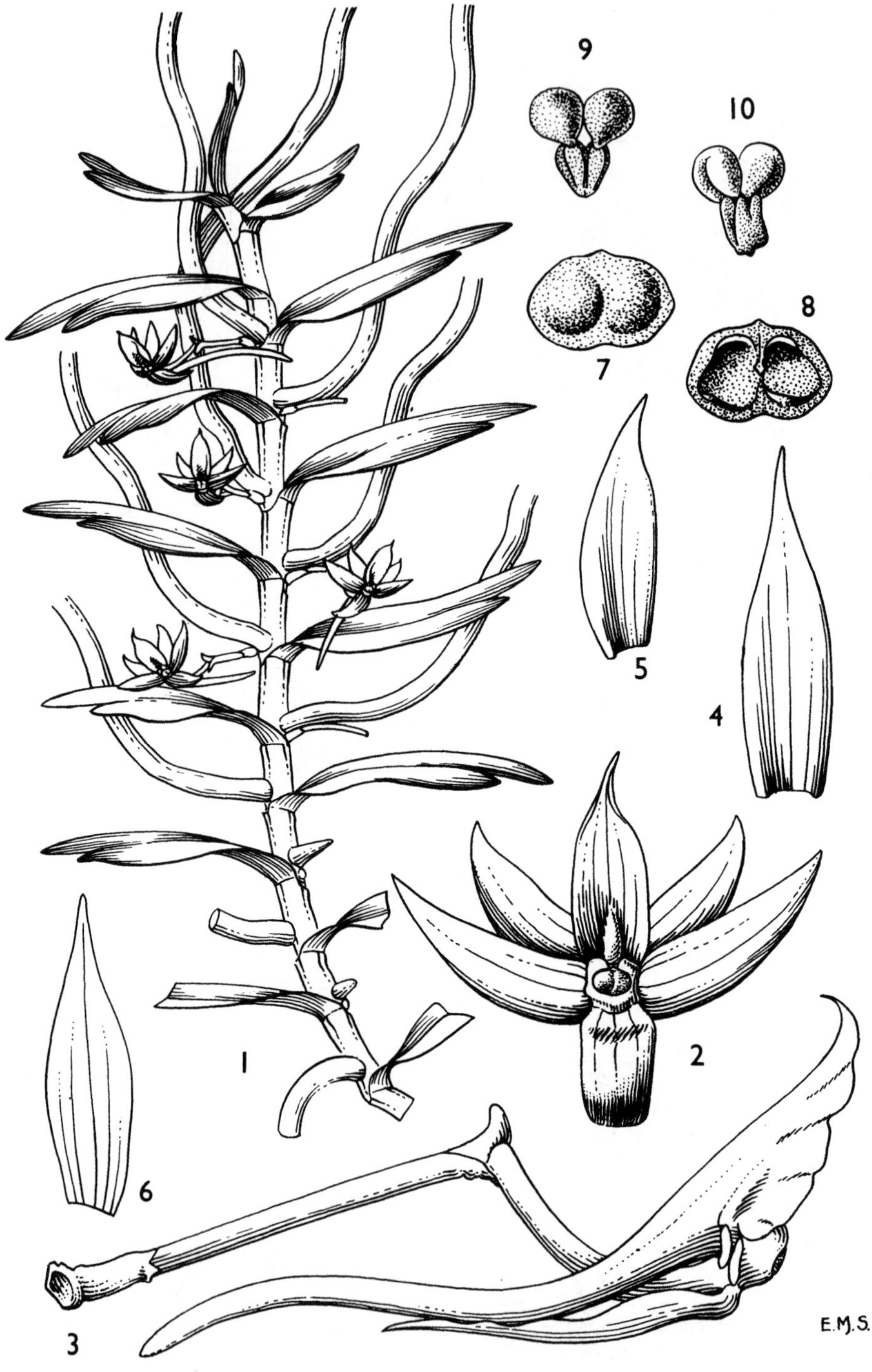

FIG. 117. *ANGRAECUM ERECTUM*—**1**, habit, × 1; **2**, flower, from front, × 4; **3**, flower from side, lateral sepal and petal removed, × 4; **4**, dorsal sepal, × 4; **5**, lateral sepal, × 4; **6**, petal, × 4; **7**, anther from above, × 12; **8**, anther, from beneath, × 12; **9**, pollinarium from above, × 12; **10**, pollinarium, from beneath, × 12. 1, from *Moreau* 579; 2–10, from Cult. Oxford Bot. Gdn. 169-554. Drawn by Miss Margaret Stones. Reproduced from Hooker's Icones Plantarum.

linear, unequally obtusely bilobed, 3.8–6 cm. long, 0.4–0.7 cm. wide, twisted at the base and articulated to verrucose leaf-sheaths, 5–8 mm. long. Inflorescences (1–)2-flowered; peduncle 2.5–2.7 cm. long; bracts triangular-ovate, acute, 1.5–5 mm. long. Flowers stellate, white; pedicel and ovary 18–20 mm. long. Sepals linear-lanceolate, acuminate, 17–31 mm. long, 2–4 mm. wide. Petals linear, long-acuminate, 21–31 mm. long, 1–3 mm. wide. Lip concave, ovate-subcircular, 9–11 mm. long, 10–11 mm. wide, with a 3.5–9 mm. long apicule; spur decurved at the base, upcurved at the apex, slenderly cylindrical, 3.5–5 cm. long. Column 2–3 mm. long. Fruit 2–3 cm. long. Pl. 5.

KENYA. Kiambu District: Kikuyu Forest, Karamenu R., 20 Sept. 1959, *Archer*!; Masai District: Namanga, Oldonyo Orok, 19 Oct. 1970, *Archer* 647!
TANZANIA. Pare District: S. Pare Mts., fl. cult. 16 Oct. 1942, *Greenway* in *Moreau* 176!; Ulanga District: Uzungwa, Idegenda, Ng'insula, 3 Nov. 1980, *Leedal* 6154!
DISTR. **K** 4, 6; **T** 3, 6, 7; Malawi, Mozambique, Zimbabwe and South Africa
HAB. In montane forest; 1250–2400 m.

SYN. *Angorkis conchifera* (Lindl.) O. Kuntze, Rev. Gen. Pl. 2: 651 (1891)
Angraecum verrucosum Rendle in J.B. 33: 250 (1895). Type: Malawi, Mt. Mulanje, *A. Whyte* (BM, holo.)
Mystacidium verrucosum (Rendle) Rolfe in F.T.A. 7: 171 (1897)
Angraecum scabripes Kraenzl. in E.J. 33: 73 (1902). Type: Tanzania, W. Usambara Mts., Kwai, *Eick* 416 (B, holo.)

16. **A. infundibulare** *Lindl.* in J.L.S. 6: 136 (1862); Schltr. in Beih. Bot. Centr. 36: 172 (1918); Dandy in Exell, Vasc. Pl. São Tomé: 331 (1944); Piers, Orch. E. Afr., ed. 2: 239 (1968); J. Stewart in U.K.W.F.: 785 (1974). Type: Principe, *Barter* 2005 (K, holo.!)

A pendent epiphyte with bilaterally compressed, rarely branching stems, up to 60 cm. long, 4–7 mm. in diameter. Leaves distichous, narrowly elliptic- ligulate or oblanceolate, unequally obliquely roundly or obtusely bilobed at the apex, 11–21.5 cm. long, 1.5–3 cm. wide, twisted at the base and articulated to leaf-sheaths 2.2–5 cm. long. Inflorescences 1-flowered; peduncle 3–5.5 cm. long; bracts elliptic-ovate, obtuse, 6 mm. long. Flowers white, tinged with green on the spur, suffused with yellow inside lip, fragrant; pedicel and ovary 5.5–6.5 cm. long. Sepals linear-lanceolate, acuminate, 6–8.5 cm. long, 0.3–1 cm. wide. Petals linear, acuminate, slightly shorter than the sepals, 5.5–6 cm. long, 4–5 mm. wide. Lip concave, ovate-elliptic to oblong-ovate, 6.5–8.5 cm. long, 5–5.6 cm. wide, with a long apicule up to 12 mm. long; spur funnel-shaped at the base, geniculate then cylindrical above, sigmoid, incurved or pendent, 16–21 cm. long, 1.5 cm. wide across base. Column fleshy, 5 mm. long. Fruit 5.5 cm. long. Pl. 5. Fig. 118, p. 503.

UGANDA. Bunyoro District: Budongo, June 1935, *Eggeling* 2040!; Masaka District: Sese Is. Bubekke, 4 July 1935, *A.S. Thomas* 1382!; Mengo District: Nagunga, Nov. 1915, *Dummer* 2663!
KENYA. Kavirondo District: Mumias area, *Evans* 31 (*fide* Stewart, 1974)
DISTR. **U** 2, 4; **K** 5; Nigeria, Cameroon, Principe, Ethiopia
HAB. In rain-forest; 1150–1300 m.

SYN. *Angorkis infundibularis* (Lindl.) O. Kuntze, Rev. Gen. Pl. 2: 651 (1891)
Mystacidium infundibulare (Lindl.) Rolfe in F.T.A. 7: 170 (1897)

17. **A. spectabile** *Summerh.* in K.B. 13: 274 (1958); Piers, Orch. E. Afr., ed. 2: 244 (1968). Type: Tanzania, Bukoba District, Lake Victoria, *C.G. Rogers* 456 (K, holo.!)

An erect terrestrial or epiphyte with slightly flattened stems, 24–30 cm. long, 5–6 mm. in diameter; roots 2.5 mm. in diameter. Leaves distichous, elliptic, unequally roundly bilobed, 7–10 cm. long, 2.1–3 cm. wide, twisted at the base and articulated to 2–2.2 cm. long leaf-sheaths. Inflorescences 1-flowered, up to 10 cm. long; peduncle 2.8–3.2 cm. long; bracts ovate, obtuse, 5–6 mm. long. Flowers green shaded with white; pedicel and ovary 3.5 cm. long. Sepals narrowly lanceolate, acuminate, 4.3–5 cm. long, 0.9–1.1 cm. wide. Petals obliquely lanceolate, acuminate, 4–4.3 cm. long, 0.6 cm. wide. Lip concave, broadly obovate or obcordate, obscurely 3-lobed at the apex, 3.2–3.4 cm. long, 4.2–4.4 cm. wide, with an apicule 2.5 mm. long; side lobes rounded; spur funnel-shaped at the base, geniculate in the middle, cylindrical in the apical part, 6.2–6.5 cm. long, 2 cm. wide across the mouth. Column 5 mm. long.

TANZANIA. Bukoba District: by Lake Victoria, July 1933, *C.G. Rogers* 456! & fl. cult. 6 June 1945, *A.S Thomas* 4122!
DISTR. **T** 1; not known elsewhere
HAB. On rocks by the lake; 1150–1200 m.

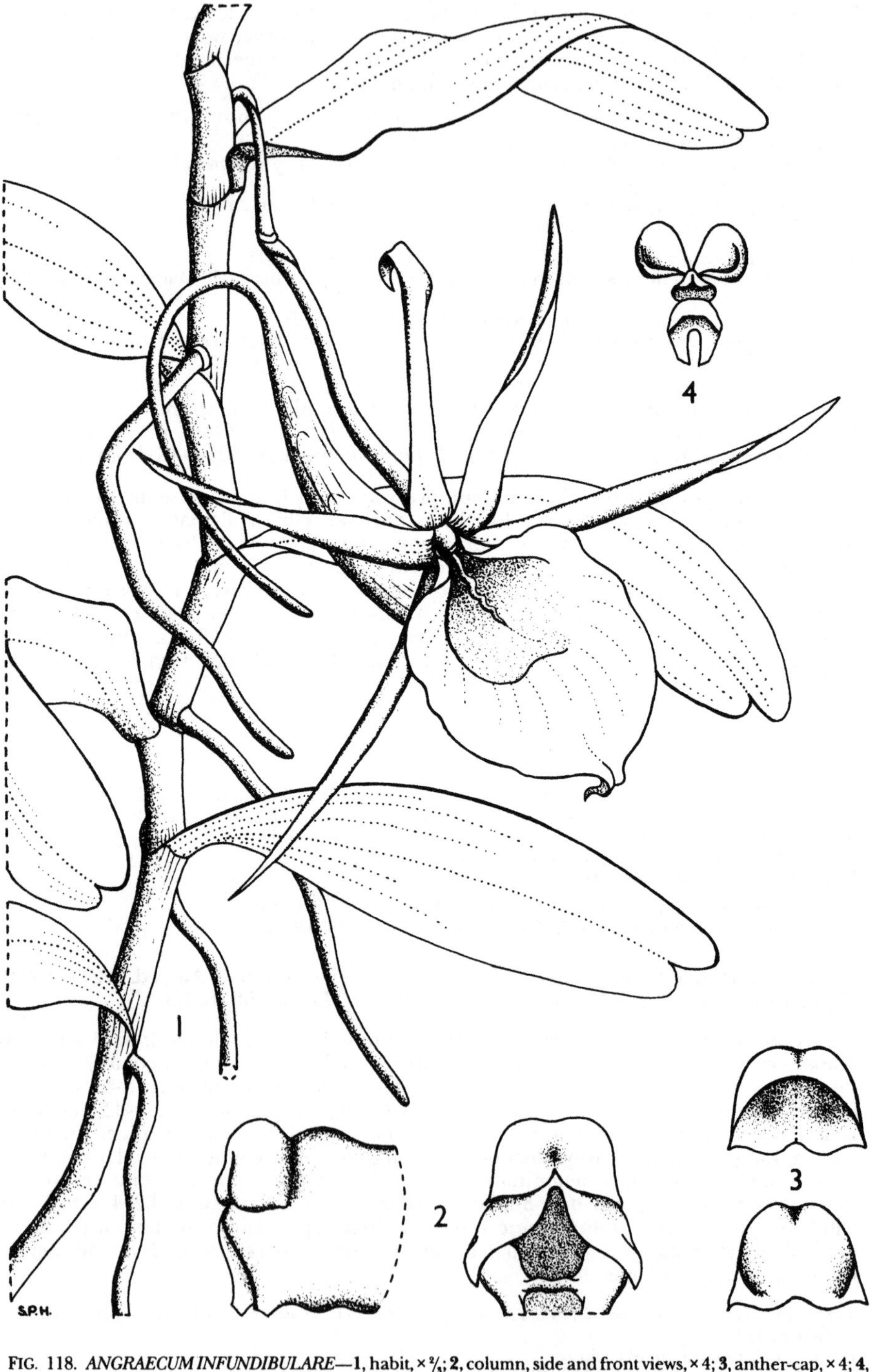

FIG. 118. *ANGRAECUM INFUNDIBULARE*—**1**, habit, × ⅔; **2**, column, side and front views, × 4; **3**, anther-cap, × 4; **4**, pollinium, × 4. 1, from *Thomas* 1382 and *Meyer* K72; 2–4, from *Meyer* K72. Drawn by Susan Hillier.

18. **A. affine** *Schltr.* in E.J. 38: 19 (1905). Type: Cameroon, Ngoko, *Schlechter* 12744 (B, holo.)

A pendent or scandent epiphyte with sigmoid stems 15–25 cm. long, 3 mm. in diameter. Leaves ligulate-oblanceolate, unequally obliquely roundly bilobed at the apex, 10–15.7 cm. long, 1.3–2.1 cm. wide, twisted at the base and articulated to leaf-sheaths 1.5–2 cm. long. Inflorescences 1-several, 3–7 cm. long, 3–4-flowered; peduncle 2.2–2.8 cm. long; bracts ovate, acuminate, 4–5 mm. long. Flowers white; pedicel and ovary 12–13 mm. long, emerging 4–6 mm. above the axil of the bract. Sepals lanceolate, acute or acuminate, 9–14 mm. long, 2–4.7 mm. wide. Petals obliquely lanceolate, acute, 9–11.2 mm. long, 2–3.9 mm. wide. Lip concave, lanceolate, acuminate, 8–10.5 mm. long, 4–5.7 mm. wide, with a longitudinal linear callus in the basal part of the lip; spur cylindrical, slightly dilated at the base and apex, 18–24 mm. long. Column 1.5–1.7 mm. long.

UGANDA. Kigezi District: Impenetrable Forest, Apr. 1948, *Purseglove* 2668!
DISTR. **U** 2; Cameroon, Bioko [Fernando Po] and Congo Brazzaville
HAB. In rain-forest; 1500–1600 m.

SYN. *A. ligulatum* Summerh. in K.B. 13: 276 (1958); Piers, Orch. E. Afr., ed. 2: 243 (1968). Type: Uganda, Kigezi District, Impenetrable Forest, *Purseglove* 2668 (K, holo.!)

19. **A. firthii** *Summerh.* in K.B. 13: 267 (1958); Piers, Orch. E. Afr., ed. 2: 233 (1968); J. Stewart in U.K.W.F.: 783 (1974). Type: Kenya, Kisumu-Londiani District, near Fort Ternan, *Firth* (K, holo.!)

An epiphyte with an elongate, bilaterally compressed stem 7–15 or more cm. long. Leaves distichous, oblanceolate or oblong, unequally roundly or obtusely bilobed at the apex, 9–11 cm. long, 1.6–2.3 cm. wide, twisted at the base. Inflorescences 7.5–11.5 cm. long, 3–5-flowered; peduncle 3–3.5 cm. long; rhachis slightly fractiflex; bracts broadly ovate, obtuse, 1.5–2 mm. long. Flowers yellow-green; pedicel and ovary 13–16 mm. long, articulated 1 mm. above the base of the pedicel. Dorsal sepal lanceolate, long-acuminate, 13.5–19 mm. long, 3.7–3.8 mm. wide; lateral sepals similar, 13.9–17.7 mm. long, 2.9–3.2 mm. wide. Petals lanceolate, acuminate, 11.5–13 mm. long, 2–2.5 mm. wide. Lip 3-lobed, 12–13 mm. long, 4.5–5.6 mm. wide; side lobes erect, rounded; mid-lobe recurved, linear, acuminate, 6.5 mm. long; spur sigmoid, 14.5–16 mm. long. Column 1.3–2 mm. long.

UGANDA. Bunyoro District: Budongo, June 1937, *Eggeling* 3358!
KENYA. N. Kavirondo District; Kakamega, 8 June 1938, *Firth*!; Kisumu-Londiani District: near Fort Ternan, June 1952, *Firth*!
DISTR. **U** 2; **K** 5; Cameroon
HAB. In rain-forest; 1450–1600 m.

20. **A. reygaertii** *De Wild.* in B.J.B.B. 5: 190 (1916). Type: Zaire, Mobwasa, *Reygaert* 1104 (BR, holo.)

A pendent epiphyte with ancipitous stems up to 15 cm. long or more. Leaves distichous, ligulate to oblanceolate, unequally obtusely bilobed at the apex, 9.5–16 cm. long, 2–3 cm. wide, twisted at base and articulated to 1.5–3.2 cm. long sheaths. Inflorescences 3–6-flowered, up to 18 cm. long; peduncle 4–11 cm. long; bracts amplexicaul, obconical, obtuse, 4–5 mm. long. Flowers white; pedicel and ovary 1.2–1.4 cm. long. Sepals linear-lanceolate to elliptic-lanceolate, acuminate, 2.6–2.8 cm. long, 3–6 mm. wide. Petals linear-lanceolate, acuminate, 2.2–2.7 cm. long, 2.5–5 mm. wide. Lip concave, elliptic-lanceolate, acute, 1.5–1.7 cm. long, 6 mm. wide, with a longitudinal ridge-like callus in basal half; spur sigmoid, slenderly cylindrical, 6.5–7.8 cm. long. Column 3–4 mm. long. Pl. 5.

UGANDA. Mengo District: R. Nile, near Kalagala Falls, fl. cult. Mar. 1964, *C. Leakey* 39! & Bugerere County, fl. cult. 12 Apr. 1975, *Meyer*!
DISTR. **U** 4; Cameroon and Zaire
HAB. In riverine forest; 1200–1250 m.

21. **A. eburneum** *Bory*, Voy. 1: 359, t. 19 (1804); Senghas & Cribb in Die Orchidee 30(1): CXXI–CXXII (1979). Type: Réunion, *Bory de St.-Vincent* (P, holo.)

A large robust epiphyte or lithophyte with an erect stem 60–100 cm. long, 1.5–2 cm. in diameter; roots thick, fleshy, like boot-laces, 3.5–5 mm. in diameter. Leaves 6–10 or more, distichous, fleshy-coriaceous, obovate-ligulate, unequally roundly bilobed at the apex,

16–50(–70) cm. long, 2.9–7 cm. wide. Inflorescences suberect or ascending, 8–20-flowered, 30–75 cm. long; peduncle stout, 12–29 cm. long, bearing 1–3 well-spaced sheaths along length; bracts amplexicaul, broadly ovate, obtuse, 0.9–1.5 cm. long, dark brown. Flowers with a white lip, yellow-green at the mouth of the spur and with green sepals, petals and column, pleasantly scented at night; pedicel and ovary 16–18 mm. long, with the ovary twisted through 180°. Sepals reflexed, linear-lanceolate, acuminate, 3.3–4.5 cm. long, 5–7(–10) mm. wide. Petals linear-lanceolate, acuminate, 3.2–4.2(–4.5) cm. long, 5–7 mm. wide. Lip erect, transversely oblong to broadly ovate, (2–)2.4–3(–4.5) cm. long, 2.8–4.2 cm. wide, with a terminal apicule 5–7 mm. long; callus an entire ridge tapering from the base to the apex in the middle of the lip; spur cylindrical, pendent, 4–5.8(–7) cm. long. Column fleshy, 5 mm. long, green.

subsp. **giryamae** (*Rendle*) *Senghas & Cribb* in Die Orchidee 30(1): CXXI (1979). Type: Kenya, Kilifi District, Giriama [Giryama] and Kwale District, Shimba Hills, *W. E. Taylor* (BM, holo.!)

Usually an epiphyte but also a lithophyte. Flowers with a transversely oblong, obscurely 3-lobed lip 2.4–3.3 cm. long, 2.8–4.2 cm. wide, with a short apicule up to 5 mm. long and spur 4–5.8 cm. long.

KENYA. Kwale District: between Samburu and Mackinnon Road, near Taru, 5 Sept. 1953, *Drummond & Hemsley* 4181! & N. of Kwale, 13 July 1937, *Greenway* 4969!; Mombasa, on rocky island, Aug. 1936, *Dale* in *F.D.* 3566!

TANZANIA. Moshi District: Kilimanjaro, *Johnston*!; Lushoto District: Usambara Mts., Magila, Sept. 1877, *Kirk!*; Tanga District: Moa, fl. cult. 13 Aug. 1944, *Moreau* 209!; Zanzibar I., Sir John Kirk's garden, fl. cult. 16 Aug. 1944, *Boscawen in Moreau* 209B!; Mangapwani, 11 June 1950, *Oxtoby* 8! & 1 June 1960, *Faulkner* 2578!; Pemba I., Ras Kigomasha [Kigomacha], June 1909, *de Legh*!

DISTR. **K** 7; **T** 2, 3; **Z**; **P**; not known elsewhere

HAB. On shrubs in thickets on outcropping rocks, on mangroves, on rocks by sea and on cliff-faces; sea-level–350(–1700) m.

SYN. *A. giryamae* Rendle in J.L.S. 30: 388 (1895); Rolfe in F.T.A. 7: 143 (1897); Summerh. in K.B. 13: 277 (1958); Piers, Orch. E. Afr., ed. 2: 234 (1968)

NOTE. *A. eburneum* is a widespread species found in East Africa, the Comores, Madagascar, the Mascarene Islands, Seychelles and Aldabra. Four subspecies have been recognised by Senghas & Cribb (1979). Subsp. *giryamae*, the only one found in continental Africa, is distinguished from the other subspecies by its transversely oblong lip and shorter spur always less than 6 cm. long.

22. **A. keniae** *Kraenzl.* in E.J. 17: 59 (1893); Summerh. in K.B. 13: 271 (1958); Piers, Orch. E. Afr., ed. 2: 242 (1968). Type: Kenya, N. Nyeri District, Ndoro, *von Hoehnel* 24 (B, holo.)

An epiphyte with elongate ancipitous stems up to 30 cm. long; roots many, clustered, unbranched. Leaves subremote, equitant (conduplicate and overlapping), shaped like a scalpel, linear, acute, 3–5 cm. long, 2 mm. wide. Inflorescences much shorter than the leaves, 1–few-flowered; bracts minute, triangular. Flowers 1.8 cm. across. Sepals lanceolate, acuminate. Petals similar but narrower. Lip similar but a little shorter; spur filiform, 4 times longer than the ovary, 15 cm. long. Column very short. Fruit 2 cm. long, 8 mm. in diameter.

KENYA. N. Nyeri District: Ndoro, foot of Mt. Kenya, *von Hoehnel* 24

DISTR. **K** 4; known only from the type

HAB. Not known

NOTE. The above description has been drawn from Kraenzlin's original. No further collections of this species have apparently been made. Kraenzlin's descriptions were often poor and inaccurate and, therefore, some doubt remains as to the exact identity of this species.

51. JUMELLEA*

Schltr., Die Orchideen: 609 (1914)

Epiphytic or lithophytic, entirely glabrous monopodial herbs. Stems abbreviated or elongated, simple or occasionally branched. Leaves linear to ligulate, equally or unequally bilobed at apex, coriaceous or fleshy, erect to spreading. Inflorescence 1(–2)-flowered, axillary, peduncle abbreviated; bracts small, mostly imbricate, scarious. Flowers resupinate, white, small or medium-sized, often scented; ovary with a long filiform

* By J.J. Wood

pedicel. Sepals and petals free, narrow, acute to acuminate, often undulate at the base, spreading. Dorsal sepal usually recurved. Petals connivent with base of lip. Lip entire, spurred at base, flat or conduplicate, rhombic, narrowly elliptic or ovate, acute to acuminate, attenuate or constricted at the base; disc usually carinate; spur usually long, filiform, retrorse or antrorse. Column short, porrect, with 2 conspicuous stelidia and a short foot; rostellum short, emarginate; anther cucullate, truncate or retuse, not beaked; pollinia 2, globose or broadly ellipsoid, each attached by a short stipes to a large viscidium. Capsule narrowly cylindrical, dehiscing by means of a single fissure.

A genus of about 45 species centred in Madagascar and the Mascarene Is., with 2 species (and three other doubtful records) in East Africa and 1 species in South Africa.

Stems 18–30(–40) cm. long; leaves 4–8(–9.5) × 1–1.3 cm., unequally bilobed; spur (1.8–)2–3 cm. long; pedicellate ovary 3.5–4 cm. long; sepals (1–)3–4 mm. broad . . . 1. *J. filicornoides*
Stems 30–80 cm. long; leaves (7–)10–13 × 1.5–2 cm., usually equally bilobed; spur 8–9 cm. long; pedicellate ovary 2.5–3 cm. long; sepals 4–6 mm. broad . . . . 2. *J. usambarensis*

1. **J. filicornoides** (*De Wild.*) *Schltr.* in Beih. Bot. Centr. 36: 152 (1918); Morris, Epiph. Orch. Malawi: 81 (1970); Harrison, Epiph. Orch. S. Afr.: 77 (1972). Type: Mozambique, Morrumbala, *Luja* (BR, holo.)

An erect epiphytic or lithophytic herb. Stem simple, covered in scarious persistent leaf-sheaths below, leafy above, 18–30(–40) cm. long, slightly pendant. Leaves 6–12(–15), ligulate or oblong-ligulate, unequally bilobed, fleshy, coriaceous, 4–8(–9.5) cm. long, 1–1.3 cm. broad. Inflorescence 1-flowered, suberect; peduncle 1.5–3.5 cm. long; bracts 3–4, tubular, obtuse to acute, scarious, lowermost imbricate, uppermost subtending the flower, 4–15 mm. long. Flowers white with spur, column and anther pale green, clove-scented, particularly at night; pedicel with ovary slender, often S-shaped, 3.5–4 cm. long. Sepals narrowly oblong-elliptic or ligulate, usually undulate at base, obtuse, dorsal erect to reflexed, laterals deflexed and recurved, (10–)13–20 mm. long, (1–)3–4 mm. broad. Petals narrowly elliptic or linear-ligulate, attenuate, undulate at base, acute, deflexed and recurved, (10–)12–18 mm. long, (1–)2–3 mm. broad. Lip rhombic, narrowly constricted at base, acute, conduplicate, carinate towards base, (11–)16–20 mm. long, (3–)5–6 mm. broad; spur cylindrical, filiform, bent at base, acute, descending, (1.8–)2–3 cm long. Column 2 mm. long, with 2 anvil-shaped stelidia 1 mm. long; anther cucullate, retuse, 1.5 mm. long and broad. Pl. 5.

KENYA. Kilifi District: Mangea Hill, 25 Apr. 1989, *Luke & Robertson* 1826!
TANZANIA. Morogoro District: Nguru Mts., Maskati, 1 Aug. 1942, *Moreau* 362! & 363!; Ulanga District: Sali, Ngongo Mt., 23 Jan. 1979, *Cribb, Grey-Wilson & Mwasumbi* 11138!
DISTR. **K** 7; **T** 6; Mozambique, Malawi, Zimbabwe, South Africa (Natal, Transvaal)
HAB. Riverine forest, savannah trees and on rocks; 700–1800 m.

SYN. *Angraecum filicornoides* De Wild., Pl. Nov. Herb. Hort. Then. 1, t. 21 (1904)
Mystacidium walleri Rolfe in F.T.A. 7: 172 (1897). Type: Mozambique, Moramballa Mt., *Waller* (K, holo.!)

2. **J. usambarensis** *J.J. Wood* in K.B. 37: 77 (1982). Type: Tanzania, Lushoto District, between Magamba and Gare Mission, *Eggeling* 391 (K, holo.!)

An erect epiphytic herb. Stems clustered, simple, covered in persistent scarious leaf-sheaths below, leafy distally, rooting at nodes, 30–80 cm. long. Leaves oblong-ligulate, equally or slightly unequally bilobed at apex, coriaceous, mid-nerve prominent below, (7–)10–13 cm. long, 1.5–2 cm. wide, sheaths 1.5–2 cm. long. Inflorescence 1-flowered, porrect; pedicel abbreviated, concealed by bracts; bracts 3–5, ovate, obtuse, scarious, imbricate, 3–7(–10) mm. long. Flowers white with a green spur and brown anther, scented at night; pedicel with ovary slender, curved or gently sigmoid, 2.5–3 cm. long. Tepals spreading, slightly undulate at base. Dorsal sepal ovate-elliptic, obtuse, 15–16 mm. long, 6 mm. broad; lateral sepals oblong-ligulate, obtuse, 17 mm. long, 4 mm. broad. Petals linear-ligulate to linear-spathulate, obtuse to acute, 14–15 mm. long, 2.5–3 mm. broad. Lip rhombic, acute, conduplicate, disc carinate towards base, recurved, 16–17 mm. long, 6 mm. broad; spur cylindrical, filiform, geniculate and slightly dilated near base, often coiled into a loop, subacute, 8–9 cm. long. Column oblong, with 2 triangular acute stelidia,

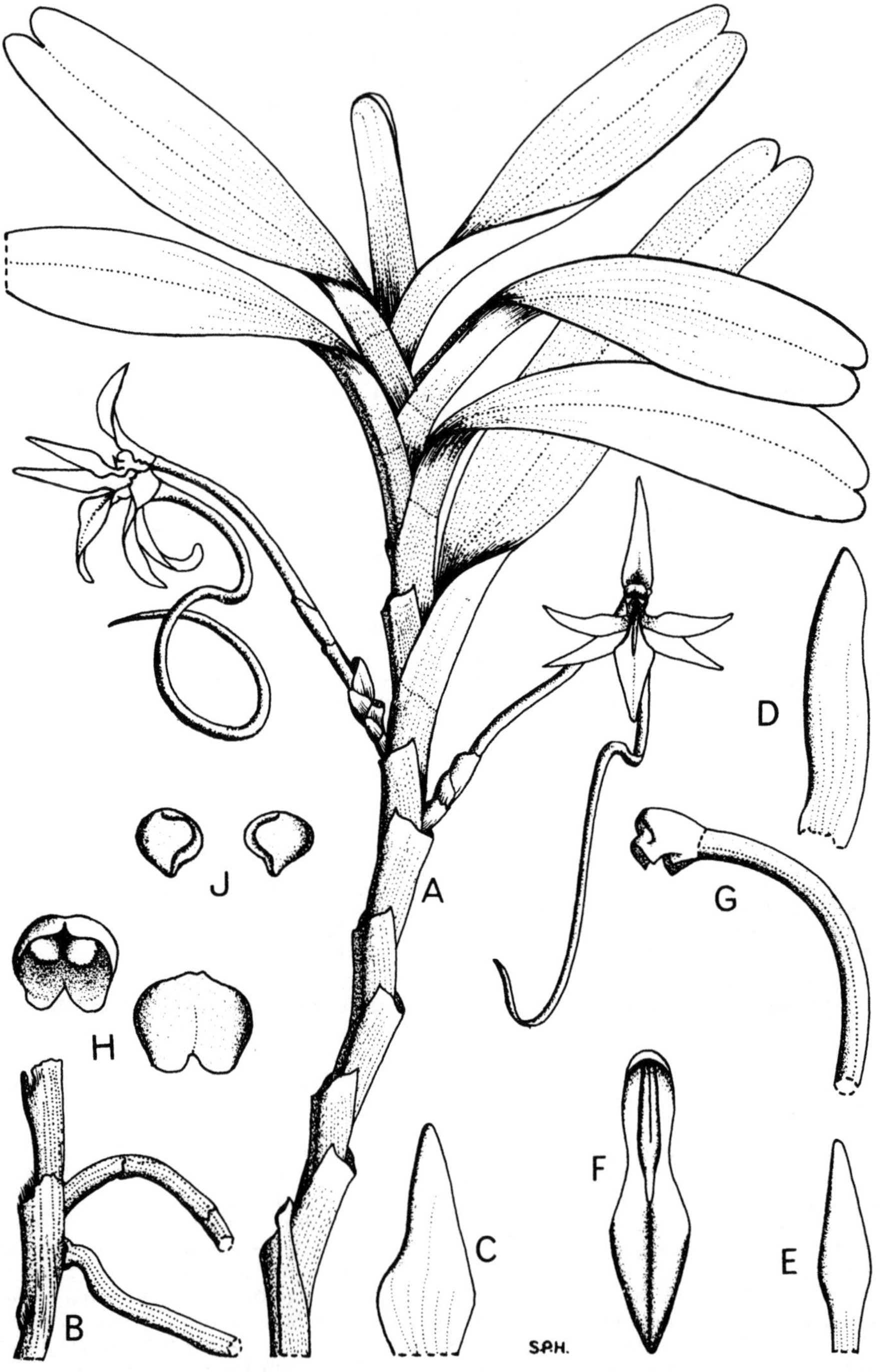

FIG. 119. *JUMELLEA USAMBARENSIS*—**A**, habit, × ⅔; **B**, roots, × ⅔; **C**, dorsal sepal, × 2; **D**, lateral sepal, × 2; **E**, petal, × 2; **F**, lip, front view, × 2; **G**, column, × 2; **H**, anther-cap, front and back views, × 6; **J**, pollinia, × 6. Drawn from *Eggeling* 391 by Susan Hillier.

4 mm. long, 3.5 mm. broad; anther cucullate, emarginate, 2 mm. long, 2.5 mm broad; pollinia ovoid, 1 mm. long and broad. Fig. 119.

KENYA. Masai District: Maparashu Hills, 3 Dec. 1981, *M.G. Gilbert* 6823!
TANZANIA. Lushoto District: between Magamba and Gare Mission, 17 Sept. 1942, *Eggeling* 391! & Hekulo [Herkulu] Forest, *Protzen*!
DISTR. **K** 6; **T** 3; Malawi
HAB. Rain-forest, epiphytic on *Ocotea usambarensis*, etc.; ± 1600–1900 m.

SYN. *Jumellea sp.* sensu Morris, Epiph. Orch. Malawi: 82 (1970)

NOTE. The following three undetermined taxa, reputedly from East Africa, are preserved in the Kew spirit collection. Their occurrence in the area remains doubtful.

3. **J. sp. A**

Stems to 60 cm. long. Leaves ligulate, unequally bilobed, coriaceous, 14–17 cm. long, 1.7–1.9 cm. broad. Inflorescence 1-flowered; pedicel elongate, not entirely concealed by bracts; bracts 3 basal, imbricate, 1 apical, ovate, obtuse, 4–12 mm. long. Flowers white with green spur and column and pale brown anther. Pedicel with ovary slender, 10.5 cm. long. Sepals and petals undulate at base. Dorsal sepal elliptic-ligulate, acute, reflexed, 22 mm. long, 5 mm. broad (at base); laterals elliptic-ligulate, acute, gently curved, deflexed, 26 mm. long, 4 mm. broad. Petals linear-ligulate, acute, 24 mm. long, 3 mm. broad. Lip rhombic, acute, conduplicate, disc carinate towards base, 25 mm. long, 7 mm. broad; spur cylindrical, filiform, geniculate and slightly dilated near base, subacute, 10.5 cm. long.

KENYA. (?) cult. *L.M. Mason* via RHS No. 1091, EN. 270-75.02508!

4. **J. sp. B**

Pedicel elongated, not entirely concealed by bracts, 7.2 cm. long; bracts 3 below, the 2 lowermost imbricate, 1 apical, ovate, obtuse, 6–10 mm. long. Flowers white with pale brown anther. Ovary 5 cm. long. Sepals and petals not undulate at base. Dorsal sepal ovate-elliptic, subacute, 15 mm. long, 5 mm. broad; laterals elliptic-ligulate, subacute, 18 mm. long, 4.5 mm. broad. Petals linear-elliptic, acute, 17 mm. long, 4 mm. broad. Lip rhombic, acute, conduplicate, disc carinate towards base, 26 mm. long, 8.5 mm. broad; spur cylindrical, filiform, curved and slightly dilated towards base, obtuse, ± 5 cm. long.

UGANDA. (?) cult. *J.A. Burgess* K.31, EN. 600-62!

5. **J. sp. C**

Leaves to 21 cm. long, 1 cm. broad. Pedicel elongated, wiry, flexuous, not entirely concealed by bracts, minutely ramentaceous, 5 cm. long; bracts 2 below, imbricate, 1 apical, ovate, obtuse, minutely ramentaceous, ± 7 mm. long. Flowers white, sepals flushed green on reverse, column green and anther cream. Ovary ± 6.7 cm. long. Sepals and petals undulate at base. Sepals linear-ligulate, dorsal acute, reflexed, 28 mm. long, 4 mm. broad; laterals subulate, 30 mm. long, 4 mm. broad. Petals linear-spathulate, acute, 27 mm. long, 3.5 mm. broad. Lip rhombic, acuminate, conduplicate, disc carinate at base, 25 mm. long, 9 mm. broad; spur cylindrical, filiform, obtuse, geniculate and somewhat dilated near base, forming a loop, ± 11.5 cm. long.

EAST TROPICAL AFRICA. *B.L. Perkins* 4, EN. 568-66!

52. **RHAESTERIA**

Summerh. in K.B. 20: 191 (1966)

Dwarf monopodial epiphytic herbs. Stems elongate, leafy throughout length; roots emitted along length of stem opposite the leaves. Leaves distichous, conduplicate, twisted at the base to lie in one plane and articulated to persistent sheathing leaf-bases. Inflorescences axillary, 1–several from the upper nodes, few-flowered. Flowers very small. Sepals and petals connate in their basal halves, free above, subsimilar. Lip free, concave, enclosing the column at its base, spurred; spur with a broad mouth. Column very short, fleshy; rostellum anchor-shaped, fleshy, pendent; pollinia 2, globose; stipites very short; viscidia 2.

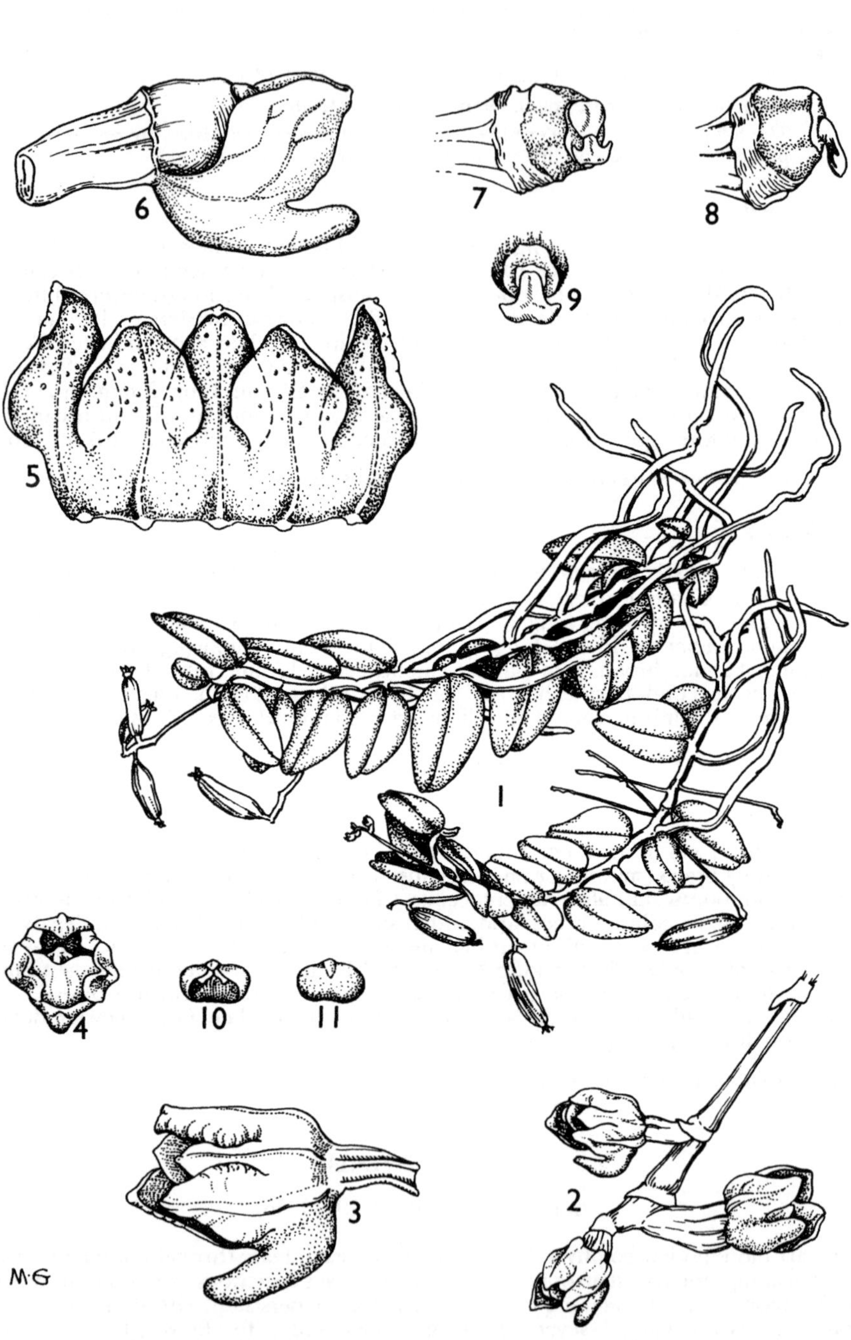

FIG. 120. *RHAESTERIA EGGELINGII*—**1**, habit, × 1; **2**, inflorescence, × 6; **3**, **4**, flower, side and front views, × 8; **5**, joined sepals and petals, × 10; **6**, flower, with sepals and petals removed showing column, lip and spur, × 10; **7**, column with anther-cap, × 10; **8**, column with anther-cap removed showing rostellum, side view, × 10; **9**, same, front view, × 10; **10**, anther-cap, from inside, × 10; **11**, anther-cap, from outside, × 10. Drawn from *Eggeling* 5849 by Miss M.A. Grierson.

A monotypic genus with a single species known only from Uganda.

R. eggelingii *Summerh.* in K.B. 20: 191 (1966). Type: Uganda, Kigezi District, Impenetrable Forest, *Eggeling* 5849 (K, holo.!)

A dwarf epiphyte with a simple or little-branching spreading or pendent stem, 6–8 cm. long, 1 mm. in diameter; roots slender, elongate, 1 mm. in diameter, grey. Leaves subfleshy, ovate or ovate-elliptic, unequally subacutely bilobed at the apex, 1–1.6 cm. long, 0.5–0.9 cm. wide. Inflorescences fasciculate or successive, slightly longer than the leaves, 2–3-flowered; peduncle filiform, 1–1.5 cm. long; rhachis similar but slightly fractiflex; bracts very small, triangular, acute. Flowers yellowish; pedicel and ovary 2–3 mm. long, furfuraceous. Sepals ovate, obtuse, 2.5–3 mm. long, 1.4 mm. wide, joined in their basal halves to the petals. Petals ovate, obtuse, 2.5 mm. long, 1.5 mm. wide. Lip cymbiform, concave, 2.5 mm. long, 3 mm. wide; spur incurved strongly, conical from a broad mouth, 2.5 mm. long. Column ± 1 mm. long. Fruit ellipsoidal, 9 mm. long, 3 mm. in diameter. Fig. 120.

UGANDA. Kigezi District: Impenetrable Forest, Nov. 1948, *Eggeling* 5849!
DISTR. U 2; known only from the type collection
HAB. On small savanna tree in grassland on outskirts of the forest; 1300–1500 m.

53. MICROCOELIA

Lindl., Gen. Sp. Orch. Pl.: 60 (1830); Summerh. in Bot. Mus. Leafl. Harv. Univ. 11: 140 (1943) & 250 (1945); L. Jonsson in Symb. Bot. Upsal. 23(4): 63 (1981)

Dicranotaenia Finet in Bull. Soc. Bot. Fr. 54, Mém. 9: 47 (1907)
Encheiridion Summerh. in Bot. Mus. Leafl. Harv. Univ. 11: 161 (1943)

Leafless epiphytic or rarely lithophytic herbs with short stems. Roots firmly or loosely attached to the substrate, often dense, terete or less commonly dorso-ventrally flattened, smooth or rarely verrucose, unbranched or with a few branches, usually elongate. Scales on stem protecting the stem apex, acute to rostrate. Inflorescences few–many, axillary, racemose, concentrated in the apical part of the stem, few–many-flowered; peduncle long or short; rhachis terete or angular, smooth or with processes; bracts sheathing or not. Flowers small to minute, ± sessile or pedicellate, usually white variously tinged with green, brown or pink on spur and other segments. Sepals and petals free, subsimilar. Lip entire or obscurely 3-lobed, free, usually with fleshy calli at the base either side of the mouth of the spur; spur globose, cylindrical or variously swollen. Column fleshy; androclinium short to long; anther-cap hemispherical, often elongated at the apex; pollinia 2, subglobose to pyriform; stipes linear to oblanceolate, entire or bifid at the apex; viscidium linear or oblong, short to long; rostellum bifid, short to as long as the column.

A genus of some 27 species from Madagascar and tropical and southern Africa.

The genus has been most recently revised by L. Jonsson (1981) in Symbolae Botanicae Upsalienses 23(4). The major difference between this treatment and that of Summerhayes (1945) has been the inclusion of the genus *Encheiridion* in *Microcoelia*. In its root anatomy and seed morphology, *Encheiridion* agrees well with *Microcoelia*. The major morphological distinction, the very pronounced rostellar arms and the correspondingly elongated stipes and anther-cap of *Encheiridion* have been interpreted by Jonsson in terms of a more or less continuous gradation in the length of these features between that genus and *Microcoelia*. Because of this, no infrageneric categories have been proposed by him.

Spur globose, 1 mm. in diameter . . . . 1. *M. exilis*
Spur conical, filiform or variously inflated at base and/or apex, never globose:
  Roots verrucose . . . . 8. *M. megalorrhiza*
  Roots smooth:
    Rostellar arms very prominent, 4–5 mm. long; apical margin of lip denticulate; roots strongly dorso-ventrally flattened . . . . 14. *M. macrorrhynchia*
    Rostellar arms not prominent, less than 2 mm. long; apical margin of lip entire or erose; roots usually terete:

Spur 18.5–20 mm. long, cylindrical, dilated at the apex . . . 11. *M. microglossa*
Spur less than 15 mm. long:
Spur bulbous or strongly dilated at the apex:
Spur straight or recurved; lip less than 3 mm. long . . . 7. *M. physophora*
Spur incurved; lip more than 4 mm. long:
Spur strongly dilated in the basal two-thirds, incurved only in the apical part; lip ovate-elliptic, with a dark green mark in the centre . . . . 12. *M. bulbocalcarata*
Spur inflated at the apex only; lip obovate, lacking a dark green marking:
Lip rounded at the apex, with an erose margin; spur strongly dilated and ellipsoidal at apex: stipes bifid at the apex . . . . 10. *M. konduensis*
Lip subacute at apex, with entire margins; spur conically inflated at the apex; stipes entire . . . . 9. *M. koehleri*
Spur not inflated at the apex:
Lip less than half the length of the spur; spur pendent, straight, inflated in the basal half, 7–11.5 mm. long . . . . 13. *M. caespitosa*
Lip as long as the spur; spur straight or incurved, cylindrical, usually from a broad mouth but not inflated in the basal half:
Lowermost floral bracts not sheathing:
Inflorescences up to 3 cm. long, up to 8-flowered; lip 1.3–2.1 mm. long; scale leaves 1–11.3 mm. long . . . . . . 2. *M. moreauae*
Inflorescences usually more than 5 cm. long, up to 40-flowered; lip 2–3.2 mm. long; scale leaves 4–4.5 mm. long . . . . 4. *M. stolzii*
Lowermost floral bracts sheathing:
Lip narrowly ovate, acute . . . . . . 3. *M. smithii*
Lip oblong or pandurate, obtuse or truncate:
Sepals less than 3.5 mm. long; lip 2.2–3.3 mm. long; spur 2.1–3.1 mm. long, conical at the base, cylindrical above . . . 5. *M. globulosa*
Sepals 3.7–7.3 mm. long; lip 4.9–7.6 mm. long; spur cylindrical from the base, 3.3–5.6 mm. long . . . . 6. *M. obovata*

1. **M. exilis** *Lindl.*, Gen. Sp. Orch. Pl.: 61 (1830); Summerh. in Bot. Mus. Leafl. Harv. Univ. 11: 142 (1943); Williamson, Orch. SC. Africa: 184 (1977); Ball in S. Afr. Epiphytic Orch. 118 (1978); L. Jonsson in Symb. Bot. Upsal. 23(4): 66 (1981). Type: Madagascar, without exact locality, *Forbes* (K, holo.!, BM, iso.!)

An epiphytic or rarely a lithophytic plant with a short to long stem, up to 45 mm. long, 2–3 mm. in diameter; roots many, slender to robust, characteristically twisted and much branched, ± terete, smooth, 1–2 mm. in diameter. Scale-leaves rostrate to acuminate, up to 4.5 mm. long, 5–8-nerved. Inflorescences erect to pendent, 10–20, up to 25 cm. long, densely up to 50(–80)-flowered; peduncle up to 3 cm. long; rhachis slightly angular; bracts not sheathing at base, triangular to ovate, acute, 0.6–1.2 mm. long. Flowers minute, white with a yellow-green tip to the spur; pedicel and ovary 0.5–1 mm. long. Dorsal sepal obovate, obtuse to subacute, 1–1.5 mm. long, 0.4–0.9 mm. wide; lateral sepals obliquely ovate, obtuse to subacute, 1–1.5 mm. long, 0.4–0.8 mm. wide. Petals obovate to ovate, obtuse, 0.6–1.2 mm. long, 0.3–0.7 mm. wide. Lip concave, ovate to semiorbicular, acute, 0.6–1.1 mm. long, 0.4–0.8 mm. wide; spur globose, slightly compressed, 0.6–0.9 mm. long, 0.5–1 mm. in diameter. Column 0.2–0.5 mm. long.

UGANDA. Kigezi District: Rwenzoli [Queen Elizabeth] National Park, Ishasha R., Mar. 1968, *Lock* 45!
KENYA. Machakos District: Masongaleni, Mar. 1964, *Archer* 421!; Kwale District: Lunga-Lunga to Msambweni road, between Umba and Mweni Rivers, Aug. 1953, *Drummond & Hemsley* 3890!; Tana River District: Karawa, Oct. 1961, *Polhill & Paulo* 636!

TANZANIA. Musoma District: Mara R. Guard Post, Oct. 1961, *Greenway* 10229!; Lushoto District: E. Usambara Mts., Amani, Aug. 1942, *Moreau* 180!; Rungwe District: Songwe valley, Feb. 1976, *Cribb & Grey-Wilson* 10825!

DISTR. **U** 2; **K** 4, 7; **T** 1, 3, 6–8; Zaire, Zambia, Malawi, Mozambique, Zimbabwe, South Africa and Madagascar

HAB. In riverine forest and woodland, also in secondary forest and in plantations, mostly on twigs and small branches of trees and bushes; sea-level–1800 m.

SYN. *Angraecum chiloschistae* Reichb.f. in Linnaea 20: 678 (1847). Type: South Africa, Durban, *Gueinzius* 263 (W, holo.!, K, iso.!)
Gussonea exilis (Lindl.) Ridley in J.L.S. 21: 493 (1885)
Epidorkis exilis (Lindl.) O. Kuntze, Rev. Gen. Pl. 2: 660 (1891)
Mystacidium exile (Lindl.) Th. Dur. & Schinz, Consp. Fl. Afr. 5: 52 (1895)
Rhaphidorhynchus chiloschistae (Reichb.f.) Finet in Bull. Soc. Bot. Fr. 54, Mém. 9: 35 (1907)
Gussonea chiloschistae (Reichb.f.) Schltr. in Beih. Bot. Centr. 36: 9 (1918)

2. **M. moreauae** *L. Jonsson* in Symb. Bot. Upsal. 23(4): 71 (1981). Type: Tanzania, E. Usambara Mts., Amani, *Moreau* 31 (K, holo.!, EA, iso.!)

An epiphyte with a short stem 3–6 mm. long, 1–1.2 mm. in diameter. Roots few, slender, tufted, undulating and usually not branched, terete, smooth, up to 12 cm. long, 0.5–1 mm. in diameter. Scale-leaves acute, 3–5-nerved, up to 1.3 mm. long. Inflorescences erect, spreading, 1–4, up to 30 mm. long, each with up to 8 flowers; peduncle up to 8 mm. long; rhachis flexuose, angular, smooth; bracts not sheathing, ovate, acute, to subacute, 0.5–1 mm. long. Flowers small, white; pedicel 1.4 mm. long; ovary 0.9–1.3 mm. long. Tepals similar, slightly recurved apically. Dorsal sepal oblong, obtuse, 1.1–2.3 mm. long, 0.6–0.9 mm. wide; lateral sepals similar but oblique, 1.6–2.4 mm. long, 0.6–0.8 mm. wide. Petals oblong, obtuse, 1.4–2.1 mm. long, 0.5–0.8 mm. wide. Lip concave, oblong, truncate or obtuse, 1.3–2.1 mm. long, 0.6–1 mm. wide; spur perpendicular, geniculately incurved, conically inflated at the base, obtusely cylindrical at the apex, 1.5–1.9 mm. long, up to 0.3 mm. in diameter. Column 0.3–0.5 mm. long.

KENYA. Nairobi District: Langata Forest, Sept. 1950, *Piers* 21!; Meru District: Imenti Forest, Aug. 1951, *Piers* 81!; Masai District: Ngong Forest Road, Oct. 1974, *Gillett* 20750!

TANZANIA. Lushoto District: E. Usambara Mts., Amani, Dec. 1928, *Greenway* 1065! & Dec. 1941 *Moreau* 31!

DISTR. **K** 4, 6; **T** 3; Malawi and Zimbabwe

HAB. In rain-forest, dry evergreen forest and riverine forest, usually on finer twigs of bushes and understorey trees; 670–1800 m.

3. **M. smithii** (*Rolfe*) *Summerh.* in Bot. Mus. Leafl. Harv. Univ. 11: 148 (1943); Piers, Orch. E. Afr., ed. 2: 276 (1968); L. Jonsson in Symb. Bot. Upsal. 23(4): 73 (1981). Type: Tanzania, Lushoto/Tanga District, Umba steppe, *C.S. Smith* (K, holo.!)

A small epiphytic plant. Stem short, 3–9 mm. long, 1–2 mm. wide. Roots few, radially spreading, densely crowded, unbranched, terete, smooth, up to 15 cm. long, 1–1.5 mm. in diameter. Scale-leaves rostrate, up to 2 mm. long, with ± 3 nerves. Inflorescences erect, spreading, 7–12, up to 50 mm. long, with up to 12 flowers; peduncle up to 15 mm. long; rhachis flexuose, terete, smooth; bracts loosely sheathing, acute to subacute, up to 1.6 mm. long. Flowers white, spur with greenish apex; pedicel minute, up to 0.5 mm. long; ovary 1.3–2.2 mm. long. Dorsal sepal narrowly ovate to obovate, acute to subacute, 2.0–2.7 mm. long, 0.7–0.9 mm. wide; lateral sepals similar but slightly asymmetric, 2.3–2.7 mm. long, 0.6–0.8 mm. wide. Petals slightly asymmetric, narrowly oblong-lanceolate, acute, 2.0–2.6 mm. long, 0.6–0.8 mm. wide. Lip navicular, narrowly ovate, acute, folded, 2.3–2.8 mm. long, 0.6–0.9 mm. wide; spur perpendicular to slightly decurved, conical to cylindrical, obtuse, 1.6–2.2 mm. long, 0.3–0.4 mm. wide at the apex. Column 0.4–0.6 mm. long.

KENYA. Kwale District: Shimba Hills, 1887, *W. E. Taylor*!

TANZANIA. Lushoto District: SW. slopes of Usambara Mts., Mashewa, Feb. 1893, *Holst* 3549!; Handeni District: Kwa Nkono, Aug. 1969, *Archbold* 1092!; Pangani District: Msubugwe Forest, fl. cult. Aug. 1956, *Milne-Redhead & Taylor* 7320!

DISTR. **K** 7; **T** 3, 6; not known elsewhere

HAB. In dry evergreen forest, on *Euphorbia* sp.; sea level–500 m.

SYN. *Angraecum smithii* Rolfe in K.B. 1895: 37 (1895)
Gussonea smithii (Rolfe) Schltr. in Beih. Bot. Centr. 36: 94 (1918)

4. **M. stolzii** (*Schltr.*) *Summerh.* in Bot. Mus. Leafl. Harv. Univ. 11: 149 (1943); Williamson, Orch. SC. Afr.: 184 (1977); L. Jonsson in Symb. Bot. Upsal. 23(4): 75 (1981). Type: Tanzania, Njombe District, Madehani, *Stolz* 2350 (B, holo.†, K, lecto.!, BM, BR, P, iso.)

An epiphyte with stem up to 45 mm. long, 1.5–2 mm. wide. Roots many, with few branches, twisted, terete, smooth, up to 30 cm. long, 0.5–1.5 mm. in diameter. Scale-leaves acuminate, with 4–6 nerves, up to 4–4.5 mm. long. Inflorescences erect or pendulous, spreading, up to 120 mm. long, up to 5 simultaneously, each with up to 40 flowers; peduncle up to 10 mm. long; rhachis straight to indistinctly flexuose; bracts not sheathing, ovate, acute, up to 2 mm. long. Flowers up to 9.4 mm. long, white, spur-apex darkish green; pedicel minute, up to 2 mm. long; ovary 1.1–2.1 mm. long. Dorsal sepal 2.3–3.4 mm. long, 1.0–1.6 mm. wide, slightly convex, obovate to ovate, obtuse to subacute, apiculate with a single thickened nerve; lateral sepals similar but oblique, 2.0–3.5 mm. long, 1.0–1.5 mm. wide. Petals similar, 2.3–3.3 mm. long, 1.0–1.5 mm. wide. Lip pandurate, narrowly obovate to ovate, obtuse to subacute, 2.0–3.2 mm. long, 1.0–1.6 mm. wide; spur conical, tapering into a variously pointed cylindrical obtuse apex, slightly to markedly incurved or when undeveloped pointing backwards, 1.4–2.5 mm. long. Column 0.3–0.7 mm. long. Stipes characteristically spathulate. Fig. 121.

KENYA. Nairobi District: surroundings of Nairobi, 1938, *Copley* 5!; Masai District: Nguruman Escarpment, Endosapia R., Dec. 1970, *Archer* 652!; Teita District: Kasigau, S. slope, above Bungule, Dec. 1970, *Faden, Evans & Msafiri* 979!

TANZANIA. Mbulu District: Pienaars Heights, Jan. 1962, *Polhill & Paulo* 1077!; Lushoto District: E. Usambara Mts., Amani, Dec. 1941, *Moreau* 112!; Rungwe District: Livingstone Mts. Forest Reserve, Dec. 1959, *Procter* 1615! & Feb. 1976, *Cribb & Grey-Wilson* 10628!

DISTR. **K** 3, 4, 6, 7; **T** 2, 3, 6–8; Zambia, Malawi, Zimbabwe, Mozambique

HAB. In upland rain-forest, dry evergreen and riverine forest, occasionally in woodland and plantations, on the smaller branches and trunks of bushes and understorey trees; 600–2300 m.

SYN. *Gussonea stolzii* Schltr. in E.J. 53: 595 (1915)
Microcoelia eriocosma Summerh. in Bot. Mus. Leafl. Harv. Univ. 11: 251 (1945). Type: Tanzania, Lushoto District, Bomole, *Moreau* 111 (K, holo.!)

5. **M. globulosa** (*Hochst.*) *L. Jonsson* in Symb. Bot. Upsal. 23(4): 78 (1981). Type: Ethiopia, Sana, *Schimper* 1565 (K, lecto.!, BM, BR, FT, G, H, P, W, isolecto.)

An epiphytic plant with a short to long stem, up to 65 mm. long, 2–3 mm. diameter. Roots few–many, characteristically undulating and twisting, terete, smooth, up to 35 cm. long, 1–2 mm. in diameter. Scale-leaves acuminate to obtuse, up to 3.5 mm. long, with 5–7 nerves. Inflorescences stiff, erect or descending, 5–10, up to 90 mm. long, each with up to 15 flowers; peduncle 15 mm. long; rhachis flexuose, variously furrowed and excavated with hook-shaped processes below the bracts; bracts sheathing at base, ovate, acute to subacuminate, up to 2.5 mm. long. Flowers 10 mm. long, white, base of perianth and ovary yellow-green, tip of spur orange-brown; pedicel forming a distinct angle with the ovary, up to 3 mm. long; ovary 0.3–0.8 mm. long. Dorsal sepal slightly convex, ovate, acute, ± apiculate, with a distinct outer central groove, 2.2–3.1 mm. long, 1.0–1.4 mm. wide; lateral sepals ovate to narrowly ovate, acute, ± apiculate, 2.5–3.5 mm. long, 0.9–1.3 mm. wide. Petals obovate to elliptic, ± apiculate, narrowly thickened along central nerve, 2.1–3 mm. long, 0.8–1.2 mm. wide. Lip pandurate, apically folded with a distinct cushion-like thickening at each side of spur-mouth, 2.2–3.3 mm. long, 1.2–1.8 mm. wide; spur characteristically incurved, base conical, tapering into a geniculate cylindrical part, apex obtuse and slightly inflexed, 2.1–3.3 mm. long. Column 0.3–0.6 mm. long.

UGANDA. Bunyoro District: Budongo Forest, fl. cult. Feb. 1944, *Eggeling* 5493!; Ankole District: SW. Mbarara, Kabale road, Ruizi R., fl. cult. Jan. 1949, *Eggeling* 5869!; Mengo District: ± 40 km. NNW. of Kampala, Feb. 1962, *C. Leakey* 10!

KENYA. Ravine District: near Eldama Ravine, Jan. 1940, *Tweedie* 498!; N. Kavirondo District: Kakamega Forest, fl. cult. Mar. 1944, *Carroll* in *Moreau* 667!; Masai District: Mara R., Ngebai Langata, Apr. 1961, *Glover, Gwynne & Samuel* 370!

TANZANIA. Musoma District: Serengeti National Park, W. end of Mara R., Dec. 1971, *Greenway* 14940!; Tanga District: Moa, Oct. 1941, *Moreau* 108!; Mbeya/Chunya District: Unyiha, Mlowo R., Oct. 1976, *Leedal* 3859!

DISTR. **U** 2, 4; **K** 3, 5, 6; **T** 1, 3, 4, 7; Nigeria east to Ethiopia and south to Angola, Zambia, Malawi and Zimbabwe

HAB. On margins of evergreen forest, relict rain-forest, riverine forest, coastal bushland and secondary growth; sea-level–1950 m.

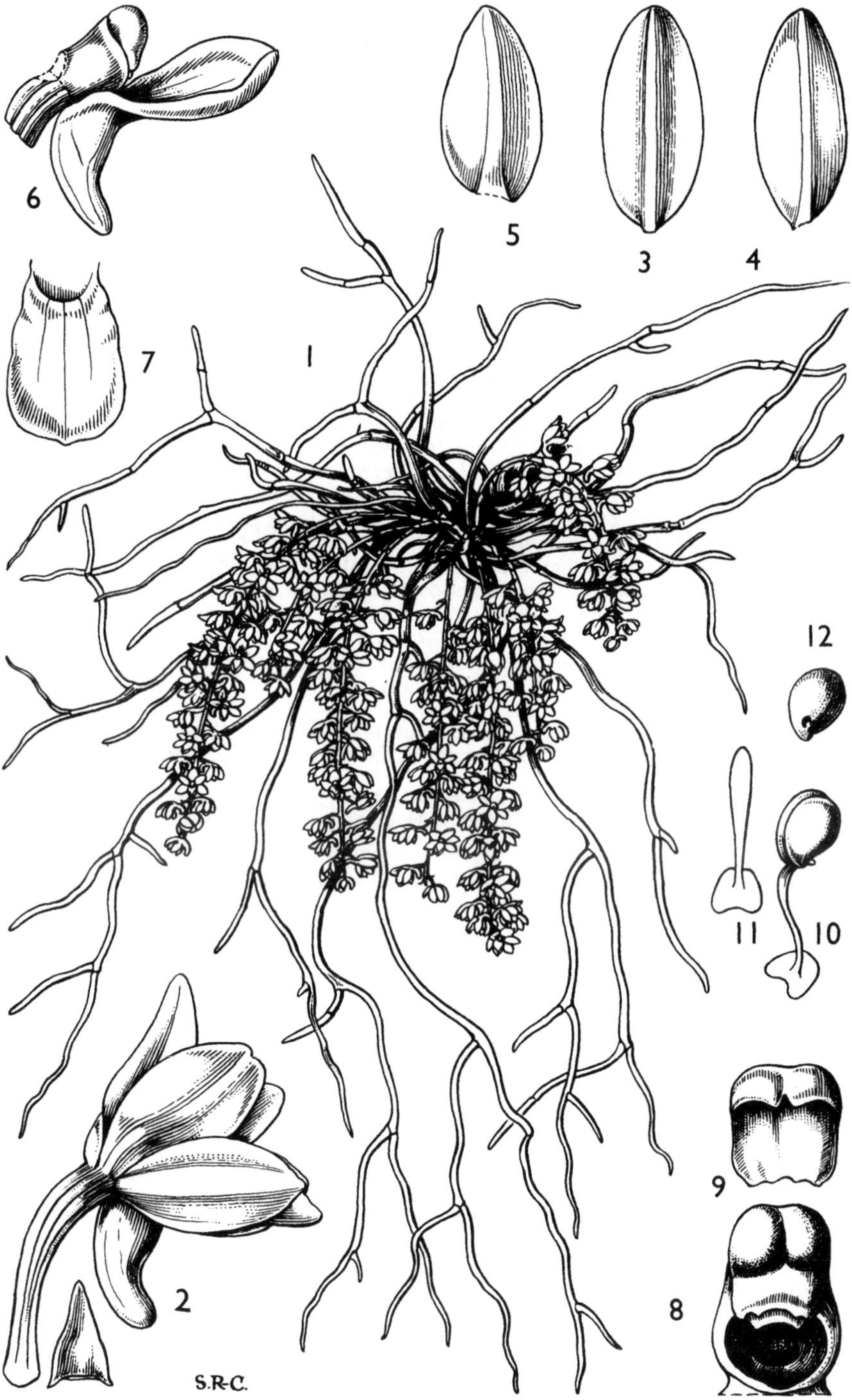

FIG. 121. *MICROCOELIA STOLZII*—**1**, habit, × 1; **2**, flower and bract, × 8; **3**, dorsal sepal × 8; **4**, lateral sepal, × 8; **5**, petal, × 8; **6**, column and lip, side view, × 8; **7**, lip lamina, spread out, × 8; **8**, column, front view, × 16; **9**, anther, from inside, × 16; **10**, pollinarium, × 24; **11**, stipes and viscidium, front view, × 24; **12**, pollinium, side view, × 24. 1, from *Moreau* 111; 2–12, from *Moreau* 105. Drawn by Miss Stella Ross-Craig. Reproduced from Hooker's Icones Plantarum.

SYN. *Angraecum globulosum* Hochst., Pl. Exs. Schimp., Sect. 3., No. 1565 (1844)
A. guyonianum Reichb.f. in Linnaea 22: 865 (1849). Type: Ethiopia, Tigré, Sana, *Schimper* sect. II, s.n. (W, holo.!, P, iso.)
Aeranthus guyonianus (Reichb.f.) Reichb.f. in Flora 48: 190 (1865)
Gussonea globulosa (Hochst.) Ridl. in J.L.S. 21: 491, in obs. (1885)
Epidorkis guyoniana (Reichb.f.) O. Kuntze, Rev. Gen. Pl. 2: 660 (1891)
Mystacidium globulosum (Hochst.) Th. Dur. & Schinz, Consp. Fl. Afr. 5: 53 (1895)
Microcoelia guyoniana (Reichb.f.) Summerh. in Bot. Mus. Leafl. Harv. Univ. 11: 144 (1943)

6. **M. obovata** *Summerh.* in Bot. Mus. Leafl. Harv. Univ. 11: 253 (1945); Piers, Orch. E. Afr., ed. 2: 276 (1968); L. Jonsson in Symb. Bot. Upsal. 23(4): 83 (1981). Type: Tanzania, Lushoto/Tanga District, Lwengera valley, *Moreau* 445a (K, holo.!)

An epiphytic plant with a stem up to 30 mm. long, 3–5 mm. in diameter. Roots characteristically undulating, apically with a rigid appearance, terete, smooth, up to 35 cm. long. Scale-leaves acute to acuminate, with 5–7 nerves, up to 5 mm. long. Inflorescences erect or descending, up to 90 mm. long, up to 10 simultaneously, each with up to 20 flowers; peduncle up to 15 mm. long; rhachis straight to indistinctly flexuose, terete to faintly angled with rows of developed verruculae; bracts sheathing towards apex, acute to subacute, up to 2.5 mm. long. Flowers up to 14 mm. long, white, rust-brown at base of tepals; pedicel up to 3.2 mm. long; ovary 1.6–3.2 mm. long. Dorsal sepal slightly convex, obovate to oblong, obtuse, apiculate, 3.7–6 mm. long, 1.2–2 mm. wide; lateral sepals slightly assymmetric, oblong, acute, apiculate, apically reflexed, 5.2–7.3 mm. long, 1.7–2.3 mm. wide. Petals asymmetric, oblong to obovate, apiculate, apically reflexed, 3.9–6.4 mm. long, 1.3–1.8 mm. wide. Lip indistinctly pandurate, broadly obovate, folded, with a thickening on each side of spur mouth, 4.9–7.6 mm. long, 3.1–5.1 mm. wide; spur slender, conical, tapering into a cylindrical obtuse apex, distinctly incurved, 3.3–5.6 mm. long. Column dorsally slightly lobed, 1.1–1.6 mm. long.

KENYA. Embu District: Tana R., Kindaruma Dam, July 1972, *Archer* 684!; Teita District: Maungu Hills, W. slopes of Nyangala Hill, Dec. 1969, *Archer* 660! & Taveta, near Lumi R., Dec. 1961, *Polhill & Paulo* 977!
TANZANIA. Lushoto District: W. Usambara Mts., Mazumbai, Jan. 1976, *Cribb & Grey Wilson* 10206! & E. Usambara Mts., Kigongoi, Dec. 1942, *Moreau* 445!; Tanga District: Daluni R., bridge near Tanga, Feb. 1976, *L. Tanner* 4!
DISTR. **K** 4, 7; **T** 3; Mozambique
HAB. In woodland, wooded grassland and riverine forest, growing in exposed places often in the crowns of trees; sea-level–1100 m.

7. **M. physophora** (*Reichb.f.*) *Summerh.* in Bot. Mus. Leafl. Harv. Univ. 11: 148 (1943); L. Jonsson in Symb. Bot. Upsal. 23(4): 91 (1981). Type: Madagascar, *Hildebrandt* 3255 (W, holo., K, P, iso.!)

An epiphytic plant with a short stem, 15–20 mm. long, 2–3 mm. in diameter. Roots few–many, triangular in cross-section, up to 35 cm. long, 3 mm. in diameter. Scale-leaves rostrate, up to 5 mm. long, with ± 5 nerves. Inflorescences erect, spreading, up to 25 mm. long, up to 6 simultaneously, each with up to 15 flowers; peduncle distinct, up to 9 mm. long; rhachis straight to flexuose, angular, smooth; bracts distinctly sheathing, acute to subacute, up to 2 mm. long. Flowers whitish green; pedicel 5–9 mm. long; ovary 2.3–4.8 mm. long. Dorsal sepal hooded, ovate, ± acute, slightly thickened along central nerve, 1.7–2.6 mm. long, 1–1.7 mm. wide; lateral sepals asymmetric, slightly convex, ovate, acute, subapiculate, slightly thickened along central nerve, connate below ovary, 2.2–3.1 mm. long, 1.6–2.2 mm. wide. Petals ovate, obtuse, 1.4–2.5 mm. long, 0.7–1.1 mm. wide. Lip 3-lobed, 1.1–2.7 mm. long, 0.4–1.1 mm. wide; mid-lobe tongue-shaped, obtuse; side lobes indistinct, rounded, extending over spur-mouth; spur perpendicular or occasionally apically curved towards ovary, cylindrical, tapering into a distinct inflated, rounded apex, 7.5–9.5 mm. long. Column 0.7–1.1 mm. long.

KENYA. Kwale District: Gongoni Forest, Mar. 1958, *J.B. Smart*!
TANZANIA. Tanga District: Sawa, 1954, *Faulkner* 1510!; Uzaramo District: NW. Dar es Salaam, Kibaha, Nov. 1971 & Nov. 1972, *Flock* 154! & W. Dar es Salaam, Kigogo to Mwisho cross-roads bus stop, Nov. 1973, *Wingfield* 2492!; Zanzibar I., Chwaka, Dec. 1930, *Vaughan* 1747!
DISTR. **K** 7; **T** 3, 6; **Z**; Madagascar
HAB. In coastal bushland and woodland; sea-level–500 m.

SYN. *Angraecum physophorum* Reichb.f., Otia Bot. Hamb. 2: 77 (1881)
Gussonea physophora (Reichb.f.) Ridl. in J.L.S. 21: 492 (1885)
Angorkis physophora (Reichb.f.) O. Kuntze, Rev. Gen. Pl. 2: 651 (1891)
Epidorkis physophora (Reichb.f.) O. Kuntze, Rev. Gen. Pl. 2: 660 (1891)
Mystacidium physophorum (Reichb.f.) Th. Dur. & Schinz, Consp. Fl. Afr. 5: 54 (1895)

8. **M. megalorrhiza** (*Reichb.f.*) *Summerh.* in Bot. Mus. Leafl. Harv. Univ. 11: 158 (1943); L. Jonsson in Symb. Bot. Upsal. 23(4): 109 (1981). Type: Malawi, Shire R., *Waller* (K, holo.!)

An epiphytic plant. Stem up to 35 mm. long, 3–4.5 mm. in diameter. Roots long, mostly few, terete, distinctly verrucose, up to 30 cm. long, 1.5–2.5 mm. in diameter. Inflorescences erect or pendulous, spreading, up to 130 mm. long, up to 10 simultaneously, each with up to 20 flowers; peduncle up to 10 mm. long; rhachis straight to indistinctly flexuose, terete and slightly excavated; bracts sheathing, acute, 2 mm. long. Flowers up to 12 mm. long, whitish with base of tepals and apex of spur fading to salmon-pink; pedicel up to 2 mm. long; ovary 3–3.9 mm. long. Dorsal sepal narrowly elliptical to narrowly oblong, acuminate with slightly bulging area, 5.7–7 mm. long, 1.8–2.4 mm. wide; lateral sepals slightly asymmetric, narrowly oblong, acuminate, 6.6–8.7 mm. long, 1.6–2.4 mm. wide. Petals narrowly elliptic, acuminate, apex slightly erose, 5.6–7.4 mm. long, 1.8–2.4 mm. wide. Lip large, subspathulate, obscurely 3-lobed; mid-lobe deeply concave, ovate, acuminate, apically erose; side lobes narrow, oblong, thickened towards spur-mouth, 6.5–8.1 mm. long, 2.2–3.5 mm. wide; spur cylindrical, semicircularly curved, tapering to an obtuse apex, 6.2–7.5 mm. long, 0.7–1 mm. in diameter. Column with a central trigonal pillar-like thickening, 1.8–2.7 mm. long.

TANZANIA. Lushoto District: E. Usambara Mts., between Kizara and Daluni, Dec. 1941, *Moreau* 118!; Handeni District: ± 33 km. SSE. of Korogwe, June 1970, *Mwasumbi, Harris & Tadros* 10713!; Tanga District: Moa, Oct. 1941, *Moreau* 98!
DISTR. **T** 3; Malawi
HAB. In dry woodland, usually on the finer branches in the crowns of large trees; sea-level–550 m.

SYN. *Angraecum megalorrhizum* Reichb.f., Otia Bot. Hamb. 2: 117 (1881)
Angorkis megalorrhiza (Reichb.f.) O. Kuntze, Rev. Gen. Pl. 2: 651 (1891)
Epidorkis megalorrhiza (Reichb.f.) O. Kuntze, Rev. Gen. Pl. 2: 660 (1891)
Gussonea megalorrhiza (Reichb.f.) Schltr. in Beih. Bot. Centr. 36: 93 (1918)

9. **M. koehleri** (*Schltr.*) *Summerh.* in Bot. Mus. Leafl. Harv. Univ. 11: 158 (1943) & in Hook., Ic. Pl. 35, t. 3465 (1947); Piers, Orch. E. Afr., ed. 2: 272 (1968); J. Stewart in U.K.W.F.: 785 (1974); Williamson, Orch. SC. Afr.: 184 (1977). Type: Tanzania, Usambara Mts., Lushoto, *Koehler* (B, holo.)

An epiphytic plant. Stem up to 40 mm. long, 2–3 mm. in diameter. Roots many, often forming dense conical-shaped tufts, terete to slightly flattened, smooth, up to 20 cm. long, 1–2 mm. in diameter. Scale-leaves acute to acuminate, with ± 5 nerves, up to 4 mm. long. Inflorescences erect, spreading, up to 70 mm. long, up to 7 simultaneously, each with up to 15 flowers; peduncle 10 mm. long; rhachis subflexuose, terete, apically angular; bracts sheathing, acute, 1–1.6 mm. long. Flowers up to 16 mm. long, whitish, spur apex orange-brown; pedicel up to 2 mm. long; ovary curved and twisted, 1.2–2.3 mm. long. Dorsal sepal narrowly oblong, acuminate, subapiculate, with a slightly bulging central area, 5.5–7.3 mm. long, 1.5–2.2 mm. wide; lateral sepals slightly asymmetric, narrowly oblong, acuminate, apiculate, 6.4–8.4 mm. long, 1.2–1.9 mm. wide. Petals slightly asymmetric, ovate to narrowly ovate, apiculate, apically erose, 5.2–6.2 mm. long, 1.6–2 mm. wide. Lip large, ± entire, broadly ovate, acute to acuminate or obtuse to subapiculate, with upfolded erose margins, 5.8–7 mm. long, 4.1–5.1 mm. wide; spur variously incurved, cylindrical, tapering to a conical inflation at apex, up to 1.7 mm. long. Column 1.7–2.5 mm. long. Fig. 122.

UGANDA. Bunyoro District: Butiaba [Buitiba] Flats, fl. cult. Mar. 1944, *Eggeling* 5495! & Budongo Forest, Feb. 1943, *Eggeling* 5234!; Mengo District: Bulemezi, Kangavwe Forest, Apr. 1915, *Maitland* 121!
KENYA. Kisumu area, Mar. 1937, *C. van Someren* 37! & May 1945, *Copley* 126!; near Lake Victoria, Mar. 1938, *Firth*!
TANZANIA. Lushoto District: W. Usambara Mts., Jan. 1946, *Moreau* 672!; Mbeya District: Santilya, May 1970, *Leedal* 445!
DISTR. **U** 2, 4; **K** 5; **T** 3, 7; Nigeria, Zaire, Rwanda, Zambia
HAB. In lowland forest, riverine forest, dry evergreen forest and occasionally into woodland in mist zone, on twigs and finer branches of understorey and canopy trees; 200–1500 m.

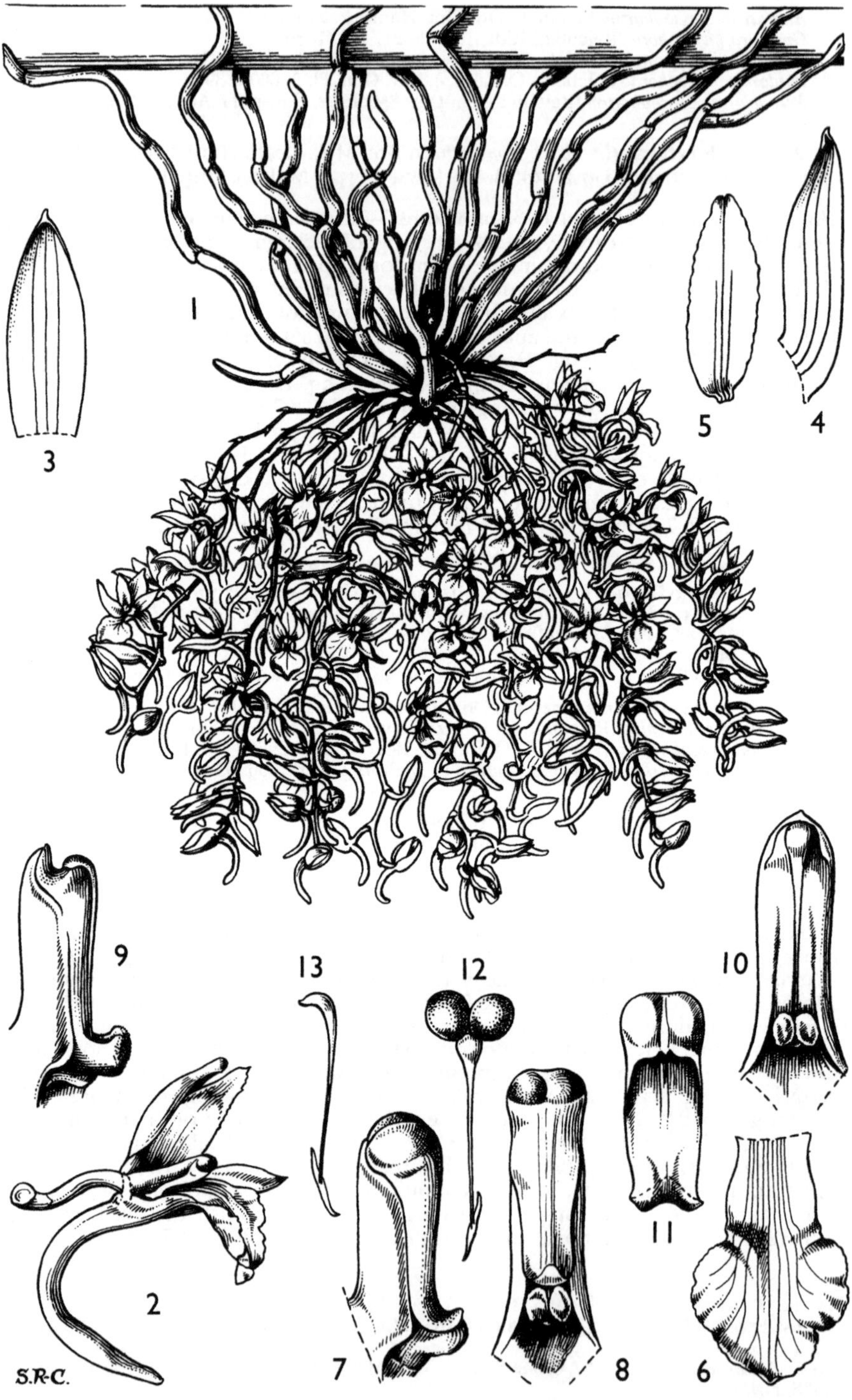

FIG. 122. *MICROCOELIA KOEHLERI*—**1**, habit, × 1; **2**, flower, lateral view, one lateral sepal and petal removed, × 4; **3**, dorsal sepal, × 6; **4**, lateral sepal, × 6; **5**, petal, × 6; **6**, lip lamina, spread out, × 6; **7**, **8**, column, side and front views, × 12; **9**, **10**, column, side and front views, anther and pollinia removed, × 12; **11**, anther, from inside, × 12; **12**, pollinarium, front view, × 12; **13**, same, side view, pollinia removed, × 12. Drawn by Miss Stella Ross-Craig. Reproduced from Hooker's Icones Plantarum.

SYN. *Angraecum koehleri* Schltr. in E.J. 38: 162 (1906)
Gussonea friesii Schltr. in R.E. Fries, Wiss. Ergebn. Schwed. Rhod.-Kongo-Exped. 1911–12: 250 (1916). Type: Zambia, between Mbala and Kasanga, *Fries* 1342 (UPS, holo.!)
G. koehleri (Schltr.) Schltr. in Beih. Bot. Centr. 36: 93 (1918)
Microcoelia friesii (Schltr.) Summerh. in Orch. Rev. 47: 21, in obs. (1939)
M. pachystemma Summerh. in Orch. Rev. 47: 21 (1939). Type: Kenya, near Lake Victoria, *Firth* (K, holo.!)

10. **M. konduensis** (*De Wild.*) *Summerh.* in Bot. Mus. Leafl. Harv. Univ. 11: 154 (1943); L. Jonsson in Symb. Bot. Upsal. 23(4): 114 (1981). Type: Zaire, W. Lusambo, *E. & M. Laurent* (BR, holo.)

An epiphytic plant. Stem short, up to 10 mm. long, 2–3 mm. in diameter. Roots many, mostly twisted and usually with many branches, terete, smooth, up to 50 cm. long, 0.5–1.5 mm. in diameter. Scale-leaves acuminate to rostrate, 5-nerved, 3–5 mm. long. Inflorescences descending, spreading, up to 150 mm. long, up to 15 simultaneously, each with up to 15 flowers; peduncle 10 mm. long; rhachis straight to indistinctly flexuose, stiff, terete; bracts sheathing, acute to subacuminate, up to 2.5 mm. long. Flowers up to 11 mm. long, whitish, sepals tinged in orange-brown around base, lip tinged green around spur-mouth; pedicel curved and twisted, up to 2.3 mm. long; ovary curved and twisted, 1.3–2 mm. long. Dorsal sepal hooded, oblong to elliptic, acuminate, keeled apically, 3.1–5.3 mm. long, 1.1–2 mm. wide; lateral sepals asymmetric, narrowly ovate to obovate, convex, acute, apiculate, 4–6.6 mm. long, 1.3–2.3 mm. wide. Petals slightly asymmetric, narrowly ovate, acute, subapiculate, 2.2–6.1 mm. long, 0.8–1.6 mm. wide. Lip large, ± entire, spathulate or orbicular to broadly transversely elliptic, obtuse to slightly emarginate, erose, 4.5–7.9 mm. long, 2.2–6 mm. wide; spur distinctly incurved, obtuse, cylindrically inflated apically, 2–4.8 mm. long. Column 1–2.3 mm. long.

UGANDA. Mengo District: Mabira Forest, June 1908, *E. Brown* 433!
DISTR. U 4; Ivory Coast, Ghana, Benin, Nigeria, Cameroon, Zaire
HAB. In rain-forest and in plantations, on finer branches and twigs in the canopies of trees; ± 1050–1100 m.

SYN. *Angraecum konduensis* De Wild., Not. Pl. Utiles Congo 1: 321 (1904)
Dicranotaenia dahomeensis Finet in Bull. Soc. Bot. Fr. 54, Mém. 9: 47 (1907). Type: Benin, *Le Testu* 125 (P, holo., UPS, drawing)
Gussonea konduensis (De Wild.) Schltr. in Beih. Bot. Centr. 36: 93 (1918)
G. dahomeensis (Finet) Schltr. in Beih. Bot. Centr. 36: 92 (1918)
Microcoelia dahomeensis (Finet) Summerh. in F.W.T.A. 2: 454 (1936)

11. **M. microglossa** *Summerh.* in F.W.T.A. 2: 454 (1936) & in K.B. 1936: 231 (1936) & in Bot. Mus. Leafl. Harv. Univ. 11: 147 (1943) & in Hook., Ic. Pl. 35, t. 3466 (1947) & in F.W.T.A., ed. 2, 3: 259 (1968); L. Jonsson in Symb. Bot. Upsal. 23(4): 119 (1981). Type: Nigeria, NE. Calabar, *Talbot* (K, holo.!)

An epiphytic plant. Stem up to 32 mm. long, 2–3 mm. in diameter. Roots twisting and with a few branches, terete, smooth, up to 44 cm. long, 1–2 mm. in diameter. Scale-leaves acuminate, with 5–7 nerves, up to 3.6 mm. long. Inflorescences erect, spreading, up to 50 mm. long, up to 4 simultaneously, each with up to 10 flowers; peduncle up to 10 mm. long; rhachis terete, smooth; bracts sheathing, acute to subacute, up to 2 mm. long. Pedicel distinct, up to 5.5 mm. long; ovary 1.9–2.6 mm. long. Flowers up to 11 mm. long, whitish with a brown median band on the sepals; pedicel distinct, up to 5.5 mm. long; ovary 1.9–2.6 mm. long. Dorsal sepal hooded, ovate, acute to subacute, 2.6–3 mm. long, 1.4–2.5 mm. wide; lateral sepals asymmetric, slightly convex, ovate, acute to subapiculate, connate below ovary, 3.6–3.7 mm. long, 2.3–3 mm. wide. Petals ovate, acute, 2.3–2.5 mm. long, 1.5–1.6 mm. wide. Lip short, obscurely 3-lobed, 2.3–2.5 mm. long, 1.6–2.1 mm. wide; mid-lobe ovate-triangular in outline, porrect to decurved; side lobes indistinct; spur pendent, long, slightly curved, cylindrical, 18.6–19.9 mm. long, tapering into a distinctly inflated, conical-shaped apex. Column 1–1.4 mm. long.

UGANDA. Bunyoro District: Budongo Forest, Nov. 1939, *Eggeling* 3836! & Oct. 1943, *Eggeling* 5450!; Toro District: E. slope of Ruwenzori Mts., Bwamba, Feb. 1939, *A.S. Thomas* 2788! & July 1939, *A.S. Thomas* 3042!
DISTR. U 2; Nigeria, Cameroon, Zaire
HAB. In dense rain-forest, on understorey bushes and trees; 470–1070 m.

12. **M. bulbocalcarata** *L. Jonsson* in Symb. Bot. Upsal. 23(4): 121 (1981). Type: Uganda, Kigezi District, Impenetrable Forest, *Meyer* K56 (K, holo.!)

An epiphytic plant with a short stem, up to 12 mm. long, 3 mm. in diameter. Roots few, terete, smooth, up to 53 cm. long, 1.5–2 mm. in diameter. Scale-leaves acuminate, with 7–8 nerves, up to 5 mm. long. Inflorescences erect, spreading, up to 15 mm. long, up to 4 simultaneously, each with up to 6 flowers; peduncle up to 5 mm. long; rhachis flexuose, terete, smooth; bracts sheathing, acute, up to 1.6 mm. long. Flowers up to 15.4 mm. long, whitish with a dark green patch on the lip; base of sepals and apex of dorsal sepal pale green; pedicel distinct, smooth, up to 5.3 mm. long; ovary up to 3.2 mm. long. Dorsal sepal hooded, elliptic, acute, subapiculate, 4.8–6.3 mm. long, 2.6–3.7 mm. wide; lateral sepals asymmetric, elliptic, acute, slightly convex, shortly connate at base, 5.2–7 mm. long, 3.2 mm. wide. Petals asymmetric, obovate to ovate, obtuse, slightly convex, 4.5–6 mm. long, 3–3.8 mm. wide. Lip oblong, acute, obscurely 3-lobed, 4.2–5 mm. long, 2.6–3.7 mm. wide; mid-lobe triangular, horizontal to decurved; side lobes indistinct; spur long, cylindrical, distinctly inflated, 7.7–10.5 mm. long tapering into a bulbous apical inflation, apex incurved. Column 1–1.3 mm. long. Pl. 7.

UGANDA. Kigezi District: Impenetrable Forest, Dec. 1967, *C. Leakey* 1000! & near Ihihizo R., fl. cult. Nov. 1977, *Meyer* K56!
DISTR. U 2; adjacent Rwanda
HAB. In dense upland forest, on undergrowth trees and shrubs; 1680–1950 m.

13. **M. caespitosa** (*Rolfe*) *Summerh.* in F.W.T.A. 2: 454 (1936), in Bot. Mus. Leafl. Harv. Univ. 11: 151 (1943) & in F.W.T.A., ed. 2, 3: 259 (1968); Sanford in Mitt. Bot. Staats. München 10: 296 (1971); L. Jonsson in Symb. Bot. Upsal. 23(4): 123 (1981). Type: Cameroon, *Bates* 393 (K, holo.!, BM, iso.)

An epiphytic plant with a short, often stout stem, 10–31 mm. long, 2.5–5 mm. in diameter. Roots few–many, undulating and twisting, branched, terete, smooth, up to 90 cm. long, 1.5–2.5 mm. wide. Scale-leaves acuminate, with 6–7 nerves, up to 5–7 mm. long. Inflorescences erect, spreading, up to 30 mm. long, up to 10 simultaneously, each with up to 22 flowers; peduncle up to 10 mm. long; rhachis straight to indistinctly flexuose, terete; bracts sheathing, acute to subacute, up to 2 mm. long. Flowers up to 17.5 mm. long, whitish or fading into greenish white; pedicel 4.7–8.2 mm. long; ovary 1.1–2.9 mm. long. Dorsal sepal hooded, oblong, obtuse to subacute, 2.5–3.5 mm. long, 1–1.9 mm. wide; lateral sepals asymmetrical, ovate, obtuse to subapiculate, convex, 2.7–4 mm. long, 2–2.6 mm. wide. Petals obovate, obtuse or subacute, 2.3–3 mm. long, 1.2–1.8 mm. wide. Lip small, obscurely 3-lobed, 2.5–3.6 mm. long, 1.1–1.9 mm. wide; mid-lobe triangular-shaped, flat or cup-shaped, straight or deflexed; side lobes indistinct with well-developed thickening on each side of the spur; spur long, pendent, basal part constricted, central part inflated, tapering into an obtuse apex, 7.4–11.3 mm. long. Column 1.2–2.2 mm. long.

UGANDA. Bunyoro District: Siba area, fl. cult. Sept. 1943, *Eggeling* 5238!
DISTR. U 2; Sierra Leone to Cameroon, Gabon, Central African Republic, Zaire
HAB. In rain-forest; 1000–1100 m.

SYN. *Angraecum caespitosum* Rolfe in F.T.A. 7: 150 (1897)
A. crinale De Wild., Not. Pl. Utiles Congo 1: 320 (1904). Type: Zaire, Bandundu, *E. & M. Laurent* (BR, holo., K, iso.!)
A. micropetalum Schltr. in E.J. 38: 23 (1905). Types: Cameroon, NW. Ngoko R., Dja R., *Schlechter* 12799 (B, syn.†, BR, lecto.), near Bakossi, *Schlechter* 12786 (B, syn.†) & *Schlechter* 12892 (B, syn.†)
A. andersonii Rolfe in K.B. 1912: 134 (1912). Type: Ghana, without exact locality, *Anderson* (K, holo.!)
A. bieleri De Wild. in B.J.B.B. 5: 182 (1916). Types: Zaire, Upper Lopori R., *Bieler* 6! (BR, lecto.) & Eala, *Laurent* 1780 (BR, syn.)
Gussonea caespitosa (Rolfe) Schltr. in Beih. Bot. Centr. 36: 91 (1918)
G. crinalis (De Wild.) Schltr. in Beih. Bot. Centr. 36: 91 (1918)
G. micropetala (Schltr.) Schltr. in Beih. Bot. Centr. 36: 91 (1918)
G. bieleri (De. Wild.) Schltr. in Beih. Bot. Centr. 36: 91 (1918)
Microcoelia micropetala (Schltr.) Summerh. in Bot. Mus. Leafl. Harv. Univ. 11: 155 (1943)
M. bieleri (De Wild.) Summerh. in Bot. Mus. Leafl. Harv. Univ. 11: 151 (1943)

14. **M. macrorrhynchia** (*Schltr.*) *Summerh.* in F.W.T.A. 2: 454 (1936); L. Jonsson in Symb. Bot. Upsal. 23(4): 131 (1981). Types: Cameroon, near Ngoko, Dja R., *Schlechter* 12785 (B, holo.†) & Dja R., Bitye, near Ndu, *Bates* 1429 (K, neo.!)

An epiphytic plant with a small stem, up to 15 mm. long, 1.5–3 mm. in diameter. Roots few, undulating, flattened to substrate to terete, smooth, up to 70 mm. long, 1–3 mm. in diameter. Scale-leaves acute to acuminate, with 5–7 nerves, up to 4 mm. long. Inflorescences erect to pendulous, up to 10 cm. long, up to 5 simultaneously, each with up to 18 flowers; peduncle up to 10 mm. long; rhachis straight or indistinctly flexuose, terete; bracts sheathing, acute, 0.7–1.9 mm. long. Flowers up to 11.5 mm. long; dorsal sepal brown; lateral sepals and petals white; lip white with yellow-green spot at base; spur green-brown; pedicel straight to distinctly twisted, up to 2.3 mm. long; ovary 1–1.9 mm. long. Dorsal sepal hood-shaped, ovate, acute, 3–4.9 mm. long, 1–2 mm. wide; lateral sepals asymmetric, ovate, acute, apex recurved, 3.2–5.7 mm. long, 1.5–2.4 mm. wide. Petals asymmetric, ovate, obtuse to subacute, erose, 3–4.5 mm. long, 1.6–3.4 mm. wide. Lip indistinctly 3-lobed, 4.5–6 mm. long, 3.6–5.8 mm. wide; mid-lobe spathulate, semicircular in outline, distinctly dentate, with minutely erose margins, apically folded and distinctly decurved, giving an emarginate appearance; side lobes erect, oblong, wedge-shaped, almost closing the spur-mouth; spur long, pendulous, incurved at late anthesis, tapering apically into a distinct conical inflation, 5–6.5 mm. long. Column 1.0–2.0 mm. long. Pl. 7.

UGANDA. Bunyoro District: Budongo Forest, fl. cult. July 1950, *Eggeling* 5213!
DISTR. U 2; Liberia to Cameroon, Gabon, Central African Republic, Zaire, Malawi and Zambia
HAB. In rain-forest and in riverine forest, also in secondary forest and plantations, in exposed positions on twigs and small branches; 1000–1100 m.

SYN. *Angraecum macrorrhynchium* Schltr. in E.J. 38: 22 (1905)
Gussonea macrorrhynchia (Schltr.) Schltr. in Beih. Bot. Centr. 36: 93 (1918)
Encheiridion macrorrhynchium (Schltr.) Summerh. in Bot. Mus. Leafl. Harv. Univ. 11: 162 (1943)

54. DIAPHANANTHE

Schltr., Die Orchideen: 593 (1914) & in Beih. Bot. Centr. 36: 95 (1918); Summerh. in Bot. Mus. Leafl. Harv. Univ. 12: 95 (1945) & in K.B. 14: 140 (1960)

Rhipidoglossum Schltr. in Beih. Bot. Centr. 36: 80 (1918); Summerh. in Blumea, Suppl. 1: 80 (1937); Garay in Bot. Mus. Leafl. Harv. Univ. 23: 195 (1972); Senghas in Schltr., Die Orchideen, ed. 3, 1: 1110 (1986)
Sarcorhynchus Schltr. in Beih. Bot. Centr. 36: 104 (1918)

Epiphytic or rarely lithophytic herbs with short or long monopodial stems, usually unbranched and covered by sheathing leaf-bases; roots elongate, emerging all along stem through the leaf-bases, often prominent and rarely branched. Leaves distichous, coriaceous or fleshy, rarely thin-textured, unequally bilobed at apex, articulated at base to a sheathing leaf-base, often twisted at base to lie in one plane. Inflorescences 1–many, emerging through the sheathing leaf-bases in upper part of stem, usually several–many-flowered; bracts usually amplexicaul, rarely prominent. Flowers white, pale green or yellow, rarely pinkish, usually translucent, rarely showy. Sepals free, subsimilar. Petals free. Lip entire or obscurely lobed, spurred, usually with a tooth-like or transverse callus in the mouth of the spur. Column porrect; stipites 2; viscidia 1 or 2; rostellum pendent or reflexed, linear or tapering and bifid to pendent, clavate and obscurely 3-lobed.

A genus of about 50 species confined to continental, and mostly tropical, Africa.

In this account I have followed the generic treatment of Summerhayes (1960) in including the genus *Rhipidoglossum* Schltr. as a section within *Diaphananthe*. This has recently been challenged by Garay (1972) and by Senghas (1986) who have preferred to recognise *Rhipidoglossum* as distinct. However, the reasoning of Summerhayes that several species have characteristics that bridge the gap between the two genera seems as valid now as then.

Senghas has also retained *Sarcorhynchus* as distinct but I agree with Rasmussen in Norw. Journ. Bot. 21: 229 (1974) that it falls within the range of variability of *Diaphananthe*.

Schlechter (1918) recognised three sections in *Diaphananthe*, but Summerhayes united these in sect. *Diaphananthe* (species 1–7 here), recognised by the pollinarium in which the pollinia are each attached by a separate stipes to a common viscidium, the rostellum being bifid and usually tapering. He placed all of the other species (8–27 here) in sect. *Rhipidoglossum*, characterised by each pollinium attached by a stipes to a separate viscidium and the rostellum being obscurely 3-lobed with the midlobe prominent, pendent and fleshy. In section *Diaphananthe* the callus at the base of the spur is usually a prominent tooth, while in sect. *Rhipidoglossum* it is often obscure and in some species absent. Rasmussen included all of those species previously referred to *Sarcorhynchus* in the latter section.

Flowers borne in whorls of 2–4 at the nodes of the inflorescence . . . . 7. *D. fragrantissima*
Flowers borne singly at the nodes of the inflorescence:
Bracts of the inflorescence prominent, more than 6 mm. long and wide:
Sepals more than 15 mm. long; lip 18–20 mm. long, 20–23 mm. wide; bracts 10–15 mm. long, papery, creamy white . . . . 8. *D. kamerunensis*
Sepals 15 mm. or less long; lip 7.5–15 mm. long, 7–18 mm. wide; bracts 6–10 mm. long, dark brown . . . . 1. *D. pellucida*
Bracts of the inflorescence inconspicuous, usually less than 3 mm. long:
Spur more than 15 mm. long:
Flowers white; spur ± straight from a broad mouth; lip 4.5–6.5 mm. long, 6.2–7 mm. wide . . . . . . 14. *D. tenuicalcar*
Flowers pale yellow; spur incurved from a narrow mouth; lip 11–12 mm. long, 10–17 mm. wide 12. *D. stolzii*
Spur less than 15 mm. long:
Spur very short, less than 3 mm. long, almost globular:
Leaves oblong-oblanceolate, 1–2 cm. wide; flowers white; lip broader than long, 2.7–3.8 mm. long, 3.5–4.7 mm. wide; spur papillate-pubescent in mouth . . . . 18. *D. densiflora*
Leaves linear, less than 6 mm. wide; lip as long as or longer than wide; spur glabrous within:
Leaves less than 3.5 cm. long, scarcely falcate, rounded at apex; inflorescence 1.5 cm. long, 2–3-flowered; lip 4–4.5 mm. long, 3–3.3 mm. wide; rostellum spathulate . . . . . . 25. *D. orientalis*
Leaves more than 3.5 cm. long, falcate, acute; inflorescences 2–3.5 cm. long, 3–8-flowered; lip 2 mm. long and wide; rostellum shortly clavate . . . . 22. *D. adoxa*
Spur more than 3 mm. long, fusiform or cylindric:
Pollinia attached by stipites to a single gland or viscidium:
Lip ovate; spur clavate to cylindrical, more than 1 cm. long; stems 1.5–5 cm. long:
Spur cylindrical, 15 mm. long; lip 8–9 mm. long, 4–5 mm. wide . . . . 2. *D. bueae*
Spur slightly swollen towards apex, 11–12 mm. long; lip 3.4–3.8 mm. long, 1.7–2.3 mm. wide 3. *D. rohrii*
Lip oblong, pandurate or subcircular:
Leaves less than 12 cm. long, elliptic-ovate to elliptic-oblong; lip pandurate or oblong, notched at apex . . . . 4. *D. bidens*
Leaves more than 14 cm. long, linear to oblanceolate; lip broadly ovate or subcircular, obtuse or shortly apiculate:
Lip subcircular or elliptic; leaf linear, falcate, 1–1.8 cm. wide . . . . 6. *D. lorifolia*
Lip obscurely 3-lobed, broadly ovate; leaf oblanceolate, weakly falcate, 2.5–2.8 cm. wide . . . . 5. *D. eggelingii*
Pollinia attached by stipites to 2 separate glands:
Leaves less than 2 mm. wide:
Spur slender, incurved, cylindrical, ± 7 mm. long . . . . 17. *D. microphylla*
Spur clavate-fusiform, pendent, 3.8–4 mm. long 26. *D. tanneri*
Leaves more than 2 mm. wide:
Lip lanceolate, acute . . . . 24. *D. stellata*
Lip neither lanceolate nor acute:

Leaves ovate-elliptic, less than 3 times as long as wide; lip pandurate, deeply bilobed at apex . . . . 16. *D. bilobata*
Leaves linear to oblanceolate, more than 5 times as long as broad:
Lip broader than long, usually fan-shaped or transversely elliptic:
Lip 6–9.5 mm. long, 7–9.5 mm. wide; spur 8.5–14 mm. long, with a prominent tooth in the mouth of the spur 9. *D. pulchella*
Lip less than 6 mm. long and 7 mm. wide:
Lip more than 5 mm. long and 5.5 mm. wide; spur 8–9 mm. long . . . 10. *D. ugandensis*
Lip less than 4.5 mm. long and 5.5 mm. wide; spur less than 7 mm. long:
Bracts 3–6 mm. apart on inflorescence-axis; lip 3.3–4.5 mm. long, 3.2–5.7 mm. wide, broadest across the apical half . . . . . 19. *D. xanthopollinia*
Bracts 1–2 mm. apart; lip 1.4–3.5 mm. long, 2.5–4.5 mm. wide, broadest across the middle . . . . 21. *D. rutila*
Lip as long as broad or longer, ovate to oblong:
Inflorescences less than 3.5 cm. long; lip ovate, acute . . . . 23. *D. montana*
Inflorescences more than 5 cm. long; lip ovate to oblong or fan-shaped:
Spur 13–15 mm. long, filiform; stem very short . . . . 27. *D. meliantha*
Spur less than 7 mm. long; stem usually 5 or more cm. long:
Lip deeply notched at apex; stem ± 5 cm. long; peduncle ± 4 cm. long 11. *D. cuneata*
Lip obtuse or shallowly notched; stem usually more than 10 cm. long; peduncle 1 cm. long or less:
Spur 8–8.5 mm. long . . . . 13. *D. schimperiana*
Spur less than 7 mm. long:
Lip obovate to fan-shaped; spur 4.5–6.8 mm. long; leaves 5–10 mm. wide; flowers 5–10, pale yellow or green 15. *D. subsimplex*
Lip ovate; spur 4–4.5 mm. long; leaves 3–5 mm. wide; flowers up to 20, cream or buff . . . . 20. *D. ovalis*

1. **D. pellucida** (*Lindl.*) *Schltr.*, Die Orchideen: 593 (1914) & in Beih. Bot. Centr. 36: 97 (1918); Summerh. in F.W.T.A., ed. 2, 3: 261 (1968). Types: Nigeria, Brass, *Barter* 37 & Onitscha, *Barter* 1757 (K, syn.!)

An epiphyte with a short stem up to 12 cm. long and 1 cm. diameter. Leaves distichous in upper half of stem, oblanceolate, subacute and unequally bilobed at apex, 18–40 cm. long, 2–9 cm. wide, with prominent net-venation, articulated to rough overlapping leaf-bases 2–3 cm. long. Inflorescences 1–7, pendent, 20–50 cm. long, up to 50- or more-flowered; peduncle 2.5–8 cm. long; bracts broadly ovate, obtuse, amplexicaul, 6–10 mm. long and wide. Flowers pale yellow, lacking a scent; pedicel and ovary 2–3 mm. long. Dorsal sepal ovate, subacute, 7.8–13.5 mm. long, 2.8–6.2 mm. wide; lateral sepals lanceolate, subacute, 8–16.5 mm. long, 2.7–7.5 mm. wide. Petals linear to ovate, acute, 7.5–17.5 mm. long, 2.6–6.5 mm. wide, erose. Lip geniculate at base, obscurely 3-lobed, 7.5–15 mm. long, 7–18 mm. wide, with a lacerate margin, with a tooth in the mouth of the spur which is bordered on each side by a fleshy raised rim; spur pendent, narrowly fusiform, inflated, 5–14.5 mm. long. Column 2.5–4.5 mm. long; stipites 2; viscidium 1; rostellum tapering, bifid. Fruit ellipsoidal, 10–20 mm. long. Fig. 123.

UGANDA. Bunyoro District: Budongo Forest, Wambabya [Luanbabya], 27 Dec. 1909, *Dawe* 799!; Toro District: 26 Nov. 1921, *Snowden* 733!; Victoria Nyanza region, 14 July 1914, *Maitland* 193!
DISTR. **U** 2, 4; Sierra Leone to Cameroon and Zaire
HAB. In rain-forest; 900–1500 m.

SYN. *Angraecum pellucidum* Lindl. in J.L.S. 6: 134 (1862)
Listrostachys pellucida (Lindl.) Reichb. f. in Walp., Ann. Bot. 6: 908 (1864)

2. **D. bueae** (*Schltr.*) *Schltr.* in Beih. Bot. Centr. 36: 96 (1918); Summerh. in F.W.T.A., ed. 2, 3: 261 (1968). Type: Cameroon, near Buea, *Diestel* (B, holo., K, record of holo.!)

An epiphyte with a short stem ± 1.5 cm. long, 5 mm. diameter. Leaves ± 5, slightly falcate, linear, unequally bilobed at apex, 4–7 cm. long, 0.5–1.5 mm. wide, twisted at base to lie in one plane, articulate to a short 2–3 mm. long sheath. Inflorescences 5–7 cm. long, laxly many-flowered; peduncle 6–12 mm. long; bracts amplexicaul, 1 mm. long. Flowers pale yellow; pedicel and ovary 5–8 mm. long. Dorsal sepal elliptic-lanceolate, subacute, 6–8.6 mm. long, 2.7–4 mm. wide; lateral sepals elliptic, obtuse, 5.6–8.7 mm. long, 2.6–3.7 mm. wide. Petals linear to obovate, obtuse, 5–8 mm. long, 1.3–2.2 mm. wide, erose. Lip ovate, subacute, 8–9 mm. long, 4–4.5 mm. wide, with an obscure tooth in the mouth of the spur; spur swollen and slightly lobed at apex, 15–15.5 mm. long. Column 3.5 mm. long; stipites 2, usually coherent, ciliate; viscidium 1, orange-red; rostellum pointed, entire. Fruit ellipsoidal, 9–11 mm. long.

UGANDA. Ankole District: Kalinzu Forest, 18 Feb. 1975, *H. Meyer* 129!
DISTR. **U** 2; Ivory Coast, Cameroon
HAB. Rain-forest; 1500–1600 m.

SYN. *Angraecum bueae* Schltr. in E.J. 38: 159 (1906)

NOTE. This agrees well in most respects with the type material but has a somewhat less clavate, longer spur.

3. **D. rohrii** (*Reichb.f.*) *Summerh.* in K.B. 14: 140 (1960). Type: Ethiopia, without exact locality, *Roth* (W, holo., K, iso.!)

An epiphyte with an irregular short stem 2–5 cm. long, 5–7 mm. diameter. Leaves 2–7, somewhat falcate, linear to obovate, slightly unequally bilobed at the apex, 5–15 cm. long, 0.7–2.2 cm. wide, articulated to a 5–10 mm. long base. Inflorescences 5–17 cm. long, laxly many-flowered; penduncle 3–4 cm. long; rhachis zig-zag; bracts amplexicaul, 2–8 mm. long. Flowers waxy, yellow; pedicel and ovary 4 mm. long, scabrid. Dorsal sepal oblanceolate, subacute, 3.6–4 mm. long, 1.5–2 mm. wide; lateral sepals linear to obovate, obtuse, 3.7–4 mm. long, 1–1.4 mm. wide. Petals linear, obtuse, 3.4–4.3 mm. long, 1 mm. wide. Lip ovate, acute, 3.4–3.8 mm. long, 1.7–2.3 mm. wide, with an obscure tooth in the mouth of the spur; spur incurved, clavate, 11–12 mm. long. Column 2.4 mm. long; stipites 2, obovate, coherent, ciliate; viscidium 1; rostellum pointed, pendent. Fruit ellipsoidal, 7–10 mm. long.

UGANDA. Toro District: Bwamba Pass, 13 Aug. 1938, *A.S. Thomas* 2358!; Mbale District: Elgon, 10 Oct. 1962, *C. Leakey* 25! & Bulambuli, 14 July 1924, *Snowden* 921!
KENYA. Trans-Nzoia/N. Kavirondo District: Elgon, May 1960, *Tweedie* 2015!; Trans-Nzoia District: Kitale, Mar. 1938, *C. van Someren* 36!; Naivasha District: Kinangop, Apr. 1938, *Chandler* 2386!
TANZANIA. Masai District: Ngorongoro Crater, 23 Sept. 1977, *Raynal* 19229! & 23 May 1983, *Chuwa* 2537!
DISTR. **U** 2, 3; **K** 3; **T** 2; Liberia, Ivory Coast, Ghana, Togo, Cameroon, São Tomé, Bioko [Fernando Po], Ethiopia, Zaire, Burundi and Angola
HAB. Montane forest; 2100–3000 m.

SYN. *Angraecum rohrii* Reichb.f., Otia Bot. Hamb. 2: 117 (1881)
Angorchis rohrii (Reichb.f.) O. Kuntze, Rev. Gen. Pl. 2: 651 (1891)
Angraecum quintasii Rolfe in F.T.A. 7: 141 (1897). Type: São Tomé, *Quintas* 6 (K, holo.!)
Diaphananthe quintasii (Rolfe) Schltr. in Beih. Bot. Centr. 36: 99 (1918); Summerh. in F.W.T.A., ed. 2, 3: 261 (1968); Piers, Orch. E. Afr., ed. 2: 267 (1968); U.K.W.F.: 788, fig. (1974)

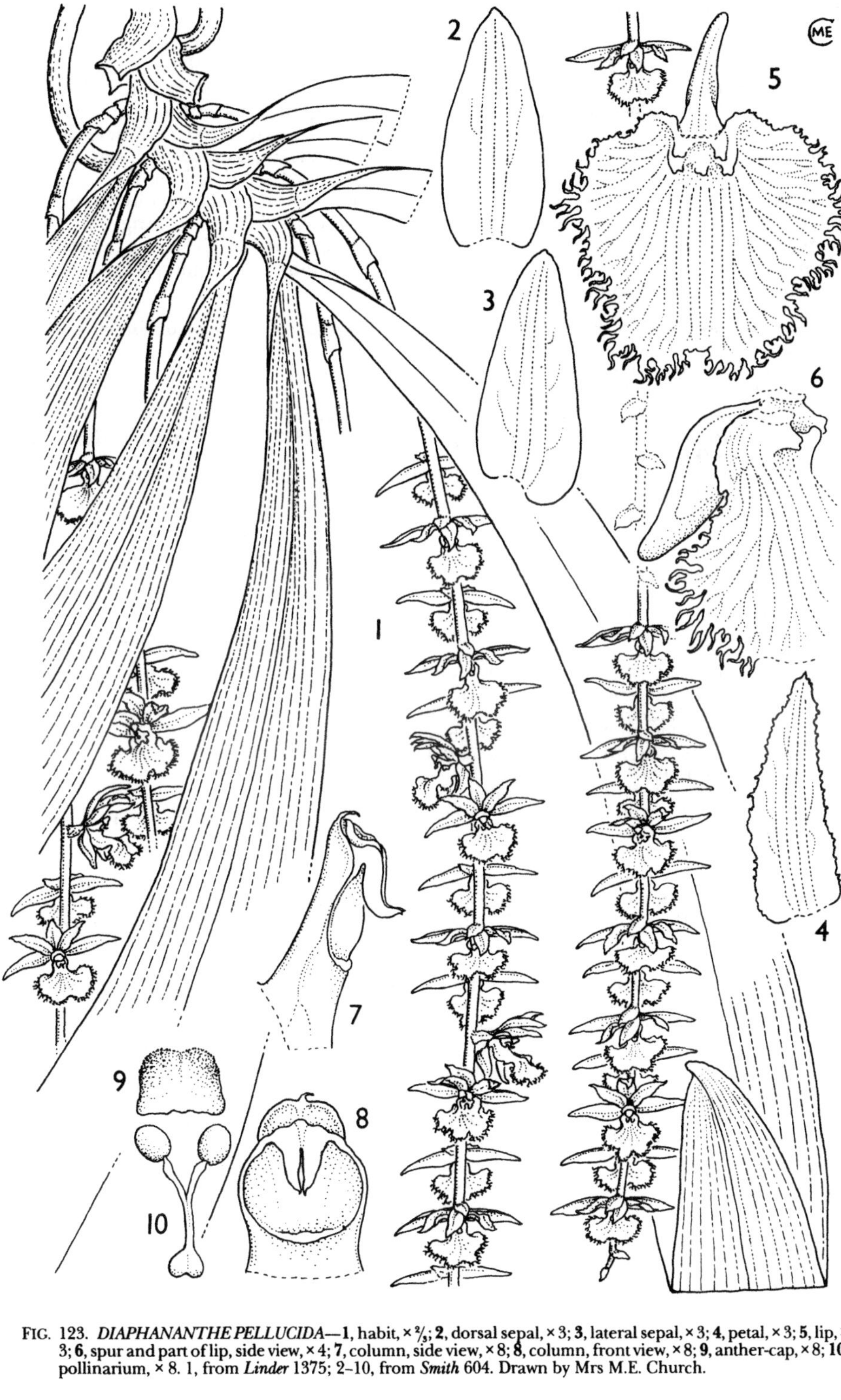

FIG. 123. *DIAPHANANTHE PELLUCIDA*—**1**, habit, × ⅔; **2**, dorsal sepal, × 3; **3**, lateral sepal, × 3; **4**, petal, × 3; **5**, lip, × 3; **6**, spur and part of lip, side view, × 4; **7**, column, side view, × 8; **8**, column, front view, × 8; **9**, anther-cap, × 8; **10**, pollinarium, × 8. 1, from *Linder* 1375; 2–10, from *Smith* 604. Drawn by Mrs M.E. Church.

4. **D. bidens** (*Sw.*) *Schltr.*, Die Orchideen: 593 (1914); Summerh. in F.W.T.A., ed. 2, 3: 261 (1968); Piers, Orch. E. Afr., ed. 2: 263 (1968). Type: Sierra Leone, without exact locality, *Afzelius* (S, holo.)

An epiphyte with an elongate, slender, pendent stem up to 50 cm. long, 2–6.5 mm. diameter. Leaves up to 22, distichous, twisted at base to lie in one plane, ovate-elliptic, acute, to an unequally bilobed apex, 6.5–17 cm. long, 1.1–3.1 cm. wide, articulated at base to a longitudinally ridged sheathing base 10–25 mm. long. Inflorescences all along upper part of stem, 4–22 cm. long, up to twice length of leaf, up to 25-flowered; peduncle short, ± 5 mm. long, covered in sheathing bracts; bracts amplexicaul, conical, 1–1.5 mm. long. Flowers whitish; pedicel and ovary 2–5 mm. long. Dorsal sepal oblong-lanceolate or triangular, acute, 2.5–7 mm. long, 1.2–2.4 mm. wide; lateral sepals broadly triangular, 2.5–6.5 mm. long, 1–2.8 mm. wide. Petals lanceolate to ovate, 2.5–6 mm. long, 0.6–2 mm. wide. Lip pandurate, retuse to apiculate, 3–7 mm. long, 2.6–6 mm. wide, with a tooth in the mouth of the spur; spur inflated, 3–7.3 mm. long. Column 1.5–2 mm. long; stipites 2; viscidium 1. Fruit ellipsoidal, 6–16 mm. long.

UGANDA. Bunyoro District: Budongo Forest, Dec. 1942, *Eggeling* 5791!; Mengo District: Entebbe Botanic Garden, 23 Aug. 1946, *A.S. Thomas* 4514!
DISTR. U 2, 4; Sierra Leone to Cameroon, Zaire and Angola
HAB. Rain-forest; 1100–1300 m.

SYN. *Limodorum bidens* Sw. in K. Svenska Vetensk.-Acad. Handl. 21: 243 (1800)
Angraecum ashantense Lindl. in Bot. Reg. 29, Misc.: 56 (1843). Type: Ghana, Ashanti, cult. *Loddiges* (K, holo.!)
A. monodon Lindl. in Paxt., Fl. Gard. 2: 102 (1851). Type: Gabon, without exact locality, cult. *Pescatore* (K, holo.!)
A. bakeri Kraenzl. in Mitt. Deutsch. Schutz. 2: 159 (1889). Type: Cameroon, without exact locality, *Braun* (B, holo.†)
Mystacidium productum Kraenzl. in E.J. 22: 30 (1895). Type: Cameroon, between Victoria and Bimbia, *Preuss* 1227 (B, holo.†)
Listrostachys mystacioides Kraenzl. in E.J. 28: 170 (1900). Type: Cameroon, without exact locality, cult. Berlin Botanic Garden (B, holo.†)
L. longissima Kraenzl. in E.J. 48: 400 (1912). Type: Guinea Bissau, Beba, near Resia, *Tessman* 510 (B, holo.†)
Angraecum subfalcifolium De Wild. in B.J.B.B. 5: 192 (1916). Types: Zaire, Dundusana, *Mortehan* 909 & Moma Region, *Jespersen* (BR, syn.)
Diaphanthe ashantensis (Lindl.) Schltr. in Beih. Bot. Centr. 36: 98 (1918)
D. monodon Schltr. in Beih. Bot. Centr. 36: 98 (1918)
D. mystacioides (Kraenzl.) Schltr. in Beih. Bot. Centr. 36: 99 (1918)
D. producta (Kraenzl.) Schltr. in Beih. Bot. Centr. 36: 99 (1918)
D. subfalcifolia (De Wild.) Schltr. in Beih. Bot. Centr. 36: 100 (1918)
Mystacidium duemmerianum Kraenzl. in Viert. Nat. Ges. Zürich 74: 100 (1929). Type: Uganda, Mengo District, Mabira Forest, *Dummer* 4437 (K, iso.!)

5. **D. eggelingii** *Cribb*, sp. nov. affinis *D. lorifoliae* Summerh. sed caulo crassiore, foliis oblanceolatis, leviter falcatis, floribus pergamaceo-albis, labello ovato plus minusque trilobato satis differt. Typus: Uganda, Kigezi District, Impenetrable Forest, *Eggeling* 4144 (K, holo.!)

An epiphyte with an elongated pendent stem, ± 10 mm. diameter. Leaves very fleshy, oblong or oblong-lanceolate, unequally bilobed at apex, 18–19.5 cm. long, 2.5–2.8 cm. wide, twisted at base to lie in one plane, articulated to a sheathing leaf-base ± 2.5 cm. long. Inflorescences several in upper part of stem, 1–2 per node, 8–16 cm. long, 7–16-flowered; peduncle 10–15 mm. long; bracts amplexicaul, conical, 3 mm. long. Flowers parchment white; pedicel and ovary 3–4 mm. long. Dorsal sepal lanceolate, acute, 8–9.5 mm. long, 3–3.6 mm. wide; lateral sepals falcate-ovate, subacute, 7.5–9 mm. long, 3–3.8 mm. wide. Petals falcate-elliptic, acute, 7.5–7.6 mm. long, 3–3.3 mm. wide. Lip obscurely 3-lobed, subcordate, apiculate, 7–7.5 mm. long, 8–8.6 mm. wide, erose, with a tooth in the mouth of spur; spur pendent, fusiform-inflated, 8–9 mm. long. Column 3 mm. long; stipites 2; viscidium 1; rostellum tapering, reflexed. Fig. 124.

UGANDA. Kigezi District: Impenetrable Forest, Nyebeya, Oct. 1940, *Eggeling* 4144!
DISTR. U 2; known only form the type gathering
HAB. Rain-forest; ± 1000 m.

6. **D. lorifolia** *Summerh.* in K.B. 11: 235 (1956); Piers, Orch. E. Afr., ed. 2: 266 (1968); J. Stewart in U.K.W.F.: 788 (1974). Type: Kenya, SW. Mau Forest, *Gardner* in *F.D.* 3625 (K, holo.!)

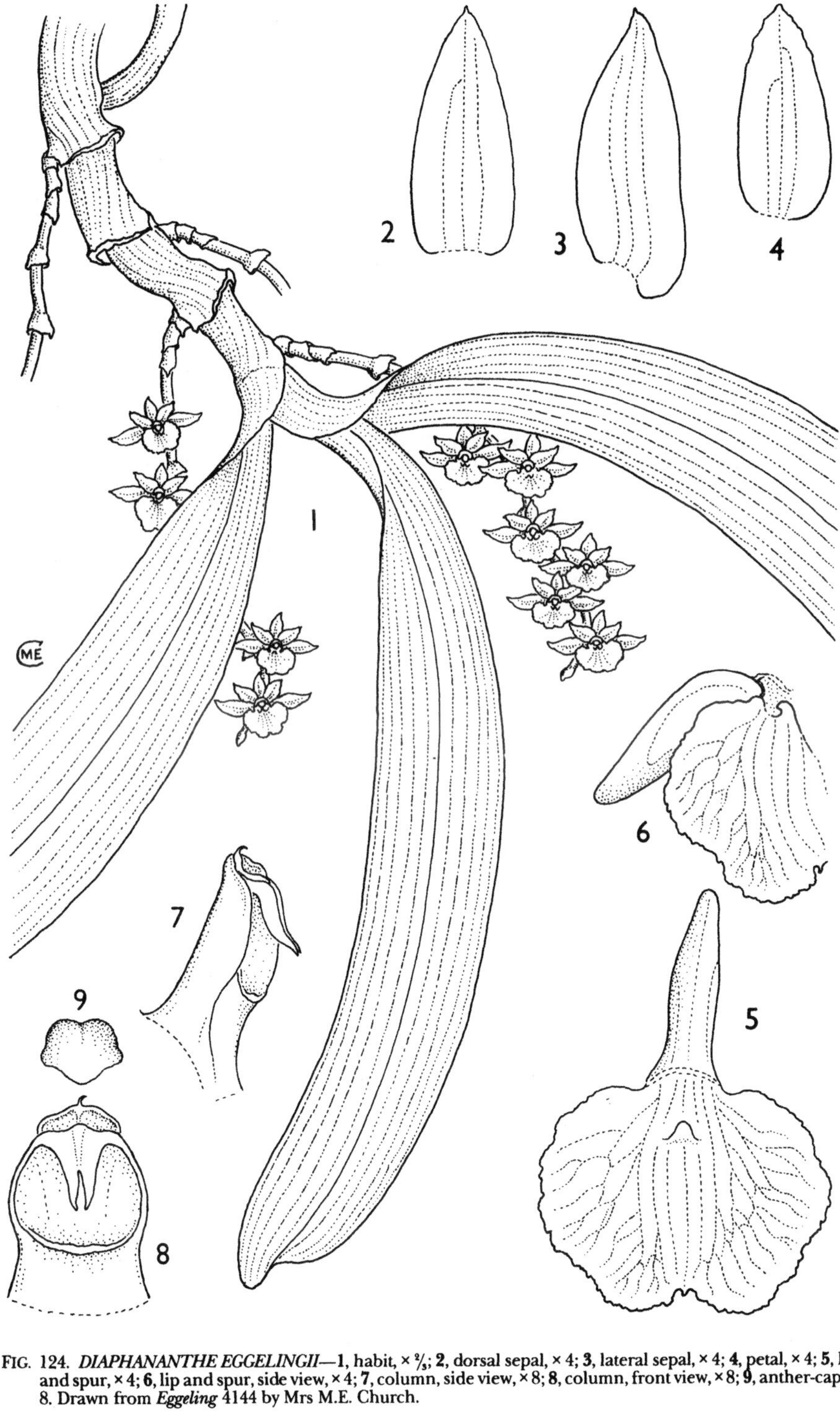

FIG. 124. *DIAPHANANTHE EGGELINGII*—**1**, habit, × ⅔; **2**, dorsal sepal, × 4; **3**, lateral sepal, × 4; **4**, petal, × 4; **5**, lip and spur, × 4; **6**, lip and spur, side view, × 4; **7**, column, side view, × 8; **8**, column, front view, × 8; **9**, anther-cap, × 8. Drawn from *Eggeling* 4144 by Mrs M.E. Church.

An epiphyte with long hanging stems 8–27 cm. long, ± 5 mm. diameter, leafy in upper part. Leaves distichous, falcate, linear, unequally obtusely bilobed at apex, 14–26 cm. long, 1–1.8 cm. wide, twisted at base to lie in one plane, articulated at base to a rough sheath 1.5–2 cm. long. Inflorescences pendent, 1–several, 7–15 cm. long, laxly many-flowered; peduncle 1–2.5 cm. long; bracts usually amplexicaul, broadly ovate, 1–1.5 mm. long. Flowers pale yellowish, white or pinkish; pedicel and ovary 3–4 mm. long. Dorsal sepal lanceolate or narrowly elliptic, acute, 5.2–8 mm. long, 2–3 mm. wide; lateral sepals oblong-ovate, acute, 6.5–9 mm. long, 2.3–3.2 mm. wide. Petals lanceolate to oblong, acute, 5.4–6.6 mm. long, 1.8–2.6 mm. wide, slightly erose. Lip entire to obscurely 3-lobed, circular to broadly ovate, shortly apiculate, 5.3–6.3 mm. long, 6–8.3 mm. wide, erose, with a small tooth in the mouth of the spur; spur slightly incurved, narrowly fusiform, 5–9.5 mm. long. Column 2.5–3 mm. long; stipites 2; viscidium 1; rostellum shortly ligulate, recurved.

UGANDA. Elgon, Sipi, 6 Apr. 1924, *Snowden* 969!
KENYA. Nandi District: Kaimosi, June 1962, *Tweedie* 588!; Kericho District: Kiptiget, Dec. 1941, *Copley* 57! & SW. Mau Forest, Feb. 1937, *Gardner* in *F.D.* 3625!
TANZANIA. Arusha District: Meru Forest, 6 Dec. 1956, *Saska* 92!; Lushoto District: W. Usambara Mts., Mazumbai, 2 Apr. 1981, *L. Tanner* 259!; Morogoro District: Uluguru Mts., near Morogoro, 19 Nov. 1932, *Schlieben* 3001!
DISTR. **U** 3; **K** 3, 5, 6; **T** 2, 3, 6; Sudan, Ethiopia
HAB. In montane and riverine forest; (300–)1300–2200 m.

7. **D. fragrantissima** (*Reichb.f.*) *Schltr.* in Beih. Bot. Centr. 36: 100 (1918); Summerh. in Bot. Mus. Leafl. Harv. Univ. 12: 101 (1945); Piers, Orch. E. Afr., ed. 2: 265 (1968); J. Stewart in U.K.W.F.: 788 (1974); Williamson, Orch. SC. Afr.: 186 (1977); la Croix et al., Malawi Orch. 1: 68 (1983). Type: Angola, Pungo Adongo, *Welwitsch* 702 (W, holo., K, iso.!)

An epiphyte or rarely a lithophyte with long pendulous stems 7–50 cm. long, 5–10 mm. diameter, covered in leaf-sheaths. Leaves up to 9, distichous, falcate, linear-oblanceolate, unequally bilobed at apex, 10–44 cm. long, 1–3.5 cm. wide, twisted at base, articulated to a sheathing base 1.2–3 cm. long. Inflorescences pendulous, 1–several, up to 60 cm. long, up to 60-flowered; peduncle 2–20 cm. long; bracts amplexicaul, 3–5 mm. long. Flowers usually borne in whorls of up to 4, greenish yellow, diurnally scented; pedicel and ovary 1–2.25 mm. long. Dorsal sepal linear-triangular, acute, 7–10.5 mm. long, 2–2.2 mm. wide; lateral sepals slightly falcate, linear, acute, 9.7–11.2 mm. long, 2–3 mm. wide. Petals linear to obovate, acuminate, 8.3–8.5 mm. long, 2–2.2 mm. wide. Lip rectangular, 3-lobed at apex, 10–13 mm. long, 6.5–7 mm. wide, with erose lateral margins, with a marked tooth at the mouth of the spur; mid-lobe elongated and acuminate; spur geniculate at base, fusiform, 6–9 mm. long. Column 2 mm. long, rostellum deeply bifid; stipites 2; viscidium 1.

UGANDA. Bunyoro District: Mururongo, Sept. 1935, *Eggeling* 2228!; Mbale District: Busia, Oct. 1933, *Jack* 6009!; Masaka District: NW. side of Lake Nabugabo, 9 Oct. 1953, *Drummond & Hemsley* 4719!
KENYA. S. Kavirondo District: Kiya R., Feb. 1960, *Glover* 12162!; Teita District: Maungu Hill, 28 Mar. 1970, *Archer* 603!
TANZANIA. Biharamulo District: Banga, May 1950, *Watkins* 443!; Tanga District: E. Usambara Mts., Sigi, 26 May 1942, *Moreau* 42!; Ulanga District: [Liando], 9 Apr. 1932, *Schlieben* 2058!
DISTR. **U** 2–4; **K** 5, 7; **T** 1, 3, 6; Cameroon, Zaire, Rwanda, Burundi, Sudan, Ethiopia, Zambia, Angola, Mozambique and Zimbabwe
HAB. In forest, bushland and rocky country; 150–1400 m.

SYN. *Listrostachys fragrantissima* Reichb.f. in Flora 48: 190 (1865)
L. kirkii Rolfe in F.T.A. 7: 164 (1897). Type: Tanzania, Usambara Mts., Magila, *Kirk* (K, holo.!)
L. falcata De Wild., Not. Pl. Utiles Congo 1: 147 (1903). Type: Zaire, Ojuma Valley, *Gentil* (BR, holo.)
L. fimbriata Rolfe in K.B. 1906: 115 (1906). Type: Uganda, Mengo District, Entebbe, *Mahon* 5 (K, holo.!)
Angraecum muansae Kraenzl. in Orchis 2: 99 (1908). Type: Tanzania, Mwanza [Muansa], *Uhlig* 119 (B, holo.)
Diaphananthe falcata (De Wild.) Schltr. in Beih. Bot. Centr. 36: 100 (1918)
D. fimbriata (Rolfe) Schltr. in Beih. Bot. Centr. 36: 100 (1918)
D. kirkii (Rolfe) Schltr. in Beih. Bot. Centr. 36: 101 (1918)
D. muansae (Kraenzl.) Schltr. in Beih. Bot. Centr. 36: 101 (1918)

8. **D. kamerunensis** (*Schltr.*) *Schltr.* in Orchis 8: 137 (1914); Summerh. in F.W.T.A., ed. 2, 3: 261 (1968); Piers, Orch. E. Afr., ed. 2: 265 (1968); Williamson, Orch. SC. Afr.: 190 (1977). Types: Cameroon, near Neu-Tegel, *Winkler* 209, near Buea, *Schlechter & Diestel* s.n. & near Ediki, *Schlechter* 15768 (B, syn.)

FIG. 125. *DIAPHANANTHE KAMERUNENSIS*—**1**, habit, × ⅔; **2**, flower, × 1; **3**, dorsal sepal, × 2; **4**, lateral sepal, × 2; **5**, petal, × 2; **6**, lip, × 2; **7**, column, side view, × 6; **8**, column, side view with anther-cap and pollinia removed, × 6; **9**, column, front view, × 6; **10**, anther-cap, × 6; **11**, pollinaria, × 6. Drawn from *Baker* s.n. by Mrs M.E. Church.

A large pendent epiphyte with a short stout stem 6–8 cm. long, 1.1–1.8 cm. diameter. Leaves 3–7, falcate, oblanceolate, unequally roundly bilobed at apex, 20–35 cm. long, 3–5.5 cm. wide, twisted at base to lie in one plane, articulated to a sheathing base 1–2 cm. long. Inflorescences pendent, 1–several, 10–12 cm. long, laxly 8–10-flowered; peduncle 3.5–4.5 cm. long, almost covered by loose bracts; bracts large, papery, broadly ovate, 10–15 mm. long and wide. Flowers pale translucent green; pedicel and ovary 2–2.4 mm. long. Dorsal sepal ovate, shortly apiculate or obtuse, 15–16 mm. long, 8.5–10.7 mm. wide; lateral sepals obliquely lanceolate, obtuse, 20–23 mm. long, 8–9.5 mm. wide. Petals subcircular, obtuse, 15–17 mm. long, 12–14 mm. wide, with strongly erose-lacerate margins. Lip flabellate, distantly 3-lobed, 18–20 mm. long, 22–24 mm. wide, erose on apical margin, with a tooth in the mouth of the spur; side lobes obliquely oblong; mid-lobe shortly triangular; spur incurved-pendent, 12–14 mm. long. Column 4 mm. long; stipites 2; viscidia 2; rostellum pendent, 3-lobed. Fig. 125.

UGANDA. Ankole District: Kalinzu Forest, June 1938, *Eggeling* 3646! & Lutoto Forest, 16 Aug. 1938 (fr.), *Pole-Evans & Erens* 1758!
DISTR. **U** 2; Cameroon, Zaire and Zambia
HAB. Rain-forest; ± 1000 m.

SYN. *Angraecum kamerunense* Schltr. in E.J. 38: 161 (1906)

9. **D. pulchella** *Summerh.* in Bot. Mus. Leafl. Harv. Univ. 12: 103 (1945) & in Hook., Ic. Pl. 35, t. 3486 (1950); Piers, Orch. E. Afr., ed. 2: 267 (1968); J. Stewart in U.K.W.F.: 788 (1974); Williamson, Orch. SC. Afr.: 187 (1977); la Croix et al., Malawi Orch. 1: 70 (1983). Type: Tanzania, E. Usambara Mts., Amani, *Moreau* 64 (K, holo.!)

An epiphyte or rarely a lithophyte with a 3–13 cm. long stem, 3–4 mm. diameter. Leaves distichous, leathery, oblanceolate-falcate, unequally bilobed at apex, 3–10 cm. long, 0.4–1 cm. wide, articulated to a 4–7 mm. long sheath. Inflorescences pendulous, 1–several, up to 25-flowered, as long as or longer than the leaves; peduncle 1.5–4 cm. long; bracts triangular, acute, 2 mm. long. Flowers pale yellow or cream-coloured, fragrant; pedicel and ovary 2–3 mm. long. Dorsal sepal ovate or obovate, obtuse or acuminate, 4.6–7.7 mm. long, 2.5–3.7 mm. wide; lateral sepals obliquely elliptic, obtuse, 5.7–11 mm. long, 2–3.6 mm. wide. Petals ovate to elliptic, acute, subacute or obtuse, 4.3–7.6 mm. long, 3.1–5.3 mm. wide, slightly erose, ± geniculate at the base. Lip flabellate to rectangular, obscurely 2–4-lobed, 6–9.5 mm. long, 7–9.6 mm. wide, deeply notched at apex, with a tooth in the mouth of the spur; spur slightly fusiform, porrect, 8.6–14 mm. long. Column 2–3 mm. long; stipites 2; viscidia 2; rostellum decurved.

var. **pulchella**

Inflorescence 6–15 cm. long, up to 14-flowered. Lateral sepals 5.7–9.5 mm. long.

UGANDA. Mbale District: Elgon, Suam Gorge, June 1962, *C. Leakey* 1!
KENYA. Trans-Nzoia District: Elgon, Suam saw-mills, May 1941, *Tweedie* 570!; Machakos District: Mua Hills, Apr. 1952, *Piers* in *E.A.H.* 64/53!; Masai District: Ngong Hills, Nov. 1938, *C. van Someren* 52!
TANZANIA. Lushoto District: W. Usambara Mts., Mazumbai, Mar. 1943, *Moreau* 611! & Mkusu valley, between Mkusi and Kifungilo, 23 Apr. 1953, *Drummond & Hemsley* 2208!; Ufipa District: Lake Sundu, 8 Dec. 1958, *Richards* 10254!; Songea District: Matagoro Hills, 22 Feb. 1956, *Milne-Redhead & Taylor* 8873!
DISTR. **U** 1–3; **K** 3–6; **T** 2–4, 7, 8; Zaire, Zambia and Malawi
HAB. In riverine forest, bushland, woodland and montane forest with *Podocarpus*; 700–2500 m.

var. **geniculata** *Summerh.* in Bot. Mus. Leafl. Harv. Univ. 12: 105 (1945). Type: Uganda, Bunyoro District, Bugumolo, *Eggeling* 2153 (K, holo.!)

Differs in usually having a much longer inflorescence, 10–25 cm. long, bearing more, larger flowers with longer lateral sepals, 9–11 mm. long and geniculate at the base; petals with a distinctly geniculate angle at the base; and a longer spur usually 10–14 mm. long.

UGANDA. Bunyoro District: Hoima, Aug. 1940, *Purseglove* 983!; Toro District: Bwamba, Aug. 1937, *Eggeling* 3392!; Mengo District: Entebbe Botanical Garden, 23 Aug. 1946, *A.S. Thomas* 4513!
DISTR. **U** 2, 4; Cameroon and Zaire
HAB. In rain-forest; 1200–1500 m.

10. **D. ugandensis** (*Rendle*) *Summerh.* in Bot. Mus. Leafl. Harv. Univ. 12: 111 (1945). Type: Uganda, Kigezi District, Rukiga [Ruchigga], *Bagshawe* 425 (K, holo.!)

An epiphyte with an elongate stem, 8–9 cm. long. Leaves linear, unequally bilobed at apex, 6–12 cm. long, 0.5–0.8 cm. wide, articulated to a sheathing base 8 mm. long. Inflorescences 1-several, 6–18 cm. long, sublaxly many-flowered; peduncle ± 1 cm. long; bracts sheathing, 1–1.5 mm. long. Flowers yellowish green to pinkish; pedicel and ovary 3 mm. long. Dorsal sepal oblong-elliptic, obtuse, 4–5 mm. long; lateral sepals obliquely oblong-elliptic, obtuse, 4.5–6 mm. long. Petals elliptic, subacute or obtuse, 3.7–5 mm. long, 2.8–3.3 mm. wide. Lip obovate or broadly elliptic, notched at apex, 5.25–6 mm. long, 5.5–6.75 mm. wide, erose, with a tooth in the mouth of the spur; spur 8–9 mm. long. Column 1.8–2 mm. long; stipites 2; viscidia 2; rostellum pendent, subclavate.

UGANDA. Bunyoro District: Budongo Forest, Nov. 1944, *Eggeling* 5538!; Kigezi District: Rukiga [Ruchigga], 21 Nov. 1903, *Bagshawe* 425!
DISTR. U 2; Zaire
HAB. Rain-forest; 1500–2000 m.

SYN. *Mystacidium ugandense* Rendle in J.L.S. 37: 220 (1905)

11. **D. cuneata** *Summerh.* in K.B. 14: 141 (1960) & in Hook., Ic. Pl. 37, t. 3671 (1969). Type: Uganda, Bunyoro District, Budongo, *Burgess* (K, holo.!)

An epiphyte with a short stem ± 5 cm. long, 3–4 mm. diameter. Leaves ± 5, linear to oblanceolate, unequally bilobed at rounded apex, 3–4 cm. long, 5–7 mm. wide, twisted at base and articulated to a longitudinally ridged leaf-bases 5 mm. long. Inflorescence laxly 6–8-flowered, 9 cm. long; peduncle 4 cm. long; bracts conical, somewhat amplexicaul, 1.5–2 mm. long. Flowers pale yellowish green; pedicel and ovary 3 mm. long, emerging from rhachis 2–3 mm. above axil of bract. Dorsal sepal broadly elliptic or subcircular, obtuse, 2.3–2.5 mm. long, 2 mm. wide; lateral sepals falcate, elliptic, obtuse, 3–3.5 mm. long, 1.5–2 mm. wide, fused behind spur at base. Petals circular, obtuse, 2 mm. long and wide. Lip flabellate, bilobed, 3.5–4 mm. long, 2–2.5 mm. wide, erose; spur pendent, incurved towards apex, cylindrical, 6–6.5 mm. long. Column 1.5 mm. long; stipites 2, with papillose margins; viscidia 2; rostellum clavate, pendent.

UGANDA. Bunyoro District: Budongo Forest, Queen's Mile, fl. Kew 20 Sept. 1977, *Meyer* K.41! & 25 Mar. 1979, *Pfennig* 1323!
DISTR. U 2; Cameroon
HAB. In rain-forest; ± 1000 m.

12. **D. stolzii** *Schltr.* in E.J. 53: 600 (1915); la Croix et al., Malawi Orch. 1: 74 (1983). Type: Tanzania, Rungwe District, Kyimbila, *Stolz* 714 (K, iso.!)

An epiphyte with long pendent, sometimes branching stems, 20–150 cm. long, 3–5 mm. diameter. Leaves distichous, in upper half of stem, oblong-obovate, unequally roundly bilobed at apex, 2.6–6 cm. long, 1–2 cm. wide, twisted at base and articulated to a 7–10 mm. long bilaterally flattened sheath. Inflorescences as long as or shorter than the leaves, 3–6.5 cm. long, densely many-flowered; peduncle 8–15 mm. long; bracts amplexicaul, conical, 2 mm. long. Flowers pale yellow, fragrant; pedicel and ovary 7.5–10 mm. long. Dorsal sepal lanceolate-elliptic, acute or obtuse, 8–9 mm. long, 3.2–4 mm. wide; lateral sepals falcate, oblong-elliptic to lanceolate, acute or obtuse, 8.8–14 mm. long, 2.3–3 mm. wide. Petals obliquely triangular-ovate, acute, 6.7–9 mm. long, 4.2–7 mm. wide. Lip flabellate, sometimes obscurely 2–4-lobed, emarginate to apiculate, 11–12 mm. long, 10–17 mm. wide, erose on apical margins, with a prominent tooth in the mouth of the spur; spur cylindrical from a conical base, incurved or sigmoid, 15–21 mm. long. Column 3 mm. long; stipites 2; viscidia 2; rostellum pendent, clavate. Pl. 6.

TANZANIA. Tanga District: E. Usambara Mts., Amani, 29 July 1941, *Moreau* 29! & 20 May 1950, *Verdcourt* 206!; Ulanga District: Muhulu Mts., 26 Feb. 1942, *Schlieben* 1845!
DISTR. T 2, 3, 6, 7; Malawi, Zimbabwe
HAB. In montane forest and in moist *Brachystegia glaucescens* woodland; 200–1700 m.

13. **D. schimperiana** (*A. Rich.*) *Summerh.* in Bot. Mus. Leafl. Harv. Univ. 12: 106, 109 (1945) & in K.B. 14: 142 (1960). Type: Ethiopia, Mt. Taber, near Gerar, *Schimper* 1318 (P, holo., K, iso.!)

A pendent epiphyte or rarely a lithophyte with stems up to 40 cm. or more long, 3–5 mm. diameter; roots smooth, 2–3 mm. diameter. Leaves falcate, narrowly oblong to oblanceolate, unequally and roundly bilobed at apex, 6.5–14 cm. long, 1.3–2.5 cm. wide, twisted at base to lie in one plane, articulated to a 1.4–1.7 cm. long sheathing base. Inflorescences 5.5–8.5 cm. long, laxly 5–10-flowered; peduncle 1–1.5 cm. long; bracts obconical, acute or apiculate, 2–3.5 mm. long. Flowers translucent white or cream; pedicel and ovary 2–4 mm. long. Dorsal sepal ovate, acute, 5–5.6 mm. long, 2.8–3.4 mm. wide; lateral sepals falcate, lanceolate, acute, 7–7.5 mm. long, 1.8–2 mm. wide. Petals obliquely triangular-ovate, acuminate, 5–5.7 mm. long, 3–3.6 mm. wide, erose. Lip obscurely 3-lobed in basal half, oblong-ovate in outline, emarginate to obtuse and shortly apiculate at apex, 6–7.7 mm. long, 6.6–7.5 mm. wide, with a tooth-like callus in the mouth of the spur; spur incurved, slightly dilated in middle, 8–9 mm. long. Column 1.5 mm. long; rostellum pendent, subclavate; stipites 2; viscidia 2.

UGANDA. Acholi District: Lomwaga Mt., June 1942, *Eggeling* 5068!; Karamoja District: Mt. Morongole, July 1965, *J. Wilson* 1642!
DISTR. U 1; Sudan and Ethiopia
HAB. Epiphyte and on rocks in montane forest; 2400–2850 m.

SYN. *Dendrobium schimperianum* A. Rich., Tent. Fl. Abyss. 2: 282 (1850)
Angraecum schimperianum (A. Rich.) Reichb. f. in Walp., Ann. Bot. Syst. 3: 573 (1852)

14. **D. tenuicalcar** *Summerh.* in Bot. Mus. Leafl. Harv. Univ. 12: 109 (1945) & in Hook., Ic. Pl. 36, t. 3567 (1956); Piers, Orch. E. Afr., ed. 2: 269 (1968); J. Stewart in U.K.W.F.: 790 (1974). Type: Uganda, Karamoja District, Napak, *A.S. Thomas* 3645 (K, holo.!)

An epiphyte often growing in masses, with spreading or pendulous stems, 9–26 cm. long, 1–3 mm. diameter. Leaves 3–8, towards apex of stems, falcate, linear-lanceolate, obliquely bilobed at acute apex, 1–6 cm. long, 7–13 mm. wide, twisted at base and articulated to a sheathing leaf-base 5–11 mm. long. Inflorescences 1–many, shorter than or equal to leaf, 1–3 cm. long, up to 7-flowered; peduncle 5–7 mm. long; bracts amplexicaul, conical, 2 mm. long. Flowers white, sweetly scented; pedicel and ovary 4–12 mm. long. Dorsal sepal oblong-elliptic, obtuse, 4–4.5 mm. long, 2–2.8 mm. wide; lateral sepals obliquely ovate or elliptic, obtuse, 4.2–5.1 mm. long, 2–2.5 mm. wide. Petals obliquely ovate, obtuse or acute, 4.2–4.8 mm. long, 2.5 mm. wide. Lip flabellate, obscurely 2–4-lobed on apical margin, retuse, 4.5–6.5 mm. long, 6.2–7.6 mm. wide, with a tooth in the mouth of the spur; spur slender, cylindrical from a broad mouth, 15–25 mm. long. Column 2 mm. long; stipites 2; viscidia 2; rostellum pendent, clavate.

UGANDA. Karamoja District: Napak, June 1950, *Eggeling* 5970! & May 1940, *A.S. Thomas* 3645!
KENYA. Kisumu-Londiani District: Tinderet Forest Reserve, June 1949, *Maas Geesteranus* 5229!
DISTR. U 1; **K** 5; Ethiopia
HAB. In and at the edge of montane forest and in wooded grassland; 2100–2400 m.

15. **D. subsimplex** *Summerh.* in Bot. Mus. Leafl. Harv. Univ. 12: 107 (1945) & in Hook., Ic. Pl. 35, t. 3487 (1950); Piers, Orch. E. Afr., ed. 2: 268 (1968); J. Stewart in U.K.W.F.: 788 (1974); la Croix et al., Malawi Orch. 1: 74 (1983). Type: Kenya, N. Nyeri District, Nanyuki, *Moreau* 578 (K, holo.!)

An epiphyte with 10–35 cm. long pendent stems, 3–4 mm. diameter. Leaves in apical half of stem, distichous, linear,.unequally roundly bilobed at apex, 2.5–4.5 cm. long, 5–10 mm. wide, twisted at base and articulated to a 7–13 mm. long sheath. Inflorescences 1–several, 5–13-flowered, 1.5–8 cm. long; peduncle 2–4 mm. long; rhachis zigzag; bracts triangular, acute, 1–2.5 mm. long. Flowers pale yellow or pale green; pedicel and ovary 2.5–3 mm. long. Dorsal sepal oblong to ovate-elliptic, obtuse or rounded, 2.7–4 mm. long, 1.5–2 mm. wide; lateral sepals oblong to oblong-elliptic, obtuse, 3.3–4.5 mm. long, 1.3–1.6 mm. wide. Petals ovate, obtuse or acute, 2.8–3.5 mm. long, 1.7–2.7 mm. wide. Lip flabellate or obovate, obtuse or retuse, 2.5–3.6 mm. long, 2.8–3.5 mm. wide, with an obscure tooth in the mouth of the spur; spur slightly incurved, cylindrical to narrowly fusiform, 4.5–6.8 mm. long. Column 1.2–1.5 mm. long; stipites 2; viscidia 2; rostellum pendent, clavate.

UGANDA. Karamoja District: Napak, 28 May 1940, *A.S. Thomas* 3644! & June 1950, *Eggeling* 5881!
KENYA. Trans-Nzoia District: Suam R. valley, Nov. 1934, *Tweedie* 289!; Naivasha, Sterndale, May 1943, *Andrews* in *Bally* 4496!; N. Nyeri District: Nanyuki, June 1943, *Moreau* 578!
TANZANIA. Masai District: Ol Doinyo Ngiro, fl. cult. 16 Mar. 1944, *Elliott* in *Moreau* 778!; Moshi District: W. side of Mt. Kilimanjaro, Homa R., fl. cult. 29 Mar. 1977, *Alkema* in *Piers*!; Kilosa District: Ukaguru Mts., Kwamba Mt., 2 Feb. 1976, *Cribb & Grey-Wilson* 10529!; Njombe, 6 Apr. 1986, *Petersson, Hedren & Kibuwa* 706!
DISTR. U 1; **K** 3–6; **T** 2, 3, 6, 7; Malawi and Zimbabwe
HAB. In montane forest and on forest margins; 1700–2600 m.

16. **D. bilobata** (*Summerh.*) *Rasmussen* in Norw. Journ. Bot. 21: 229 (1974). Type: Uganda, Kigezi District, Rutenga, *Eggeling* 4212 (K, holo.!)

An epiphyte or rarely a lithophyte with elongated, pendent stems up to 50 cm. long, 2.5–5 mm. diameter. Leaves up to 16 in upper half of stem, distichous, twisted at base to lie in one plane, oblong-elliptic or obovate, unequally bilobed at apex, 3.5–8 cm. long, 1.5–3 cm. wide, articulated to longitudinally veined sheaths 1–1.8 cm. long. Inflorescences 1–several, often 2–3 per node in upper part of stem, 2–8.5 cm. long, up to twice length of leaf; peduncle very short, covered by bracts; bracts very close, 2–3 mm. apart, amplexicaul, conical, acute, 1–2 mm. long. Flowers white, pale cream or green; pedicel and ovary 1–1.5 mm. long. Dorsal sepal oblong, obtuse, 3.8–4.7 mm. long, 2.1–3.4 mm. wide; lateral sepals oblique, oblong-elliptic, obtuse, 5–5.3 mm. long, 2.1–2.9 mm. wide. Petals obovate to elliptic, obtuse, 3.6–3.9 mm. long, 2.1–2.9 mm. wide. Lip obscurely 3-lobed, 4.7–5.5 mm. long, 3.6–3.7 mm. wide, with an obscurely bilobed tooth in the mouth of the spur; side lobes narrowly elliptic; mid-lobe deflexed, transversely elliptic, deeply incised at apex so that mid-lobe appears bilobed; spur dependent, slightly incurved, cylindrical, 7.8–8.3 mm. long. Column 2 mm. long; stipites 2; viscidia 2; rostellum somewhat spathulate.

UGANDA. Kigezi District: Rutenga, Apr. 1948, *Purseglove* 2622!; Ankole District: Mbarara, Apr. 1944, *Eggeling* 5496!
DISTR. **U** 2; Zaire, Rwanda
HAB. In rain-forest and thickets; 1280–2130 m.

SYN. *Sarcorhynchus bilobatus* Summerh. in Bot. Mus. Leafl. Harv. Univ. 12: 111 (1945)

17. **D. microphylla** (*Summerh.*) *Summerh.* in K.B. 14: 143 (1960). Type: Tanzania, Rungwe District, Ngulu Mt., *Stolz* 2554 (K, holo.!)

An epiphyte with slender stems, 3–20 cm. long, 2 mm. diameter, often growing in tangled masses; roots more prominent than the leaves, 2–10 cm. long, 0.5–2 mm. diameter. Leaves in apical part of stem, up to 6, linear to oblanceolate, rounded and unequally bilobed at apex, 6–15 mm. long, 2–4 mm. wide, articulated at base to a 4–5 mm. long leaf-sheath. Inflorescences 1–many, secund, 2–3 cm. long, up to 5-flowered; peduncle 1–1.2 cm. long; bracts amplexicaul, conical, 1 mm. long. Flowers greenish yellow, strongly fragrant; pedicel and ovary 4–5 mm. long, covered by scurfy hairs. Dorsal sepal oblong-elliptic, obtuse, 3 mm. long, 1.4–1.5 mm. wide; lateral sepals falcate, narrowly oblong, subacute, 3.5–4 mm. long, 1–1.5 mm. wide. Petals obliquely oblong-elliptic, obtuse, 2.5–3 mm. long, 1.5–1.6 mm. wide. Lip entire, elliptic, obtuse, 2.5–3 mm. long, 0.8–1 mm. wide; spur strongly incurved, slenderly cylindrical, ± 7 mm. long. Column 1 mm. long; stipites 2; viscidia 2; rostellum pendent, rather bulbous.

TANZANIA. Iringa District: Image Mt., 4 Mar. 1962, *Polhill & Paulo* 1683!; Rungwe District: Luteba, Mwakaleli, 20 Mar. 1975, *Leedal* 2617! & Bundali Mts., Ngulu Mt., 3 Mar. 1914, *Stolz* 2554!
DISTR. **T** 7; N. Malawi
HAB. In montane mist-forest of *Podocarpus*, *Dombeya*, *Myrica*, etc., also occasionally in old coffee plantations; 1400–2300 m.

SYN. *Rhipidoglossum microphyllum* Summerh. in Bot. Mus. Leafl. Harv. Univ. 12: 95 (1945)

18. **D. densiflora** (*Summerh.*) *Summerh.* in K.B. 14: 143 (1960); Piers, Orch. E. Afr., ed. 2: 263 (1968); Williamson, Orch. SC. Afr.: 187 (1977). Type: Gabon, Mbigou, on Mboumi R., *Le Testu* 5767 (K, holo.!)

An epiphyte with a long pendent stem, 10–43 cm. long, 3–4 mm. diameter. Leaves distichous, many in upper half of stem, elliptic to oblong or oblanceolate, acute and obliquely unequally bilobed at apex, 4–10 cm. long, 1–2 cm. wide, twisted at base and articulated to a slightly compressed leaf-sheath 7–15 mm. long. Inflorescences solitary or paired at nodes, 1–many, 1–3 cm. long, densely many-flowered; peduncle 5 mm. long or less; bracts amplexicaul, 1–1.5 mm. long. Flowers waxy, white, campanulate; pedicel and ovary 2.5–3 mm. long. Dorsal sepal oblong-elliptic, obtuse, 2.2–3 mm. long, 1.7–2 mm. wide; lateral sepals obliquely elliptic, obtuse, 3–3.5 mm. long, 1.3–1.7 mm. wide. Petals circular, truncate or obtuse, 2–2.75 mm. long and wide. Lip entire, broadly orbicular or transversely elliptic, retuse, 2.7–3.75 mm. long, 3.5–4.7 mm. wide, with a low transverse callus at the mouth of the spur; spur ellipsoidal-clavate, 2–3 mm. long, shortly pubescent in the mouth. Column 1–1.2 mm. long; stipites 2; viscidia 2; rostellum pendent, shortly spathulate.

UGANDA. Bunyoro District: Budongo Forest, Sept. 1933, *Eggeling* 1431! & Bugoma Forest, *Dawe* 759!; Mengo District: Kyagwe, Sept. 1921, *Lankester* 29! & Mabira Forest, Sept.–Oct. 1920, *Dummer* 4438!
DISTR. U 2, 4; Liberia, Cameroon, Gabon, Zaire, Angola and Zambia
HAB. In rain-forest; 1050–1200 m.

SYN. *Rhipidoglossum densiflorum* Summerh. in Bot. Mus. Leafl. Harv. Univ. 12: 91 (1945) & in Hook., Ic. Pl. 35, t. 3485 (1950)

19. **D. xanthopollinia** (*Reichb.f.*) *Summerh.* in K.B. 14: 143 (1960); Piers, Orch. E. Afr., ed. 2: 269 (1968); J. Stewart in U.K.W.F.: 788 (1974); Williamson, Orch. SC. Afr.: 189 (1977); la Croix et al., Malawi Orch. 1: 74 (1983). Type: Angola, Pungo Andongo, *Welwitsch* 678 (W, holo.)

An epiphyte usually with long stems, 6–27 cm. long, 3–6 mm. diameter. Leaves up to 25 in upper half of stem, linear to oblanceolate, unequally roundly bilobed at apex, 3–15 cm. long, 5–7 mm. wide, articulated to an 8–15 mm. long leaf-base. Inflorescences 1–many, 3–9 cm. long, densely many-flowered; peduncle 5–10 mm. long; bracts ± amplexicaul, conical, obtuse, 1–2 mm. long. Flowers creamy yellow, diurnally sweetly scented; pedicel and ovary 1.5–2.5 mm. long. Dorsal sepal oblong-elliptic to ovate, obtuse, 2.8–3.5 mm. long, 1.7–2.7 mm. wide; lateral sepals obliquely elliptic, obtuse, 3.8–5 mm. long, 1.7–2.4 mm. wide. Petals circular to broadly ovate, obtuse, 2.4–3.6 mm. long, 2–3 mm. wide. Lip flabellate, obscurely 2–4-lobed on apical margin, emarginate, 3.3–4.5 mm. long, 3.2–5.7 mm. wide, erose on apical margin, with a low obscure transverse callus at the mouth of the spur; spur incurved, cylindrical, 5–7 mm. long. Column 1.2–2 mm. long; stipites 2; viscidia 2; rostellum pendent, clavate.

UGANDA. Bunyoro District: Bugoma, Aug. 1935, *Eggeling* 2187! & Budongo, Aug. 1940, *Eggeling* 4047!; Mengo District: Kiagwe, Kolo Forest, Sept. 1932, *Eggeling* 529!
KENYA. N. Kavirondo District: Kakamega Forest, Sept. 1943, *Carroll* in *Tweedie* 619!; Teita District; Maungu Hills, Nyangala Hill, 2 May 1970, *Archer* 608!
TANZANIA. Tanga District: E. Usambara Mts., Amani, Sept. 1941, *Moreau* 71! & Longusa, Dec. 1951, *Paulo* in *E.A.H.* 11126!; Iringa District: Mwanihana Forest, Apr. 1981, *Rodgers* 919! & 991!
DISTR. U 2, 4; **K** 5, 7; **T** 3, 7; Zaire, Angola, Zambia, Malawi, Mozambique, Zimbabwe and South Africa (Natal)
HAB. In lower montane forest; 600–1800 m.

SYN. *Aeranthus xanthopollinius* Reichb.f. in Flora 48: 190 (1865)
A. erythropollinius Reichb.f. in Flora 48: 190 (1865). Type: Angola, Golungo, *Welwitsch* 665 (W, holo.)
A. gerrardii Reichb.f. in Flora 50: 117 (1867). Type: South Africa, Natal, Port Natal, *Gerrard* (W, holo.)
Mystacidium gerrardii (Reichb.f.) Bolus in J.L.S. 25: 187 (1889)
M. xanthopollinium (Reichb.f.) Th. Dur. & Schinz, Consp. Fl. Afr. 5: 55 (1892)
Angraecum gerrardii (Reichb.f.) Bolus, Ic. Orch. Austr. Afr. 1, t. 7 (1893)
Mystacidium erythropollinium (Reichb.f.) Th. Dur. & Schinz, Consp. Fl. Afr. 5: 52 (1895)
Listrostachys scheffleriana Kraenzl. in E.J. 33: 75 (1902). Type: Tanzania, Usambara Mts., Derema, *Scheffler* 124 (B, holo.)
Mystacidium mahonii Rolfe in K.B. 1906: 116 (1906). Type: Uganda, Mengo District, Entebbe, *Mahon* 7 (K, holo.!)
Rhipidoglossum xanthopollinium (Reichb.f.) Schltr. in Beih. Bot. Centr. 36: 81 (1918)
R. gerrardii (Reichb.f.) Schltr. in Beih. Bot. Centr. 36: 80 (1918)

20. **D. ovalis** *Summerh.* in K.B. 16: 313 (1962). Type: Zaire, Kivu, Kahusi Gap, *Stauffer* 1026 (K, holo.!)

An epiphyte with a pendent stem, 6–30 cm. long, 1.5–2 mm. diameter. Leaves 2–10 in upper part of stem, falcate, linear, unequally bilobed at obtuse apex, 2.5–8.5 cm. long, 3–5 mm. wide, twisted at base and articulated to a sheathing base 7-10 mm. long. Inflorescences 1–several, 5–9 cm. long, laxly up to 20-flowered; peduncle 6–10 mm. long; bracts amplexicaul, conical, 1 mm. long. Flowers cream-coloured or buff; pedicel and ovary 2 mm. long, rugulose. Dorsal sepal elliptic, obtuse, 2.5–3 mm. long, 1.8–2 mm. wide; lateral sepals oblong-elliptic, obtuse, 3 mm. long, 1.4–1.6 mm. wide. Petals oblong to elliptic, obtuse, 1.8–2 mm. long, 1.6–1.7 mm. wide. Lip concave, entire or obscurely 3-lobed, ovate, obtuse, 2.3–3 mm. long, 2.4–2.7 mm. wide, ecallose; spur incurved, cylindrical, 4–4.5 mm. long. Column 1 mm. long; stipites 2; viscidia 2; rostellum pendent, clavate. Fig. 126.

UGANDA. Kigezi District: Rutenga, Mar. 1962, *C. Leakey* 18!

FIG. 126. *DIAPHANANTHE OVALIS*—**1**, habit, × 1; **2**, part of flowering stem, × 5; **3**, dorsal sepal, × 6; **4**, lateral sepal, × 6; **5**, petal, × 6; **6**, column, with part of lip cut away, × 18; **7**, pollinarium, × 40. Drawn from *Stauffer* 1026 by Miss M.A. Grierson.

DISTR. **U** 2; adjacent Zaire
HAB. In montane forest; 2100–2300 m.

21. **D. rutila** (*Reichb.f.*) *Summerh.* in K.B. 14: 143 (1960); Piers, Orch. E. Afr., ed. 2: 268 (1968); J. Stewart in U.K.W.F.: 788 (1974); Williamson, Orch. SC. Afr.: 188 (1977); la Croix et al., Malawi Orch. 1: 72 (1983). Type: W. Africa, *von Laubach* (W, holo.)

A pendent or spreading epiphyte with stems 3–40 cm. long, 2–6 mm. diameter, occasionally branched. Leaves in upper part of stem, 3–10, linear to obovate or narrowly elliptic, unequally roundly bilobed at apex, 3–12 cm. long, 0.5–2.2 cm. wide, twisted at base and articulated to a sheathing base 7–10 mm. long. Inflorescences 1-many, pendent, 5–19 cm. long, as long as or longer than the leaves, densely up to 25-flowered or more; peduncle 12–30 mm. long; bracts amplexicaul, conical, 1–2 mm. long. Flowers dull pink with a purplish tinge; pedicel and ovary 1–2 mm. long. Dorsal sepal elliptic or obovate, obtuse, 1.5–3.3 mm. long, 1–2.5 mm. wide; lateral sepals slightly falcate-linear, obtuse, 2–4.3 mm. long, 1.2–2.1 mm. wide. Petals circular or broadly ovate, obtuse, 1.5–2.8 mm. long, 1.1–3.5 mm. wide. Lip flabellate, retuse at apex, 1.4–3.7 mm. long, 2.5–4.7 mm. wide; spur incurved, cylindrical, 4.3–7 mm. long. Column 0.8–1.2 mm. long; viscidia 2; stipites 2; rostellum pendent, swollen.

UGANDA. Bunyoro District: Budongo Forest, Sept. 1933, *Eggeling* 1356! & 1370!; Ankole District: Mt. Lutoto, *Snowden* 1643!; Mengo District: Mabira Forest, Dec. 1908, *E. Brown* 471! & *Dummer* 2985!
KENYA. Nyeri, 13 Nov. 1938, *Bally* 8603!; N. Kavirondo District: Kakamega Forest, May 1933, *Dale* in *F.D.* 3108!; Masai District: Ngong Escarpment, May 1931, *Napier* 1302!
TANZANIA. Tanga District: Amani, Aug. 1941, *Moreau* 33!; Rungwe District: Iviji, 9 Apr. 1972, *Leedal* 1116!; Lindi District: Rondo [Muera] Plateau, 20 May 1935, *Schlieben* 6532!
DISTR. **U** 2, 4; **K** 3–7; **T** 3, 4, 7, 8; Sierra Leone, Guinée and Liberia to Cameroon, Gabon, Equatorial Guinea, Central African Republic and Zaire, Angola, Malawi, Mozambique and Zimbabwe
HAB. In riverine and montane forest, also occasionally in old coffee plantations; 600–2200 m.

SYN. *Aeranthus rutilus* Reichb.f. in Flora 68: 382 (1885)
Listrostachys rutilus (Reichb.f.) Ridl. in Bol. Soc. Brot. 5: 200 (1887)
Mystacidium rutilum (Reichb.f.) Th. Dur. & Schinz, Consp. Fl. Afr. 5: 54 (1895)
Listrostachys multiflora Rolfe in F.T.A. 7: 162 (1897). Type: Tanzania, Pangani Valley, *Kirk* (K, holo.!)
L. gabonensis Rolfe in F.T.A. 7: 161 (1897). Type: Gabon, without exact locality, cult. Jardin des Plantes, Paris (K, holo.!)
Angraecum woodianum Schltr. in E.J. 26: 343 (1899). Types: Malawi, without exact locality, *Buchanan* in *J.M. Wood* 6027 (B, holo.)
Listrostachys margaritae De Wild., Not. Pl. Utiles Congo 1: 150 (1903). Type: Zaire, Basenga, *Gentil* 30 (BR, holo.)
Rhipidoglossum rutilum (Reichb.f.) Schltr. in Beih. Bot. Centr. 36: 81 (1918)
R. woodianum (Schltr.) Schltr. in Beih. Bot. Centr. 36: 81 (1918)
Diaphananthe margaritae (De Wild.) Schltr. in Beih. Bot. Centr. 36: 98 (1918)
Chamaeangis schliebenii Mansf. in N.B.G.B. 11: 809 (1933). Type: Tanzania, Ulanga District, Mahenge, *Schlieben* 2168 (B, holo.)

22. **D. adoxa** *Rasmussen* in Norw. Journ. Bot. 21: 229 (1974). Type: Ethiopia, Kaffa Province, *DEBL* 73-52 (C, holo.)

An epiphyte with 5–25 cm. long pendent stems, 2–2.5 mm. diameter, often forming dense clumps; roots prominent, much longer than leaves. Leaves 5–12 in upper part of stem, falcate, linear, unequally bilobed at apex, 3.5–12 cm. long, 1.5–5 mm. wide, twisted at base and articulated to a 7–10 mm. long leaf-sheath. Inflorescences 1–several, 2–3.5 cm. long, 3–8-flowered; peduncle 2–7 mm. long; bracts amplexicaul, 2 mm. long. Flowers green or pale yellowish green; pedicel and ovary up to 2 mm. long. Dorsal sepal lanceolate, acute, 2–2.5 mm. long, 1 mm. wide; lateral sepals lanceolate, acute, 3–3.2 mm. long, 1 mm. wide. Petals ovate, obtuse or rounded, 2.1 mm. long, 1.2 mm. wide. Lip entire, ovate, obtuse, 2 mm. long and wide, ecallose; spur clavate, 1.5–2 mm. long. Column 0.7 mm. long; stipites 2; viscidia 2; rostellum pendent, clavate.

UGANDA. Kigezi, Sept. 1936, *Eggeling* 3238!
KENYA. Nairobi District: Duke of York School, 15 May 1963, *Start* in *E.A.H.* 175/63!
DISTR. **U** 2; **K** 4; Ethiopia
HAB. In upland evergreen forest; 1300–1500(–2300 in Ethiopia) m.

SYN. *Rhipidoglossum adoxum* (Rasmussen) Senghas in Schltr., Die Orchideen, ed. 3, 1: 1110 (1986)

NOTE. The Ugandan collection agrees well with the type in its habit but possesses only immature buds. Its identification is, therefore, somewhat tentative.

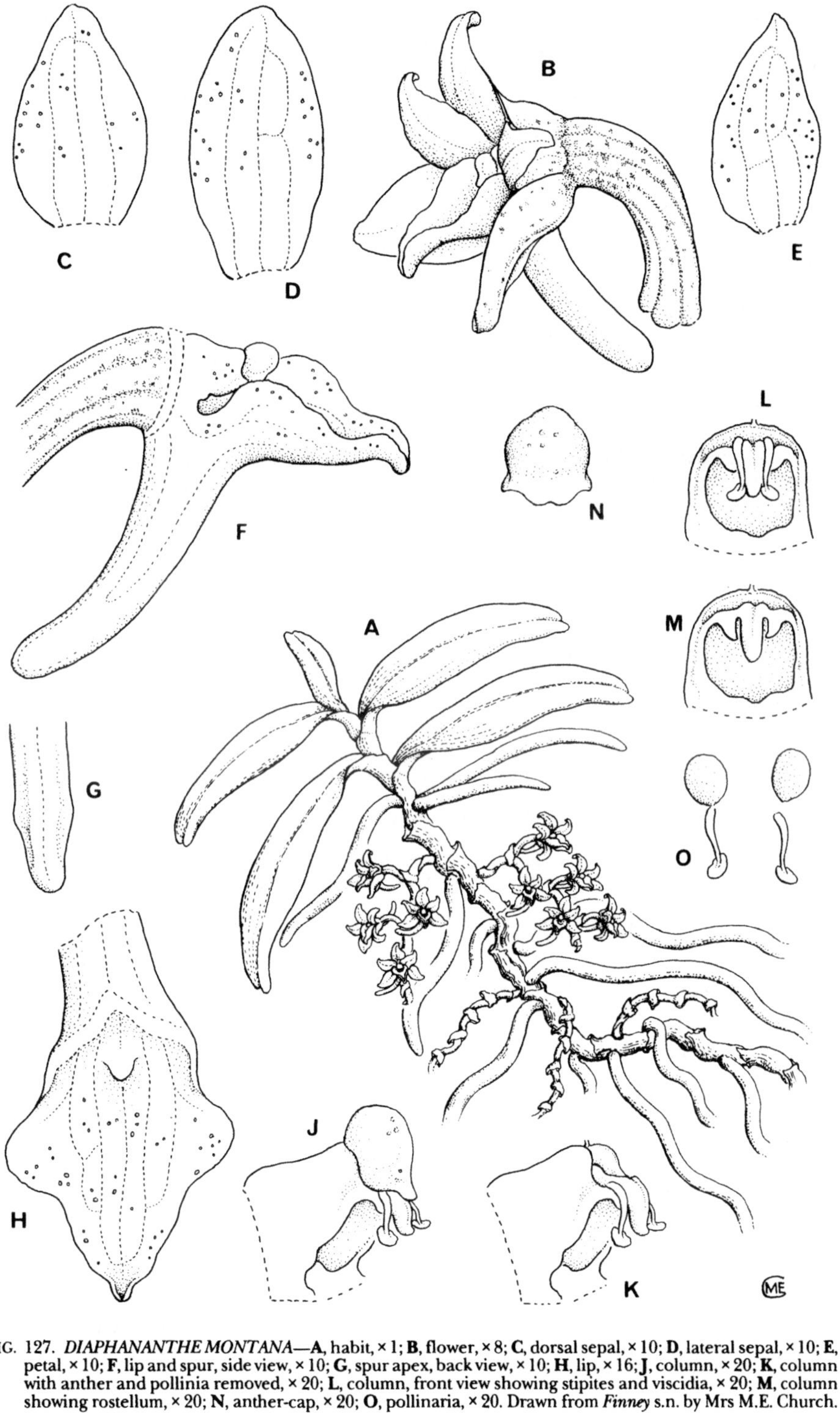

FIG. 127. *DIAPHANANTHE MONTANA*—**A**, habit, × 1; **B**, flower, × 8; **C**, dorsal sepal, × 10; **D**, lateral sepal, × 10; **E**, petal, × 10; **F**, lip and spur, side view, × 10; **G**, spur apex, back view, × 10; **H**, lip, × 16; **J**, column, × 20; **K**, column with anther and pollinia removed, × 20; **L**, column, front view showing stipites and viscidia, × 20; **M**, column showing rostellum, × 20; **N**, anther-cap, × 20; **O**, pollinaria, × 20. Drawn from *Finney* s.n. by Mrs M.E. Church.

23. **D. montana** (*Piers*) *Cribb & J. Stewart* in K.B. 40: 411 (1985). Type: Kenya, Masai District, Ngong Hills, *J. Stewart* 827 (EA, holo.!, K, iso.!)

A dwarf epiphyte with a 5–10 cm. long stem, rather obscured by a dense mass of roots. Leaves linear or narrowly elliptic, unequally bilobed at apex, 1.5–6.5 cm. long, 2–6 mm. wide, twisted at base and articulated to a 4–8 mm. long leaf-sheath. Inflorescences 1-several, 1.5–3.5 cm. long, up to 9-flowered; peduncle very short, completely covered by sterile bracts; bracts amplexicaul, conical, 3–3.5 mm. long. Flowers greenish yellow; pedicel and ovary 3–5.5 mm. long. Dorsal sepal ovate, obtuse, 3.6–4.2 mm. long, 1.9–2 mm. wide; lateral sepals elliptic or ovate, obtuse, 4.6–5.2 mm. long, 1.6–2 mm. wide. Petals ovate, acute, 3.6–3.7 mm. long, 1.5–1.6 mm. wide. Lip concave, ovate, entire or very obscurely 3-lobed, acute, 3.2–3.3 mm. long, 2.6–3.7 mm. wide, with an obscure callus at the base; spur cylindric, slightly recurved or straight, 5–5.5 mm. long. Column 1–1.2 mm. long; stipites 2; viscidia 2; rostellum pendent, tooth-like. Fig. 127, p.537.

KENYA. W. Suk/Elgeyo Districts: Cherangani Hills, 18 June 1969, *J. Stewart* 1010!; Masai District: Ngong Hills, 7 Dec. 1984, *Khayota* 69! & Ol Doinyo Orok, 19 Oct. 1970, *Archer* 648!
DISTR. **K** 2/3, 6; not known elsewhere
HAB. On trees in open woodland and in montane forest; 2300–2750 m.

SYN. *Angraecum montanum* Piers in Die Orchidee 20: 248 (1960)
Rhipidoglossum montanum (Piers) Senghas in Schltr., Die Orchideen, ed. 3, 1: 1111 (1986)

24. **D. stellata** *Cribb*, sp. nov. e aliis specibus generis planta distincta minuta, inflorescentia pendenti clavata densa, floribus stellatis parvis, tepalis lanceolatis acutis, calcare recto parallelo ad ovarium bene distinguenda. Typus: Tanzania, Iringa District, Mufindi, Luisenga R., *J. & J. Lovett* 2065 (K, holo.!)

A pendent to erect epiphyte with a 2–15 cm. long stem, 3 mm. diameter, somewhat laterally compressed; roots branching, silvery green, to 20 cm. long, 3–4 mm. diameter. Leaves distichous, narrowly oblong, falcate, unequally bilobed at apex, 3.5–8.5 cm. long, 7–9 mm. wide, twisted at base to lie in one plane, articulated to a sheathing base 5 mm. long. Inflorescences 4–7-flowered, pendent or subpendent, 2–4 cm. long; peduncle terete, tapering to base, 1.5–2 cm. long; bracts ovate, acute, 1–2 mm. long. Flowers spirally arranged in a head, yellow-green, very fragrant; pedicel and ovary 5 mm. long. Dorsal sepal elliptic-lanceolate, acute, 4–5 mm. long, 2.2 mm. wide; lateral sepals similar. Petals lanceolate, acute, 3.8–4 mm. long, 1.2–1.3 mm. wide. Lip recurved, lanceolate to obscurely 3-lobed, acute, 4.2–5 mm. long, 2.5 mm. wide, with a low transverse callus at the base; spur parallel to ovary, straight, cylindrical, 8–13 mm. long. Column 1 mm. long; stipites 2; viscidia 2, elliptic, yellow; rostellum pendent-recurved, anchor-shaped. Fig. 128.

TANZANIA. Iringa District: Mufindi, Kigogo Forest, May 1954, *Eggeling* 6815! & Luisenga R., 25 May 1970, *Paget-Wilkes* 838! & 9 June 1971, *Paget-Wilkes* 902!; Njombe District: Madaba, fl. Kew 24 Nov. 1986, *de Leyser*!
DISTR. **T** 7; not known elsewhere
HAB. In montane forest; 1800–2400 m.

25. **D. orientalis** (*Mansf.*) *Rasmussen* in Norw. Journ. Bot. 21: 229 (1974). Type: Tanzania, Uluguru Mts., Mesumba [Messumba], *Schlieben* 4178 (B, holo.)

A dwarf epiphyte with a rarely branching stem, ± 10 cm. long, 1.5–2 mm. diameter. Leaves 5–9, linear-oblong or oblong-ligulate, slightly falcate, unequally obtusely bilobed at apex, 2–3.5 cm. long, 3–6 mm. wide, twisted at base and articulated to a leaf-sheath 5 mm. long, drying almost black. Infloresences 1–few, 2–3-flowered, ± 1.5 cm. long; peduncle ± 5 mm. long; bracts ovate, acute, 1–1.5 mm. long. Flowers minute, yellowish green; pedicel and ovary 1–2 mm. long. Dorsal sepal lanceolate-elliptic, acute, 3.5–3.7 mm. long, 1.2 mm. wide; lateral sepals oblong-lanceolate, acute, 3.5–3.9 mm. long, 1.3 mm. wide. Petals lanceolate, acute or acuminate, 4 mm. long, 1.4 mm. wide. Lip concave, entire, ovate-elliptic, obtuse, 4–4.5 mm. long, 3–3.3 mm. wide, ecallose; spur pendent, cylindrical-clavate, 1.3–1.8 mm. long. Column 1 mm. long; stipites 2; viscidia 2; rostellum spathulate, dilated at apex.

TANZANIA. Morogoro District: Uluguru Mts., Mesumba [Messumba], 19 July 1933, *Schlieben* 4178!
DISTR. **T** 6; known only from the type gathering
HAB. In montane mist-forest; 1900 m.

SYN. *Sarcorhynchus orientalis* Mansf. in N.B.G.B. 13: 412 (1936)

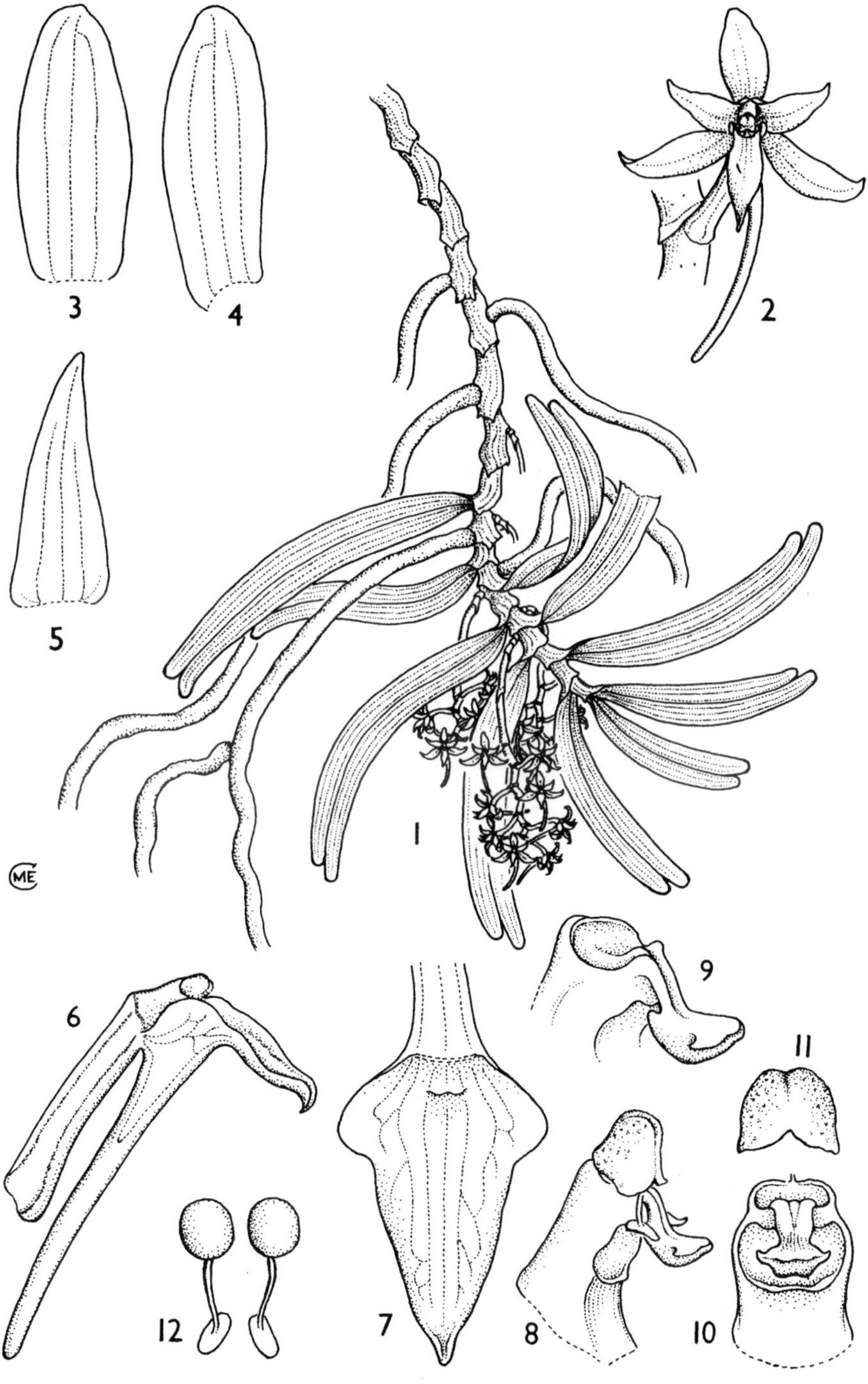

FIG. 128. *DIAPHANANTHE STELLATA*—**1**, habit, × ⅔; **2**, flower, × 4; **3**, dorsal sepal, × 8; **4**, lateral sepal, × 8; **5**, petal, × 8; **6**, lip and spur, × 6; **7**, lip, spread out, × 10; **8**, column, side view, × 16; **9**, column apex, × 20; **10**, column, front view, × 16; **11**, anther-cap, × 16; **12**, pollinaria, × 16. Drawn from *de Leyser* E.N. 155-85-02060 by Mrs M.E. Church.

26. **D. tanneri** *Cribb* in K.B. 40: 411 (1985). Type: Tanzania, W. Usambara Mts., Bungu, *G. Tait* in *Moreau* 174 (K, holo.!)

A dwarf epiphyte often forming a tangled mass of stems and roots; stems erect to pendent, up to 20 cm. long, 2 mm. diameter, bearing prominent branching roots all along stem. Leaves fleshy, not at all obvious, linear, rounded at unequally bilobed apex, up to 4.5 cm. long, 1.5 mm. wide, articulated to an 8–10 mm. long leaf-sheath. Inflorescences 1–several, 4–7-flowered, 2–2.5 cm. long; peduncle 6–10 mm. long; rhachis somewhat zig-zag; bracts ovate, acute, 1 mm. long. Flowers tiny, glistening green; pedicel and ovary 2–2.5 mm. long. Dorsal sepal elliptic, rounded to subacute, 3.5 mm. long, 1.5–1.8 mm. wide; lateral sepals spreading, elliptic, rounded, 3–3.7 mm. long, 1.3–1.6 mm. wide. Petals elliptic-ovate, acute, 3–3.3 mm. long, 1.2–1.7 mm. wide. Lip entire, recurved, ovate, acute, 2.2–2.5 mm. long, 1.5 mm. wide, with an obscure bilobed callus at the base; spur swollen towards apex, pendent-recurved, 3.8–4 mm. long. Column 1 mm. long; stipites 2; viscidia 2; rostellum pendent, peg-like.

TANZANIA. Lushoto District: W. Usambara Mts., Mazumbai Forest, 11 Jan. 1976, *Cribb & Grey-Wilson* 10175! & Bungu, fl. in cult. 9 Aug. 1943 & 14 Apr. 1944, *G. Tait* in *Moreau* 174!

SYN. *Rhipidoglossum tanneri* (Cribb) Senghas in Schltr., Die Orchideen, ed. 3, 1: 1111 (1986)

27. **D. meliantha** *Cribb* in K.B. 34: 335 (1979). Type: Tanzania, Uluguru Mts., Lukwangule Plateau, *Cribb & Grey-Wilson* 10463 (K, holo.!)

A small epiphyte with a pendent or spreading stem, 3–5 cm. long. Leaves 2–6 clustered towards apex of stem, falcate, linear-ligulate, unequally bilobed at rounded apex, 2–8 cm. long, 3–7 mm. wide, twisted at base to lie in one plane and articulated to a 4–6 mm. long sheathing base. Inflorescences 1–few, pendent, 5–10-flowered, 3.5–19 cm. long; peduncle 3–6 cm. long; bracts triangular, acute, 1–2 mm. long. Flowers pale green turning butterscotch yellow, unpleasantly scented; pedicel and ovary 6–8 mm. long. Dorsal sepal oblong-ovate, obtuse, 4.5–5 mm. long, 3 mm. wide; lateral sepals obliquely narrowly ovate, decurved, obtuse, 5–5.5 mm. long, 2.5 mm. wide. Petals orbicular, obtuse, 3 mm. long and wide. Lip decurved, entire, oblong, obtuse to emarginate, 2.5–3.5 mm. long, 2–3 mm. wide, ecallose; spur incurved, cylindrical, 13–15 mm. long. Column 1 mm. long; stipites 2; viscidia 2; rostellum anchor-shaped at apex, fleshy, pendent-reflexed.

TANZANIA. Morogoro District: Uluguru Mts., Lukwangule Plateau, 19 Jan. 1943, *Moreau* 470! & Sept. 1970, *Thulin & Mhoro* 1027!; Iringa District: Mufindi, Kigogo Forest, 30 Mar. 1986, *Lovett* in *Keeley* 62! & Luisenga Forest, 16 Mar. 1962, *Polhill & Paulo* 1780!; Rungwe District: Ipyana, Kandete, 22 Mar. 1975, *Leedal* 2621!

DISTR. **T** 6, 7; not known elsewhere

HAB. In montane mist-forest patches; 1500–2400 m.

SYN. *Rhipidoglossum melianthum* (Cribb) Senghas in Schltr., Die Orchideen, ed. 3, 1: 1110 (1986)

55. **MARGELLIANTHA**

Cribb in K.B. 34: 329 (1979)

Small monopodial, epiphytic herbs with short stems covered in persistent leaf-bases. Leaves distichous, coriaceous, ± falcate, linear or narrowly oblanceolate, unequally bilobed at the apex, twisted at the base to lie in one plane, articulated at the base to a persistent sheathing leaf-base. Inflorescences spreading, racemose, few-flowered, axillary, emerging through the leaf-bases; peduncle and rhachis slender, terete, wiry. Flowers campanulate, white to yellow or pale green. Sepals and petals free; dorsal sepal concave; lateral sepals falcate, concave. Petals elliptic or subcircular. Lip concave, entire, lacking a callus, spurred at the base; spur clavate to globose, with a narrow mouth. Column short, fleshy, sometimes with an obscure short foot; stigma concave, with a raised posterior margin; rostellum 3-lobed, thin-textured, mid-lobe longer than the lateral lobes; pollinia 2, globose; stipites 2, linear; viscidia 2, circular or elliptic.

A small genus of 4 species in tropical Africa.

Spur almost globose, 2.5–4 mm. in diameter; bracts somewhat inflated 3.5 mm. long, 3 mm. wide; flowers creamy yellow . . . 1. *M. globularis*
Spur clavate, 6–7 mm. long; bracts inconspicuous, less than 2.5 mm. long and wide; flowers white with a green anther-cap or pale greenish yellow:
Flowers glistening white with a green anther-cap; spur incurved, 2 mm. in diameter at apex; lateral sepals falcate; sepals more than 4.5 mm. long; lip more than 5 mm. long, 6–8 mm. wide . . . . 2. *M. leedalii*
Flowers pale greenish yellow; spur straight, relatively slender, less than 1.2 mm. in diameter; lateral sepals not markedly falcate; sepals 3–3.5 mm. long; lip 4 mm. long, 5.5 mm. wide . . . . 3. *M. clavata*

1. **M. globularis** *Cribb* in K.B. 34: 331 (1979). Type: Tanzania, Uluguru Mts., *Wallace* 172 (K, holo.!)

A small pendent or spreading epiphyte with a short stem up to 3 cm. long, 4 mm. in diameter; roots from basal part of the stem, 1–1.5 mm. in diameter. Leaves 4–5, distichous, falcate, ligulate, unequally obtusely bilobed at the apex, 4–10 cm. long, 0.6–0.8 cm. wide. Inflorescences 1-few, 5–6-flowered, up to 6 cm. long; peduncle wiry, up to 3 cm. long; bracts amplexicaul, ovate-triangular, 3.5 mm. long, 3 mm. wide. Flowers campanulate, pale creamy yellow; pedicel and ovary slender, 11–12 mm. long. Dorsal sepal oblong-ovate, obtuse, 4 mm. long, 2 mm. wide; lateral sepals obliquely falcate, ovate, slightly L-shaped, acute, 5 mm. long, 2.5 mm. wide. Petals subcircular, obtuse, 4–5 mm. long, 4 mm. wide, with obscurely erose margins. Lip flabellate or broadly obovate, emarginate, 5 mm. long, 6 mm. wide; spur subglobose-clavate, 4 mm. long, 2.5–3 mm. wide. Column 1 mm. long.

TANZANIA. Morogoro District: Uluguru Mts., 25 Sept. 1932, *Wallace* 172!
DISTR. **T** 6; known only from the type gathering
HAB. In montane forest; 1250 m.

2. **M. leedalii** *Cribb* in K.B. 34: 333 (1979). Type: Tanzania, Uluguru Mts., *Cribb & Grey-Wilson* 10380 (K, holo.!)

A dwarf pendent to spreading epiphyte with a short stem 1.5–4 cm. long, 2–2.5 mm. in diameter; roots basal, 1–2 mm. in diameter. Leaves 3–6, distichous, slightly falcate, ligulate, unequally obtusely or acutely bilobed at the apex, 4–8.2 cm. long, 0.6–0.8 cm. wide, twisted at the base. Inflorescences 3–8-flowered, up to 6.5 cm. long; peduncle very finely wiry, 2–3 cm. long; bracts cucullate, ovate, acute, 2–2.5 mm. long. Flowers glistening white with a green anther-cap; pedicel and ovary slender, 7–10 mm. long. Dorsal sepal ovate, obtuse, 4–5 mm. long, 3 mm. wide; lateral sepals falcate, almost L-shaped, obtuse, 4–5 mm. long, 3 mm. wide. Petals obovate to subcircular, obtuse, 3.5–4 mm. long, 3 mm. wide, with erose margins. Lip flabellate to broadly obcordate, emarginate, 5–5.6 mm. long, 6–8 mm. wide, erose on apical margin; spur slightly incurved, clavate, 6–7 mm. long, 2 mm. in diameter. Column 2–3 mm. long, with an obscure foot ± 1 mm. long. Fig. 129, p. 542.

KENYA. Teita District: Kasigau Hill, 13 June 1967, *Archer* 548!
TANZANIA. Morogoro District: Uluguru Mts., Bondwa, Jan. 1976, *Cribb & Grey-Wilson* 10380! & Lupanga Mt., 26 Dec. 1931, *B.D. Burtt* 3490!; Iringa District: Image Mt., *Polhill & Paulo* 1620!; Rungwe Mt., Feb. 1976, *Cribb & Grey-Wilson* 10642!
DISTR. **K** 7; **T** 6, 7; not known elsewhere
HAB. In montane or mist-forest, on mossy branches of trees and shrubs; 1550–2400m.

3. **M. clavata** *Cribb* in K.B. 34: 331 (1979). Type: Tanzania, W. Usambara Mts., Magamba Forest Reserve, *Eggeling* in *Moreau* 400 (K, holo.!)

A dwarf pendent to suberect epiphyte with a short stem up to 5 cm. long, 3 mm. in diameter; roots 1–2 mm. in diameter. Leaves distichous, narrowly oblong, slightly falcate, unequally obtusely bilobed at the apex, 4–5 cm. long, 0.8–1 cm. wide, twisted at the base. Inflorescences up to 2.5 cm. long, 2–5-flowered; peduncle very slender, wiry, 5 mm. long; bracts ovate, subacute, 1–1.5 mm. long. Flowers pale yellow-green; pedicel and ovary very slender, 7 mm. long. Dorsal sepal oblong-elliptic, rounded, 3.5 mm. long, 1.5 mm. wide;

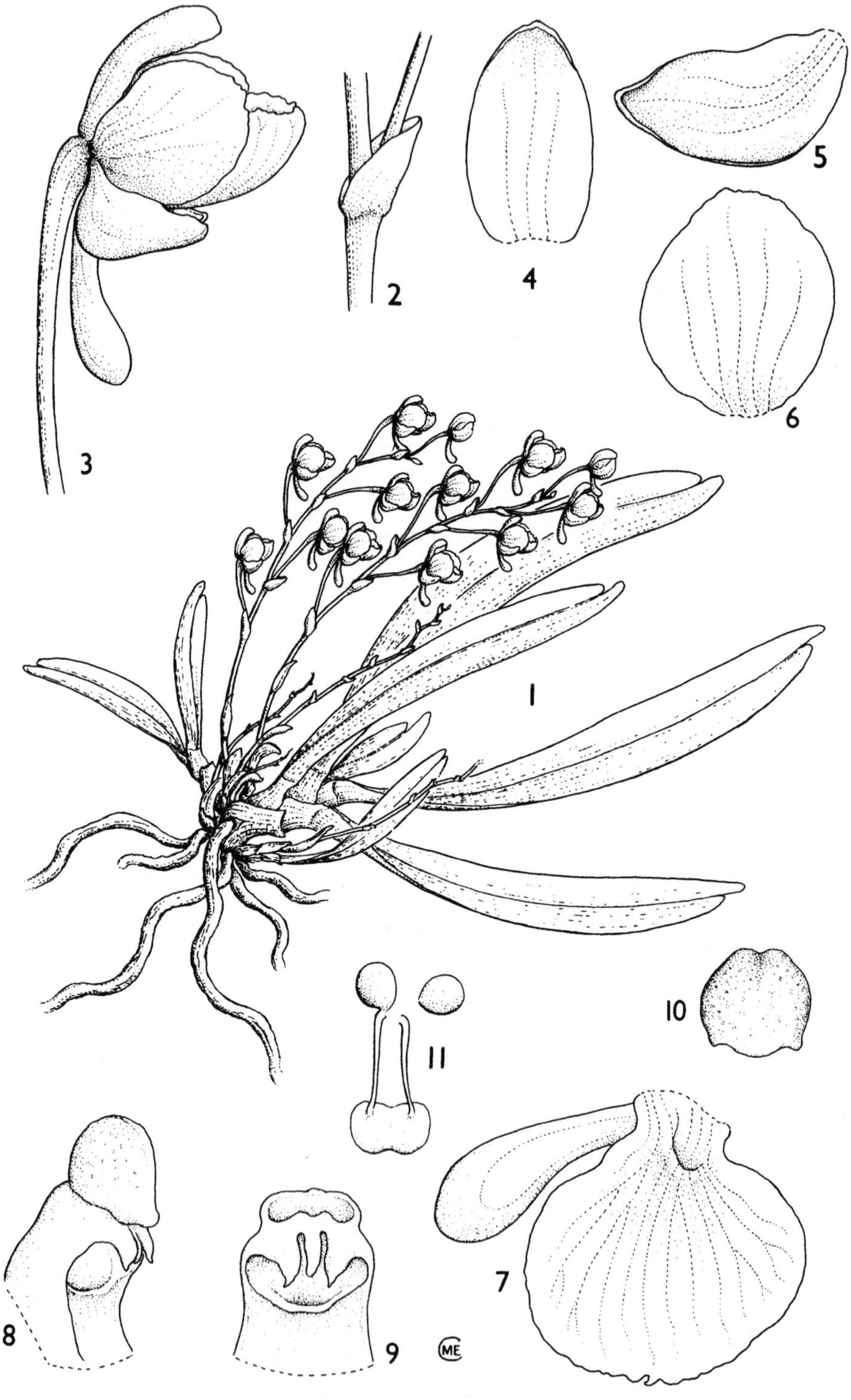

FIG. 129. *MARGELLIANTHA LEEDALII*—**1**, habit, × 1; **2**, bract, × 6; **3**, flower, × 6; **4**, dorsal sepal, × 8; **5**, lateral sepal, × 8; **6**, petal, × 8; **7**, lip, × 8; **8**, column, side view, × 24; **9**, column, front view, × 24: **10**, anther-cap, × 20; **11**, pollinia, stipites and viscidia, × 20. Drawn from *Archer* 548 by Mrs M.E. Church.

lateral sepals elliptic-obovate, rounded, 3–4 mm. long, 1.5 mm. wide. Petals elliptic or almost circular, obtuse, 2.5–3 mm. long, 2.5 mm. wide. Lip broadly elliptic, 4 mm. long, 5.5–6 mm. wide; spur straight, clavate, 6–7 mm. long, 1–1.2 mm. in diameter. Column 1–1.5 mm. long.

TANZANIA. Lushoto District: W. Usambara Mts., Magamba Forest Reserve, fl. cult. 6 Mar. 1943, *Eggeling* in *Moreau* 400! & Mazumbai, Kwagrotto, fl. cult. Mar. 1982, *L. Tanner* 278!
DISTR. T 3; not known elsewhere
HAB. In montane forest, on branches of small trees on ridge; 1800–1950 m.

56. BOLUSIELLA*

Schltr. in Beih. Bot. Centr. 36: 105 (1918)

Erect epiphytic or occasionally lithophytic, entirely glabrous monopodial herbs. Stems abbreviated, simple, leafy above, with numerous tufted roots below. Leaves equitant, imbricate, arranged in a vertical plane forming a fan, narrowly elliptic or ensiform to linear-ligulate, obtuse to acute, rigid and articulated to a sheathing base, fleshy, sometimes sulcate-canaliculate and falcate-recurved. Inflorescence exceeding leaves, densely or laxly many-flowered, arising from axils of old leaf-sheaths; peduncle shorter than or equalling rhachis, rarely longer; bracts small or large, imbricate and partially concealing flowers, membranous, grey to blackish brown or olive. Flowers resupinate, very small, white. Tepals free, oblong or narrowly elliptic, obtuse to acuminate, subequal, spreading. Lip entire or obscurely 3-lobed, spurred, oblong or ovate-elliptic, obtuse to acuminate, sometimes recurved; spur cylindrical, ellipsoid or conical, obtuse, usually shorter than lip. Column oblong, fleshy, constricted above, without a foot; rostellum subulate, hamate; anther quadrate, cucullate; pollinia 2, oblong or ellipsoid, each attached by a linear stipes to a large ovate viscidium.

A genus of 5 species distributed in tropical and southern Africa.

Leaves deeply sulcate on upper surface, often falcate-recurved in side view, 2–4 mm. broad . . . . 1. *B. iridifolia*
Leaves not deeply sulcate on upper surface, mostly straight or rarely falcate-recurved in side view, (3–)5–12 mm. broad:
Sterile bracts well-spaced on peduncle, never imbricate; fertile bracts small, 1–3 mm. long, well spaced on rhachis; leaves (1.5–)2.5–6(–8) cm. long, tapering to an acute or subacute apex . . . . 2. *B. talbotii*
Sterile bracts imbricate and often completely enclosing the peduncle; fertile bracts large, 3–4 mm. long, imbricate; leaves 1–3.5 cm. long, apex obtuse . . . . 3. *B. maudiae*

1. **B. iridifolia** (*Rolfe*) *Schltr.* in Beih. Bot. Centr. 36: 106 (1918); Robyns & Tournay, F.P.N.A. 3: 524 (1955); Piers, Orch. E. Afr., ed. 2: 249 (1968): Summerh. in F.W.T.A., ed. 2, 3: 263 (1968); J. Stewart in U.K.W.F.: 790 (1974). Type: Angola, Golungo Alto, *Welwitsch* 679 (K, holo.!)

An epiphytic or lithophytic herb. Stem short, 0.5–2 cm. long. Leaves 4–6(–10), ensiform, obtuse, falcate-recurved or straight, upper surface sulcate, 1–4.5 cm. long, 1.5–5 mm. broad. Inflorescence dense, 2.2–6 cm. long; bracts ovate, acute to acuminate, brown or black, shorter than or longer than flowers, 1–4 mm. long, well-spaced. Flowers white; pedicel with ovary 2 mm. long. Tepals oblong, obtuse, 1.1–3 mm. long, 0.7–1.4 mm. broad. Lip oblong-ligulate, obtuse, slightly sigmoid, 1–5 mm. long, 0.7–1.1 mm. broad; spur cylindrical-ellipsoid and inflated or conical and saccate, obtuse, strongly incurved at right-angles to lip or ± straight, 0.5–2 mm. long. Column 0.5–0.9 mm. long.

* By J.J. Wood

KEY TO INFRASPECIFIC VARIANTS

Spur cylindrical-ellipsoid, ± dilated, pendent or strongly incurved at right-angles to the lip, more than 0.8 mm. long; bracts brown, mostly shorter than flowers . . . . . subsp. **iridifolia**
Spur conical-saccate, ± straight, scarcely dependent, ± 0.5 mm. long; bracts black, mostly longer than flowers . . . subsp. **picea**

subsp. **iridifolia**

Stem 0.5–2 cm. long. Leaves usually falcate-recurved, (1.5–)2–4.5 cm. long, 2–4 mm. broad. Inflorescence 3.5–6 cm. long; bracts brown, shorter than flowers, 1–2(–4) mm. long. Tepals 2–3 mm. long. lip 2–2.5(–5) mm. long; spur cylindrical-ellipsoid, ± dilated, pendent or strongly incurved at right-angles to lip, 1.7–2 mm. long. Fig. 130/1–8.

UGANDA. Bunyoro District: Budongo Forest, Dec. 1942 (fr.) & Sept. 1943 (fl.), *Eggeling* 5193!; Mengo District: Mabira Forest, 1905, *Dawe* 211!; Masaka District: 2 km. NE. of Nkoma, June 1971, *Lye* 6227!
KENYA. Uasin Gishu District: Kipkarren R., June 1962, *Piers* (62) 26!; Machakos/Masai District: Chyulu Centre, July 1938, *Bally* 7854!; Masai District: Ngong Hills, Apr. 1952, *Archer* 52!
TANZANIA. Arusha District: Ngurdoto National Park, Tulusia [Tululusie] R., Nov. 1965, *Greenway & Kanuri* 12328!; Lushoto/Tanga Districts: Sigi Segoma, Nov. 1943, *Moreau* 736!; Iringa District: Sao Hill, July 1959, *Watermeyer* 101!
DISTR. **U**2, 4; **K**3, 4, 6; **T**2, 3, 7; Ivory Coast, Ghana, Equatorial Guinea, Cameroon, Zaire, Angola, Grand Comore
HAB. Forest and savanna trees; sea-level–2300 m.

SYN. *Listrostachys iridifolius* Rolfe in F.T.A. 7: 167 (1897)

subsp. **picea** *Cribb* in K.B. 32: 180 (1977); Ball, S. Afr. Epiphytic Orch.: 66 (1978). Type: Zambia, Northern Province, Mafinga Mts., *Williamson* 1940 (K, holo.!)

Stem 1.5–2 cm. long. Leaves straight or slightly falcate, 1–2.4 cm. long, 3–5 mm. broad. Inflorescence 2.2–5 cm. long. Bracts black or dark brown, longer than flowers, 2–3(–4) mm. long. Tepals 1.1–1.2 mm. long. Lip 1 mm. long; spur conical, saccate, straighter than in subsp. *iridifolia*, never incurved, 0.5 mm. long. Fig. 130/9.

KENYA. Machakos/Masai Districts: Chyulu Hills, May 1938, *Bally* 8402!
TANZANIA. Lushoto District: Mkusu valley between Mkusi and Kifungilo, Apr. 1953, *Drummond & Hemsley* 2206!; Mbeya District: Umalila, Idunda, Mar. 1975, *Leedal* 2627!; Njombe District: Kipengere Mts., Nyarere R., Jan. 1957, *Richards* 7644A!
DISTR. **K** 4/6; **T**3, 7; Malawi, Zambia, Zimbabwe
HAB. Montane forest, amongst lichens on dwarf trees in scrub forest, on rocks in river; 1500–2400 m.

SYN. *B. sp.* sensu Williamson, Orch. SC. Afr.: 190, t. 1 (1977)

2. **B. talbotii** (*Rendle*) *Summerh.* in F.W.T.A. 2: 456 (1936) & in F.W.T.A., ed. 2, 3: 264 (1968). Type: S. Nigeria, Oban, *Talbot* 941 (BM, holo., K, drawing!)

An epiphytic herb. Stem very short, 0.3–1.5 cm. long. Leaves 3–6(–12), ensiform, acute or subacute, rarely obtuse, straight, fleshy, not sulcate on upper surface, (1.5–)2.5–6(–8) cm. long, 0.5–0.8(–1.2) cm. broad. Inflorescence lax, 3–11 cm. long; bracts ovate, acute, ochraceous, edged olive-brown, shorter than flowers, 1–3 mm. long, well spaced, never imbricate. Flowers white, pink in centre and often with a green spur; pedicel with ovary 1 mm. long. Tepals narrowly elliptic or oblong, obtuse to apiculate, 1.7–4 mm. long, 0.6–0.9 mm. broad. Lip obscurely 3-lobed, 1.6–2.5 mm. long, 0.85–1 mm. broad; side lobes rounded; mid-lobe triangular to narrowly elliptic, subacute or acute; spur conical, obtuse, (1.5–)1.7–1.8 mm. long. Column 0.6 mm. long.

TANZANIA. Lushoto District: E. Usambara Mts., Amani, Apr. 1941, *Moreau* 41! & 18 Jan. 1976, *Cribb & Grey-Wilson* 10262!; Ulanga District: Sali, Muhulu Forest Reserve, Jan. 1979, *Cribb, Grey-Wilson & Mwasumbi* 11186!; Iringa District: Mufindi, Lulando Forest, fl. cult. Feb. 1988, *de Leyser* 209!
DISTR. **T**3, 6, 7; Sierra Leone, Liberia, Ivory Coast, Ghana, Togo, Nigeria, Bioko, Equatorial Guinea, Annobon
HAB. Forest and coffee plantations; 900–1350 m.

SYN. *Angraecum talbotii* Rendle, Cat. Pl. Talb.: 108, t. 15/6, 7 (1913)

3. **B. maudiae** (*Bolus*) *Schltr.* in Beih. Bot. Centr. 36: 106 (1918); Schelpe, Introd. S. Afr. Orch.: 52 (1966); Harrison, Epiphytic Orch. S. Afr.: 45 (1972); Ball, S. Afr. Epiphytic Orch.:

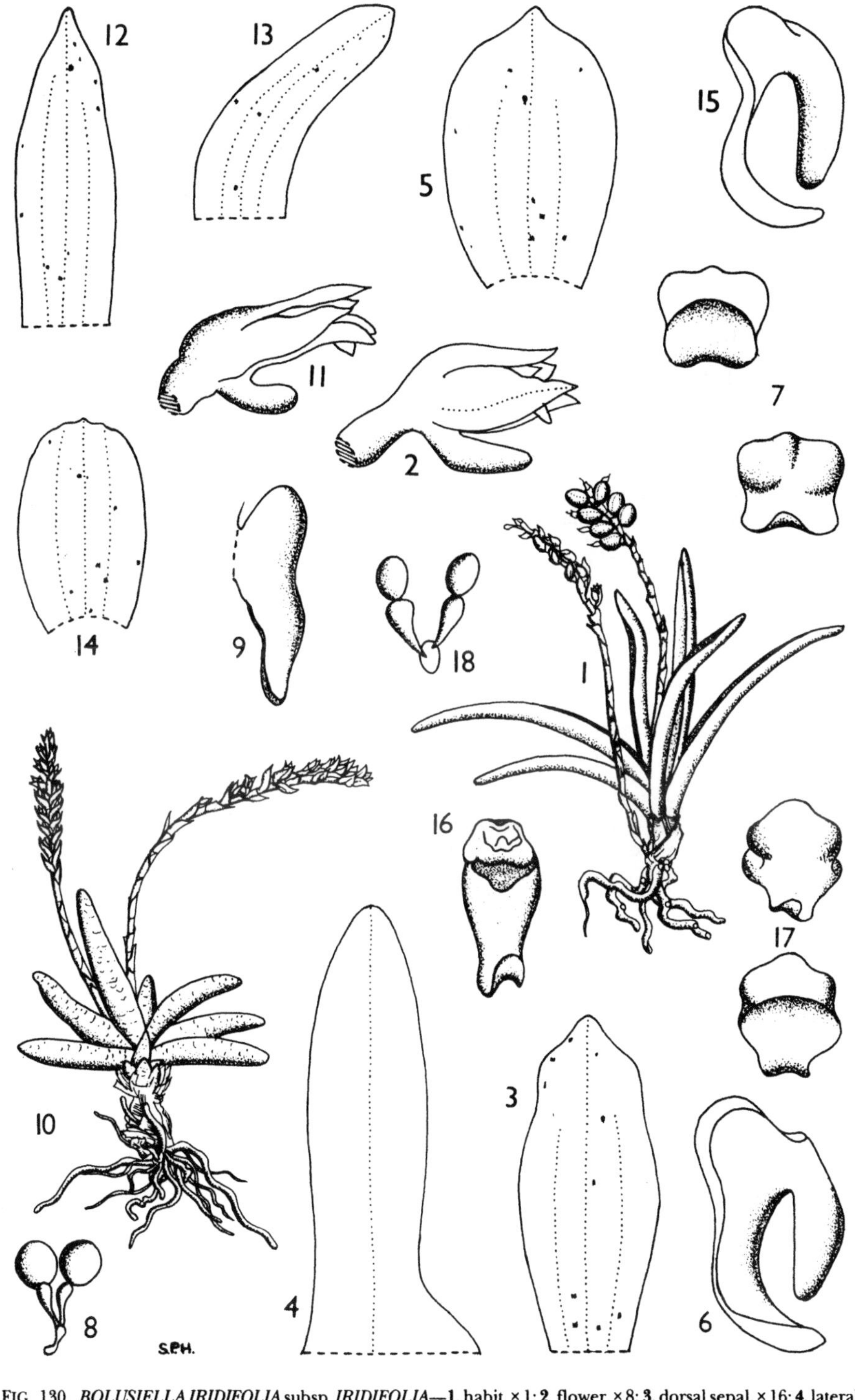

FIG. 130. *BOLUSIELLA IRIDIFOLIA* subsp. *IRIDIFOLIA*—**1**, habit, × 1; **2**, flower, × 8; **3**, dorsal sepal, × 16; **4**, lateral sepal, × 16; **5**, petal, × 16; **6**, lip and spur, × 14; **7**, anther-cap, × 8; **8**, pollinarium, × 26. *B. IRIDIFOLIA* subsp. *PICEA*—**9**, lip and spur, × 14. *B. MAUDIAE* - **10**, habit, × 1; **11**, flower, × 8; **12**, dorsal sepal, × 16; **13**, lateral sepal, × 16; **14**, petal, × 16; **15**, lip and spur, × 14; **16**, column, × 16; **17**, anther-cap, × 16; **18**, pollinarium, × 26. 1–8, from *Eggeling* 5193; 9, from *L. Tanner* 149; 10–18, from *Meyer* K77. Drawn by Susan Hillier.

68 (1978). Type: South Africa, Natal, Zululand, Etshowe, *Mrs. Charles Saunders* in *Bolus Herb.* 6270 (BOL, holo., K, iso.!)

An epiphytic herb. Stem 0.5–2(–4) cm. long. Leaves (2–)5–8, ensiform, obtuse, straight or falcate-recurved, coriaceous, surfaces convex, not sulcate on upper surface, 1–3.5 cm. long, 0.3–1 cm. broad. Inflorescence dense, (1.5–)2.5–6 cm. long; bracts ovate, acute to acuminate, olive or bronze, equalling flowers, 3–4 mm. long, closely imbricate. Flowers with a pale olive spur, faintly scented; pedicel with ovary 1.5–2 mm. long. Tepals 3–4 mm. long, 1–1.5 mm. broad, obtuse to acute; dorsal sepal narrowly elliptic to oblong, lateral sepals oblong, rounded-auriculate and briefly adnate at base; petals oblong. Lip obscurely 3-lobed, 2–3 mm. long, 0.8–1 mm. broad; side lobes rounded; mid-lobe oblong-ligulate, acute to acuminate; spur cylindrical, slightly inflated, obtuse, incurved, 1.5–2 mm. long. Column 0.6–0.8 mm. long. Fig. 130/10–18.

UGANDA. Toro District: Kibale Forest, June 1939, *A.S. Thomas* 3004!; Bunyoro District: Budongo Forest, Aug. 1940, *Eggeling* 4045!; Mengo District: Kipayo, Aug. 1914, *Dummer* 996!
KENYA. Nandi District: near Kaimosi, Sept. 1934, *Tweedie* 230! & Yala R., along the bridge, June 1953, *Williams & Piers* 615!; N. Kavirondo District: Kakamega Forest, Sept. 1949, *Maas Geesteranus* 6263!
TANZANIA. Lushoto District: 9.5 km. E. of Amani, Sigi R., fl. cult. Sept. 1944, *Moreau* 395!
DISTR. **U**2, 4; **K**3, 5; **T**3; Ivory Coast, Ghana, Malawi, Zambia, South Africa (Natal)
HAB. Forest; 900–1900 m.

SYN. *Angraecum maudiae* Bolus, Ic. Orch. S. Afr. 1, t. 9 (1893)
Listrostachys imbricata Rolfe in K.B. 1910: 161 (1910). Type: Ghana, without exact locality, *Anderson* (K, holo.!)
Bolusiella imbricata (Rolfe) Schltr. in Beih. Bot. Centr. 36: 106 (1918); Moreau in Journ. E. Afr. Nat. Hist. Soc. 17: 21 (1943); Tweedie in Journ. E. Afr. Nat. Hist. Soc. 24: 44 (1964); Piers, Orch. E. Afr., ed. 2: 247 (1968); Summerh. in F.W.T.A., ed. 2, 3: 263 (1968); J. Stewart in U.K.W.F.: 790 (1974); Williamson, Orch. SC. Afr.: 190 (1977)

57. CHAMAEANGIS

Schltr. in Beih. Bot. Centr. 36: 107 (1918)

Small to medium-sized erect to pendent monopodial epiphytic herbs. Stems covered in persistent sheathing leaf-bases, short to long. Leaves distichous, sometimes borne in a fan, fleshy or coriaceous, linear to oblanceolate, unequally bilobed at the apex. Inflorescences 1–several, ± as long as the leaves or a little longer, densely many-flowered, usually 2–several-flowered at each node. Flowers small, green, yellow or orange. Sepals and petals free, subsimilar, spreading. Lip ± entire, ecallose, spurred at the base; spur cylindrical to clavate. Column short, fleshy, lacking a foot; rostellum short, bifid; pollinia 2, globose; stipites 1 or 2; viscidia 1 or 2, ovate or circular.

A genus of some 15 species in tropical Africa, Madagascar and the Comores.

Spur abruptly dilated at the apex; leaves acute at the obliquely bilobed apex . . . . 2. *C. vesicata*
Spur cylindrical to the apex, at the most slightly swollen there; leaves roundly or obtusely bilobed at the apex:
 Stem elongate, mostly more that 20 cm. long, leafy throughout; sepals and petals very short, less than 2 mm. long; spur 9.5 mm. long or less . . . . . . 1. *C. odoratissima*
 Stem short, less than 15 cm. long, usually much shorter, leafy in apical part only; sepals and petals usually more than 2.5 mm. long; spur usually more than 10 mm. long 3. *C. sarcophylla*

1. **C. odoratissima** (*Reichb.f.*) *Schltr.* in Beih. Bot. Centr. 36: 108 (1918); Summerh. in K.B. 6: 473 (1952); Piers, Orch. E. Afr., ed. 2: 251 (1968); J. Stewart in U.K.W.F.: 790 (1974). Type: no locality or collector cited, *Herb. Reichb.f.* No. 7916 (W, holo., K, record of holo.!)

A pendent epiphyte with an elongate stem 20–45 cm. or more long, 4–6 mm. in diameter. Leaves fleshy, slightly falcate, oblanceolate, unequally obtusely bilobed at the apex, 10–24 cm. long, 1.9–3.1 cm. wide, articulated to leaf-bases 1.3–1.7 cm. long.

Inflorescence 8–28 cm. long, densely many-flowered in whorls of 2–6-flowers; bracts 5 mm. apart, amplexicaul, 1 mm. long. Flowers yellow-green to yellow; pedicel and ovary 1–2 mm. long. Dorsal sepal elliptic, obtuse, 1–1.8 mm. long, 1.5 mm. wide; lateral sepals obliquely oblong-elliptic, obtuse, 1.5–2.1 mm. long, 1–1.2 mm. wide. Petals elliptic or circular, obtuse, 1–1.4 mm. long and wide. Lip concave, ovate, obtuse, 2 mm. long, 1.3 mm. wide; spur incurved, cylindrical, very little inflated towards apex, 5–9.5 mm. long. Column 0.7 mm. long.

UGANDA. Toro District: Kibale Forest, Fort Portal, Aug. 1936, *Eggeling* 3090!; Mbale District: N. Bugisu, Buteza to Bubungi path, 22 Sept. 1954, *Norman* 236!; Mengo District: Bulemezi County, May 1931, *Sandford* 2164!
KENYA. N. Kavirondo District: Elgon, Oct. 1931, *Tweedie* 24!
TANZANIA. Lushoto District: E. Usambara Mts., Amani, 2 Apr. 1941, *Moreau* 35! & 7 Apr. 1910, *Braun* in *Herb. Amani* 3210!; Morogoro District: Nguru Mts., Maskati track, fl. cult. 16 Apr. 1943, *Moreau* 367!; Ulanga District: Muhulu Mts., 26 Feb. 1932, *Schlieben* 1846!
DISTR. U2–4; **K**3/5; **T**3, 6; Sierra Leone and Liberia to the Central African Republic, Zaire and Rwanda, Malawi and Angola
HAB. In primary and secondary forest on trees such as *Bridelia micrantha* and *Sapium ellipticum*; 900–2150 m.

SYN. *Angraecum odoratissimum* Reichb.f. in Bonplandia 4: 326 (1856)
Listrostachys urostachya Kraenzl. in E.J. 22: 22 (1895). Type: Tanzania, Usambara Mts., *Holst* (B, holo., K, record of holo.!)
Chamaeangis urostachya (Kraenzl.) Schltr. in Beih. Bot. Centr. 36: 109 (1918)

2. **C. vesicata** (*Lindl*). *Schltr.* in Beih. Bot. Centr. 36: 109 (1918); Summerh. in K.B. 13: 86, 279 (1958); Piers, Orch. E. Afr., ed. 2: 253 (1968); J. Stewart in U.K.W.F.: 790 (1974). Type: Ghana, without exact locality, cult. *Loddiges* (K, holo.!)

A pendent epiphyte with a short stem less than 8 cm. long. Leaves 3–10, very coriaceous, falcate, linear, acuminate at the obliquely bilobed apex, 20–40 cm. long, 0.7–1.6 cm. wide, twisted at the base to lie in one plane or rarely conduplicate. Inflorescences pendent, 7–30 cm. long, densely many-flowered, 2 flowers at each node; peduncle 2.6–4.5 cm. long; bracts amplexicaul, 1.5–2 mm. long. Flowers lime-green to yellow or orange, diurnally scented of *Buddleia globosa*; pedicel and ovary 4–5 mm. long. Dorsal sepal ovate, acute, 2–3 mm. long, 1.5–2.1 mm. wide; lateral sepals ovate, acute, 2.5–4 mm. long. Petals lanceolate, acute, 2.5–3 mm. long, 1–1.2 mm. wide. Lip convex, deflexed, ovate, subacute, 2.5–3 mm. long, 1 mm. wide; spur incurved, clavate, abruptly dilated at the apex, 7–12 mm. long. Column 1–1.5 mm. long.

UGANDA. Kigezi District: Impenetrable Forest, Nyebeya, Oct. 1940, *Eggeling* 4187; Bunyoro District: Budongo Forest, Sept. 1935, *Hancock* 14A! & Sept. 1935, *Eggeling* 2103!; W. Ankole District: Kalinzu Forest, Aug. 1936, *Eggeling* 3223!
KENYA. Baringo District: Kabarnet, Oct. 1962, *Tweedie* 2481!; Nairobi District: Mbagathi, *Dowson* 567! & Karura Forest, *Smart* E25!; Masai District: Ngong Hills, *C. van Someren* 10!
TANZANIA. Mbulu District: headwaters of the Endabash R., fl. cult. 8 Mar. 1944, *Moreau* 309!; Buha District: 88 km. from Kasulu on the Kibondo road, 3 Apr. 1946, *Abdallah* in *Moreau* 843!
DISTR. U2; **K**3, 4, 6; **T**2, 4; Guinée and Sierra Leone to Zaire and Rwanda
HAB. In riverine and in montane forest on *Juniperus procera*, etc.; 1100–1800 m.

SYN. *Angraecum vesicatum* Lindl. in Bot. Reg. 29, Misc.: 6 (1843)

3. **C. sarcophylla** *Schltr.* in E.J. 53: 597 (1915). Types: Tanzania, Rungwe District, near Kyimbila, *Stolz* 639 (B, syn., K, isosyn.!) & Ukinga Mts., *Stolz* 2080 (B, syn.)

A pendent or spreading epiphyte with short stems 2–15 cm. long; roots branching. Leaves fleshy-coriaceous, falcate, linear, unequally obliquely subacutely or roundly bilobed at the apex, 7–33 cm. long, 0.3–2.5 cm. wide, twisted at the base and articulated to imbricate leaf-bases. Inflorescences 1–several, pendent, 3–19 cm. long, densely many-flowered, with 2–4 flowers per node; peduncle fleshy, 2–6 cm. long, bearing 3–4 sheathing bracts; bracts broadly ovate, amplexicaul, 1.5–4 mm. long, 3–6 mm. apart. Flowers pale peach-coloured, orange or orange-brown with a white column; pedicel and ovary 3–4.6 mm. long. Dorsal sepal oblong- to ovate-elliptic, 2–3.7 mm. long, 1.2–2.3 mm. wide; lateral sepals recurved, obliquely ovate or oblong-elliptic, obtuse, 2.2–3.6 mm. long, 0.7–2.1 mm. wide. Petals recurved, ovate or lanceolate, obtuse, 2–3.6 mm. long, 1.5–2 mm. wide. Lip ovate-oblong or ovate-elliptic, obtuse or obscurely 3-toothed at the apex, 2.2–3.5 mm. long, 1.8–3 mm. wide; spur pendent, cylindrical, slightly swollen in apical half, 9–19 mm. long. Column 0.5–0.9 mm. long. Fig. 131.

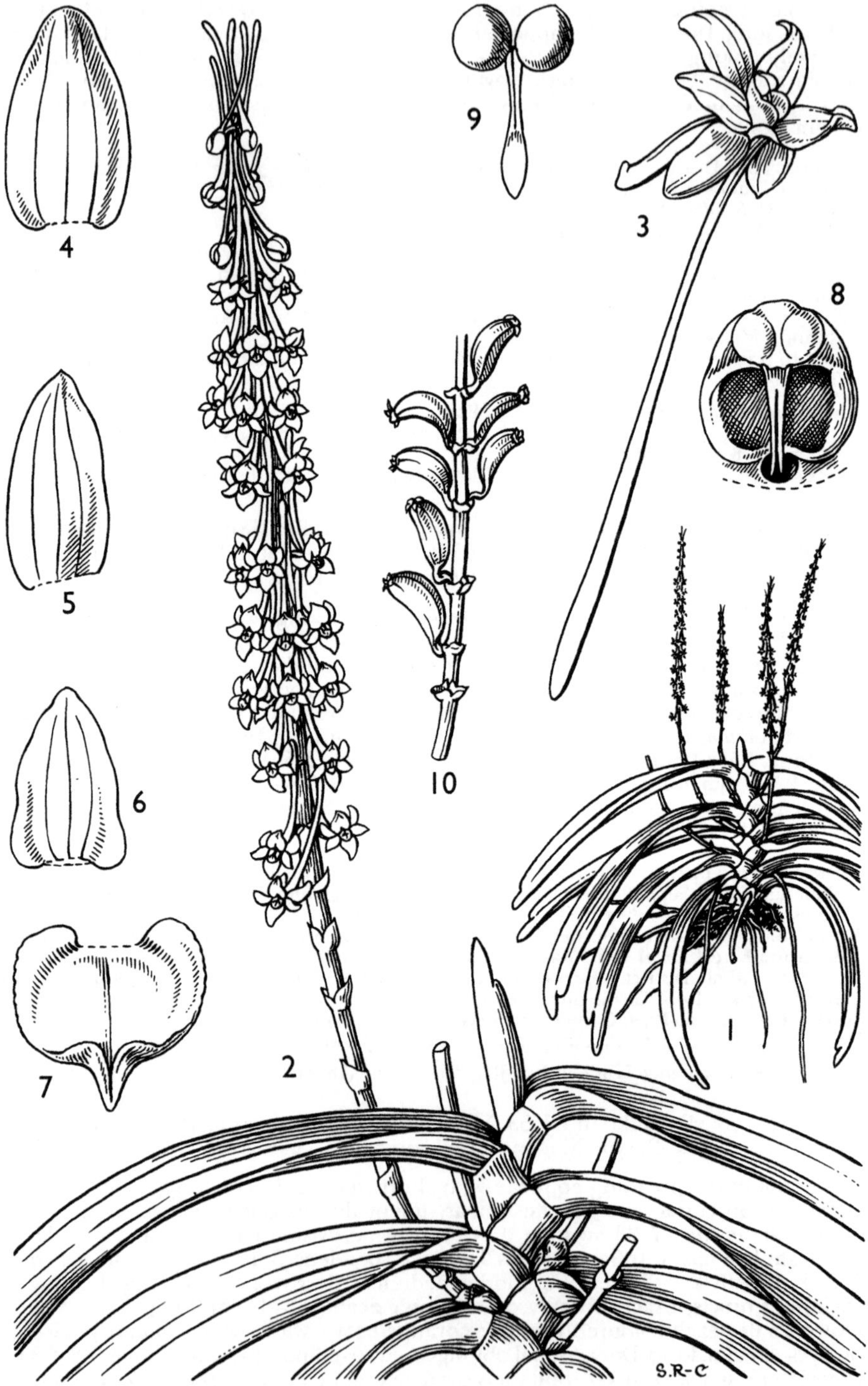

FIG. 131. *CHAMAEANGIS SARCOPHYLLA*—**1**, habit, × ¼; **2**, inflorescence, × 1; **3**, flower, × 4; **4**, dorsal sepal, × 8; **5**, lateral sepal, × 8; **6**, petal, × 8; **7**, lip, × 8; **8**, column front view, pollinia, stipes and gland removed, × 12; **9**, pollinarium, × 16; **10**, fruits, × 1. 1–7, from *Maas Geesteranus* 5344 and *Bally* 196; 8–10, from *Tweedie* 312. Drawn by Miss Stella Ross-Craig.

UGANDA. Mbale District: Elgon, 10 May 1924, *Snowden* 880!
KENYA. Uasin Gishu District: S. Elgon, Sosiani R., Mar. 1941, *Tweedie* 559!; Kisumu/Londiani District: Tinderet Forest Reserve, 2 July 1949, *Maas Geesteranus* 5344!; Masai District: Ngong Hills, 7 Dec. 1985, *Khayota* 65!
TANZANIA. Arusha District: Mt. Meru, 21 Apr. 1940, *Vaughan* 3031!; Lushoto District: W. Usambara Mts., Mazumbai, 23 Apr. 1981, *L. Tanner* 260A!; Iringa District: Mufindi, Upper Kigogo R., 16 Feb. 1979, *Cribb, Grey-Wilson & Mwasumbi* 11449! & 17 Mar. 1962, *Polhill & Paulo* 1794!
DISTR. U3: K3–6; T2, 3, 6, 7; Zaire, Rwanda and Malawi
HAB. In montane forest, especially over streams; 1500–2400 m.

SYN. *C. orientalis* Summerh. in K.B. 6: 473 (1952). Type: Kenya, Machakos/Masai Districts, Chyulu Hills, *Bally* in *C. van Someren* 19 (K, holo.!)

58. **AERANGIS***

Reichb.f. in Flora 48: 190 (1865); J. Stewart in K.B. 34: 239 (1979)

Radinocion Ridl. in J. Henriques, Bol. Soc. Brot. 5: 200 (1887)
Rhaphidorhynchus Finet in Mém. Soc. Bot. Fr. 9: 32 (1907), pro parte

Epiphytic herbs with short or elongated woody stems bearing numerous elongated aerial roots in the lower part. Stems usually unbranched, covered with the remains of overlapping leaf-bases, bearing few–several leaves apically. Leaves in 2 rows, usually thick and fleshy in the living state, sometimes leathery during the dry season, sheathing at the base, much longer than broad and usually wider in the upper half, unequally bilobed at the apex. Inflorescence lateral, a short or elongated raceme, rarely branched, few–many-flowered. Flowers resupinate, white or variously tinted with green or brown. Sepals and petals free, spreading or reflexed. Lip entire, often similar to the sepals and petals, spurred at the base. Column short and stout or somewhat elongated and more slender, often narrowed towards the base and enlarged at the level of the stigma; androclinium straight or sloping, the anther-cap sometimes beaked; rostellum entire, elongated, deflexed or porrect; pollinia 2, sessile on a single stipes; viscidium variously shaped; stigma an oval or rhomboid sticky depression. Ovary elongated, straight or curved. Capsule cylindric or ellipsoid, often much elongated.

A genus of approximately 50 species of which at least 26 occur in Africa and the remainder in Madagascar and the Comoro Is. One species is recorded from both East Africa and Sri Lanka. Several of the more widely distributed species have been known in cultivation for many years.

Flowers with spur more than 10 cm. long:
  Lip bearing 2 or more raised ridges or crests at its base and in the mouth of the spur:
    Lip pandurate, 8–15 mm. wide, spur 20–24 cm. long . . . 1. *A. kotschyana*
    Lip oblong or obovate, 5–9 mm. wide, spur 11–16 mm. long . . . 2. *A. verdickii*
  Lip smooth at the base and in the mouth of the spur:
    Leaf-venation markedly reticulate, darker coloured and raised above the upper surface of the leaf; column short and thick, not more than 5 mm. long:
      Leaves greyish green; dorsal sepal not more than 10 mm. long . . . 3. *A. somalensis*
      Leaves dark green; dorsal sepal at least 13 mm. long, usually 15–20 mm. long . . . 4. *A. coriacea*
    Leaf-venation not conspicuously coloured; column slender at the base, 6–9 mm. long:
      Woody stem usually more than 15 cm. long; leaves ± parallel-sided . . . 5. *A. thomsonii*
      Woody stem usually less than 15 cm. long; leaves wider in the upper half . . . 6. *A. brachycarpa*
Flowers with spur less than 8 cm. long:

* By J. Stewart

Lip narrow and elongated, at least 3 times as long as wide, or nearly so:
  Spur more than 4 cm. long, straight, or only slightly curved:
    Lip elliptic, lanceolate, ovate or obovate, extending into a long-acuminate tip:
      Sepals, petals and lip of ± equal length, the lateral sepals reflexed at first then curving forwards — 7. *A. confusa*
      Sepals longer than petals and lip; all perianth parts spreading or reflexed:
        Dorsal sepal more than 33 mm. long . . . . — 8. *A. carnea*
        Dorsal sepal less than 32 mm. long:
          Lip 7–8 mm. wide . . . . — 9. *A. kirkii*
          Lip 5 mm. wide . . . . — 10. *A. gravenreuthii*
    Lip elliptic, lanceolate, oblanceolate or narrowly ovate, apex acute, cuspidate or apiculate but never acuminate:
      Flowers with pedicels attached to the rhachis at the base of each bract:
        Plants usually with conspicuous stems; column 4 mm. long or more:
          Dorsal sepal 13–15 mm. long; spur usually 6–7.5 cm. long; column 5–8 mm. long . . . — 11. *A. collumcygni*
          Dorsal sepal 10–12 mm. long; spur usually 4–5.5 cm. long; column 4 mm. long . . . . — 12. *A. jacksonii*
        Plants with very short stems; column less than 4 mm. long:
          Inflorescence bearing 4–15 flowers; dorsal sepal 7–13 mm. long; spur usually 6–8 cm. long; column 2–4 mm. long . . . . — 13. *A. mystacidii*
          Inflorescence bearing 1–7 flowers; dorsal sepal 5–8 mm. long; spur usually 4–6 cm. long; column 1.5–2 mm. long . . . . . . — 14. *A. appendiculata*
      Flowers with pedicels attached to short projections of the rhachis, a few mm. above the base of each bract . . . . — 15. *A. montana*
  Spur 4 cm. long or less, with the tip curved upward in a distinct hook . . . . — 16. *A calantha*
Lip wide or narrow, but short, not more than twice as long as wide:
  Lip less than 6 mm. wide:
    Flowers crowded on the rhachis, less than 3 mm. apart — 17. *A. oligantha*
    Flowers more than 3 mm. apart, or solitary:
      Spur more than 40 mm. long . . . . — 13. *A mystacidii*
      Spur less than 35 mm. long:
        Spur 10–25 mm. long; column 2 mm. long — 18. *A. ugandensis*
        Spur 3–7 mm. long; column 1 mm. long . . . — 19. *A. hologlottis*
  Lip more than 7 mm. wide:
    Leaves linear-oblanceolate, widest near the apex; petals elliptic; lip suborbicular or flabellate . . . . — 20. *A. alcicornis*
    Leaves linear-ligulate or linear-lanceolate, narrowing at the apex; petals obovate or broadly oblanceolate; lip obovate . . . . — 21. *A. luteoalba*

1. **A. kotschyana** (*Reichb.f.*) *Schltr.* in Beih. Bot. Centr. 36: 118 (1918); F.P.S. 3: 312 (1956); F.P.U.: 219 (1962); Tweedie in Journ. E. Afr. Nat. Hist. Soc. 24: 36, 38 (1964); Hawkes, Encycl. Cult. Orch.: 35 (1965); Summerh. in F.W.T.A., ed. 2, 3: 265 (1968); Piers, Orch. E. Afr., ed. 2: 212 (1968); J. Stewart in U.K.W.F.: 793 (1974) & in K.B. 34: 252 (1979). Type: Sudan, Fassoglu, *Kotschy* 449 (W, holo.!)

Stem stout and woody, upright, up to 15 mm. in diameter and 20 cm. long, bearing numerous greyish roots 7–9 mm. in diameter. Leaves 3–20, distichous, very close together, obovate-oblong, 6–20(–30) cm. long, 2–8 cm. wide, base attenuate, apex unequally or subequally bilobed, the longer lobe subobtuse or acute, up to 12 mm. long, the shorter lobe rounded or absent, fleshy or leathery in texture, the margins undulate. Inflorescences axillary, often several, arching or pendent racemes bearing up to 20 flowers; peduncle terete, up to 12 cm. long; bracts triangular, ovate, up to 8 mm. long. Flowers in 2 rows, 2–3 cm. apart, white, often tinged with salmon pink at the centre and in the tips of the perianth and spur; pedicellate ovary pinkish green, 2–3 cm. long. Dorsal sepal erect, or arching forward over the column, elliptic-lanceolate, apiculate, winged on the back near the tip, (12–)20–25 mm. long, 9 mm. wide; lateral sepals spreading, lanceolate, winged on the back near the tip, apiculate, (14–)20–28 mm. long, 7 mm. wide. Petals strongly reflexed, oblanceolate, apiculate, 10–23 mm. long, 8 mm. wide. Lip deflexed, subpandurate, apiculate, winged on the back near the tip, margins strongly recurved in the lower half, two prominent crests on the upper surface in front of the mouth of the spur, (10–)15–20 mm. long, 8–15 mm. wide; spur pendulous, thickened in its lower third and curved or twisted just below the middle, becoming more twisted as the flowers age, 13–22 cm. long. Column thick, 3–6 mm. long, 5 mm. wide; rostellum porrect, straight or slightly curved; pollinia ovoid; stipes widened in the upper part; viscidium oblanceolate. Capsule 3–5 cm. long, up to 1.5 cm. in diameter. Pl. 6.

UGANDA. Karamoja, 1 May 1958, *Symes* 355!; Toro District: Bwamba Pass, Fort Portal side, July 1969, *J. Stewart* 1189!; Mengo District: Kampala, 28 Sept. 1926, *Snowden* 1011!
KENYA. Mombasa District: Rabai, Jan. 1886, *W.E. Taylor*!
TANZANIA. Lushoto District: 3 km. E. of Amani, 9 June 1941, *Moreau* 67!; Kigoma District: Gombe Stream Reserve, Kakombe valley, 10 Jan. 1964, *Pirozynski* 240!; Lindi District, June 1927, *District Agricultural Officer* 3!
DISTR. **U** 1–4; **K** 7; **T** 2–4, 6, 8; **Z**; Guinea, Nigeria, Central African Republic, Zaire, Rwanda, Burundi, Sudan, Ethiopia, Malawi, Mozambique, Zambia and Zimbabwe
HAB. Epiphytic on old trees in humid wooded grassland, woodland, and sometimes in forests; sea-level to 1500 m.

SYN. *Angraecum kotschyanum* Reichb.f. in Oest. Bot. Zeitsch. 14: 338 (1864); Schweinf., Beitr. Fl. Aeth.: 292 (1867); Rolfe in F.T.A. 7: 137 (1897)
A. grantii Baker in Trans. Linn. Soc., Bot. 29: 153 (1875); Rolfe in F.T.A. 7: 138 (1897). Type: Sudan, Madi, *Grant* 716 (K, holo.!)
A. kotschyi Reichb.f. in Gard. Chron., n.s., 14: 456 (1880); Veitch, Man. Orch. Pl. 7: 132 (1891); Th. Dur. & Schinz, Consp. Fl. Afr. 5: 44 (1895); Williams, Orchid Grower's Manual: 98 (1894); Hooker in Curtis, Bot. Mag. 121, t. 7442 (1895), *nom. superfl.* Type as for *A. kotschyanum* above
Aerangis kotschyi (Reichb.f.) Reichb.f. in Otia Bot. Hamb. 2: 78 (1881); Schltr., Die Orchideen, ed. 2: 590 (1927)
Angraecum semipedale Rendle in J.L.S. 30: 389 (1895); Kraenzl. in P.O.A. C: 158 (1895). Type: Kenya, Mombasa District, Rabai, *W.E. Taylor* (BM, holo.!)
Rhaphidorhynchus kotschyi (Reichb.f) Finet in Mém. Soc. Bot. Fr. 9: 36 (1907)
Aerangis grantii (Baker) Schltr. in Beih. Bot. Centr. 36: 116 (1918)

2. **A. verdickii** (*De Wild.*) *Schltr.* in Beih. Bot. Centr. 36: 121 (1918); Williamson, Orch. SC. Afr.: 194 (1977); J. Stewart in K.B. 34: 256 (1979). Type: Zaire, Lukafu, *Verdick* 328, pro parte, 2 foliis elliptico-obovatis majoribus exclusis (BR, holo.!)

Stem woody, up to 2 cm. in diameter, to 5 cm. long but usually very short, with numerous greyish roots 5–9 mm. in diameter and very elongated. Leaves 2–6, distichous, oblong-ligulate, 5–20 cm. long, 2–5 cm. wide, slightly narrowed towards the base, margins undulate, the apex bilobed, the lobes rounded or acute, equal or subequal, 2–11(–20) mm. long, thick and coriaceous or almost succulent in texture, greyish green, often with a narrow band of red at the margin, often split along the midvein in herbarium specimens. Inflorescences arising below the leaves bearing 4–12 flowers, up to 40 cm. long, usually ± 20 cm.; peduncle terete, up to 6 cm. long, 5 mm. in diameter, bearing several tubular sheaths; rhachis similar, almost straight; bracts broadly ovate, apiculate, up to 10 mm. long. Flowers arranged in 1–2 rows, 1.5–3.5 cm. apart; perianth-parts reflexed or spreading, white, rarely faintly tinted with pink in the ovary and spur; pedicel with ovary slender and curved, 2–3.8 cm. long. Dorsal sepal slightly concave, arching forwards over the column, ovate-oblong, apiculate, 11–15 mm. long, 4–8 mm. wide; lateral sepals spreading or deflexed, often falcate, narrowly oblong-ligulate with a slightly thickened apicule, 14–21 mm. long, 3–5 mm. wide. Petals reflexed, obovate-oblong, apiculate, 12–18 mm. long, 3.5–8 mm. wide. Lip oblong-obovate, 11–18 mm. long, 5–9 mm. wide, bearing 2 distinct ridges at its base in front of the spur opening and sometimes with additional ridges on either side of these, often deflexed towards the spur, apiculate; spur slender, curved, often slightly bent, thickened and flattened in the apical half, 12–16(–20) cm. long. Column terete, enlarged towards the apex, 3–5 mm. long; rostellum deflexed, porrect at the tip, wider than the stipes; anther-cap with an obtuse beak; pollinia reniform, 1 mm. long; stipes slender, widened in the upper third; viscidium ± ovate.

TANZANIA. Mwanza District: Igalukiro, 25 June 1951, *R. Tanner* 493!; Ufipa District: Lake Kwela, 30 Nov. 1954, *Richards* 3475!; Kondoa District: Kolo, 11 Jan. 1962, *Polhill & Paulo* 1129!

DISTR. **T** 1, 4, 5, 7, 8; Zaire, Mozambique, Malawi, Zambia, Zimbabwe, Angola, and South Africa (Transvaal)

HAB. On branches and trunks of trees, or on rocks, in rather dry woodland and montane scrub in areas where there is a pronounced dry season; 300–2600 m.

SYN. *Angraecum verdickii* De Wild. in Ann. Mus. Congo, Bot., sér. 4, 1: 21 (1902), pro majore parte
A. augustum Rolfe in Orch. Rev. 16: 246 (1908); Gard. Chron., ser. 3, 45: 89, Tab. Suppl. (1909). Type: Zimbabwe, *Old* s.n. (K, icon.!)
A. erythrurum Kraenzl. in Viert. Nat. Ges. Zürich 74: 109 (1929). Type: South Africa, Masatane, *Junod* 1300 (Z, holo.!)
Aerangis schliebenii Mansf. in Schlieben in Gartenfl. 86: 41, 60, in obs. (1937), nomen. Based on Tanzania, Lindi, *Schlieben* 6419 (B, LISC, PRE), *non A. schliebenii* Mansf. in N.B.G.B. 12: 704 (1935) e descr., typus non visus
[*A. kotschyana* sensu Schelpe, Introd. S. Afr. Orch.: 44 (1966); Harrison, Epiphytic Orch. S. Afr.: 49 (1972), *non* Schltr.]
A. erythyrurum (Kraenzl.) Garay in Bot. Mus. Leafl. Harv. Univ. 23: 157 (1972)

NOTE. The size of the perianth-parts and the length of the spur in this species are very variable in size from year to year, even on the same plant.

3. **A. somalensis** (*Schltr.*) *Schltr.* in Beih. Bot. Centr. 36: 120 (1918); J. Stewart in K.B. 34: 261 (1979). Type: Ethiopia or ? Somalia, *Ruspoli & Riva*, without number or locality (not seen)

Stem short, upright, bearing numerous greyish, flexuous roots, 5–7 mm. in diameter. Leaves 2–6, distichous, oblong-ligulate, 2–11 cm. long, 1.3–3.4 cm. wide, leathery, greyish green, with a markedly reticulate venation darker coloured and raised on the upper surface, conspicuous even in the dried state, unequally or subequally bilobed at the apex, the lobes subacute or rounded, 5–15 mm. long, margins thickened and undulate, sometimes the whole leaf with a pink or reddish cast, others pale green. Inflorescences arising below the leaves, usually several, 10–20 cm. long, bearing 4–17 flowers; bracts brownish, ovate, up to 5 mm. long. Flowers white, sometimes tinged with pink, 1–2 cm. apart; pedicellate ovary 1.5–2 cm. long. Dorsal sepal arching forward over the column, ovate, concave, apiculate, 8–10 mm. long, 4–6 mm. wide; lateral sepals oblique or reflexed, oblong-ligulate, concave, apiculate, 9–14 mm. long, 3–5 mm. wide. Petals spreading, oblong-apiculate, 8–11 mm. long, 3–5.5 mm. wide. Lip oblong-ligulate, the margins reflexed in the lower half, apiculate, 9–13 mm. long, 4–7 mm. wide; spur straight or slightly curved, sometimes slightly inflated in the lower half but narrowing again towards the tip, 10–15 cm. long. Column rather thick, to 4 mm. high; anther-cap slightly beaked; pollinia globose, 0.8 mm. long; stipes slender but enlarged in the upper third; viscidium ovoid, 1–1.5 mm. long, with outer margins rolled over the pointed rostellum. Fig. 132.

KENYA. Northern Frontier Province: Moyale, 16 Apr. 1952, *Gillett* 12796!; Machakos District: Sultan Hamud, 20 Oct. 1969, *J. Stewart* 1168!; Masai District: Emali Hill, Oct.–Nov. 1952, *C. van Someren* 109!

TANZANIA. Mbeya District: Umalila, Dec. 1969, *Leedal* 338!

DISTR. **K** 1, 4, 6, 7; **T** 7; Ethiopia, Zimbabwe and South Africa (Transvaal)

HAB. On trees and at the base of branching shrubs in relict patches of woodland along streams and near rock outcrops in essentially dry areas; 1000–1200 m.

SYN. *Angraecum somalense* Schltr. in E.J. 38: 163 (1906)
Aerangis sp. (aff. *Aerangis somalensis* (Schltr.) Schltr.) sensu Piers, Orch. E. Afr., ed. 2: 220 (1968)
[*A. verdickii sensu* J. Stewart in Amer. Orch. Soc. Bull. 42(6): 528 (1973) & in U.K.W.F.: 793 (1974), *non* (De Wild.) Schltr.]

4. **A. coriacea** *Summerh.* in Gard. Chron., ser. 3, 131: 96 (1952) & in Bot. Mag. 171, t. 297 (1957); Tweedie in Journ. E. Afr. Nat. Hist. Soc. 24: 36, 37 (1964); Piers, Orch. E. Afr., ed. 2: 207 (1968); J. Stewart in U.K.W.F.: 793 (1974) & in K.B. 34: 262 (1979). Type: Kenya, Nairobi District, Kikuyu–Dagoretti, *C. van Someren* 27 (K, holo.!)

Stem woody, up to 10 mm. in diameter, erect or prostrate, up to 40 cm. long, closely leafy above and covered with old leaf-bases in the lower part from which stout greyish roots up

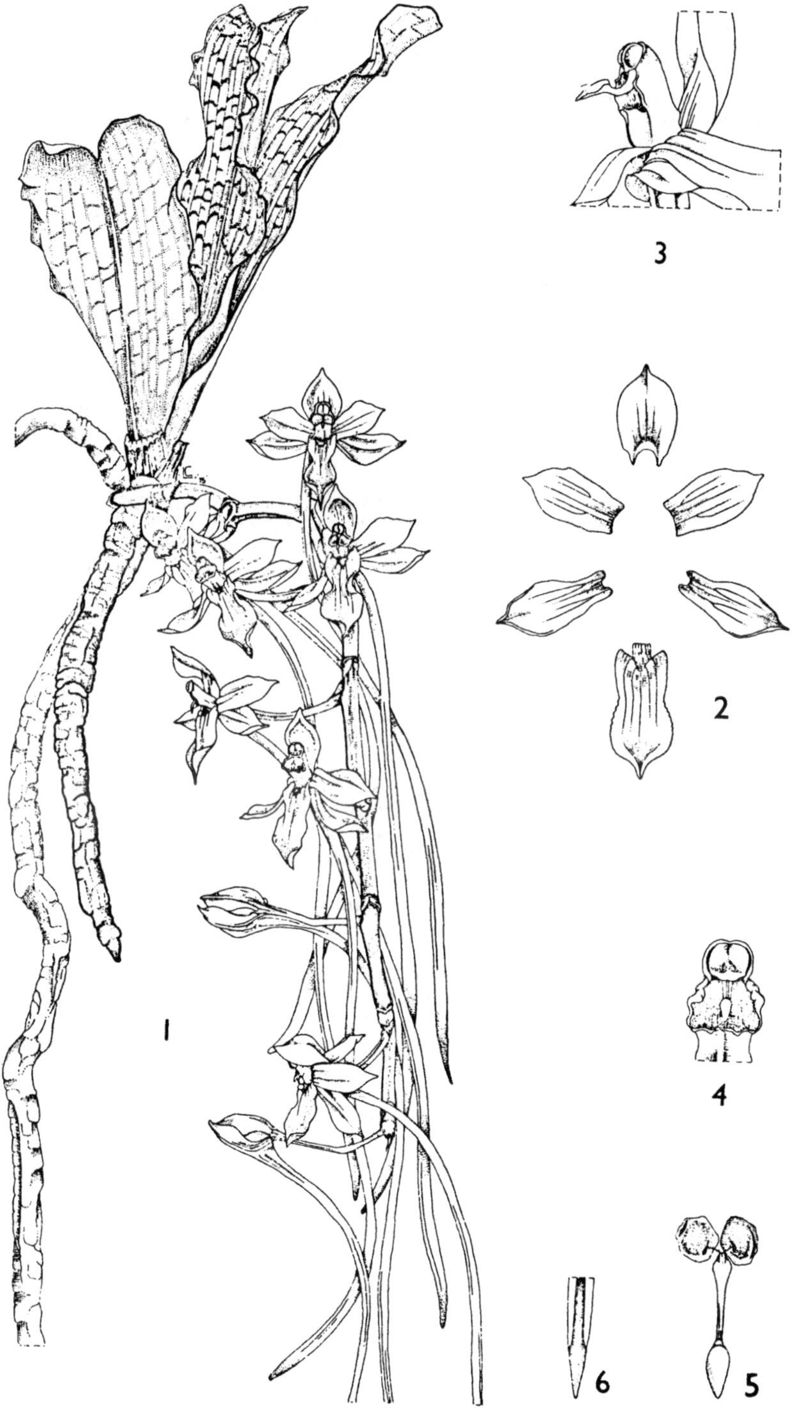

FIG. 132. *AERANGIS SOMALENSIS*—**1**, habit, × ⅔; **2**, perianth dissected, × 1⅓; **3**, column, and base of perianth, viewed from side and below, × 2⅔; **4**, column, front view, × 3⅓; **5**, pollinarium, × 5; **6**, rostellum, × 5. Drawn from *J. Stewart* 1168 by Miss L.S. Cowan.

to 5 mm. in diameter arise. Leaves distichous, 4–many, ligulate or oblanceolate-ligulate, up to 22 cm. long, 2–4.5 cm. wide, only slightly widened in the upper part, gradually narrowed below, unequally bilobed at the apex, the longer lobe rounded and incurved, up to 1 cm. long, the shorter rounded or absent, fleshy in texture, deep green with the transverse veins showing up clearly in a darker colour, margins sometimes undulate. Inflorescences arising from the axils of the upper leaves, arching racemes up to 40 cm. long bearing 4–22 flowers; peduncle terete, up to 15 cm. long, 3–6 mm. in diameter, bearing several tight sheaths; rhachis slightly zigzag, gradually tapering, tinged reddish; bracts spreading from a short sheathing base, broadly ovate, up to 1 cm. long. Flowers spreading, 1–3 cm. apart, white with a rose-pink or green diffusion at the base of the tepals, the spur pinkish or greenish; pedicel with ovary 2–3 cm. long. Dorsal sepal erect or slightly reflexed, lanceolate-elliptic, margins recurved, 13–20 mm. long, 7–10 mm. wide; lateral sepals strongly reflexed, apiculate, oblanceolate-ligulate, margins recurved, 14–19 mm. long, 5–7 mm. wide. Petals reflexed, oblanceolate-oblong, margins recurved, 13–17 mm. long, 5–8 mm. wide. Lip deflexed, lanceolate or oblong-lanceolate, margins recurved in the middle, acute at the reflexed apex, 14–18 mm. long, 6–10 mm. wide; spur pendulous, flexuous, narrowly cylindrical but widened and somewhat flattened in the apical half, 11–17 cm. long. Column thick, ± 4 mm. long; androclinium gently sloping; rostellum deflexed but curved upwards towards the apex; pollinia ellipsoid; stipes slender, slightly widened in the upper part; viscidium elliptic-oblong; stigma transversely oblong.

KENYA. Northern Frontier Province: Karisia Hills, Feb. 1969, *J. Stewart* 1203!; Meru District: Mt. Kenya, 1952, *Schelpe*!; Embu District: Rupengazi R. near Embu, *Piers* 41!
TANZANIA. Arusha District: Mt. Meru, 21 Apr. 1940, *Vaughan* 3030!; Moshi District: Weru-Weru gorge, 28 June 1942, *Wallace* in *Moreau* 125a!
DISTR. **K** 1, 4; **T** 2; not known elsewhere
HAB. Shady places, often near rivers, usually rather low down on trunks in highland forests; 1300–2300 m.

5. **A. thomsonii** (*Rolfe*) *Schltr.* in Beih. Bot. Centr. 36: 121 (1918); W.F.K.: 141, fig. 114 (1948); Tweedie in Journ. E. Afr. Nat. Hist. Soc. 24: 36, 38 (1964); Piers, Orch. E. Afr., ed. 2: 207 (1968); J. Stewart in U.K.W.F.: 792 (1974) & in K.B. 34: 264 (1979). Type: Kenya, Laikipia, *Thomson* 131 (K, holo.!)

Stem woody, up to 10 mm. in diameter, 10–100 cm. long, usually upright, at least in the terminal leafy part, covered with old leaf-bases in the lower part from which whitish roots up to 9 mm. in diameter arise. Leaves 8–20, alternate, distichous, 1–3 cm. apart, ligulate, 8–28 cm. long, 1.5–4.5 cm. wide, apex unequally lobed, the lobes rounded or obtuse, the longer 5–10 mm. long, the shorter up to 5 mm. long or absent, fleshy or leathery in texture, deep green, margins entire. Inflorescences borne at the nodes, arching racemes bearing 4–10 flowers; peduncle terete, up to 9 cm. long, 6 mm. in diameter, bearing 2–3 tubular sheaths; rhachis ± straight, green, to 20 cm. long, with a short projection 2–3 mm. long in the axil of each bract; bracts very broadly triangular-ovate, obtuse or apiculate, cucullate, 10–15 mm. long. Flowers held erect in 2 rows, 1.5–3.5 cm. apart, white; pedicel with ovary green, 3–6 cm. long. Dorsal sepal erect, sometimes curving forward at the tip, lanceolate-elliptic, cuspidate, 22–30 mm. long, 7–9 mm. wide; lateral sepals strongly reflexed, lanceolate-elliptic, acuminate, winged on the back in the upper half, 25–32 mm. long, 5–6 mm. wide. Petals reflexed, oblique, lanceolate-elliptic, acute, 20–25 mm. long, 6–8 mm. wide. Lip reflexed, elliptic-lanceolate, acuminate, margins recurved below, inrolled above, 20–25 mm. long, 7–8 mm. wide; spur pendulous, flexuous, cylindrical but widened and flattened in its terminal half, 10–15 cm. long. Column erect, widened towards the apex, 6–8 mm. long; anther-cap conical, rounded; pollinia oblong; stipes very slender, 3 mm. long; viscidium round. Fig. 133.

UGANDA. Mbale District: Elgon, 28 Oct. 1923, *Snowden* 809!
KENYA. Turkana District: SE. of Kachagalau Hill, 22 Apr. 1959, *Osmaston* 4454!; Nakuru District: near Nyahururu [Thomson's Falls], 23 Oct. 1931, *Pierce* in *Napier* 1486!; Kiambu District: Gatamayu Forest, 30 Mar. 1961, *Lucas & Polhill* 79!
TANZANIA. Masai District: Burko Forest Reserve, 18 Feb. 1967, *Carmichael* 1350!; Mbulu District: SE. slope of Mt. Hanang, Nangwa, 6 Feb. 1946, *Greenway* 7617!; Lushoto District: Shume, 14 Mar. 1944, *Moreau* 189!
DISTR. **U** 3; **K** 1–7; **T** 2, 3; Ethiopia
HAB. Shady places, usually rather low down on trunks and branches in highland forest; 1600–2600 m.

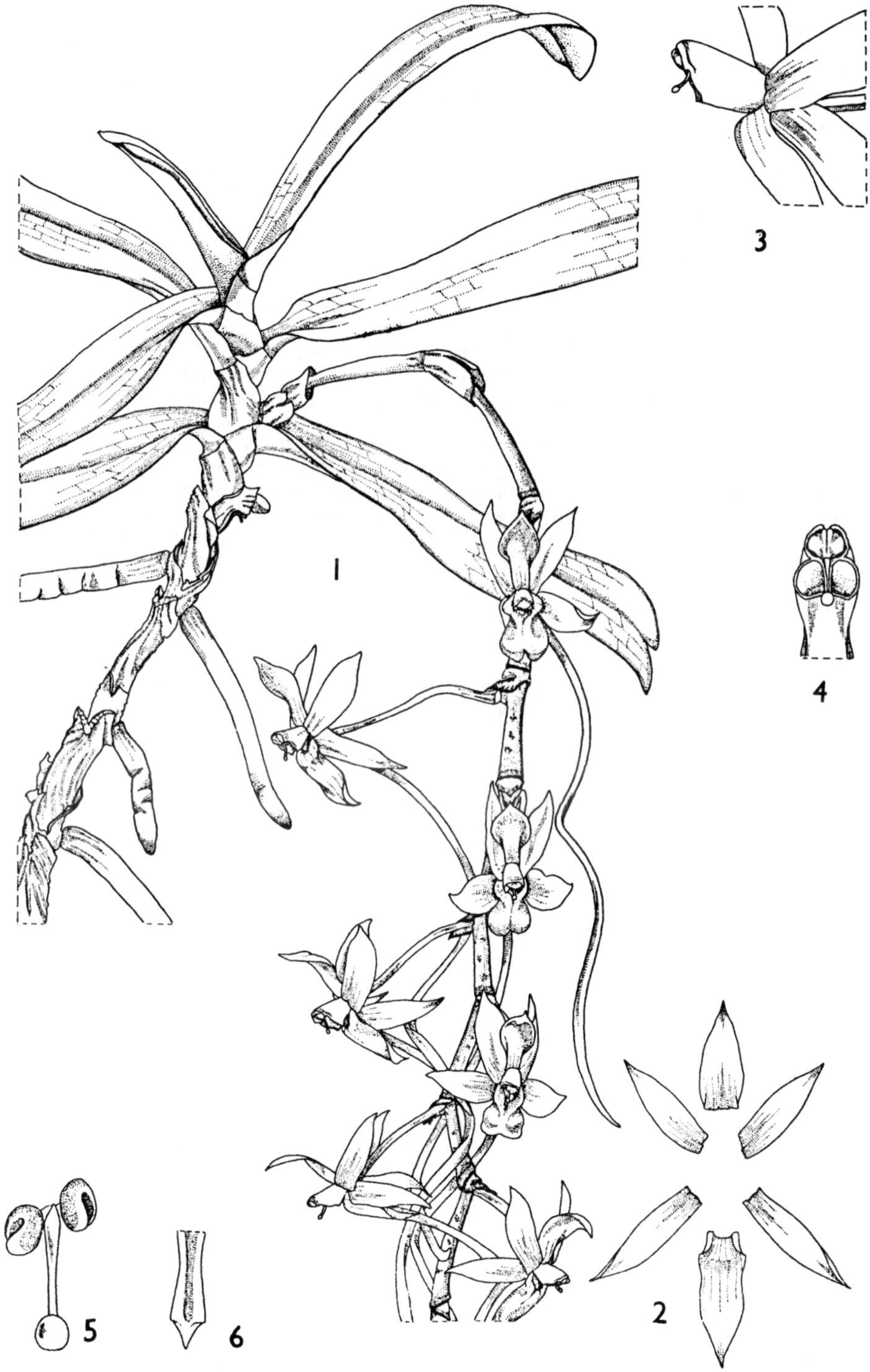

FIG. 133. *AERANGIS THOMSONII*—**1**, habit, × ⅔; **2**, perianth dissected, × ⅔; **3**, column and base of perianth, side view, × 2; **4**, column, front view, × 2; **5**, pollinarium, × 7; **6**, rostellum, × 7. Drawn from *Greenway & Kanuri* 13844 by Miss L.S. Cowan.

SYN. *Angraecum thomsonii* Rolfe in F.T.A. 7: 143 (1897)
Aerangis friesiorum Schltr. in N.B.G.B. 9: 22 (1924). Type: Kenya, Aberdare Mts., *Fries* 2561 (UPS, holo.!)

6. **A. brachycarpa** (*A. Rich.*) *Th. Dur. & Schinz*, Consp. Fl. Afr. 5: 50 (1895); Schltr. in Beih. Bot. Centr. 36: 115 (1918); J. Stewart in K.B. 34: 244 (1979). Type: Ethiopia, *Schimper* 1314 (P, holo.!, K, iso.!)

Stem woody, upright or pendent, up to 20 cm. long but usually much shorter, bearing greyish roots 5–6 mm. in diameter. Leaves 4–12, alternate, usually distichous, 5–12 mm. apart, obovate or spathulate, up to 25 cm. long, 2–6 cm. wide near the apex, with a sheathing base and unequally bilobed apex, the lobes rounded or acute, the longer up to 12 mm. long, fleshy or leathery in texture, dark green and often black-dotted, margins entire. Inflorescences axillary racemes, arching or pendent, up to 40 cm. long, bearing 2–12 flowers; peduncle terete up to 10 cm. long, 3 mm. in diameter, bearing 2–3 tight sheaths; rhachis slightly zigzag, gradually tapering, greyish green, black-dotted; bracts triangular, acute, up to 9 mm. long, enclosing a small outgrowth of the rhachis 1–2 mm. long. Flowers in 2 rows, 1.5–3.5 cm. apart, pale green when first opening, becoming white, and often with a pink or brownish tinge in the petals and spur, tepals often with the margins recurved so that they appear very narrow; pedicel with ovary straight or arching, 3.5–7 cm. long. Tepals all narrowly lanceolate, acuminate, 20–45 mm. long, 4–8 mm. wide near the base; dorsal sepal erect, lateral sepals and petals becoming sharply reflexed within a few days of the flowers opening. Lip deflexed, 5–10 mm. wide near the base, narrowing gradually or abruptly at a point about one-third from the tip to a slender acumen; spur straight or slightly curved, pendulous, gradually tapering towards the tip which is sometimes minutely bifid, 12–20 cm. long. Column slender, terete below, dilated above, 6–8 mm. long; androclinium sloping; rostellum porrect; pollinia oboval-kidney shaped, viscidium heart-shaped. Fruit a slender capsule up to 20 cm. long.

UGANDA. Acholi District: Lotuturu [Lututura], June 1942, *Eggeling* 5061!; Karamoja District: Napak, June 1950, *Eggeling* 5882!
KENYA. Nairobi District: Bahati, 12 May 1933, *C.G. Rogers* 454!; Embu, 18 July 1924, *Lamb* 988/24!; Masai District; near Narok, Siyabei gorge, 27 Apr. 1970, *J. Stewart* 1188!
TANZANIA. Moshi District: N. slope of Kilimanjaro, Ngare Rongai, 13 May 1942, *Moreau* 13!; Kigoma, 29 Nov. 1959, *Harley* 124!
DISTR. **U** 1; **K** 3, 4, 6; **T** 2; 4, 7; Ethiopia, Malawi, Zambia, Angola
HAB. In dense shade, usually rather low down on tree trunks, branches, or in the forking bases of bushes in highland forests; 1600–2300 m.

SYN. *Dendrobium ? brachycarpum* A. Rich., Voyage en Abyssinie 5: 282 (1850); Reichb.f. in Flora 48: 191 (1865)
Angraecum brachycarpum (A. Rich.) Reichb.f. in Walp., Ann. Bot. 3: 573 (1852), in note; Schweinf., Beitr. Fl. Aeth.: 292 (1867); Rolfe in F.T.A. 7: 135 (1897)
Aerangis flabellifolia Reichb.f. in Flora 48: 191 (1865); Th. Dur. & Schinz, Consp. Fl. Afr. 5: 50 (1895); Schltr. in Beih. Bot. Centr. 36: 116 (1918); J. Stewart in U.K.W.F.: 793 (1974). Type: Angola, Golungo Alto, *Welwitsch* 657 (BM, holo.!, K, LISU, iso.!)
Angraecum rohlfsianum Kraenzl. in Bot. Zeit. 40: 341 (1882) & in Orchis 2(4): 38, fig. 5 (1908); Th. Dur. & Schinz, Consp. Fl. Afr. 5: 45 (1895); Kraenzl., Xenia Orch. 3: 76, t. 240 (1900); Rolfe in F.T.A. 7: 137 (1897). Type: Ethiopia, Wadi Woina, *Steckert* (not seen)
A. flabellifolium (Reichb.f.) Rolfe in F.T.A. 7: 135 (1897); Rendle in Hiern, Cat. Afr. Pl. Welw. 2(1): 8 (1899)
Rhaphidorhynchus rohlfsianus (Kraenzl.) Finet, in Mém. Soc. Bot. Fr. 9: 38 (1907)
Aerangis rohlfsiana (Kraenzl.) Schltr. in Beih. Bot. Centr. 36: 120 (1918)
A. friesiorum sensu Tweedie in Journ. E. Afr. Nat. Hist. Soc. 24: 36, 37 (1964); Piers, Orch. E. Afr., ed. 2: 210 (1968); J. Stewart & Campbell, Orch. Trop. Afr.: 30 (1970), *non* Schltr.]
A. sp. No. 4 sensu Williamson, Orch. SC. Afr.: 193, fig. 108/4, t. 177 (1977)

NOTE. The appearance of the flowers of this species varies with age, the sepals and petals are spreading at first but after a few days become completely reflexed. No other species in East Africa has the long, acuminate tepals and long, slender spur of this *Aerangis*.

7. **A. confusa** *J. Stewart* in K.B. 34: 273 (1979). Type: Kenya, Nairobi District, Escarpment Forest, 48 km. W. of Nairobi, *Smart* E.24 (K, holo.!, EA, iso.!)

Stem woody, up to 8 mm. in diameter, to 10 cm. long, upright, or curved with the apical part upright and leafy, with numerous very long flexuous roots borne on the lower part. Leaves 3–12, distichous, 5–10 mm. apart, obovate or oblanceolate, 5–24 cm. long, 1.5–5.5 cm. wide, widest near the apex and gradually narrowed to the sheathing base, apex

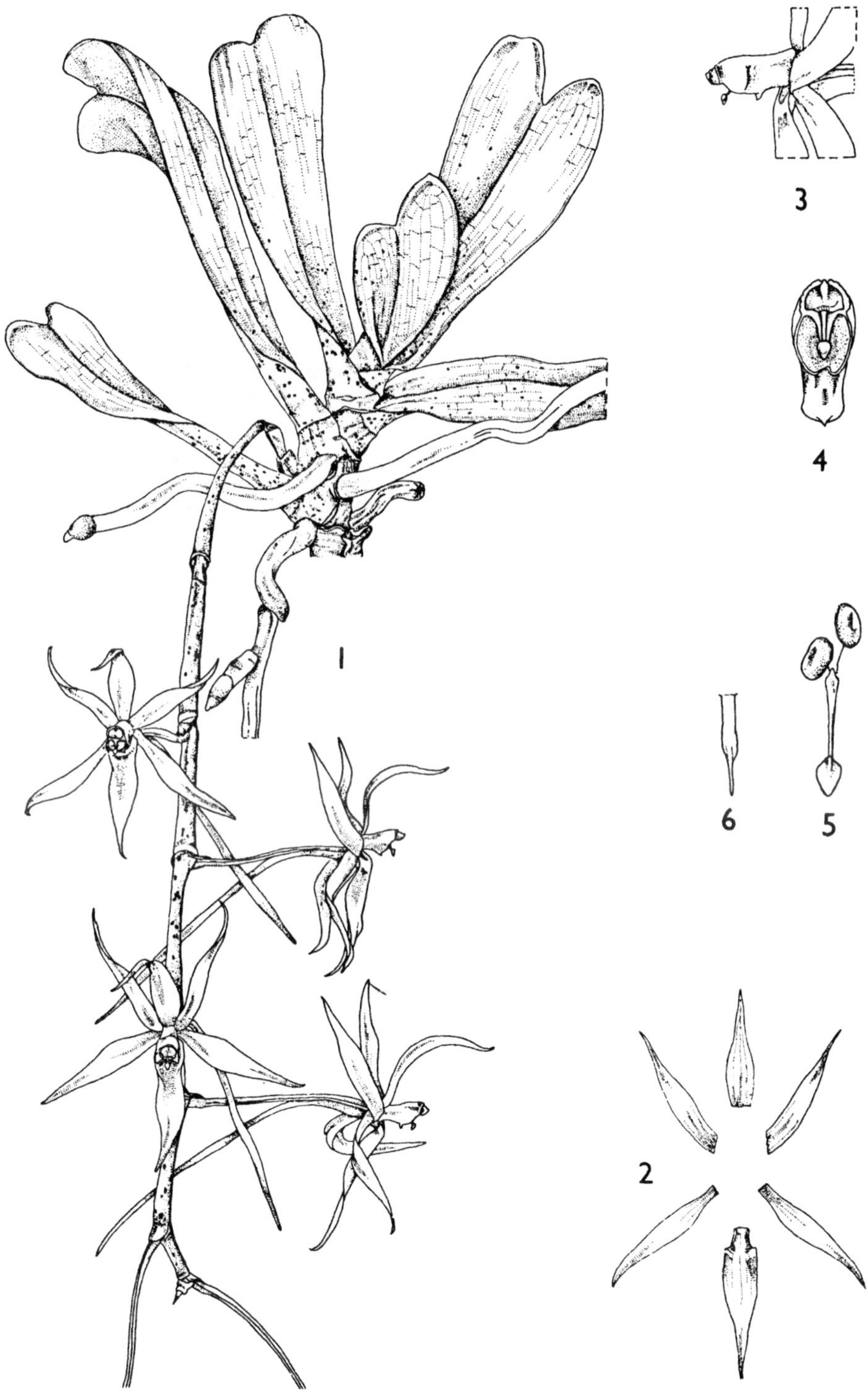

FIG. 134. *AERANGIS CONFUSA*—**1**, habit, × ⅔; **2**, perianth dissected, × ⅔; **3**, column, and base of perianth, side view, × 2; **4**, column, front view, × 3; **5**, pollinarium, × 7; **6**, rostellum, × 7. Drawn from *J. Stewart* 963 by Miss L.S. Cowan.

bilobed, the lobes equal or unequal, rounded, acute or obtuse, with or without a sinus to 10 mm. long between the lobes, leathery in texture, dark green, with minute black dots, the margins straight. Inflorescences axillary, spreading or pendulous, bearing 4–10 flowers; peduncle terete, up to 10 cm. long, 3 mm. in diameter, bearing 1–3 tubular sheaths; rhachis straight or slightly zigzag, pale pinkish or greyish green, dotted with black spots; bracts triangular-ovate, black, 4–7 mm. long. Flowers erect or curving forwards, 1–3 cm. apart, closest towards the tip of the inflorescence, greenish white at first, becoming white, with the spur, pedicel and sometimes the tips of the perianth pinkish; pedicel with ovary 2–6 cm. long. Dorsal sepal arching forwards over the column, lanceolate-elliptic, rostrate or acuminate, 15–25 mm. long, 4–7.5 mm. wide; lateral sepals first reflexed then curving forwards, lanceolate, rostrate, 18–26 mm. long, 2.5–5 mm. wide. Petals strongly reflexed, lanceolate and acuminate or cuspidate, 15–22 mm. long, 3–5 mm. wide. Lip elliptic or lanceolate, acuminate or cuspidate, or oblong-lanceolate and cuspidate like the petals, the margins strongly recurved, 15–23 mm. long, 5–7 mm. wide; spur narrow, tapering gradually from the base to apex, or slightly inflated in the lower half, 4–6 cm. long. Column straight, horizontal, terete below, inflated above, 6–8 mm. long; anther-cap with a short beak; rostellum slightly curved downwards; pollinia ellipsoid, up to 0.9 mm. long; stipes slender, 2 mm. long; viscidium ovate; stigmatic cavity oblong, facing downwards. Fig. 134.

KENYA. Naivasha District: Mt. Margaret Farm, 14 Jan. 1944, *Case* in *Moreau* 567!; Nyeri District; Uaso Nyiro R., 13 Nov. 1938, *Bally* 8461!; Nairobi District: Nairobi and Ngong, Apr. 1932, *Napier* 1328!
TANZANIA. Masai District: Kitumbeine Mt., Langata Dabash, 11 Nov. 1966, *Carmichael* 1304!
DISTR. **K** 1, 3, 4, 6; **T** 2; not known elsewhere
HAB. Shady places, usually rather low down on the trunks of small trees and bushes; often on or near hills; 1600–2100 m.

SYN. *Aerangis sp.* sensu Jex-Blake, W.F.K.: 141, fig. 115 (1948)
A. sp. (near *A. biloba* & *A. friesiorum*) sensu Piers, Orch. E. Afr., ed. 2: 220 (1968)
[*A. brachycarpa* sensu J. Stewart in Amer. Orch. Soc. Bull. 42: 528 (1973) & in U.K.W.F.: 793 (1974), *nec* Reichb.f. (*sphalmate*), *nec* Th. Dur. & Schinz]

NOTE. This species is very difficult to distinguish from *A. brachycarpa* when not in flower.

8. **A. carnea** *J. Stewart* in K.B. 34: 271 (1979). Type: Tanzania, Rungwe District, Poroto [Mporoto] Mts. near Mbeya, *Cunningham* (K, holo.!)

Stem short, up to 8 cm. long, 5 mm. in diameter, bearing numerous flexuous greyish roots 2–3 mm. in diameter below, and several leaves above. Leaves distichous, sometimes falcate, oblanceolate, up to 15 cm. long, 3 cm. wide near the apex, gradually narrowed towards the sheathing base, apex unequally bilobed, the lobes triangular-acute, 3–20 mm. long, slightly fleshy, dark green and glossy, venation distinctly reticulate in the dried state. Inflorescence spreading or pendulous, axillary, 2–7-flowered; peduncle terete, 3–8 cm. long, 1–2 mm. in diameter and bearing 2–3 tubular sheaths; rhachis ± straight, slightly zigzag between the flowers in the upper part; bracts ovate, acute, 5–7 mm. long. Flowers flesh-coloured or pinkish white with a pink spur, 1.2–3 cm. apart; pedicel with ovary 2.5–3.5 cm. long. Dorsal sepal erect or arching forward over the column, narrowly ovate, long-acuminate, 33–35 mm. long, 5–6 mm. wide; lateral sepals reflexed or spreading, oblong-ovate below, acuminate, 35–42 mm. long, 3–5 mm. wide. Petals reflexed, obovate, acuminate, 23–28 mm. long, 3.5–4 mm. wide. Lip obovate, long-acuminate, 27–30 mm. long, 5–7 mm. wide; spur slightly incurved, swollen in the apical half but narrowing towards the tip, 5.5–6.5 cm. long. Column straight or slightly curved, inflated above, terete below, 5–6 mm. long; anther small, cap nearly circular with a prominent beak at the centre; rostellum gradually tapering, extending nearly across the stigmatic surface; pollinia reniform-globose; stipes slender, 2 mm. long; viscidium elliptic-obovate, 0.6 mm. long; stigmatic surface transversely elongated, 2 mm. long, 3–4 mm. wide. Fig. 135.

TANZANIA. Mbeya District: Ikuyu, 8 km. N. of Irambo, 10 Feb. 1979, *Cribb & Grey-Wilson* 11377!; Rungwe District: Ulambya, Ishenta, Nov. 1970, *Leedal* 513 & 922!
DISTR. **T** 7; Malawi
HAB. Epiphyte of shady forests, often near rivers; 1430–1830 m.

9. **A. kirkii** (*Reichb. f.*) *Schltr.* in Beih. Bot. Centr. 36: 118 (1918); Tweedie in Journ. E. Afr. Nat. Hist. Soc. 24: 39 (1964); Piers, Orch. E. Afr., ed. 2: 210 (1968); J. Stewart in K.B. 34: 267 (1979). Type: Zanzibar I., *Kirk* (K, holo.!)

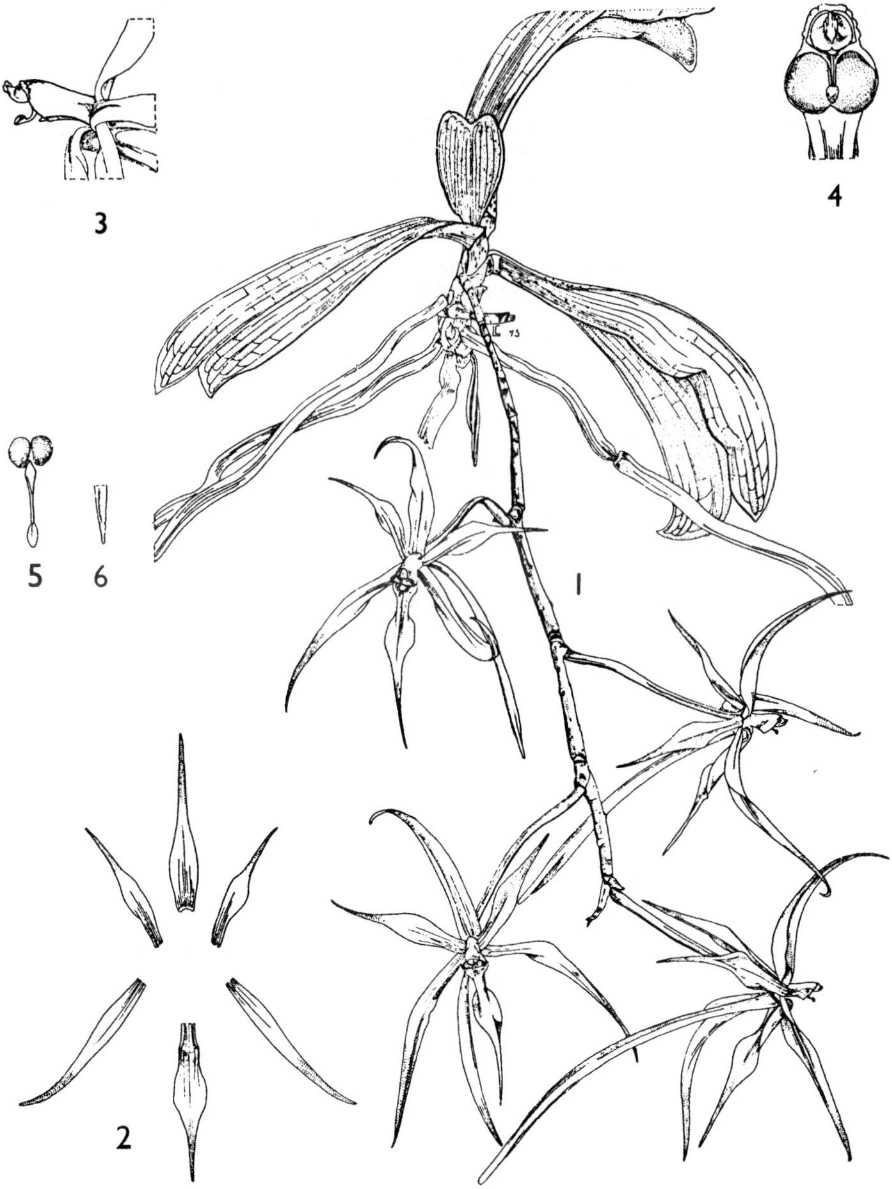

FIG. 135. *AERANGIS CARNEA*—**1**, habit, × ⅔; **2**, perianth dissected, × ⅔; **3**, column and base of perianth, side view, × 2; **4**, column, front view, × 4; **5**, pollinarium, × 4; **6**, rostellum, × 4. Drawn from *Leedal* 513 by Miss L.S. Cowan.

Stem woody, 1–5 cm. long, 3–5 mm. in diameter, covered in closely overlapping leaf-bases, with numerous flexuous greyish white roots 1–2 mm. in diameter arising from the lower half. Leaves 2–7, very close together, oblanceolate or linear-lanceolate, up to 15 cm. long, often much shorter, and up to 3 cm. across at the widest part, the apex much dilated, and bilobed, the lobes rounded or acute, unequal, 4–12 mm. long, dark greyish green and leathery or dark green and slightly fleshy. Inflorescences 1–several, arising below the leaves, bearing 2–6 flowers, usually horizontal or pendulous, up to 17 cm. long; peduncle terete, up to 6 cm. long, 1–3 mm. in diameter, bearing 1–3 tight sheaths; rhachis straight, slightly wider below each bract, greyish pink; bracts broadly triangular, acute, 4–8 mm. long. Flowers spreading, 1.5–3 cm. apart, closer together towards the tip of the inflorescence, the apical one opening first and somewhat larger than the lower ones, white, with pink tinges in the spur; pedicel with ovary 18–25 mm. long. Dorsal sepal erect, ovate-lanceolate, acuminate, 18–25 mm. long, 5–7 mm. wide; lateral sepals longer and narrower, spreading, lanceolate, acuminate, 22–28 mm. long, 4–6 mm. wide. Petals usually spreading, sometimes slightly reflexed, oblanceolate, acuminate, 16–20 mm. long, 5–6 mm. wide. Lip oblong, then acuminate or apiculate, 16–20 mm. long, 7–8 mm. wide; spur pendulous, often curving forwards, slightly inflated in the lower half, narrowing again towards the tip, 6–7.5 cm. long. Column 4 mm. long, lower surface sloping and dilated; anther-cap shortly beaked; rostellum straight, almost reaching across the mouth of the spur; pollinia globose; stipes rather broad; viscidium ovate with the margins curving over the rostellum. Capsule 4–8 cm. long, up to 5 mm. in diameter. Pl. 6.

KENYA. Northern Frontier Province: Roka, June 1937, *Dale* in *F.D.* 3791!; Kwale District: Shimba hills, Giriama Point, 25 Apr. 1968, *Magogo & Glover* 955!; Kilifi District: Arabuko, May 1929, *R.M. Graham* B.442 & in *F.D.* 2146!

TANZANIA. Tanga District: Kihuhwi R., 13 km. E. of Amani, 5 June 1941, *Moreau* 47!; Morogoro District: Turiani, Manyangu Forest, Nov. 1953, *Paulo* 188!; Pemba I., Ngezi Forest, 9 Oct. 1929, *Vaughan* 77!

DISTR. **K** 1, 7; **T** 2, 3, 6; **Z**; **P**; Mozambique

HAB. Coastal bush and riverine forest at low altitudes inland, on small trees and bushes, usually in rather dense shade; 0–450 m.

SYN. *Angraecum bilobum* Lindl. var. *kirkii* Reichb. f. in Gard. Chron., n.s., 18: 488 (1882), as '*kirkei*'; Williams, Orchid Album 4, t. 162 (1884); Veitch, Man. Orch. Pl. 7: 123 (1891); Th. Dur. & Schinz, Consp. Fl. Afr. 5: 40 (1895)

A. apiculatum Hook. var. *kirkii* (Reichb. f.) Reichb. f. in Gard. Chron., ser. 2, 24: 456 (1885), in notes

A. kirkii (Reichb. f.) Rolfe in F.T.A. 7: 136 (1897)

Aerangis biloba (Lindl.) Schltr. var. *kirkii* (Reichb. f.) Hawkes, Encycl. Cult. Orch.: 34 (1965), *non rite publ.*

NOTE. The leaves of this species are very variable in shape and may become much wider under good conditions of cultivation.

10. **A. gravenreuthii** (*Kraenzl.*) *Schltr.* in Beih. Bot. Centr. 36: 117 (1918); Summerh. in F.W.T.A., ed. 2, 3: 265 (1968); J. Stewart in K.B. 34: 277 (1979); Type: Cameroon, Buea, *Preuss* 891 (K, lecto.!)

Stem woody, up to 5 mm. in diameter, 7 cm. long, bearing several leaves 5–10 mm. apart in the apical part and numerous greyish, flexuous roots up to 3 mm. in diameter below. Leaves distichous, oblanceolate, up to 15 cm. long, 1.5–3 cm. wide, gradually narrowed towards the sheathing base, apex unequally bilobed, the lobes acute, 3–18 mm. long, dark green, reticulate venation conspicuous in the dry state, margins entire. Inflorescences pendent axillary racemes, 10–20 cm. long, bearing 2–5 flowers; peduncle terete, slender, to 8 cm. long, 1–2 cm. in diameter and bearing several loose sheaths; rhachis ± straight or curved; bracts ovate, to 5 mm. long. Flowers erect, white, sometimes with orange-pink flush, and pink spur, 1–2.2 cm. apart; pedicel with ovary 3.5 cm. long. Dorsal sepal erect or arching forward over the column, narrowly ovate in the lower half becoming long-acuminate, 20–32 mm. long, 5 mm. wide; lateral sepals spreading or reflexed, oblong-ovate below, acuminate above, 20–35 mm. long, 5–5.5 mm. wide. Petals reflexed, obovate below, acuminate above, 15–25 mm. long, 4 mm. wide. Lip oblong-ovate below, long-acuminate above, 17–25 mm. long, 5 mm. wide; spur curving forwards, somewhat inflated in the apical half (Tanzanian specimens) but narrow at the apex, 4–6(–8) cm. long. Column short, 5 mm. long, inflated above and the stigmatic surface elongated; anther-cap shortly beaked; pollinia oblong-globose; stipes short and broad; viscidium almost square, the edges clasping the blunt-ended rostellum.

TANZANIA. Morogoro District: Uluguru Mts., Bunduki, June 1953, *Eggeling* 6630!
DISTR. **T** 6; also known from Cameroun and Bioko
HAB. Forests, in deep shade; 1200–1600 m. (in Tanzania)

SYN. *Aeranthus gravenreuthii* Kraenzl. in E.J. 17: 62 (1893)
Angraecum ellisii Reichb. f. var. *occidentale* Kraenzl. in E.J. 17: 59 (1893). Type: as above
Mystacidium gravenreuthii (Kraenzl.) Rolfe in F.T.A. 7: 171 (1897)
Angraecum stella Schltr. in E.J. 38: 163 (1906). Type: Cameroon, Neu-Tegel, *Schlechter* 15794 (not seen)

11. **A. collumcygni** *Summerh.* in K.B. 1927: 418 (1927); Williamson, Orch. SC. Afr.: 193 (1977); J. Stewart in K.B. 34: 284 (1979). Type: Uganda, *Lankester* (K, holo.!)

Stem woody, up to 10 cm. long, 5 mm. in diameter, with numerous flexuous roots 2–4 mm. in diameter. Leaves 4–7, distichous, oblanceolate, 6–17 cm. long, 1.5–5.5 cm. wide, somewhat oblique or falcate, apex unequally bilobed, the longer lobe obtuse or subacute and up to 20 mm. long, the shorter rounded, 2 mm. long or absent, fleshy, dark green in colour dotted with black or dark brown especially towards the base. Inflorescences arising below the leaves or axillary, erect, horizontal or pendent racemes bearing 4–16 flowers; peduncle terete, 4–8 cm. long, bearing 2–3 tight sheaths, black-dotted; rhachis ± zigzag, greyish green heavily dotted with brown, 1–3 mm. in diameter; bracts spreading, triangular, subacute or obtuse, 3–5 mm. long. Flowers held erect or curving forwards, 0.6–3 cm. apart, opening green, becoming white after a few days with the tips of the perianth-parts usually tinted with rose or brown, the spur pale greenish brown; pedicel with ovary slender, 2–3.5 cm. long. Dorsal sepal erect or only slightly reflexed, lanceolate, acute or acuminate, 13–15 mm. long, 3.5–5 mm. wide; lateral sepals obliquely curved and reflexed, narrowly lanceolate, acuminate, slightly winged on the back, 13.5–17 mm. long, 2.5–4 mm. wide. Petals reflexed, lanceolate, acuminate or cuspidate, 9–15 mm. long, 2.5–4 mm. wide. Lip deflexed, lanceolate but with the lateral margins strongly recurved, 10–14.5 mm. long, 3.5–6 mm. wide; spur pendent, straight or slightly incurved, gradually narrowed towards the apex, 6–7.5 cm. long, 2 mm. in diameter near the mouth. Column erect, terete below with a short pointed tooth projecting into the mouth of the spur, inflated above, 6–8 mm. long; anther-cap hemispherical, with a short central beak; rostellum straight or porrect; pollinia oblong; stipes slender; viscidium cordate; stigmatic cavity oblong.

UGANDA. Bunyoro District: Budongo Forest, Nyamagita, June 1935, *Eggeling* 2076!; Mengo District: km. 16 Kampala–Masaka, Mar. 1938, *C. van Someren* 33! & Mpanga Forest, fl. in cult. 29 May 1970, *Faden* in *J. Stewart* 1192!
TANZANIA. Bukoba District: Kibugoa Forest Reserve, fl. at Kew Oct. 1960, *Procter*!
DISTR. **U** 2, 4, **T** 1; Cameroon, Central African Republic, Zaire, Zambia
HAB. Moist woodlands, rain-forest, and in neglected plantations in areas that were formerly forested; 800–1200 m.

SYN. *A. compta* Summerh. in B.J.B.B. 25: 259 (1955); Robyns & Tournay, F.P.N.A. 3: 524 (1955); Piers, Orch. E. Afr., ed. 2: 208 (1968). Type: Uganda, Bunyoro District: Budongo Forest, Nyamagita, *Eggeling* 2076 (K, holo.!, EA, iso.!)

12. **A. jacksonii** *J. Stewart* in Bot. Mag. 182, t. 754 (1978) & in K.B. 34: 288 (1979). Type: Uganda, Mengo District, ? Mpanga Forest, *Tweedie* 2601 (K, holo.!, EA, iso.!)

Stem woody, up to 20 cm. in height, 10 mm. in diameter, bearing 8–15 leaves distributed over its entire length and 6 mm. apart; roots numerous, greyish coloured, 3–4 mm. in diameter. Leaves distichous, often falcate, ligulate, linear-lanceolate or oblanceolate, up to 15 cm. long, 1–2.3 cm. wide, gradually narrowed to the sheathing base, dark green, black-dotted on both surfaces, the apex unequally bilobed, the lobes obtuse or subacute, the longer one up to 10 mm. long, the shorter often absent, margins entire. Inflorescences arising opposite to a leaf, from the same place in several successive years, horizontal or arching racemes bearing 4–8 flowers, up to 17 cm. long; peduncle terete, up to 4 cm. long, 2 mm. in diameter, pale greenish brown, black-dotted; rhachis ± straight; bracts broadly ovate, 5–6 mm. long, black at flowering time. Flowers white with a greenish or orange-pink suffusion, 8–18 mm. apart, opening at the base of the inflorescence first; pedicel with ovary greenish or brownish, 3 cm. long. Dorsal sepal erect, narrowly elliptic, acute, 10–12 mm. long, 5 mm. wide; lateral sepals reflexed, narrowly elliptic, 12–15 mm. long, 4 mm. wide. Petals reflexed, narrowly elliptic, acute, 11–13 mm. long, 4 mm. wide. Lip narrowly obovate, the lateral margins reflexed, apiculate, 13–15 mm. long, 5 mm. wide; spur deflexed or curving forwards slightly, somewhat thickened in the lower half but tapering towards the tip which is sometimes minutely bicuspidate, 4–5.5 cm. long. Column 4 mm. high, terete below, widened above; anther-cap beaked; rostellum curving downwards, across the face of the stigma; pollinia globose; stipes slender; viscidium obovate; stigmatic cavity rhomboid. Pl. 6. Fig. 136.

UGANDA. Toro District: Kibale Forest, 1974, *Meyer* 142A, B & C! & Toro, 6 Mar. 1922, *Snowden* 772!; Mengo District: origin uncertain,? Mpanga Forest, fl. in cult. May 1963 and Oct. 1969, *Tweedie* 2601!
DISTR. U 2, 4; not known elsewhere
HAB. Rain-forest; 1200–1600 m.

13. **A. mystacidii** (*Reichb. f.*) *Schltr.* in Orchis 11: 17 (1917) & in Beih. Bot. Centr. 36: 119 (1918); Schelpe, Introd. S. Afr. Orch.: 44 (1966), excl. t. 17; Morris, Epiph. Orch. Malawi: 97 (1970); Harrison, Epiph. Orch. S. Afr.: 47 (1972); Williamson, Orch. SC. Afr.: 194 (1977); J. Stewart in S. Afr. Orch. Journ. 8: 113 (1977) & in K.B. 34: 289 (1979). Type: South Africa, Natal, *Gueinzius* in *Herb. R. Romer* (W, holo.!)

Stem woody, up to 4 cm. long, 5 mm. in diameter, bearing numerous greyish roots 2–3 mm. in diameter. Leaves 4–8, distichous, oblanceolate, oblong-spathulate or shortly obovate, 4–15 cm. long, 1.5–2.5(–4.5) cm. wide, somewhat oblique, base attenuate, apex bilobed, the lobes equal or strongly unequal, acute, obtuse or rounded, somewhat leathery or fleshy in texture, margins entire. Inflorescences arising below the leaves, horizontal or pendulous, bearing 4–15(–24) flowers; peduncle terete, up to 6 cm. long, gradually thickened towards the rhachis, bearing 1–2 tight sheaths at the base and one broadly ovate sheath halfway along its length; rhachis straight, greenish grey or brownish, 5–25 cm. long, and 3 mm. in diameter; bracts triangular-ovate, subacute, 3–5 mm. long. Flowers erect or horizontal, white with a pinkish or light brown spur; pedicel with ovary 1–1.5(–2) cm. long. Dorsal sepal erect or arching forward over the column, oblong, apiculate or acute, 7–13 mm. long, 2–4(–5) mm. wide; lateral sepals oblique, deflexed on either side of the lip, elliptic-oblong, apiculate, 8–14 mm. long, 2.5–4 mm. wide. Petals nearly always strongly reflexed, oblong, subacute, 6–13 mm. long, 2–4.5 mm. wide. Lip deflexed, oblong or ovate-oblong, obtuse or apiculate, 7–12 mm. long, 3–4.5 mm. wide; spur curving fowards, rarely straight, often somewhat enlarged in the distal half, (4.5–)6–8 cm. long. Column stout and thick, terete and somewhat narrowed below, 2–4 mm. high; anther-cap rounded, sometimes minutely apiculate; rostellum curved, porrect, pollinia globose; stipes slender; viscidium narrowly elliptic.

TANZANIA. Rungwe District: Mbaka Kilambo, Mar. 1913, *Stolz* 1908!
DISTR. T 7; Mozambique, Malawi, Zambia, Zimbabwe, Swaziland and South Africa (Cape Province and Natal)
HAB. Epiphytic at low levels in forest, and rarely epilithic; in deep shade along forested streams; 700 m.

SYN. *Angraecum mystacidii* Reichb. f. in Linnaea 20: 677 (1847) & in Walp., Ann. Bot. Syst. 1: 792 (1849); Bolus in J.L.S. 25: 186 (1889); Th. Dur. & Schinz, Consp. Fl. Afr. 5: 44 (1895); Rolfe in Fl. Cap. 5(3): 69 (1912)
A. saundersiae Bolus in Hook., Ic. Pl. 18, t. 1728 (1888) & in J.L.S. 25: 186 (1889); Th. Dur. & Schinz, Consp. Fl. Afr. 5: 45 (1895); Bolus in Ic. Orch. Austr.-Afr. 2, t. 2 (1911). Type: South Africa, Natal, *Saunders* (K, holo.!)
A. pachyurum Rolfe in F.T.A. 7: 138 (1897). Type: Malawi, Zambesia region, *O'Brien* (K, holo.!)
Aerangis mystacidioides Schltr. in E.J. 53: 598 (1915). Type: Tanzania, Rungwe District, Mbaka, Kilambo, *Stolz* 1908 (K!, W, iso.)
A. pachyura (Rolfe) Schltr. in Beih. Bot. Centr. 36: 119 (1918).

14. **A. appendiculata** (*De Wild.*) *Schltr.* in Beih. Bot. Centr. 36: 114 (1918); J. Stewart in K.B. 34: 293 (1979). Type: Mozambique, Morrumbala, *Luja* 466 (BR, holo.!)

Epiphytic herb. Stem short, bearing very numerous dark greyish roots, 2–3 mm. in diameter, which radiate across the bark of the host tree or rock like those of a *Mystacidium.* Leaves 2–4, obovate, up to 8 cm. long, 2 cm. wide, unequally bilobed at the tip, the lobes subacute, dark green with a conspicuous reticulate venation. Inflorescence pendulous, 11–14 cm. long, bearing 1–7 flowers; peduncle terete, short, bearing 1–2 sheaths; rhachis slightly zigzag; bracts loose, ovate, obtuse, 3–5 mm. long. Flowers erect, white with pinkish pedicel and spur; pedicel with ovary 11–14 mm. long. Dorsal sepal erect or arching forward over the column, elliptic or obovate, subobtuse, 5–8 mm. long, 2–3.5 mm. wide; lateral sepals similar, reflexed or spreading, slightly narrower than the dorsal sepal. Petals similar to the sepals and of the same size. Lip elliptic, oblong, minutely apiculate, (4.5–)7–8 mm. long, 2.8–4.5 mm. wide; spur slender, 4–6 cm. long. Column short, erect,

FIG. 136. *AERANGIS JACKSONII*—**1**, habit, × ⅔; **2**, perianth dissected, × 1; **3**, column, and base of perianth, side view, × 3⅓; **4**, column, front view, × 3; **5**, pollinarium, × 3⅓; **6**, rostellum, × 3⅓. Drawn from *H. Meyer* 142A by Miss L.S. Cowan.

1.5–2 mm. long; rostellum very elongated, directed towards the mouth of the spur at first, porrect when the viscidium is removed; pollinia globose; stipes short, very slender; viscidium narrowly elliptic-lanceolate, the margins rolled over the rostellum, 2 mm. long.

TANZANIA. Iringa District: Mufindi escarpment forest, fl. in cult. Kew (058-85 01352) on 29 Apr. 1987, *P. Keeley*!
DISTR. T 7; Mozambique, Malawi, Zambia and Zimbabwe
HAB. On boles of trees in upland evergreen forest; 2000 m.

SYN. *Mystacidium appendiculatum* De Wild. in Pl. Nov. Hort. Then. 1: 85, t. 21/15–21 (1904) & in F.R. 3: 164 (1906)
[*Aerangis pachyura* sensu Morris in Epiph. Orch. Malawi: 99 (1970), *non* (Rolfe) Schltr.
] *A. sp.* (*Williamson* 2050) sensu Williamson, Orch. SC. Afr.: 194 (1977)

15. **A. montana** *J. Stewart* in K.B. 34: 295 (1979). Type: Tanzania, Rungwe District, Kiwira R., NW. of Rungwe Mt., *Pollock* in *Moreau* 862 (K, holo.!, EA, iso.!)

Stem short, up to 6 cm. long, 8 mm. in diameter, bearing numerous flexuous roots 2–3 mm. in diameter, and 2–6 leaves in the upper part. Leaves distichous, ± spreading, ligulate or oblanceolate-ligulate, 10–18 cm. long, 2–2.5 cm. wide, lightly curved, the apex decidedly unequally bilobed, the lobes rounded, 3–10 mm. long, the shorter lobe sometimes almost absent, fleshy, dark green. Inflorescence arising below the leaves, spreading or pendulous, simply racemose, to 20 cm. long, bearing 4–10 flowers; peduncle terete, to 5 cm. long, 2–4 mm. in diameter provided with 2 loose sheaths; rhachis ± straight, with short flower-bearing projections 3–5 mm. above the insertion of the bracts; bracts tubular, broadly cylindrical, to 4 mm. long. Flowers spreading, 1–3 cm. apart, becoming closer towards the tip of the rhachis, pale green at first, becoming white, but with the spur, pedicel and base of the tepals salmon-pink; pedicel with ovary curved, 1.5–2 cm. long. Dorsal sepal erect, narrowly ovate, apiculate, 13–16 mm. long, 5–7.2 mm. wide; lateral sepals spreading, narrowly ovate, acute or apiculate, 11–15 mm. long, 4–6 mm. wide. Petals spreading, narrowly ovate, subacute or obtuse, 11–16 mm. long, 4–6 mm. wide. Lip similar to the tepals, 11–18 mm. long, 5.5–7.5 mm. wide; spur curving forwards, slightly wider in the distal half, 5–6 cm. long. Column short and thick, to 3 mm. high; anther-cap rounded; rostellum projecting downwards and forwards; pollinia ellipsoid-globose; stipes narrow, 2 mm. long; viscidium oblanceolate; stigmatic surface transversely oblong.

TANZANIA. Mbeya District: Umalila Forests, 30 Apr. 1969, *Leedal* 215!; Rungwe District: Kiwira R., NW. of Rungwe Mt., 27 Nov. 1943, *Pollock* in *Moreau* 743; Njombe District: Kiligati, 22 Feb. 1943, *Pollock* in *Moreau* 515!
DISTR. T 7; Malawi and Zambia
HAB. Usually growing with its roots among mosses and lichens in evergreen forests on mountains; 1700–2400 m. (in Tanzania)

16. **A. calantha** (*Schltr.*) *Schltr.* in Beih. Bot. Centr. 36: 115 (1918); Summerh. in F.W.T.A., ed. 2, 3: 265 (1968); J. Stewart in K.B. 34: 297 (1979). Types: Zaire, Likilemba, *Schlechter* 12700 & Cameroon, Ngoko, *Schlechter* (not seen, presumably destroyed)

Stem very short, woody, up to 3 cm. long, 3 mm. in diameter, bearing smooth slender greyish roots 1–1.5 mm. wide in the lower part. Leaves 2–6, alternate, usually distichous, closely overlapping on the stem, linear, 3–8 cm. long, 4–8 mm. wide, often falcate, somewhat attenuated towards the base, unequally bilobed at the apex, the lobes acute, 0–3 mm. long, the shorter one sometimes absent, margins entire, thin or somewhat fleshy, dark green with a few black dots. Inflorescences arising below the leaves or in the leaf-axils, often from the same site as in preceding years, spreading or pendent, up to 11 cm. long bearing 2–5(–8) flowers; peduncle terete, up to 3 cm. long, 2 mm. in diameter, bearing 1–3 tubular sheaths; rhachis straight; bracts broadly triangular, 1–3 mm. long. Flowers held erect, secund, 1–1.5 cm. apart, white, the spur often pink; pedicel with ovary pinkish brown, 12–17 mm. long. Dorsal sepal erect, oblong-lanceolate, subacute, 7–12 mm. long, 2–3.5 mm. wide; lateral sepals similar, slightly longer and narrower, projecting forwards. Petals spreading or slightly reflexed, linear-oblong, subacute, 7–10 mm. long, 2–2.5 mm. wide. Lip oblong-lanceolate, apiculate, 7–10 mm. long, 2–3.5 mm. wide; spur slender, gradually tapering to the pointed tip which is often curved or hooked forwards in the lower 5–10 mm., 2.5–4 cm. long. Column short and thick, 1.5–2 mm. long; androclinium sloping, anther-cap beaked; rostellum curving downwards and then forwards; pollinia reniform; stipes slender, attached to the centre of the viscidium;

viscidium slipper-shaped, fitting over the tip of the rostellum; stigmatic surface ± oblong, large, occupying most of the front of the column. Fruit a cylindrical capsule, to 5 cm. long, 5 mm. in diameter.

UGANDA. Bunyoro District: Budongo Forest, Nov. 1943, fl. in cult. Mar. 1944, *Eggeling* 5461!
TANZANIA. Lushoto District: W. Usambara Mts., Magamba, fl. in cult. 13 July 1944, *Eggeling* in *Moreau* 394!; Mpanda District: Silkcub Highlands, 4 Dec. 1956, *Richards* 7138!
DISTR. U 2; **T** 3, 4; Ghana, Central African Republic, Cameroon, Equatorial Guinea, Zaire and Angola
HAB. In deep shade, usually on small twigs and branches, at low levels and in the canopy, found in primary forest and on plantation trees such as coffee, cocoa and orange where these have replaced forest; 1000–1650 m.

SYN. *Angraecum calanthum* Schltr. in E.J. 38: 20 (1905)
A. roseocalcaratum De Wild. in B.J.B.B. 5: 190 (1916). Type: Zaire, between Zobia and Buta, *Seret* 889 (BR, holo.!)
A. sankuruense De Wild. in B.J.B.B. 5: 190 (1916). Type: Zaire, Sankuru, *Ed Luja* 1910 (BR, holo.!)
Aerangis roseocalcarata (De Wild.) Schltr. in Beih. Bot. Centr. 36: 120 (1918)
A. sankuruensis (De Wild.) Schltr. in Beih. Bot. Centr. 36: 120 (1918)
A. parvula Schltr. in N.B.G.B. 7: 329 (1919). Type: Tanzania, cultivated material (presumably destroyed)

17. **A. oligantha** *Schltr.* in E.J. 53: 599 (1915); J. Stewart in K.B. 34: 302 (1979). Type: Tanzania, Rungwe Mt., *Stolz* 2399 (K, iso.!)

Stem woody up to 6 cm. long, up to 4 mm. in diameter, closely covered by overlapping leaf-bases, with numerous slender, flexuous, greyish white roots to 2 mm. in diameter. Leaves 4–8, alternate, spreading or erect, up to 4 mm. apart, oblanceolate or narrowly spathulate, up to 8 cm. long, 2 cm. wide just below the apex, base sheathing, apex bilobed, lobes acute, 4–15 mm. long, sometimes one lobe much shorter or absent, rather thin and leathery in texture, dark green, margins entire. Inflorescences axillary, or from the woody stem below the leaves, often arising from the site of previous inflorescences, usually erect, bearing few–12 flowers, 1.5–4 cm. long; peduncle very short, terete; rhachis straight; bracts triangular, acute, 1.5–3 mm. long. Flowers very close together, crowded towards the apex of the rhachis and not always resupinate, pale green when first opening, becoming white, 15–18 mm. in diameter, all tepals spreading; pedicel with ovary straight, 10–15 mm. long. Dorsal sepal lanceolate-oblong, 6–9 mm. long, ± 3 mm. wide; lateral sepals slightly falcate, 6–10 mm. long, 2 mm. wide. Petals lanceolate-elliptic, 6–9 mm. long, 3–3.5 mm. wide. Lip ovate, apiculate, 6–8.5 mm. long, 4 mm. wide; spur parallel to the pedicellate ovary, slightly swollen in the lower half, 8–13 mm. long. Column small, 2 mm. long; androclinium sloping, centre of anther-cap with a green vertical stripe; rostellum straight and directed downwards towards the spur; pollinia globose; stipes slender; viscidium broadly elliptic, or rounded and fitting over the swollen end of the rostellum. Fruit a capsule 15–30 mm. long, 3 mm. wide.

TANZANIA. Mbeya District: Isangati area, near Umalila, Nov. 1969, *Leedal* 285!; Rungwe District: Kiwira R., lower fishing camp, 24 Oct. 1947, *Greenway & Brenan* 8262! & Tukuyu area, May 1969, *Leedal* 251!
DISTR. **T** 7; not known elsewhere
HAB. On low bushes and small trees, always amongst mosses and lichens; 1800–2100 m.

18. **A. ugandensis** *Summerh.* in K.B. 1931: 390 (1931); Robyns & Tournay, F.P.N.A. 3: 525 (1955); Tweedie in Journ. E. Afr. Nat. Hist. Soc. 24: 42 (1964); Piers, Orch. E. Afr., ed. 2: 218 (1968); J. Stewart in U.K.W.F.: 792 (1974) & in K.B. 34: 304 (1979). Type: Uganda, Mbale District, Elgon, *Snowden* 879 (K, holo.!)

Stem woody, up to 6 mm. diameter, erect, at least in the terminal leafy part, up to 20 cm. long but usually much shorter, bearing greenish white roots 3–4 mm. in diameter. Leaves 4–12, distichous, cuneate, oblanceolate, 5–15 cm. long, 1–2 cm. wide, often falcate, unequally bilobed at the apex, the longer lobe rounded or obtuse up to 8 mm. long, the shorter lobe very short or absent, widest just below the apex and gradually tapering to a shortly sheathing base, dark green with scattered black dots, leathery, without a distinct midrib. Inflorescences 1–5, arising from the axils of the lower leaves, always pendent, 7–10-flowered, up to 15 cm. long; peduncle terete, up to 2 cm. long, 1–2 mm. in diameter, bearing 2–3 tubular sheaths; rhachis straight; bracts truncate, 2–3 mm. long. Flowers spreading, 5–12 mm. apart, closest towards the tip of the inflorescence, white, often tinged with green; pedicel with ovary 1–2.5 cm. long, green. Dorsal sepal erect, oblong-lanceolate, apiculate or subacuminate, 6–12 mm. long, 3–4 mm. wide; lateral sepals similar but often slightly reflexed. Petal reflexed, oblong-lanceolate, acute, 6–10 mm. long, 3 mm. wide. Lip oblong, suddenly acute or cuspidate, 6–10 mm. long, 3–5 mm. wide at the middle; spur straight, narrowly tubular, 1–2.5 cm. long. Column short, terete below, slightly dilated above, 2 mm. long; anther-cap ± conical; rostellum straight, directed downwards towards the mouth of the spur; pollinia globose; stipes slightly widened in the upper half; viscidium round and flat, at right-angles to the rostellum. Pl. 6. Fig. 137.

UGANDA. Toro District, without precise locality, fl. in cult. Mar. 1939, *Chandler* 2779!; Ankole District: Lake Lutoto, Apr. 1944, *Eggeling* 5501!; Kigezi District: Impenetrable Forest, Nov. 1948, *Eggeling* 5850!

KENYA. Nandi District: Kaimosi Farm Forest, 17 June 1956, *Williams & Piers* 601!; N. Kavirondo District: Kakamega, fl. in cult. 9 June 1944, *Carroll* in *Moreau* 660!; Kericho District: Kimugu R., 31 May 1969, *J. Stewart* 991!

DISTR. **U** 2, 3; **K** 3–5; Zaire and Burundi

HAB. In moist forest, particularly near rivers, usually rather low down on large tree trunks and often with its roots in a deep growth of moss; 1500–2000 m.

19. **A. hologlottis** (*Schltr.*) *Schltr.* in Beih. Bot. Centr. 36: 117 (1918); J. Stewart in K.B. 34: 306 (1979). Type: Sri Lanka, Peradeniya, *Schlechter* (PDA, topo.!, holotype presumably destroyed)

Stem short and stout, woody, up to 3 cm. long, 5 mm. in diameter, covered with closely overlapping old leaf-bases, bearing numerous greyish roots 1–2 mm. in diameter. Leaves 2–6, closely overlapping on the stem, linear or narrowly oblong-ligulate, 5–9 cm. long, 7–18 mm. wide, attenuated below, often falcate, bilobed at the apex, the lobes unequal or subequal, acute, 2–7 mm. long, somewhat fleshy, dark green, margins entire. Inflorescences 1–several racemes, arising below the leaves, erect or spreading, up to 18 cm. long, bearing 6–15 flowers; peduncle terete, up to 5 cm. long, 1 mm. in diameter, bearing 4–5 tight sheaths; rhachis straight; bracts broadly triangular-ovate, up to 2 mm. long. Flowers 5–8 mm. apart, white with the pedicel and spur pale greenish; pedicel with ovary 5–7 mm. long; perianth-parts all spreading. Dorsal sepal narrowly elliptic, apiculate, 6–7 mm. long, 1.5–3 mm. wide; lateral sepals similar, but narrower, 5.5–6.5 mm. long, 1.3–2 mm. wide. Petals narrowly obovate-elliptic, apiculate, 5–6 mm. long, 2–4 mm. wide. Lip obovate, apiculate, 5–6 mm. long, 2.5–3 mm. wide; spur straight, slender, 3.5–7 mm. long, 0.5 mm. wide. Column short and stout, 1 mm. high; androclinium transversely oblong; rostellum straight; pollinia globose; stipes slender; viscidium narrowly elliptic.

KENYA. Kwale District: Shimba Hills, Jan. 1963, *Castelino* in *E.A.H.* 18/63!; Kilifi District: Bamba-Ganze road, June 1964, *Archer* 522!

TANZANIA. Lushoto District: Sigi, Dec. 1942, *Moreau* 402!; Zanzibar I., Oct. 1943, *Davies* in *Moreau* 402A!

DISTR. **K** 7; **T** 3; **Z**; Mozambique and Sri Lanka

HAB. Epiphytic on twigs and small branches of trees in coastal forest; 0–500 m.

SYN. *Angraecum hologlottis* Schltr. in F.R. 3: 82 (1906)
Aerangis sp. (aff. *A. rhodosticta* (Kraenzl.) Schltr. ?) sensu Piers, Orch. E. Afr., ed. 2: 222, t. IIK (1968)

20. **A. alcicornis** (*Reichb.f.*) *Garay* in Bot. Mus. Leafl., Harv. Univ. 23: 157 (1972); J. Stewart in K.B. 34: 308 (1979). Type: Malawi, Shupanga, *Kirk* 114 (K, holo.!)

Stem woody, up to 3 cm. long, with numerous flexuous glabrous roots up to 2–3 mm. in diameter. Leaves 2–8, linear-oblanceolate to elliptic-oblanceolate, 3.5–12 cm. long, 1–2.5 cm. wide, apex usually acutely occasionally somewhat obtusely bilobed, the lobes unequal or subequal, up to 18 mm. long, dark green, or greyish green with a much darker distinct venation. Inflorescences arising below the leaves, horizontal or somewhat pendent, 10–30 cm. long, bearing 3–15 flowers; peduncle up to 10 cm. long, terete, bearing 3–6 tubular sheaths; rhachis straight; bracts shortly ovate, acute, 3–5 mm. long. Flowers spreading, 8–17 mm. apart, becoming closer together towards the tip of the inflorescence, white with pinkish tinges towards the tip of the spur; pedicel with ovary slender, 12–15 mm. long, tinged pink. Dorsal sepal erect, oblong-elliptic, apiculate, 10–18 mm. long, 4–7 mm. wide; lateral sepals similar, 10–18 mm. long, 5–6 mm. wide. Petals elliptic, narrowed towards the base, 9–15 mm. long, 5–8 mm. wide. Lip cuneate at the base, suborbicular to fan-shaped, rounded at the apex, sometimes with a short apiculus, 10–20 mm. long, 10–15

FIG. 137. *AERANGIS UGANDENSIS*—**1**, habit, × ⅔; **2**, perianth dissected, × 3; **3**, column, and base of perianth, side view, × 3; **4**, column, front view, × 5; **5**, pollinarium, × 10; **6**, rostellum, × 10. Drawn from *J. Stewart* 991 by Miss L.S. Cowan.

mm. wide; spur filiform, somewhat incurved and slightly inflated towards the tip, 2.5–4 cm. long. Column stout, 2–4 mm. high, dilated below the anther; anther semi-globose, anther-cap beaked; rostellum elongated, obtuse; pollinia ellipsoid; stipes dilated above, slender below; viscidium round or elliptic.

TANZANIA. Uzaramo District: 96 km. inland from Dar es Salaam, fl. at Kew, *Pullen-Burry*! & July 1969, *Harris* 3097!; Lindi District: Lake Lutamba, 20 Sept. 1934, *Schlieben* 5365!
DISTR. **T** 6, 8; Malawi and Mozambique
HAB. On trees, shrubs and lianes, in thicket; 200–1000 m.

SYN. *Angraecum alcicorne* Reichb.f. in Otia Bot. Hamb. 2: 117 (1881); Th. Dur. & Schinz in Consp. Fl. Afr. 5: 39 (1895); Kraenzl. in P.O.A. C: 157 (1895); Rolfe in F.T.A. 7: 136 (1897); Morris, Epiph. Orch. Malawi: 84 (1970)
Aerangis calodictyon Summerh. in K.B. 1934: 214 (1934). Type: Tanzania, Uzaramo District, 96 km. inland from Dar es Salaam, *Pullen-Burry* (K, holo.!)
A. lutambae Mansf. in N.B.G.B. 12: 704 (1935). Type: Tanzania, Lindi District, Lake Lutamba, *Schlieben* 5365 (BR, P, Z, iso.!)

21. **A. luteoalba** (*Kraenzl.*) *Schltr.* in Beih. Bot. Centr. 36: 118 (1918); J. Stewart in K.B. 34: 309 (1979). Types: Zaire, Ituri, *Stuhlmann* 2650 & Lindu, *Stuhlmann* 2720 (B, syn.†)

Stem 1–3 cm. long, often pendent with numerous thin flexuous roots which are relatively long for the size of the plant. Leaves 2–8, lying in one plane, linear-ligulate or linear-lanceolate, up to 15 cm. long, 6–15 mm. wide, sometimes falcate, unequally bilobed at the apex, the lobes rounded, the shorter lobe sometimes suppressed, fleshy and brittle in the rainy season becoming leathery in the dry season. Inflorescences arising from the stem below the leaves, arching or pendent racemes bearing few–25 flowers in the upper half or two-thirds, up to 35 cm. long; peduncle slender, terete, bearing 2–3 sheaths; rhachis similar, slightly zigzag; bracts spreading or slightly reflexed, ovate, acute, 3–4 mm. long. Flowers arranged in 2 rows all in the same plane, 7–20 mm. apart; perianth parts all spreading, white, cream, greenish, or yellowish white; column similar in colour, or bright red; pedicel with ovary slender, 1–2 cm. long. Sepals oblanceolate, acute, 10-15 mm. long, 3–7 mm. wide. Petals obovate or broadly oblanceolate, equalling or a little longer than the sepals, 5–10 mm. wide. Lip obovate or rhomboid, widest above the middle, acute, 15–20 mm. long, 7–15 mm. wide; spur slender, incurved, slightly thickened in the apical half, 2.3–4 cm. long. Column short and thick, 2–3 mm. long; androclinium slightly sloping with frilly margins; rostellum projecting straight downwards; pollinia ovoid; stipes considerably widened in the upper half; viscidium oval or round. Capsule 2–2.5 cm. long.

var. **luteoalba**

Flowers entirely cream or pale yellow.

UGANDA. Ankole District, without precise locality, *Doggett*!
DISTR. **U** 2; Zaire
HAB. In forest, on small twigs and branches of bushes, also on tree trunks and lianes; 1250–2000 m.

SYN. *Angraecum luteo-album* Kraenzl. in P.O.A.C: 158 (1895) & in E.J. 22: 27 (1895); Rolfe in F.T.A. 7: 141 (1897)
Rhaphidorhynchus luteo-albus (Kraenzl.) Finet in Mém. Soc. Bot. Fr. 9: 35 (1907)

var. **rhodosticta** (*Kraenzl.*) *J. Stewart* in K.B. 34: 310 (1979). Types: Cameroon, Yaoundé, *Zenker & Staudt* 434 (K, isosyn. !) & Ethiopia, *Ruspoli & Riva* 1360 (FT, syn.!)

Differs from typical var. *luteoalba* in having a bright red column. Pl. 6. Fig. 138.

UGANDA. Bunyoro District: Budongo, Sept. 1935, *Eggeling* 2185!; Kigezi District: Ishasha Gorge, Apr. 1948, *Purseglove* 2635!; Mengo District: Kampala, fl. in cult. Sept. 1937, *Chandler* 1904!
KENYA. Meru District: E. Mt. Kenya, 24 July 1913, *Battiscombe* 692!; N. Kavirondo District: Kakamega Forest, June 1961, *Lucas* 112! Kericho District: Kiptiget R., 1 June 1969, *J. Stewart* 996!
TANZANIA. Bukoba District: Minziro Forest, Mar. 1953, *Procter* 145!; Ufipa District: Chagwe, 20 Sept. 1921, *Lankester* 692!
DISTR. **U** 2–4; **K** 4–6; **T** 1, 4; Central African Republic, Cameroon, Zaire and Ethiopia
HAB. In forest; 1250–2200 m.

SYN. *Angraecum rhodostictum* Kraenzl. in N.B.G.B. 1: 154 (1896); Rolfe in F.T.A. 7: 140 (1897)
A. albidorubrum De Wild. in B.J.B.B. 5: 180 (1916). Type: Zaire, Suronga, *Seret* 190 (BR, holo.!)
Aerangis albidorubra (De Wild.) Schltr. in Beih. Bot. Centr. 36: 113 (1918)
A. rhodosticta (Kraenzl.) Schltr. in Beih. Bot. Centr. 36: 119 (1918); W.F.K.: 141, fig. 116 (1948); Robyns & Tournay, F.P.N.A. 3: 525, fig. 32 (1955); Summerh. in Bot. Mag. 171, t. 276 (1956); Tweedie in Journ. E. Afr. Nat. Hist. Soc. 24: 39 (1964); Hawkes, Encycl. Cult. Orchids: 35

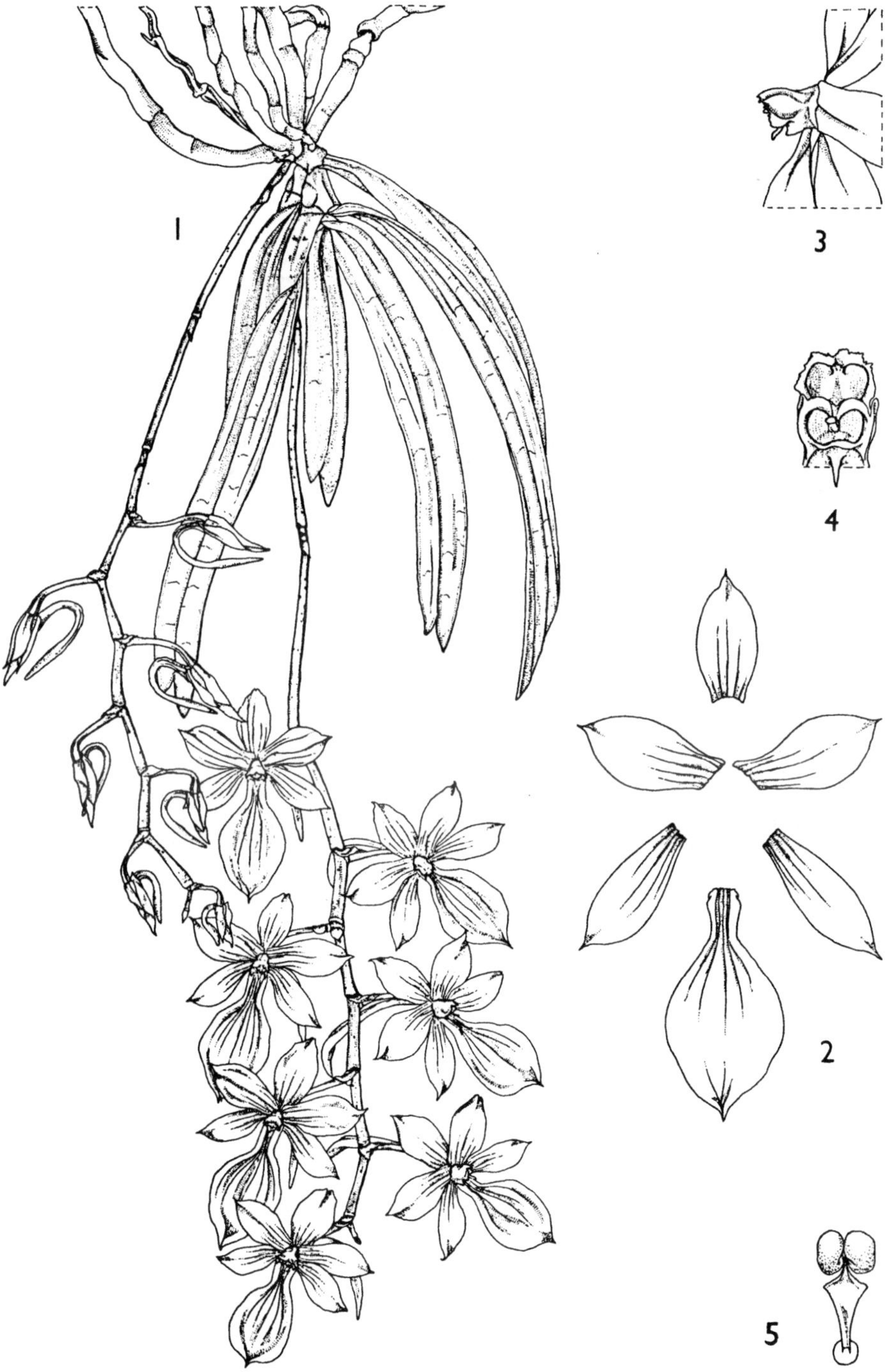

FIG. 138. *AERANGIS LUTEO-ALBA* var. *RHODOSTICTA*—**1**, habit, × 1; **2**, perianth dissected, × 1½; **3**, column, and base of perianth, side view, × 3⅓; **4**, column, front view, × 5; **5**, pollinarium, × 7. Drawn from *J. Stewart* 1201 by Miss L.S. Cowan.

(1965); Piers, Orch. E. Afr., ed. 2: 214 (1968); P.F. Hunt in K.B. 22: 492 (1968); J. Stewart & Campbell, Orch. Trop. Afr.: 32–33 (1970); P.F. Hunt & Grierson, Orchidaceae: 110–111 (1973); J. Stewart in U.K.W.F.: 793, 797 (1974)
Angraecum mirabile Hort. in Gard. Chron., ser. 3, 73: 17, fig. 7 (1923)

59. RANGAERIS

Summerh. in K.B. 1936: 227 (1936) & 4: 435 (1949); Senghas in Schltr., Die Orchideen, ed. 3: 1099 (1986)

Epiphytic or rarely lithophytic herbs with short to long stem entirely covered with persistent leaf-bases; roots emerging along the stem opposite the leaves, rather stout, unbranched or little branching. Leaves coriaceous, distichous, conduplicate, linear or oblong, unequally bilobed at the apex. Inflorescences pendent or spreading to suberect, usually laxly few–several-flowered. Flowers small to relatively large, usually white or yellowish; pedicel and ovary longer than the bract. Sepals and petals free, subsimilar, spreading to recurved. Lip entire or obscurely 3-lobed, ecallose; spur pendent, filiform, elongate from a narrow mouth. Column short to long, glabrous or puberulent, lacking a foot; rostellum bifid; anther cucullate, produced in front and truncate at the apex; pollinia 2, pyriform; stipites 2, linear or oblanceolate; viscidium relatively large, oblong or cordate.

A small genus of about 6 species confined to tropical Africa.

Summerhayes (1949) divided the genus into two sections, the typical one comprising all of the species except *R. brachyceras*, which he placed in his monotypic section *Biglandulosae*. Senghas (1986), in the third edition of Schlechter's Die Orchideen, removed this species to the newly erected genus *Cribbia*. For a discussion of this refer to the following genus.

Column 2 cm. or more long, strongly dorsally puberulent . . . 1. *R. schliebenii*
Column less than 5 mm. long, not noticeably puberulent:
  Lip entire, ovate, 8.5 mm. or less long; spur 7 cm. or less long; stems 1–6 cm. long . . . . 2. *R. muscicola*
  Lip obscurely 3-lobed, 10–24 mm. long; spur 7.5–16 cm. long; stems up to 45 cm. long . . . . 3. *R. amaniensis*

1. **R. schliebenii** (*Mansf.*) *Cribb*, comb. nov. Type: Tanzania, Ulanga District, near Mahenge, Masagati, *Schlieben* 1159a (B, holo., Z, iso.!)

An erect epiphyte with a 7–11 cm. long stem covered by persistent leaf-bases; roots in lower half of stem, 1.5–2 mm. in diameter. Leaves coriaceous, conduplicate, linear-ligulate, unequally roundly bilobed at the apex, 15–20 cm. long, 1.7–2.5 cm. wide, articulated to conduplicate leaf-bases. Inflorescences spreading, axillary, 18–27 cm. long, 4–8-flowered; bracts scarious, ovate-oblong, obtuse, 8–11 mm. long, brown. Flowers white with a greenish orange spur; pedicel and ovary 6–9 cm. long, ovary slightly pustular. Sepals recurved, linear, acute or acuminate, 3.4–4 cm. long, 3.5–4.5 mm. wide. Petals reflexed, linear, acuminate, 3.4–3.8 cm. long, 4 mm. wide. Lip recurved, narrowly elliptic, acuminate, 3–4.2 cm. long, 7–11 mm. wide, with distal margins erose-denticulate; spur filiform, pendent, wide mouthed, 16–19 cm. long. Column slender, porrect, 2–2.7 cm. long, minutely puberulent in basal half; stipites obovate, rugulose; viscidium ovate-cordate. Pl. 7. Fig. 139.

TANZANIA. Ulanga District: Masagati, June 1931, *Schlieben* 1159a! & Kwiro Forest Reserve, Jan. 1979, *Cribb, Grey-Wilson & Mwasumbi* 11087!; Songea District: Madaba waterfall, 20 May 1989, *Spurrier* 89!

DISTR. **T** 6, 8; not known elsewhere

HAB. Epiphytic in montane and riverine forest; 400–1400 m.

SYN. *Leptocentrum schliebenii* Mansf. in N.B.G.B. 12: 704 (1935); Moreau in Journ. E. Afr. Nat. Hist. Soc. 17: 23 (1943)
Barombia schliebenii (Mansf.) Cribb in S. Afr. Orch. Journ. 11: 110 (1980)

FIG. 139. *RANGAERIS SCHLIEBENII*—**1**, habit, × ⅔; **2**, inflorescence, × ⅔; **3**, flower, × 1; **4**, dorsal sepal, × 1; **5**, lateral sepal, × 1; **6**, petal, × 1; **7**, lip, × 2; **8**, column, × 6; **9**, anther-cap, × 6; **10**, pollinia, × 8. Drawn from *Cribb, Grey-Wilson & Mwasumbi* 11087 by Susan Hillier.

2. **R. muscicola** (*Reichb.f.*) *Summerh.* in F.W.T.A. 2: 450 (1936); Piers, Orch. E. Afr., ed. 2: 284 (1968); J. Stewart in U.K.W.F.: 796 (1974); Williamson, Orch. SC. Afr.: 194 (1977); J. Stewart et al., Wild Orch. S. Afr.: 265 (1982); la Croix et al., Malawi Orch. 1: 126 (1983). Type: Angola, Pungo Andongo, *Welwitsch* 699 (W, holo., K, iso.!)

An epiphyte or lithophyte with very short stems 1–6 cm. long; roots 2–3 mm. diameter. Leaves arranged in a fan, coriaceous, conduplicate, 5–11, arcuate in side view, linear, unequally roundly or obtusely bilobed at the apex, 6.5–20 cm. long, 6–13 mm. wide, articulated to imbricate spreading sheaths at the base. Inflorescences arcuate to suberect, 5–22(–42) cm. long, 5–12-flowered; peduncle 2.5–10 cm. long; bracts amplexicaul, broadly ovate, obtuse, 3–9 mm. long and wide. Flowers white with a pinkish olive spur, diurnally sweetly scented; pedicel and ovary 1.4–3.6 cm. long. Dorsal sepal ovate, acute, 7–9 mm. long, 2.3–4.2 mm. wide; lateral sepals obliquely lanceolate, acute, 7.3–9.4 mm. long, 2.3–4 mm. wide. Petals obliquely elliptic, acute, 6.5–8 mm. long, 1.5–3.3 mm. wide. Lip entire, ovate, acute, 6.7–8.5 mm. long, 4–6 mm. wide; spur filiform, pendent, 5.5–7 cm. long. Column 3–4.5 mm. long; stipites 2, united at the base; viscidium oblong.

UGANDA. Bunyoro District: Budongo Forest, May 1933, *Eggeling* 1335!; Kigezi District: Impenetrable Forest, June 1947, *Purseglove* 2469!; Mbale District: Elgon, Tracy Falls, 9 Apr. 1924, *Snowden* 874!; Mengo District: 33.5 km. on Entebbe road, Oct. 1937, *Chandler* 2013!

KENYA. Trans-Nzoia/N. Kavirondo District: Elgon, 30 Apr. 1931, *E.J. & C. Lugard* 645!; Kericho, Nov. 1929, *Gray* 1!; Masai District: Ngong, May 1938, *C. van Someren* 25!

TANZANIA. Bukoba, Nov. 1940, *Tweedie* 558!; Lushoto District: W. Usambara Mts., Magamba, 9 Dec. 1941, *Harris* in *Moreau* 90!; Iringa District: Sao Hill, fl. in cult. 14 Apr. 1943, *Moreau* 323!; Songea District: E. slope of Chandamara Hill, 23 Mar. 1956, *Milne-Redhead & Taylor* 9336!

DISTR. **U** 2–4; **K** 3, 5, 6; **T** 1, 3, 7, 8; Guinée, Sierra Leone, Liberia, Ivory Coast, Ghana, Nigeria, Cameroon, Zaire and south to South Africa (Natal)

HAB. An epiphyte in lowland and upland evergreen forest and in *Brachystegia* woodland, also a lithophyte in the latter; 1180–2200 m.

SYN. *Aeranthes muscicola* Reichb.f. in Flora 48: 190 (1865)
Angraecum englerianum Kraenzl. in E.J. 7: 333 (1886). Type: Cameroon, near Kumba, *Schlechter* 15770 (B, holo., K, record of holo.!)
Listrostachys engleriana (Kraenzl.) Kraenzl. in E.J. 19: 254 (1894), in obs.
Mystacidium muscicolum (Reichb.f.) Th. Dur. & Schinz, Consp. Fl. Afr. 5: 54 (1895)
Listrostachys muscicola (Reichb.f.) Rolfe in F.T.A. 7: 158 (1897)
Mystacidium batesii Rolfe in F.T.A. 7: 172 (1897). Type: Cameroon, Efuleu, *Bates* 382 (K, holo.!)
Angraecum batesii (Rolfe) Schltr., Westafr. Kautsch.-Exped.: 283 (1900)
Aerangis falcifolia Schltr. in E.J. 53: 598 (1915). Type: Tanzania, Rungwe District, Mulinda, *Stolz* 1960 (B, holo., K, iso.!)
A. muscicola (Reichb.f.) Schltr. in E.J. 53: 599 (1915)
A. englerianum (Kraenzl.) Schltr. in E.J. 53: 599 (1915)
Angraecum solheidii De Wild. in B.J.B.B. 5: 191 (1916). Types: Zaire, Nala, *Boone* 98, 99, 108, 109, 110 (BR, syn.); Dundusana, *Mortehan* 163 (BR, syn.) & *De Giorgi* 1018 (BR, syn.); Yambuya, *Laurent* 1028 (BR, syn.) & *Solheid* 6 & 127 (BR, syn.)
Listrostachys solheidii (De Wild.) De Wild., in B.J.B.B. 5: 191 (1916)
Aerangis mixta Schltr. in Beih. Bot. Centr. 36: 122 (1918), nom. nov. pro *Mystacidium batesii* Rolfe
A. solheidii (De Wild.) Schltr. in Beih. Bot. Centr. 36: 123 (1918)
Listrostachys floribunda Rolfe in K.B. 1918: 236 (1918). Type: Uganda, Kampala, cult. Kew 225–16 (K, holo.!)
Aerangis floribunda (Rolfe) Summerh. in K.B. 1932: 509 (1932)

3. **R. amaniensis** (*Kraenzl.*) *Summerh.* in K.B. 4: 438 (1949); Piers, Orch. E. Afr.: 438 (1949); J. Stewart in U.K.W.F.: 796 (1974). Type: Tanzania, E. Usambara Mts., Amani. *Braun* in *Herb. Amani* 1551 (B, holo., K, iso.!)

An erect to pendent epiphyte or lithophyte, often forming dense clumps. Stems up to 45 cm. long, unbranched or little branching, covered by persistent leaf-bases; roots little branched, 3–4 mm. in diameter. Leaves coriaceous, conduplicate, narrowly oblong, unequally roundly bilobed at the apex, 3.5–11.5 cm. long, 1–2.3 cm. wide, articulated to distichous, imbricate leaf-bases. Inflorescences spreading, 6–9 cm. long, 5–13-flowered; peduncle 1–1.5 cm. long; bracts 5–6.5 mm. long, drying black. Flowers very variable in size, white fading to yellow, greenish on the outer surface and spur, diurnally scented; pedicel and ovary 2.5–5.5 cm. long, scabrid. Sepals lanceolate, acuminate, 1–2.5 cm. long, 2.7–5.6 mm. wide. Petals lanceolate, acuminate, 0.9–2.1 cm. long, 2.2–5 mm. wide. Lip obscurely 3-lobed in the middle, 10–24 mm. long, 5–11 mm. wide; side lobes rounded; mid-lobe lanceolate; spur pendent, filiform, 7.5–16 cm. long. Column 4.5–5 mm. long; stipes bifid, each lobe linear; viscidium oblong. Pl. 7.

UGANDA. Karamoja District: Mt. Moroto, Apr. 1958, *J. Wilson* 436a! & 11 June 1970, *Lye* 5641! & Warr, 4 Nov. 1939, *A.S. Thomas* 3181!
KENYA. N. Frontier Province: Mt. Kulal, 8 Oct. 1947, *Bally* 5528!; Trans-Nzoia District: Nicholson's Farm, on the banks of the Nzoia R., May 1964, *Tweedie* 2816!; Masai District: Ngong Hills, 9 Aug. 1985, *Khayoto* 304!
TANZANIA. Musoma District: about 98 km. Seronera to Klein's Camp, 6 Apr. 1961, *Greenway & Myles Turner* 9998!; Masai District: S. slope of Monduli Mt., fl. cult. 16 Nov. 1943, *Moreau* 315A!; Lushoto District: W. Usambara Mts., Bumbuli, fl. cult. 18 Dec. 1942, *Moreau* 148!; Morogoro District: Mindu Mt., fl. cult. 31 Apr. 1944, *Moreau* 689!
DISTR. **U** 1; **K** 1, 3, 4, 6; **T** 1–3, 5, 6; Ethiopia, Zimbabwe
HAB. In montane forest of *Juniperus procera*, etc. and in open woodland; 1100–2600 m.

SYN. *Listrostachys amaniensis* Kraenzl. in E.J. 43: 397 (1909)
Leptocentrum amaniense (Kraenzl.) Schltr. in Beih. Bot. Centr. 36: 112 (1918)
Cyrtorchis cufodontii Chiov. In Miss. Biol. Borana, Racc. Bot.: fig. 113 (1939). Type: Ethiopia, Arero, *Cufodontis* 317 (FT, holo.)

60. CRIBBIA

Senghas in Schltr., Die Orchideen, ed. 3: 1118 (1986)

Azadehdelia Braem in Schlechteriana 1(2): 34 (1988)

Small epiphytic or lithophytic monopodial herbs; roots slender, branching, shortly puberulent. Leaves distichous, relatively thin-textured, twisted at the base to lie in one plane, articulated at the base to imbricate leaf-bases. Inflorescences 1–many, laxly few–many-flowered, racemose; bracts free. Flowers small, not opening widely, somewhat translucent. Sepals and petals free. Dorsal sepal shorter than and distinctly shaped from the laterals. Petals smaller than the sepals. Lip entire, ecallose, spurred at the base. Column terete; rostellum short, 3-lobed, mid-lobe slightly longer than the side lobes, pendent; anther-cap hemispherical, truncate in front; pollinia 2, subglobose; stipites 2, linear; viscidia 2, very small.

A genus of one or possibly two species, confined to Africa.

This genus has only recently been split from *Rangaeris* by Senghas (1986). Braem (1988) has argued that *Cribbia* Senghas is a homonym of fungal genus *Cribbea* Smith & Reid and renamed the former *Azadehdelia*. This opinion has yet to be met with any consensus and, following similar cases cited recently in Taxon, is unlikely to be upheld. I have, therefore, retained the genus *Cribbia* Senghas for the time being because there is no danger of its confusion with the fungal genus.

C. brachyceras (*Summerh.*) *Senghas* in Schltr., Die Orchideen, ed. 3: 1118 (1986). Type: Zaire, Virunga Mts., *B.D. Burtt* 3123 (K, holo.!)

A small epiphyte or rarely a lithophyte. Stem erect or ascending, up to 20 cm. long; roots many from lower part of stem, slender, flexuous, branching, 0.5–2 mm. in diameter, shortly pubescent. Leaves distichous,suberect to spreading, straight or slightly falcate, linear-oblong, unequally roundly or obtusely bilobed at the apex, 8–13 cm. long, 7–13 mm. wide, articulated to imbricate sheathing bases. Inflorescences 1–several, racemose, suberect, laxly 7–15-flowered, 4–17 cm. long; bracts lanceolate, acuminate or acute, 2.5–5 mm. long. Flowers pale yellow, somewhat translucent; pedicel and ovary 3–5 mm. long, very slender. Dorsal sepal oblong-lanceolate, acute, 5.5–7 mm. long, 1.5–2 mm. wide; lateral sepals deflexed, linear or linear-oblanceolate, acute, 6–7.5 mm. long, 1.2–1.5 mm. wide. Petals oblong or elliptic-oblong, obtuse or rounded, 5–5.5 mm. long, 1.4–2 mm. wide. Lip entire, slightly recurved in apical half, lanceolate or oblong-lanceolate, acute, 5–5.5 mm. long, 3 mm. wide, ecallose; spur decurved, cylindrical, 5–6 mm. long, very slightly inflated in apical half. Column 2 mm. long. Fig. 140.

UGANDA. Ankole District: Igara, Kalinzu Forest, 15 Jan. 1953, *Osmaston* 2792! & Lutoto, Nov. 1940, *Eggeling* 4221; Kigezi District: Impenetrable Forest, fl. cult. 29 Dec. 1958, *Burgess* in *Piers*!; Mbale District: Elgon, Bunkoko and Bulago, 5 June 1924, *Snowden* 900! & 11 Oct. 1918, *Dummer* 3686!
KENYA. Nandi District: Kaimosi, N. of Kisumu, Sept. 1934, *Tweedie* 287!; N. Kavirondo District; Kakamega Forest, fl. cult., *Carroll* in *Moreau* 792!
DISTR. **U** 2, 3; **K** 3, 5; Liberia to Zaire, Zambia and Malawi
HAB. On mossy trunks of trees and on mossy rocks in forest; 1500–2200 m.

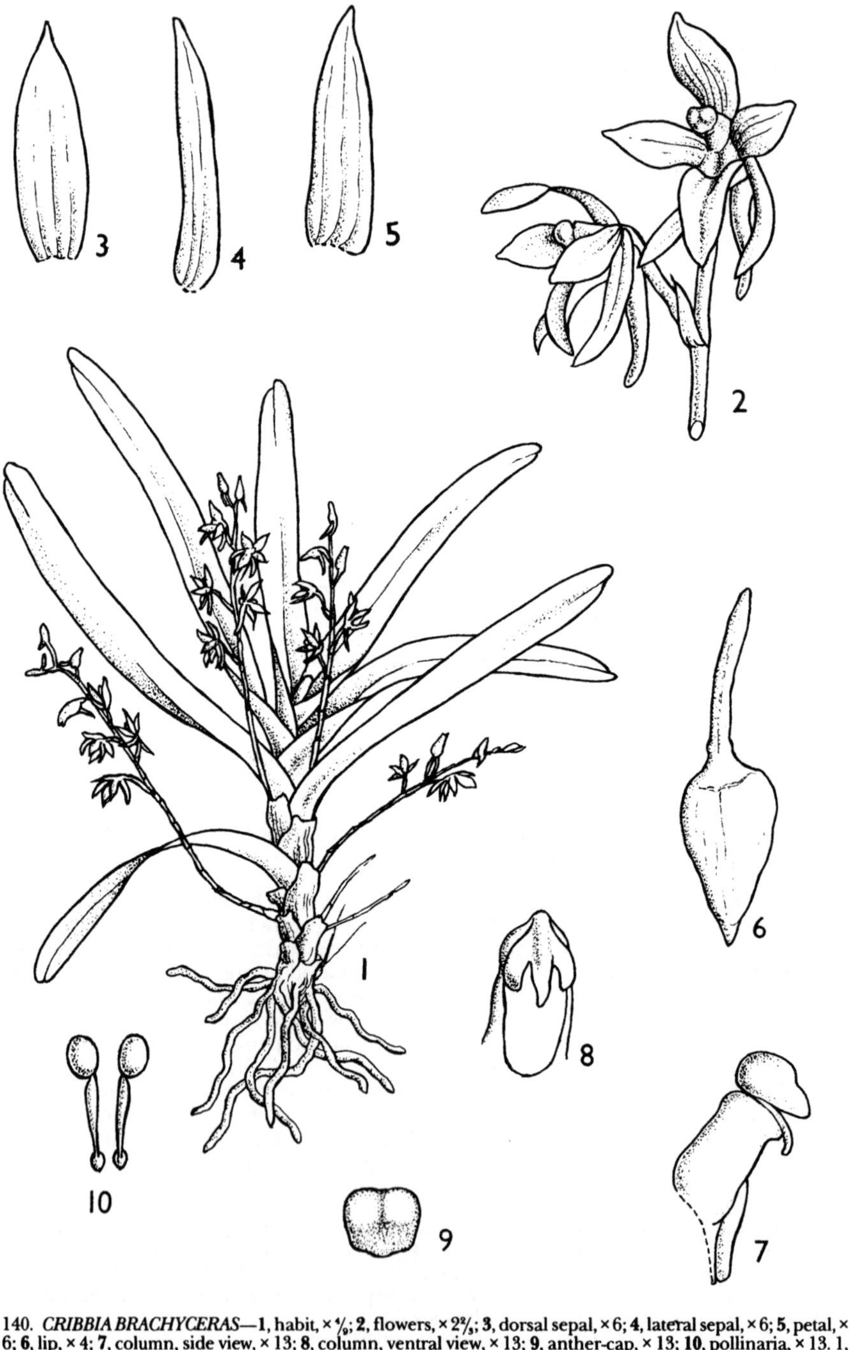

FIG. 140. *CRIBBIA BRACHYCERAS*—**1**, habit, $\times \frac{4}{9}$; **2**, flowers, $\times 2\frac{2}{3}$; **3**, dorsal sepal, × 6; **4**, lateral sepal, × 6; **5**, petal, × 6; **6**, lip, × 4; **7**, column, side view, × 13; **8**, column, ventral view, × 13; **9**, anther-cap, × 13; **10**, pollinaria, × 13. 1, from *Copley* 41; 2–10, from *Bailes* 258. Drawn by Cherry Ann Lavrih.

SYN. *Aerangis brachyceras* Summerh. in K.B. 1934: 213. (1934)
Rangaeris brachyceras (Summerh.) Summerh. in K.B. 1936: 228 (1936)
Azadehdelia brachyceras (Summerh.) Braem in Schlechteriana 1(2): 34 (1988)

61. CYRTORCHIS

Schltr., Die Orchideen: 595 (1914) & in Beih. Bot. Centr. 36: 127–133 (1918); Summerh. in K.B. 14: 143–156 (1960)

Epiphytic or rarely lithophytic herbs with short to long, erect or pendent stems, covered by distichous leaf-bases; roots emerging all along stem, elongate, branching. Leaves distichous, coriaceous or fleshy, flat or conduplicate, much longer than broad, unequally bilobed at apex, articulated at base to a sheathing base, decidous, leaving obvious sharp edges of leaf-bases free. Inflorescences 1–several, usually shorter than the leaves, few–many-flowered, axillary in upper part of stem; peduncle short, often covered by sterile sheathing bracts; bracts usually large. Flowers stellate, sweetly scented, white, sometimes with a greenish, pinkish or brownish spur. Sepals and petals subsimilar, free, linear-lanceolate, acuminate, usually recurved. Lip similar, entire or very obscurely 3-lobed, ecallose, with a long tapering spur at the base; spur sigmoid or slightly incurved. Column short, fleshy; rostellum 3-lobed, pendent, outer lobes much longer than the mid-lobe, often papillate towards apex; pollinia 2; stipites 2, oblanceolate; viscidium either hyaline or comprising an indurated saddle-shaped upper part and a hyaline lower part; anther-cap elongated at apex. Fruit a 3–6-winged capsule.

A genus of about 16 species confined to tropical and southern Africa.

The genus *Cyrtorchis* is a taxonomically difficult one. The flowers of all the species bear a distinct resemblance to each other probably because of their similar pollination syndrome, all being Hawk-moth pollinated. *Cyrtorchis* was revised by Summerhayes (1960) in the Kew Bulletin. He divided the genus on viscidium-type placing those species with a bipartite viscidium comprising an indurated and a hyaline part in sect. *Heterocollection* (now sect. *Cyrtorchis*) and the remaining species with a hyaline viscidium in sect. *Homocollection*. In the Flora of West Tropical Africa (1968) he used the viscidium-type as the major split in the key. I have preferred to use macroscopic characters in constructing the key given here but users may note that species 1–4 fall into the typical section while species 5–10 are in sect. *Homocollection*.

Leaves conduplicate (V-shaped in cross-section), recurved:
  Leaves mid- to dark green, (4.5–)6–10 cm. long; inflorescence 6–10.5 cm. long; spur 1.8–2.8 cm. long . . . . . 9. *C. praetermissa*
  Leaves grey-green, 1.5–4(–6) cm. long; inflorescence 1.5–3 cm. long; spur 1.8–2.2 cm. long . . . . 10. *C. crassifolia*
Leaves flat in cross-section, not recurved:
  Spur more than 9.5 cm. long . . . . 1. *C. chailluana*
  Spur less than 9 cm. long:
    Leaves 22–35 cm. long . . . . 6. *C. injoloensis*
    Leaves less than 22 cm. long:
      Leaves oblanceolate, 3–5 cm. wide; inflorescence 18–32 cm. long . . . . 5. *C. monteiroae*
      Leaves linear, narrowly oblong or narrowly elliptic; inflorescence less than 22 cm. long, usually much shorter:
        Pedicel and ovary more than 18 mm. long; viscidium with an indurate and hyaline part:
          Leaves suberect; bracts 5–11 mm. long . . . 3. *C. neglecta*
          Leaves spreading; bracts more that 15 mm. long:
            Leaves linear, less than 16 mm. wide; inflorescence less than 7 cm. long, 5–6-flowered . . . . 4. *C. seretii*
            Leaves oblong or narrowly elliptic, 16 mm. or more wide; inflorescence usually more than 7 cm. long, 8–14-flowered . . . . . . 2. *C. arcuata*
        Pedicel and ovary less than 15 mm. long:

Leaves 7–8.5 cm. long, 7–8 mm. wide; pedicel and ovary 3–4 mm. long . . . . 8. *C. brownii*
Leaves 7.5–12.5 cm. long, 12–25 mm. wide; pedicel and ovary 10–13 mm. long . . . . . . 7. *C. ringens*

1. **C. chailluana** (*Hook. f.*) *Schltr.*, Die Orchideen: 596 (1914) & in Beih. Bot. Centr. 36: 130 (1918); Summerh. in K.B. 14: 130 (1960) & in F.W.T.A., ed. 2, 3: 267 (1968). Types: Nigeria, Nun R., *Mann* 521 (K, syn.!) & Gabon, without exact locality, *du Chaillu* (K, syn.!)

A pendulous epiphyte with a long stem. Leaves linear-oblanceolate, unequally roundly bilobed at apex, 11–25 cm. long, 1.9–3.2 cm. wide. Inflorescences 13–23 cm. long, 7–10-flowered; peduncle ± 3 cm. long, covered by sheathing bracts; bracts cymbiform, broadly ovate, obtuse, 18–23 mm. long, 10–22 mm. apart. Flowers white, scented; pedicel and ovary 42–45 mm. long. Sepals linear-lanceolate, long-acuminate, recurved, 45–49 mm. long, 6.5–9 mm. wide. Petals similar, 40 mm. long, 5.5 mm. wide. Lip linear-lanceolate, long-acuminate, recurved, 40–42 mm. long, 5 mm. wide; spur incurved to sigmoid, 9.5–14.5 cm. long. Column 2.5 mm. long; rostellum 6.5 mm. long; viscidium bipartite. Fig. 141.

UGANDA. Masaka District: Lake Nabugabo, NW. side, 9 Oct. 1953, *Drummond & Hemsley* 4724! & 5 May 1969, *Lye* 2782! & Bugala, 24 Feb. 1945, *Greenway & Thomas* 7163!; Mengo District: Entebbe, 6 Sept. 1905, *E. Brown* 241!
DISTR. U 4; Sierra Leone to Cameroon, Gabon and Zaire
HAB. Epiphyte in forest; 1150–1250 m.

SYN. *Angraecum chailluanum* Hook. f. in Bot. Mag. 92, t. 5589 (1866)
Listrostachys chailluana (Hook. f.) Reichb.f. in Flora 68: 381 (1885)
Angorchis chailluana (Hook. f.) O. Kuntze, Rev. Gen. Pl. 2: 651 (1891)

2. **C. arcuata** (*Lindl.*) *Schltr.*, Die Orchideen: 596 (1914) & in Beih. Bot. Centr. 36: 129 (1918); Summerh. in K.B. 14: 146 (1960) & in F.W.T.A., ed. 2, 3: 267 (1968); J. Stewart in U.K.W.F.: 799 (1974); Williamson, Orch. SC. Afr.: 197 (1977); J. Stewart et al., Wild Orch. S. Afr.: 272 (1982); la Croix et al., Malawi Orch. 1: 63 (1983). Type: South Africa, Cape Province, Albany, *Drège* 4580 C (K, holo.!)

An epiphytic or lithophytic herb, often forming large masses; stems 15–30 cm. or more long. Leaves linear, narrowly oblong or narrowly elliptic, rarely oblanceolate, 12–22.5 cm. long, 1.6–3.6 cm. wide. Infloresences 6–20 cm. long, 5–14-flowered; peduncle 1.5–2 cm. long; bracts cymbiform, ovate, acute or subacute, 15–33 mm. long, 5–17 mm. apart. Flowers white with a greenish spur, sweetly scented; pedicel and ovary 20–40 mm. long. Sepals lanceolate, acuminate, recurved, 18–49 mm. long, 5–7 mm. wide. Petals similar, 14–27 mm. long, 4.5–6 mm. wide. Lip lanceolate, acuminate, recurved, 15–36 mm. long, 4–6 mm. wide; spur incurved to somewhat sigmoid, 30–70 mm. long. Column 2–2.5 mm. long; rostellum pendent, 4.5–5 mm. long; viscidium bipartite.

UGANDA. Kigezi District: Impenetrable Forest, Nyebeya, Oct. 1940, *Eggeling* 4139!; Busoga District: Jinja, 1 Jan. 1904, *E. Brown* 62!; Mengo District, without locality, Apr. 1939, *Chandler* 2787!
KENYA. Nandi District: Kaimosi, *Williams & Piers* 609!; Kericho District: Sotik, Feb. 1931, *Gardner* 2554!; Kwale District: Kaloleni, 1 Sept. 1959, *Verdcourt & Bayliss* 2413!
TANZANIA. Ngara, 20 June 1943, *Dobson* in *Moreau* 598!; Masai District: Loliondo, Ngusero Sambu, *Page-Jones* 678!; Rungwe District: Tukuyu, 9 May 1944, *Moreau* 832!
DISTR. **U** 1–4; **K** 1, 3–7; **T** 1–4, 6–8; **Z**; **P**; Sierra Leone to Zaire and Ethiopia and south to Zimbabwe and South Africa
HAB. On trees and rocks in bush, woodland and forest; sea-level to 2000 m.

SYN. *Angraecum arcuatum* Lindl. in Hook., Comp. Bot. Mag. 2: 204 (1837)
Listrostachys arcuata (Lindl.) Reichb.f. in Walp., Ann. Bot. 6: 907 (1864)
L. sedenii Reichb.f. in Gard. Chron., ser. 2, 9: 138 (1878). Type: E. Trop. Africa, cult. Veitch, *Seden* (W, holo.)
Angraecum sedenii (Reichb.f.) Nicholson, Dict. Gard. 1: 79 (1884)
Angorchis arcuata (Lindl.) O. Kuntze, Rev. Gen. Pl. 2: 651 (1891)
A. sedenii (Reichb.f.) O. Kuntze, Rev. Gen. Pl. 2: 652 (1891)
Cyrtorchis bracteata Schltr. in E.J. 53: 600 (1915). Type: Tanzania, near Kyimbila, *Stolz* 715 (B, holo.)
C. sedenii (Reichb.f.) Schltr. in Beih. Bot. Centr. 36: 132 (1918)
C. arcuata (Lindl.) Schltr. subsp. *variabilis* Summerh. in K.B. 14: 148 (1960). Type: Tanzania, Tanga District, Kiwanda, *Moreau* 20 (K, holo.!)

FIG. 141. *CYRTORCHIS CHAILLUANA*—**A**, habit, × ⅔; **B**, **C**, column and rostellum, side view and front view; **D**, anther-cap; **E**, pollinarium; **F**, **G**, viscidium from above and below. Drawn by Miss Stella Ross-Craig. Reproduced from Flora of West Tropical Africa.

NOTE. *C. arcuata* is an extremely variable species. Summerhayes (1960) revised it and distinguished three subspecies. The typical subspecies is geographically isolated from the others in southern Africa. Subsp. *variabilis* distinguished by its longer and broader leaves and larger flowers, as its name suggests is very variable in both leaf and floral morphology and size and is widespread in West Africa from Liberia to Zaire and south to Zambia and Malawi. Summerhayes assigned the majority of East African collections to this subspecies. According to him, subsp. *whytei* (Rolfe) Summerh., distinguished by its broad leaves and large flowers with a long spur, has a strangely disjunct distribution in West Africa, Malawi, where the type originates, and southern Tanzania. The West African plants are very like short-spurred forms of *C. chailluana*, while the Tanzanian specimens (*Moreau* 85 & 482) merge in flower size and spur length with some specimens he assigned to subsp. *variabilis*. Altogether, I have found Summerhayes' treatment very difficult to apply in East Africa, although it must be said that it does work in neighbouring Malawi.

On his own admission subsp. *variabilis* is difficult to distinguish from the typical variety and it seems best for the time being to forego the recognition of subspecies until a more detailed assessment can be made of the infraspecific variability of *C. arcuata*.

3. **C. neglecta** *Summerh.* in K.B. 14: 147 (1960). Type: Tanzania, Lushoto District, Kwamkoro, *Moreau* 27 (K, holo.!)

An erect epiphyte with a 20–40 cm. long, rarely branching stem. Leaves suberect, coriaceous, linear, unequally roundly bilobed at apex, 8–24 cm. long, 1.4–2.8 cm. wide. Inflorescences 3.5–9 cm. long, 4–10-flowered; peduncle 10 mm. long, covered by sheathing sterile bracts; bracts elliptic, obtuse, 5–11 mm. long, 4–10 mm. apart, brownish black. Flowers white with a salmon-coloured spur; pedicel and ovary 18–22 mm. long. Sepals lanceolate, acuminate, 12–19.5 mm. long, 4–7.6 mm. wide. Petals similar, 10–16 mm. long, 3.3–5.7 mm. wide. Lip recurved, lanceolate, acuminate, 12–16.5 mm. long, 5–6 mm. wide; spur ± straight, 3.5–5 cm. long. Column 2–2.5 mm.long; rostellum pendent, 4.5 mm. long; viscidium bipartite. Fig. 142.

TANZANIA. Lushoto District: E. Usambara Mts., Amani, 12 Jan. 1942, *Greenway* 6446! & 29 Mar. 1943, *Greenway* 6674! & W. Usambara Mts., Bumbuli, 17 Mar. 1941, *Moreau* 874!; Morogoro District: Turiani, 30 Mar. 1953, *Drummond & Hemsley* 1912!
DISTR. **T** 3, 6; Burundi
HAB. Epiphyte in riverine forest; 450–1750 m.

4. **C. seretii** (*De Wild.*) *Schltr.* in Beih. Bot. Centr. 36: 132 (1918). Type: Zaire, Suronga, *Seret* 661 (BR, holo.)

An epiphytic herb with pendent stems, 9–15 cm. long. Leaves up to 14, linear, unequally roundly or obtusely bilobed at apex, 9–17 cm. long, 9–16 mm. wide. Inflorescences 5.5–7 cm. long, 5–6-flowered; peduncle 1.5 cm. long, covered by sheaths; bracts ovate, subacute, 12–16 mm. long, 7–11 mm. apart. Flowers white with a greenish or salmon-coloured spur, sweetly fragrant; pedicel and ovary 26–28 mm. long. Sepals linear-lanceolate, acuminate, 19–31 mm. long, 4.3–6.75 mm. wide. Petals similar, 18.5–20 mm. long, 4–5 mm. wide. Lip lanceolate, acuminate, 17.5–26 mm. long; spur sigmoid, 5.5–6.3 cm. long. Column 2–3 mm. long; rostellum 6–7 mm. long; viscidium bipartite, hyaline part distinctly longer than the indurated part; anther-cap broadly beaked.

UGANDA. Bunyoro District: Bugomolo, Aug. 1935, *Eggeling* 5201!, & May 1943, *Eggeling* 5201! & 20 June 1944, *Eggeling* in *Moreau* 440! & Budongo, Biso area, May 1934, *Eggeling* 1994!
DISTR. **U** 2; Central African Republic, Zaire
HAB. In shade in primary forest; 1000–1500 m.

SYN. *Angraecum seretii* De Wild. in B.J.B.B. 5: 191 (1916)

5. **C. monteiroae** (*Reichb.f*) *Schltr.*, Die Orchideen: 596 (1914) & in Beih. Bot. Centr. 36: 132 (1918); Summerh. in K.B. 14: 153 (1960) & in F.W.T.A., ed. 2, 3: 267 (1968). Type: Angola, Bembi, *Monteiro* (W, holo., K, iso.!)

A pendent epiphyte with an elongate stem 20–80 cm. long. Leaves oblanceolate or narrowly elliptic, tapering to both ends, roundly unequally bilobed at the apex, 10–21 cm. long, 3–5 cm. wide, often rather crinkled. Inflorescences pendent, 18–32 cm. long, 10–20-flowered; peduncle slender, 5.5–8 cm. long; bracts broadly ovate, obtuse, 5–8 mm. long, 7–23 mm. apart. Flowers well spaced, white, tinged with orange-brown on the spur; pedicel and ovary 13–15 mm. long. Sepals lanceolate, acuminate, 15–18 mm. long, 4.5–5.3 mm. wide. Petals similar, 14–15 mm. long, 4–4.5 mm. wide. Lip lanceolate, acuminate, 14–15 mm. long, 5 mm. wide; spur slightly incurved, 3.5–4.5 cm. long. Column 2 mm. long; rostellum 4.5–5 mm. long; viscidium hyaline.

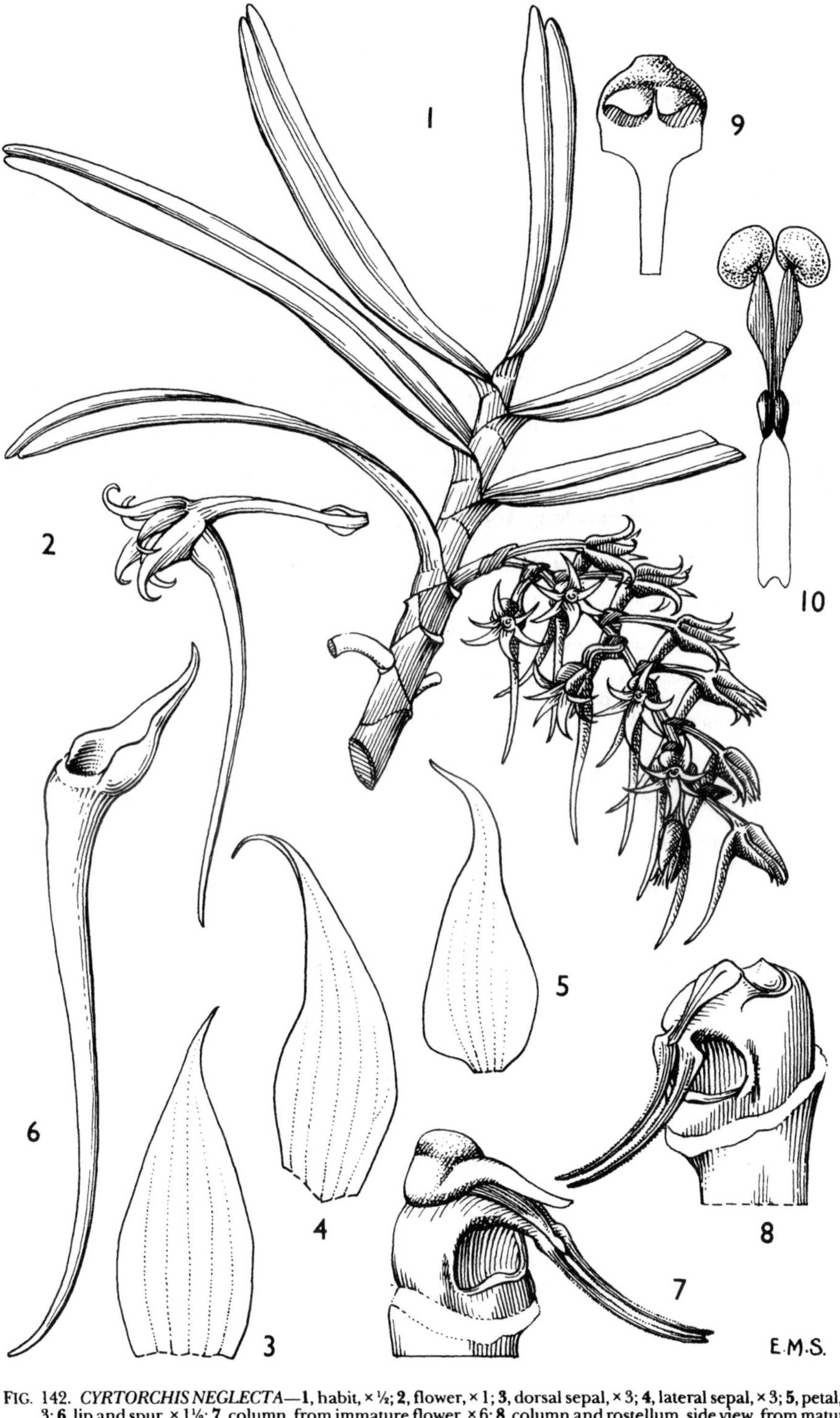

FIG. 142. *CYRTORCHIS NEGLECTA*—**1**, habit, × ½; **2**, flower, × 1; **3**, dorsal sepal, × 3; **4**, lateral sepal, × 3; **5**, petal, × 3; **6**, lip and spur, × 1½; **7**, column, from immature flower, × 6; **8**, column and rostellum, side view, from mature flower, × 6; **9**, anther-cap, × 6; **10**, pollinarium, × 6. Drawn from *Moreau* 27 by Miss Margaret Stones. Reproduced from Hooker's Icones Plantarum.

UGANDA. Masaka District: NW. side of Lake Nabugabo, 9 Oct. 1953, *Drummond & Hemsley* 4730!; Mengo District: Entebbe, June 1905, *E. Brown* 239! & 21 km. on Kampala to Entebbe road, Oct. 1932, *Eggeling* 942! & Mabira Forest, 8 Nov. 1938, *Loveridge* 22!
DISTR. U 4; Sierra Leone and Liberia to Zaire and south to Angola
HAB. Epiphyte in lakeside and riverine forest; 1150–1300 m.

SYN. *Listrostachys monteiroae* Reichb.f. in Linnaea 41: 76 (1877); Rolfe in F.T.A. 7: 156 (1897)
Angraecum antennatum Kraenzl. in E.J. 17: 61 (1893). Type: Cameroon, Barombi, *Preuss* 418 (K, iso.!)
Listrostachys droogmansiana De Wild. in Belg. Colon. 8: 425 (1902) & in Ann. Mus. Congo, Bot., sér. 5, 1: 24 (1903). Type: Zaire, Lubi rapids, *Gentil* 68 (K, record of holo.!)
Cyrtorchis droogmansiana (De Wild.) Schltr. in Beih. Bot. Centr. 36: 130 (1918)

6. **C. injoloensis** (*De Wild.*) *Schltr.* in Beih. Bot. Centr. 36: 131 (1918); Summerh. in K.B. 14: 153 (1960). Type: Zaire, Injolo, *Laurent* 1756 (BR, holo., K, record of holo.!)

An epiphyte with 6–12 cm. long pendent stems. Leaves 8–10, linear-oblong, unequally roundly bilobed at apex, 22–40 cm. long, 2.3–3 cm. wide. Inflorescences 22–25 cm. long, densely 15–22-flowered; peduncle 7.5–8.5 mm. long, covered by a sheath; bracts broadly ovate, obtuse, 8–11 mm. long, longer than the ovary, 5–15 mm. long. Flowers white; pedicel and ovary 6–7 mm. long. Sepals lanceolate, acute to acuminate, 13–15 mm. long, 3.9–4.5 mm. wide. Petals similar, 11–12 mm. long, 3–3.5 mm. wide. Lip lanceolate, acute, 11–12 mm. long, 4 mm. wide; spur straight or sigmoid, 2.5–3.2 cm. long. Column 1.5 mm. long; rostellum 3.5–4.3 mm. long; viscidium hyaline.

UGANDA. Mengo District: Entebbe, May 1906, *E. Brown* 383! & Old Entebbe road, Apr. 1962, *Leakey* 15!
TANZANIA. Pangani District: Mwembeni, Madanga, 27 Sept. 1956 (sterile), *R. Tanner* 3129!
DISTR. **U** 4; **T** 3; Zaire
HAB. Epiphyte in primary forest and in open forest patches; 30–1200 m.

SYN. *Angraecum injoloense* De Wild. in B.J.B.B. 5: 187 (1916)
A. ealaense De Wild. in B.J.B.B. 5: 185 (1916). Type: Zaire, Eala, *Laurent* 1784 (BR, holo., K, iso.!)
Cyrtorchis ealaensis (De Wild.) Schltr. in Beih. Bot. Centr. 36: 130 (1918)

7. **C. ringens** (*Reichb.f.*) *Summerh.* in K.B. 13: 87 (1958) & 14: 154 (1960) & in F.W.T.A., ed. 2, 3: 267 (1968). Type: Cameroon, without exact locality, cult. *Bouché* (W, holo.)

An epiphyte with arcuate or ascending stems up to 20 cm. long. Leaves linear, unequally roundly bilobed at apex, 7.5–12.5 cm. long, 1.2–2.5 cm. wide. Inflorescences 4–7 cm. long, densely 5–8-flowered; peduncle 1 cm. long or less, covered by sterile bracts; bracts ovate, obtuse, 3–8 mm. long, 4–7 mm. apart. Flowers white with an ochre-salmon spur, sweetly scented; pedicel and ovary 10–13 mm. long. Sepals lanceolate, acute or shortly acuminate, 9.5–11.6 mm. long, 3–5.5 mm. wide. Petals similar, 9–9.7 mm. long, 3.4–4 mm. wide. Lip lanceolate, acuminate, recurved, 9.5–10.5 mm. long, 3–4.5 mm. wide; spur incurved to straight, 2–3.3 cm. long. Column 1–1.5 mm. long; rostellum 3–5 mm. long; viscidium hyaline.

UGANDA. Kigezi District: Ihihizo R., 27 Jan. 1958, *Burgess* in *Piers* 300!
TANZANIA. Masai District: Talamai, Lolmongi Hill, 22 Feb. 1943, *Page-Jones* in *Moreau* 405! & 17 Apr. 1944, *Elliott* in *Moreau* 765!; Morogoro District: Nguru Mts., Ndege Mt., 30 Jan. 1933 (sterile), *Schlieben* 3346!
DISTR. **U** 2; **T** 2, 6; Sierra Leone to Zaire, Zambia, Malawi and Zimbabwe
HAB. In upland evergreen forest; 1200–1600 m.

SYN. *Listrostachys ringens* Reichb.f. in Gard. Chron., ser. 2, 10: 266 (1878)
Angraecum buchholzianum Kraenzl. in E.J. 7: 334 (1886). Type: Cameroon, Mungo, *Buchholz* (B, holo.)
A. bistortum Rolfe in K.B. 1893: 65 (1893). Type: Nigeria, Lagos, *Moloney* (K, holo.!)
Listrostachys buchholziana (Kraenzl.) Th. Dur. & Schinz, Consp. Fl. Afr. 5: 48 (1895)
L. hookeri Rolfe in F.T.A. 7: 154 (1897). Types: Cameroon, Cameroon Mt., *Mann* 2114 (K, syn.!) & *Johnston* 101 (K, syn.!)
L. bistorta (Rolfe) Rolfe in F.T.A. 7: 155 (1897)
L. ignoti Kraenzl. in E.J. 51: 395 (1914). Type: Cameroon, Cameroon Mt., coll. unknown (B, holo.†)
Cyrtorchis bistorta (Rolfe) Schltr. in Beih. Bot. Centr. 36: 129 (1918)
C. buchholziana (Kraenzl.) Schltr. in Beih. Bot. Centr. 36: 129 (1918)
C. hookeri (Rolfe) Schltr. in Beih. Bot. Centr. 36: 131 (1918)

8. **C. brownii** (*Rolfe*) *Schltr.* in Beih. Bot. Centr. 36: 129 (1918); Summerh. in K.B. 14: 155 (1960). Type: Uganda, Entebbe, *E. Brown* 248 (K, holo.!)

An epiphyte with stems 8–20 cm. long. Leaves very coriaceous, linear, unequally bilobed at apex, 7–8.5 cm. long, 7–8 mm. wide. Inflorescences 3–9 cm. long, densely up to 12-flowered; peduncle 2–2.5 cm. long; bracts elliptic, obtuse, 6–8 mm. long, 3–4 mm. apart. Flowers white; pedicel and ovary 3 mm. long, much shorter then the bract. Sepals lanceolate or elliptic-lanceolate, acuminate or acute, 10–13.5 mm. long, 2.4–3.3 mm. wide. Petals oblong-lanceolate, acuminate, 8.4–9.5 mm. long, 2.4–2.6 mm. wide. Lip lanceolate, acuminate, recurved, 8.5–10.5 mm. long, 2.5–3 mm. wide; spur straight, 2–2.5 cm. long. Column 1.3–1.5 mm. long; rostellum 2.5–3.2 mm. long; viscidium hyaline.

UGANDA. Kigezi District: Impenetrable Forest, Nyebeya, Oct. 1940, *Eggeling* 4176!; Mengo District: Entebbe, *E. Brown* 248!
TANZANIA. Tanga District: Sigi Segoma, 14 May 1946, *Shauri* in *Moreau* 737! & 737B!
DISTR. **U** 2, 4; **T** 3; Sierra Leone to Zaire and Malawi
HAB. Epiphyte in evergreen forest; up to 1500 m.

SYN. *Listrostachys brownii* Rolfe in K.B. 1906: 378 (1906)
Angraecum latibracteatum De Wild. in B.J.B.B. 5: 188 (1916). Type: Zaire, Eala, *Laurent* 1773 (BR, holo., K, iso.!)
Cyrtorchis latibracteata (De Wild.) Schltr. in Beih. Bot. Centr. 36: 131 (1918)

9. **C. praetermissa** *Summerh.* in K.B. 3: 279 (1948) & in 14: 155 (1960); J. Stewart in U.K.W.F.: 796 (1974); Williamson in Orch. SC. Afr.: 198 (1977); J. Stewart et al., Wild Orch. S. Afr.: 272 (1982); la Croix et al., Malawi Orch. 1: 64 (1983). Type: Zambia, Mwinilunga, *Milne-Redhead* 3146 (K, holo.!)

An erect or pendent epiphyte or lithophyte with stems up to 20 cm. long. Leaves coriaceous-fleshy, conduplicate, recurved, linear, unequally bilobed at apex, 4.5–10 cm. long, 8–12 mm. wide. Inflorescences 6–10.5 cm. long, 3–11-flowered; peduncle 8–12 mm. long; bracts ovate-triangular, obtuse, 4–5 mm. long, 3–5 mm. apart. Flowers white with a greenish spur; pedicel and ovary 9–11 mm. long, scabrid. Sepals lanceolate or oblong-lanceolate, acute or acuminate, 8–11 mm. long, 2.5–4.8 mm. wide. Petals oblong-lanceolate, acuminate, 8–8.5 mm. long, 2–3.4 mm. wide. Lip recurved, lanceolate, acuminate, 8.5–9.5 mm. long, 1.8–3 mm. wide; spur incurved, 1.8–2.8 cm. long. Column 1 mm. long; rostellum 3–4 mm. long; viscidium hyaline.

UGANDA. Bunyoro District: Budongo Forest, June 1943, *Eggeling* 5449!
KENYA. Masai District: Mara Bridge, 12 Feb. 1971, *Tait* 2!
TANZANIA. Lushoto District: Kisiwani, 29 May 1942, *Moreau* 66!; Ufipa District: Namwele escarpment, 28 Dec. 1961 (sterile), *Richards* 15793!; Kondoa District: Bukulu, 14 Jan. 1962, *Polhill & Paulo* 1172!; Njombe District, without precise locality, 29 Dec. 1931, *Lynes* DR169!
DISTR. **U** 2; **K** 6; **T** 3–5, 7; Zaire, Rwanda, Burundi, Zambia, Malawi, Zimbabwe and South Africa (Transvaal)
HAB. In woodland and on rocks; 400–1850 m.

10. **C. crassifolia** *Schltr.* in R.E. Fries, Wiss. Ergebn. Schwed. Rhod.-Kongo-Exped. 1911–12, 1: 25, t. 27 (1916); Summerh. in K.B. 14: 156 (1960); Williamson in Orch. SC. Afr.: 198 (1977); la Croix et al., Malawi Orch. 1: 65 (1983). Type: Zambia, near Mbala [Abercorn], *Fries* 1274 (B, holo., K, record of holo.!)

A dwarf epiphyte or lithophyte with a very short stem, up to 4 cm. long. Leaves very fleshy, conduplicate, recurved, oblong, unequally bilobed at apex, 1.5–4(–6) cm. long, 6–12 mm. wide, grey-green. Inflorescences 1.5–3 cm. long, 3–6-flowered; peduncle less than 1 cm. long; bracts papery, elliptic, obtuse, 3–4 mm. long. Flowers white, fading to cream, sweetly scented of hyacinths at night; pedicel and ovary 7–12 mm. long. Sepals oblong-lanceolate, acute, 7–11.5 mm. long, 2.5–3 mm. wide. Petals lanceolate, acuminate, 5.9–8.6 mm. long, 2.5–2.7 mm. wide. Lip obscurely 3-lobed, lanceolate, acuminate, recurved, 6.5–8.5 mm. long, 3.5–5.3 mm. wide; spur slightly incurved, 1.8–2.4 cm. long. Column 1–1.5 mm. long; rostellum 3.3 mm. long; viscidium hyaline.

TANZANIA. Ufipa District: Kate, 16 Aug. 1943, *Bell* in *Moreau* 649!; Mbeya/Chunya District: Unyiha, Ivula, 28 Nov. 1970, *Leedal* 529!; Rungwe District: Ipinda, *Leedal* 1353b!
DISTR. **T** 4, 7; Zambia, Malawi and Zimbabwe
HAB. Epiphyte in *Brachystegia* woodland and also on rocks in woodland; 1500–1800 m.

62. SOLENANGIS

Schltr. in Beih. Bot. Centr. 36: 133 (1918)

Small to large monopodial epiphytic herbs. Stems elongate, often scandent, either with or without leaves; roots produced all along the stems, elongate, sometimes climbing by the roots. Leaves if present distichous, fleshy or coriaceous, unequally bilobed at the apex, articulated to sheathing leaf-bases, sometimes twisted at the base to lie in one plane. Inflorescences 1–many, axillary, few–many-flowered, spreading. Flowers minute to medium-sized. Sepals and petals free, subsimilar. Lip entire to 3-lobed, usually very small, ecallose, spurred at the base; spur conical, cylindrical or clavate. Column short, fleshy, lacking a foot; rostellum short, 2-lobed; pollinia 2, ovoid; stipes 1, linear to oblanceolate; viscidium 1, linear to lunate or reniform, often large in relation to the stipes.

A small genus of 5 species in tropical Africa, Madagascar and the Mascarene Is.

Plant leafless; spur incurved, swollen at the apex . . . . 2. *S. aphylla*
Plant with leaves; spur conical or filiform:
Leaves less than 2 cm. long, less than 1 mm. wide; spur conical, 1.8–2 mm. long; lip entire . . . . 1. *S. conica*
Leaves 1.5–3 cm. long, more than 5 mm. wide; spur cylindrical, filiform, 6–7 cm. long; lip strongly 3-lobed 3. *S. wakefieldii*

1. **S. conica** (*Schltr.*) *L. Jonsson* in Bot. Notis. 132: 382 (1979). Type: Mozambique, near Beira, *Schlechter* (B, holo.)

A dwarf scandent epiphyte with stems 4–22 cm. long, 0.5 mm. in diameter; roots prominent, much longer than the leaves, 0.5–1 mm. in diameter, grey. Leaves conduplicate-iridiform, linear, acute, 13–16 mm. long, 0.5–1 mm. wide, articulated to a leaf-sheath at the base. Inflorescences 10–14 mm. long, 2–4-flowered; peduncle 7–9 mm. long; bracts amplexicaul, ovate-triangular, 0.3–0.5 mm. long. Flowers greenish white or white; pedicel and ovary 2 mm. long. Dorsal sepal elliptic, obtuse, 2.8–3 mm. long, 1.3–1.5 mm. wide; lateral sepals oblong-elliptic, obtuse, 2.8–3.2 mm. long, 1.3–1.4 mm. wide. Petals linear, subacute, 2.5–2.7 mm. long, 0.7–0.8 mm. wide. Lip concave, oblong, truncate, 3.2–3.5 mm. long, 1.5–1.8 mm. wide; spur conical, 1.8–2 mm. long. Column 0.8–1.2 mm. long; rostellum 2-lobed; viscidium broadly lunate; stipes oblanceolate. Fig. 143.

KENYA. Masai District: Ol Doinyo Orok, [Namanga Hill], 31 May 1971, *Archer* 683!
TANZANIA. Mbulu District: Nou Forest, no date given, *Moreau* 316!; Lushoto District: W. Usambara Mts., Shagai Forest, 15 May 1953, *Drummond & Hemsley* 2548! & Mazumbai, 4 June 1981, *L. Tanner* 263!; Songea District: Matengo Hills, 29 Feb. 1956, *Milne-Redhead & Taylor* 9166!
DISTR. **K** 6; **T** 2, 3, 8; Malawi, Zimbabwe and Guinea Bissau
HAB. A twig epiphyte in montane forest, on shrubs in moss and on *Podocarpus;* 1800–2250 m.

SYN. *Angraecum conicum* Schltr. in E.J. 38: 160 (1906)
Solenangis angustifolia Summerh. in K.B. 13: 280 (1958). Type: Zimbabwe, Umtali District, Engwa, *Wild* 4536 (K, holo.!, SRGH, iso.)

2. **S. aphylla** (*Thou.*) *Summerh.* in Bot. Mus. Leafl. Harv. Univ. 11: 159 (1943). Type: Mauritius, without exact locality, *Thouars* (P, holo.)

A scandent or ± pendent leafless epiphyte climbing by its hook-like roots; stems up to 40 cm. or more long, 2 mm. in diameter; roots elongate, all along the stem, branching. Inflorescences many, all along the stem, 1–2.5 cm. long, 8–16-flowered; peduncle 2–3 mm. long, brown; rhachis brown; bracts less than 1 mm. apart, triangular, 0.5 mm. long. Flowers white with rusty-red tipped sepals, faintly scented; pedicel and ovary 4–5 mm. long. Dorsal sepal elliptic, obtuse, 2.5–3 mm. long, 1.5–1.7 mm. wide; lateral sepals obliquely elliptic, obtuse, 2.5–3.2 mm. long, 1.5–1.8 mm. wide. Petals linear-oblong, acute or obtuse, 2.5–3 mm. long, 0.7–0.8 mm. wide. Lip obscurely 3-lobed, concave, 2.6–3 mm. long, 2 mm. wide; side lobes erect, obscure, rounded; mid-lobe ovate, obtuse; spur strongly incurved, swollen at the apex, 4–5 mm. long. Column 2 mm. long; rostellum 2-lobed, short; viscidium linear, small; stipes linear.

KENYA. Northern Frontier Province: Boni Forest, 2 Oct. 1947, *J. Adamson* in *Bally* 5896!; Kwale District: near Taru, 9 Sept. 1953, *Drummond & Hemsley* 4225!; Kilifi District: Arabuko, no date given, *R.M. Graham* 1637!

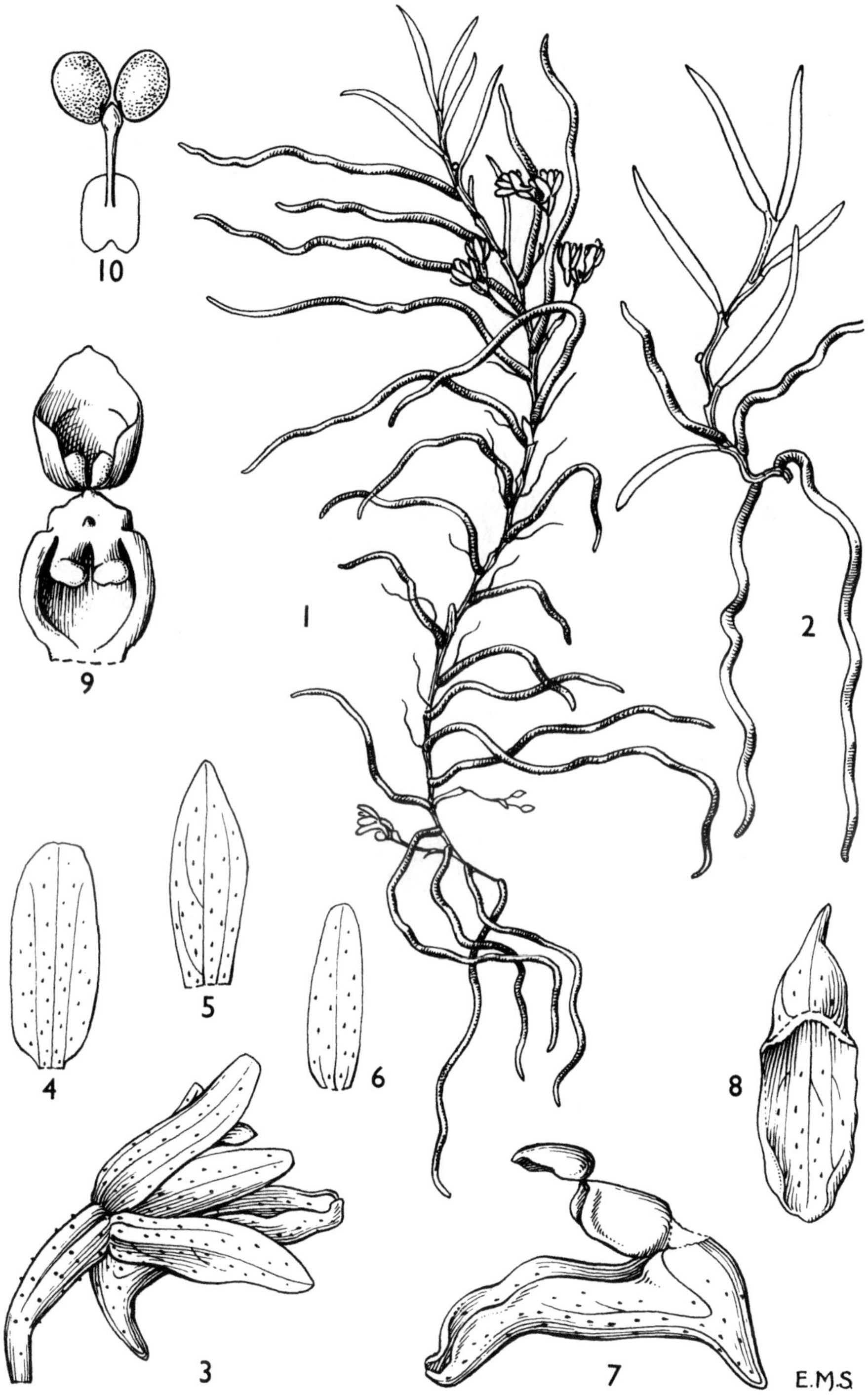

FIG. 143. *SOLENANGIS CONICA*—**1**, habit, × 1; **2**, leaves and roots, × 1; **3**, flower, × 10: **4**, dorsal sepal, × 10; **5**, lateral sepal, × 10; **6**, petal, × 10; **7**, lip, column, anther and spur, × 10; **8**, lip and spur from above, × 10; **9**, column and anther-cap, × 20, **10**, pollinarium, × 20. Drawn from *Drummond & Hemsley* 2548 by Miss Margaret Stones. Reproduced from Hooker's Icones Plantarum.

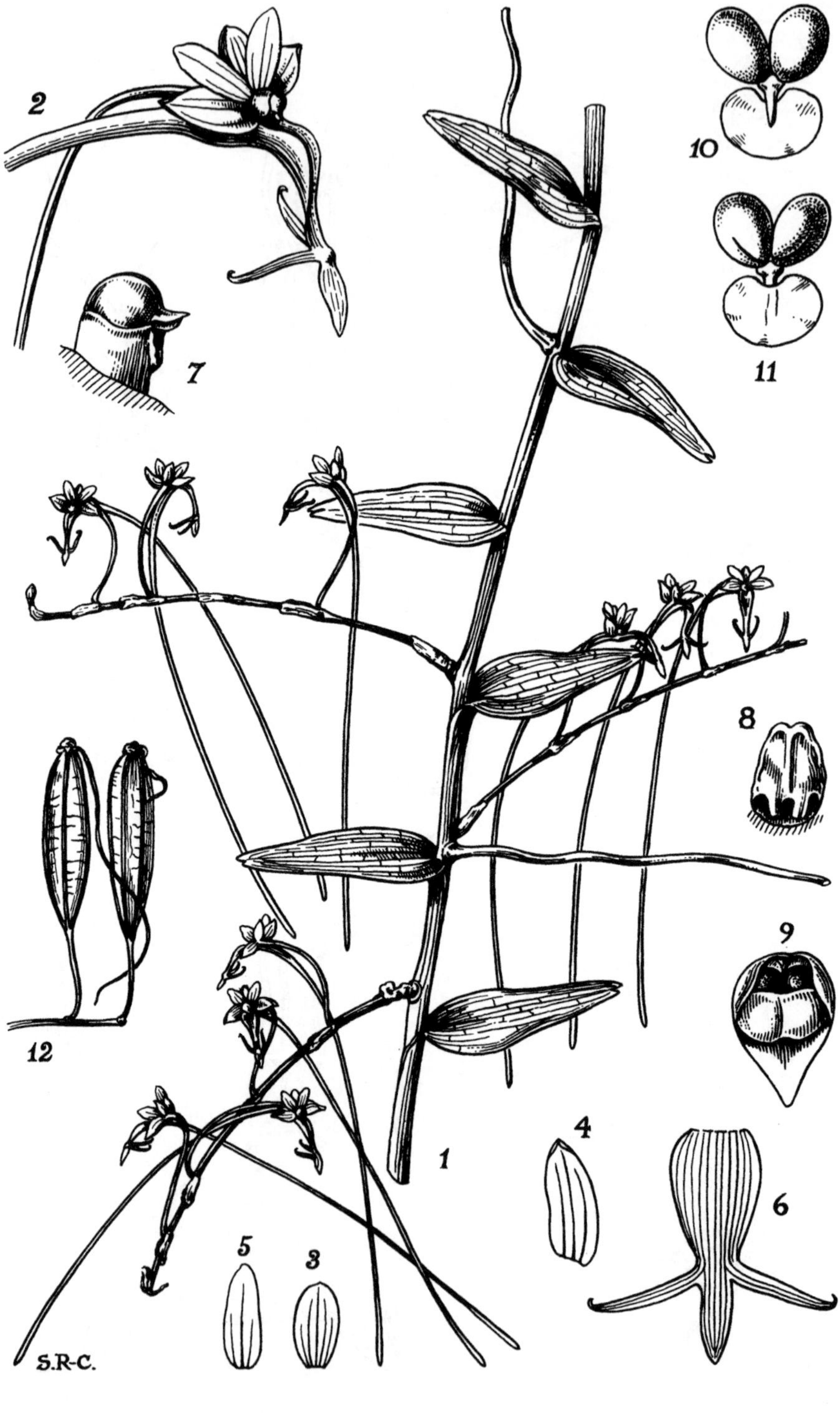

FIG. 144. *SOLENANGIS WAKEFIELDII*—**1**, habit, × 1; **2**, flower, × 4; **3**, dorsal sepal, × 4; **4**, lateral sepal, × 4; **5**, petal, × 4; **6**, lip, spread out, × 4; **7**, column, side view, × 10; **8**, column, with anther removed, × 10; **9**, anther-cap, with pollinia, × 16; **10**, **11**, pollinarium, front and back views, × 16; **12**, fruits, × 1. Drawn by Miss Stella Ross-Craig. Reproduced from Hooker's Icones Plantarum.

TANZANIA. Tanga District: Sigi, 29 Dec. 1940, *Moreau* 43/2! & Longuza, 12 Dec. 1942, *Greenway* in *Moreau* 43!; Morogoro District: Turiani, Nov. 1954, *Semsei* 1934!; Zanzibar I., Sept. 1896, *Kirk*!
DISTR. **K** 1, 7; **T** 3, 6; **Z**; Mozambique, Zimbabwe, Madagascar and Mascarene Is.
HAB. In thickets of *Acacia/Commiphora*, in scrub in riverine forest on *Barringtonia racemosa*; sea-level–900 m.

SYN. *Angraecum aphyllum* Thou., Orch. Iles Austr. Afr., t. 73 (1822)

3. **S. wakefieldii** (*Rolfe*) *Cribb & J. Stewart* in K.B. 40: 413 (1985). Type: Kenya, Mombasa, *Wakefield* (K, holo.!)

A scandent epiphyte climbing by its roots; stems up to 1 m. long, 1.5–3 mm. in diameter, glossy bright brown; roots numerous, all along the stem, flexuous, branching. Leaves 1–2.5 cm. apart, oblong-lanceolate, lanceolate or ovate-lanceolate, unequally acutely bilobed at the apex, 1.5–3 cm. long, 0.5–1.3 cm. wide, sap-green, twisted at the base and articulated to sheaths 1–1.2 cm. long. Inflorescences spreading, 4–6 cm. long, laxly 4–6-flowered; bracts amplexicaul, broadly ovate, obtuse or truncate, 3–5 mm. long. Flowers white, sweetly scented; pedicel and ovary 1–2 cm. long. Dorsal sepal oblong-elliptic, obtuse or apiculate, 3 mm. long, 1.5–2 mm. wide; lateral sepals obliquely oblong-elliptic, subacute, 4–4.5 mm. long, 1.5–1.7 mm. wide. Petals obliquely oblong-lanceolate, subacute or obtuse, 3.5 mm. long, 1.5 mm. wide. Lip 3-lobed in the apical half, narrowly clawed and concave at the base, 10 mm. long and wide; side lobes reflexed, spreading, linear, acute, 4.5–5 mm. long; mid-lobe lanceolate, acute, 3.5 mm. long, 1–1.2 mm. wide; spur filiform, pendent, 6–7 cm. long. Column 1 mm. long; rostellum bifid; viscidium large, reniform; stipes short. Pl. 7. Fig. 144.

KENYA. Mombasa, Nov. 1884, *Wakefield*!
TANZANIA. Uzaramo District: Msasani, May 1939, *Vaughan* 2805! & Msangasi R., 50 km. S. of Korogwe, by Segera–Chalinze road, Sept. 1976, *Wingfield* 2881A!; Zanzibar I., May 1933, *Vaughan* 2109!
DISTR. **K** 7; **T** 3, 6; **Z**; not known elsewhere
HAB. In coastal and lowland bushland; sea-level–300 m.

SYN. *Angraecum wakefieldii* Rolfe in F.T.A. 7: 146 (1897)
Tridactyle wakefieldii (Rolfe) Summerh. in K.B. 1936: 232 (1936) & in Hook., Ic. Pl. 35, t. 3445 (1943)

63. EURYCHONE*

Schltr. in Beih. Bot. Centr. 36: 134 (1918)

Epiphytic, entirely glabrous monopodial herbs. Stems abbreviated, simple. Leaves broadly oblong-elliptic, obliquely ovate-elliptic or liguliform, unequally bilobed at apex, spreading. Inflorescence 2–10(–12)-flowered, lax, axillary, pendulous, shorter than leaves; bracts small. Flowers white, green and purple-brown or rose, large, thin-textured, scented. Sepals and petals free, narrowly elliptic or oblong, acute, erect to spreading. Lip entire or subentire, spurred, infundibuliform, cucullate, emarginate, ecallose, spur conical below, constricted at the middle, inflated above and slightly to strongly geniculate. Column short and broad, erect, without a foot or stelidia; rostellum large, liguliform, truncate to emarginate, obscurely carinate; anther cucullate, retuse; pollinia 2, broadly ellipsoid, both attached by a long, narrowly linear stipes to a large, oblong or ovate viscidium.

A genus of 2 species in Tropical Africa.

E. rothschildiana (*O'Brien*) *Schltr.* in Beih. Bot. Centr. 36: 135 (1918); Summerh. in F.W.T.A. 2: 461 (1936); Moreau in Journ. E. Afr. Nat. Hist. Soc. 17: 22 (1943); Piers, Orch. E. Afr., ed. 2: 270 (1968); Summerh. in F.W.T.A., ed. 2, 3: 270 (1968). Type: Uganda, in the direction of the Victoria Nyanza, *H.B. Rattray*, cult. *W. Rothschild* (holo. not located)

An epiphytic herb. Stem covered in persistent leaf-sheaths, producing many long roots at the nodes, 2–8 cm. long. Leaves 3–8, broadly oblong-elliptic or obliquely ovate-elliptic,

*By J.J. Wood

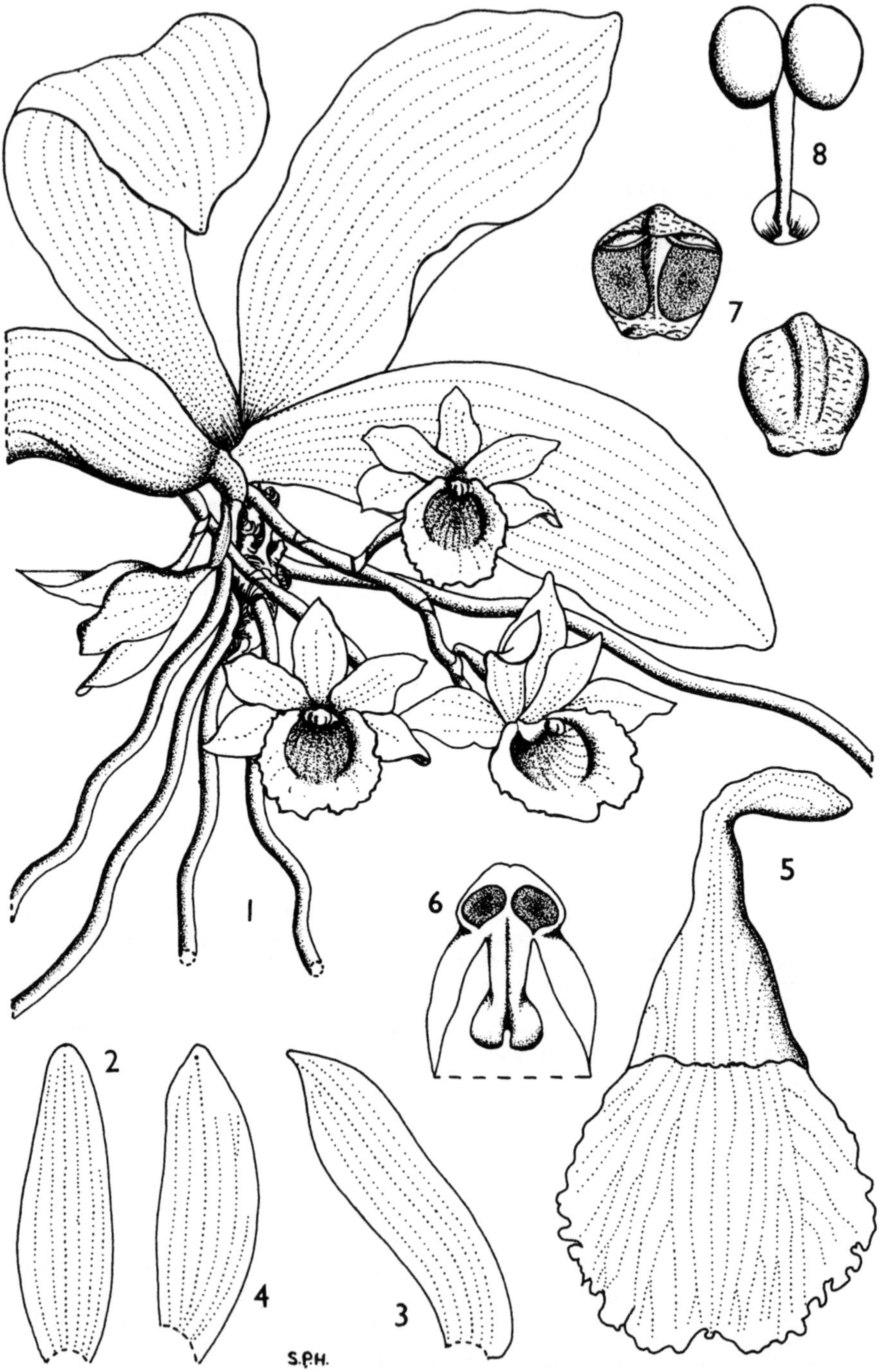

FIG. 145. *EURYCHONE ROTHSCHILDIANA*—**1**, habit, × ⅔; **2**, dorsal sepal, × 2; **3**, lateral sepal, × 2; **4**, petal × 2; **5**, lip and spur, × 2; **6**, column and rostellum, × 6; **7**, anther-cap, × 4; **8**, pollinarium, × 6. 1, from *Mason* 589 and *Schnell* 1428; 2–6, 8, from *Summerhayes* s.n.; 7, from *Segerbäck* 1233. Drawn by Susan Hillier.

unequally and subacutely bilobed at apex, margin undulate, thinly coriaceous, sheath 4–10 mm. long, lamina 6–21 cm. long, 1.5–7 cm. broad. Inflorescence laxly 2–6(–12)-flowered, deflexed or pendulous; peduncle and rhachis 3–9 cm. long; bracts triangular-ovate to ovate, obtuse, the upper closely appressed to the pedicellate ovary, 2–5 mm. long. Flowers scented of Lily-of-the-Valley (*Convallaria majalis*), with translucent white tepals, flushed and veined pale green; lip white with dark green disc and veins, with a chocolate-brown to purple area at the spur-entrance; spur greenish brown to reddish pink; column pale green; anther cream or pale greenish yellow; pedicellate ovary curved, 10–12 mm. long. Sepals and petals acute or apiculate. Dorsal sepal oblong-elliptic, 20–23(–25) mm. long, 5.5–7 mm. broad; laterals narrowly elliptic or oblong-elliptic, slightly curved, 22–25 mm. long, 6–7 mm. broad. Petals oblong or oblong-elliptic, 18–23 mm. long, 6.5–10 mm. broad. Lip ovate in outline, entire, infundibuliform, margin undulate, erose, retuse to emarginate, with a mucro in the sinus, 2–2.7 cm. long, 2–2.5 cm. broad; spur conical at base, constricted at middle, the distal portion abruptly dilated, laterally compressed and geniculate, obtuse, 2–2.5 cm. long. Column 8–10 mm. long, 6–7 mm. broad; anther 4 mm. long, 3 mm. broad. Pl. 7. Fig. 145.

UGANDA. Acholi District: Paimol, 11 May 1961, *Lind* 2997A!; Mengo District: Entebbe, 9 Aug. 1906, *E. Brown* 382! & Kajansi Forest, Mar. 1935, *Chandler* 1244!

DISTR. U 1, 4; Guinée, Sierra Leone, Liberia, Ivory Coast, Ghana, Nigeria, Bioko [Fernando Po], Zaire

HAB. Forest, in shade; 1100–1200 m.

SYN. *Angraecum rothschildianum* O'Brien in Gard. Chron., ser. 3, 34: 131 (1903)

64. ANCISTRORHYNCHUS

Finet in Mém. Soc. Bot. Fr. 9: 44 (1907); Schltr. in Beih. Bot. Centr. 36: 137 (1918); Summerh. in Bot. Mus. Leafl. Harv. Univ. 11: 203 (1944) & in F.W.T.A., ed. 2 3: 270 (1968)

Cephalangraecum Schltr. in Beih. Bot. Centr. 36: 135 (1918)

Small to medium-sized monopodial epiphytic herbs. Stems short, thick, covered by persistent sheathing leaf-bases; roots elongate, clustered towards the base of the stem. Leaves imbricate, suberect, spreading or recurved, ligulate or tapering to apex, ± unequally bilobed at the apex, lobes sometimes toothed, coriaceous or fleshy. Inflorescences axillary from the lower leaves, almost sessile, usually forming globose or ellipsoidal heads; bracts chartaceous to membranous, as long as the flowers. Flowers mostly white, often marked with green on the lip. Sepals and petals free, subsimilar. Lip oblong to orbicular, entire to 3-lobed, spurred at the base, lacking a callus; spur straight to sigmoid or geniculate, sometimes swollen at the apex and/or the mouth. Column short, fleshy; rostellum projecting down then abruptly geniculate in the middle and upcurved in the apical half, 2-lobed, each lobe falcate and acute; anther-cap hemisphaerical, produced in front into a short truncate appendage; pollinia 2, globose; stipites either 2 and subspathulate or 1 and bifid; viscidium 1, long and narrow.

A genus of about 16 species widespread in tropical Africa.

Spur ± S-shaped:
- Spur swollen in the basal half and at the apex, less than 10 mm. long:
  - Sepals and petals 12 mm. or more long; lip 13 mm. or more long; spur 9 mm. long; leaves rounded at the apex — 8. *A. laxiflorus*
  - Sepals and petals 5 mm. long or less; lip less than 7.5 mm. long; spur 6.5 mm. long or less; leaves acuminate or subacute at the apex:
    - Leaves subacute and bidentate on each lobe of the apex — 6. *A. recurvus*
    - Leaves acuminate at the apex . . . . — 9. *A. clandestinus*
- Spur slightly swollen at the apex only, 13–17 mm. long — 7. *A. refractus*

Spur ± straight or only slightly incurved:
- Spur more than 7 mm. long . . . . — 1. *A. capitatus*
- Spur less than 6 mm. long:

Spur globose or subglobose, 1–2.5 mm. across:
- Leaves evenly spaced along the stem, longest less than 4.5 cm. long . . . . 5. *A. parviflorus*
- Leaves clustered towards the apex of the stem, longest 5.5–7 cm. long . . . . 4. *A. tenuicaulis*

Spur 3–6 mm. long:
- Spur 4.5–5.8 mm. long; column 1.6–1.7 mm. long; leaves not tapering noticeably to the apex . . . . . . 2. *A. metteniae*
- Spur 3–4.3 mm. long; column 0.7–1.4 mm. long; leaves tapering to the apex . . . . 3. *A. ovatus*

1. **A. capitatus** (*Lindl.*) *Summerh.* in Bot. Mus. Leafl. Harv. Univ. 11: 205 (1944) & in F.W.T.A., ed. 2, 3: 272 (1968). Type: Nigeria, Brass R., *Barter* 1857 (K, holo.!)

A suberect to pendent epiphyte with short stems 1–10 cm. long; roots 1.5–2.5 mm. in diameter. Leaves 3–8, coriaceous, linear, unequally lobed at the apex, 16–32 cm. long, 1.4–2.1 cm. wide, each lobe 2–3-toothed, articulated at the base to leaf-bases 2–2.7 cm. long. Inflorescences capitate, densely many-flowered, up to 1.5 cm. long, 2.5 cm. across; bracts papery, oblong-lanceolate, acute, up to 1 cm. long. Flowers white with a green mark on the lip; pedicel and ovary 7–9 mm. long, scabrid-papillose. Dorsal sepal elliptic, obtuse, 5–5.5 mm. long, 2–2.3 mm. wide; lateral sepals oblong, obtuse, 5–5.6 mm. long, 2.5–2.8 mm. wide. Petals oblong, rounded, 4–5.5 mm. long, 2 mm. wide, erose. Lip concave, ovate, rounded and fleshy at the apex, 4.5–5.7 mm. long, 4.3–5 mm. wide, upcurved and undulate on the margin; spur slightly incurved, clavate, 7.4–8.5 mm. long. Column 2.5–12.8 mm. long; stipites 2, oblanceolate; viscidium 1, linear.

UGANDA. Bunyoro District: Budongo Forest, fl. cult. June 1944, *Eggeling* 5471!
DISTR. **U** 2; Sierra Leone to Zaire
HAB. In evergreen forest; 1100–1300 m.

SYN. *Angraecum capitatum* Lindl. in J.L.S. 6: 137 (1862)
Listrostachys capitata (Lindl.) Reichb.f. in Flora 48: 190 (1865)
Angorchis capitata (Lindl.) O. Kuntze, Rev. Gen. Pl. 2: 651 (1891)
Listrostachys gentilii De Wild., Not. Pl. Utiles Congo: 147 (1903). Type: Zaire, Bombe, *Gentil* (BR, holo.)
Cephalangraecum capitatum (Lindl.) Schltr. in Beih. Bot. Centr. 36: 136 (1918)
C. gentilii (De Wild.) Schltr. in Beih. Bot. Centr. 36: 136 (1918)

2. **A. metteniae** (*Kraenzl.*) *Summerh.* in Bot. Mus. Leafl. Harv. Univ. 11: 209 (1944) & in F.W.T.A., ed. 2, 3: 272 (1968). Type: Cameroon, without exact locality, *Braun* 19 (B, holo., K, iso.!)

A pendent epiphyte with a short stem 2–7 cm. long; roots branching, 2–3 mm. in diameter. Leaves linear, unequally bilobed at the apex, 16–25 cm. long, 1.1–1.6 cm. wide, each lobe 2-toothed, articulated at the base to leaf-sheaths 1–1.5 cm. long. Inflorescences densely many-flowered, 1.5–1.7 cm. long; bracts papery, ovate, acute, 6–10 mm. long, 3–5 mm. wide. Flowers white with a green mark on the lip; pedicel and ovary 4–5 mm. long, scabrid. Dorsal sepal elliptic, obtuse, 3.6–5.1 mm. long, 2.1–2.3 mm. wide; lateral sepals obliquely oblong, obtuse, 4–5.1 mm. long, 2 mm. wide. Petals elliptic-oblong, obtuse, 3.7–4.7 mm. long, 1.7–2.4 mm. wide. Lip broadly ovate-subcircular, obtuse, 3.5–4 mm. long, 4.3–4.5 mm. wide; spur clavate, straight, 3–4.3 mm. long. Column 1.6–1.7 mm. long; stipites 2, oblanceolate; viscidium linear.

UGANDA. Mengo District: near Entebbe, 1909, *Fyffe* 70!
TANZANIA. Lushoto District: E. Usambara Mts., Mlinga Mt., fl. cult. 26 Oct. 1942, *Moreau* 371!
DISTR. **U** 4; **T** 3; Sierra Leone to Nigeria and the Central African Republic and Ethiopia
HAB. In evergreen forest; 900–1300 m.

SYN. *Listrostachys metteniae* Kraenzl., Xenia Orch. 3: 122, t. 270 (1893)
L. braunii Th. Dur. & Schinz, Consp. Fl. Afr. 5: 48 (1895), *nom. superfl.* Type: based on same type as *L. metteniae*
Cephalangraecum metteniae (Kraenzl.) Schltr. in Beih. Bot. Centr. 36: 137 (1918)
C. braunii (Th. Dur. & Schinz) Summerh. in F.W.T.A. 2: 462 (1936)

3. **A. ovatus** *Summerh.* in Bot. Mus. Leafl. Harv. Univ. 11: 210 (1944) & in Hook., Ic. Pl. 35, t. 3461 (1947); Piers, Orch. E. Afr., ed. 2: 222 (1968). Type: Uganda, Bunyoro District, Budongo Forest, *Eggeling* 2048 (K, holo.!)

A pendent to ascending epiphyte with an arcuate stem up to 20 cm. long, ± 5 mm. in diameter; roots branching, 1–2 mm. in diameter. Leaves up to 12, linear, unequally bilobed at the apex, 7–19 cm. long, 0.5–1.4 cm. wide, each lobe subacute to obscurely 2-toothed, twisted at base and articulated to imbricate leaf-bases 1–2.5 cm. long. Inflorescences capitate, 1–2 cm. long and across; bracts ovate-lanceolate to ligulate, acute, 6–12 mm. long, 5 mm. wide. Flowers white; pedicel and ovary 6–8 mm. long. Dorsal sepal oblong-elliptic or obovate-elliptic, rounded, 3.3–4.4 mm. long, 1.3–2 mm. wide; lateral sepals oblique, elliptic-oblong, rounded, 3.5–4.8 mm. long, 1.3–2.1 mm. wide. Petals broadly elliptic-oblanceolate, rounded, 3.2–4.5 mm. long, 2.4–4.3 mm. wide. Lip concave, broadly ovate or ovate-orbicular, rounded, 3–4.8 mm. long, 2.4–4.3 mm. wide, with undulate margins; spur clavate, 4.6–5.8 mm. long. Column 0.8–1.4 mm. long; stipites 2, spathulate; viscidium ligulate.

UGANDA. Bunyoro District: Budongo Forest, June 1935, *Eggeling* 2048!, June 1937, *Eggeling* 3360! & May–June 1943 & May 1944, *Eggeling* 5184!
DISTR. **U** 2; Congo, Zaire
HAB. In evergreen forest; 1200–1600 m.

4. **A. tenuicaulis** *Summerh.* in K.B. 8: 590 (1954). Type: Uganda, Kigezi District, Ishasha Gorge, *Purseglove* 2034 (K, holo.!)

An epiphyte ascending or arcuate with stems 2–16 cm. long, 2 mm. in diameter; roots few-branched, 1–2 mm. in diameter. Leaves linear, unequally bilobed at the apex, 2–7.2 cm. long, 0.4–0.5 cm. wide, each lobe subacute or obscurely 2-toothed, articulated to imbricate longitudinally rugulose leaf-bases. Inflorescences capitate, 4–5 mm. long, densely many-flowered; bracts elliptic, obtuse, 2–3 mm. long. Flowers white; pedicels and ovary ± 3 mm. long. Dorsal sepal oblong-elliptic, obtuse or rounded, 1.5–3.5 mm. long, 0.9–1.3 mm. wide; lateral sepals obliquely elliptic-oblong, obtuse, 2–3.5 mm. long, 1–1.2 mm. wide. Petals elliptic, rounded, 1.7–3 mm. long, 0.7–1.4 mm. wide. Lip concave, ovate or broadly ovate, obtuse, 1.4–2.5 mm. long, 1.4–2.4 mm. wide; spur globose to shortly clavate, 1–2.1 mm. long. Column 0.6–0.7 mm. long; stipites 2, subspathulate; viscidium ligulate.

UGANDA. Bunyoro District: Budongo Forest, June 1944, *Eggeling* 5536!; Kigezi District: Ishasha Gorge, Apr. 1946, *Purseglove* 2034! & May 1950, *Purseglove* 3445!
TANZANIA. Lushoto District: E. Usambara Mts., Amani, Nov. 1903, *Warnecke* 500!; Morogoro District: Uluguru Mts., Bondwa, 25 Sept. 1932, *Schlieben* 3031A!
DISTR. **U** 2; **T** 3, 6; Zaire, Malawi
HAB. In evergreen forest; 900–1400 m.

5. **A. parviflorus** *Summerh.* in Bot. Mus. Leafl. Harv. Univ. 11: 212 (1944). Type: Tanzania, Lushoto District, Amani, *Moreau* 70 (K, holo.!)

A dwarf epiphyte with a slender stem 4–5 cm. long, 3 mm. in diameter; roots slender, less than 1 mm. in diameter. Leaves up to 10, ligulate, unequally bilobed at the apex, 3.5–4.4 cm. long, 0.6–0.7 cm. wide, dark green, twisted at the base and articulated to leaf-bases ± 1 cm. long. Inflorescences capitate, 6 mm. long and across; bracts orbicular, rounded, 3.3–3.7 mm. long, 2.9–3 mm. wide. Flowers white with an almost transparent spur; pedicel and ovary 2.5–3 mm. long. Dorsal sepal elliptic, rounded, 2.5–2.6 mm. long, 1.3–1.5 mm. wide; lateral sepals elliptic, rounded, 2.5–2.7 mm. long, 1.3–1.5 mm. wide. Petals obovate, rounded, 2.5–2.6 mm. long, 1.3–1.5 mm. wide. Lip concave, broadly ovate, obscurely 3-lobed, 1.7–1.8 mm. long, 2.22—2.4 mm. wide; spur globose, 1.2–1.3 mm. in diameter. Column 0.9 mm. long; stipites 2, spathulate; viscidium ligulate.

TANZANIA. Lushoto District: E. Usambara Mts., Amani, 24 Jan. 1941, *Moreau* 70!
DISTR. **T** 3; known only from the type collection
HAB. In evergreen forest; 850–950 m.

6. **A. recurvus** *Finet* in Mém. Soc. Bot. Fr. 9: 46 (1907); Schltr. in Beih. Bot. Centr. 36: 138 (1918); Summerh. in Bot. Mus. Leafl. Harv. Univ. 11: 213 (1944) & in F.W.T.A., ed. 2, 3: 272 (1968). Types: Gabon, Fernan Vaz, *Lecomte* (P, syn., K, record of syn.!), Lamberéné, *Thollon* (P, syn.) & Guinée, Timbo to Conakry, *Pobéguin* 788 (P, syn.)

An epiphyte with a short stem 1–6 cm. long; roots 1.5–2 mm. in diameter. Leaves linear, unequally bilobed at the apex, 14–38 cm. long, 0.8–2 cm. wide, each lobe subacute to obscurely 2-toothed, articulated to imbricate leaf-bases 1–2 cm. long. Inflorescences capitate, 1.5–3 cm. long; bracts 2–3 mm. long. Flowers white, covered on outer surface with scurfy hairs; pedicel and ovary 5–5.5 mm. long. Dorsal sepal elliptic-ovate, rounded, 3.3–3.8 mm. long, 1.8–1.9 mm. wide; lateral sepals obliquely elliptic, obtuse, 4.2–4.3 mm. long, 2.1–2.2 mm. wide. Petals oblong, rounded, 3.6–3.7 mm. long, 1.2–1.3 mm. wide. Lip 3-lobed, 4.5–7.5 mm. long, 3–4.5 mm. wide; side lobes obscure, rounded; mid-lobe recurved, ovate, acute; spur swollen at the mouth and at the apex, geniculate in the basal half and recurved towards the apex, 4.75–6 mm. long. Column 2–2.3 mm. long; stipes oblanceolate, cleft at the apex; viscidium elongate, linear.

UGANDA. Bunyoro District: Budongo Forest, Siba area, fl. cult. Apr. 1944, *Eggeling* 5503!
DISTR. **U** 2; Sierra Leone to Gabon
HAB. In rain-forest; 1100–1300 m.

7. **A. refractus** (*Kraenzl.*) *Summerh.* in Bot. Mus. Leafl. Harv. Univ. 11: 213 (1944). Type: Tanzania, Usambara Mts., Derema, *Scheffler* 123 (B, holo., K, record of holo.!)

An epiphyte with an erect very short stem up to 4 cm. long; roots branching, 1.5–3 mm. in diameter. Leaves 6–8, fleshy-coriaceous, conduplicate, linear, unequally bilobed at the apex, 6–40 cm. long, 0.8–1.2 cm. wide, articulated to imbricate leaf-bases, twisted towards the apex; leaf-bases 2–3 cm. long, acuminate at apex on each side. Inflorescences clustered, 3–3.5 cm. long; bracts ovate, acute, 7–8 mm. long. Flowers shining white; pedicel and ovary 12–14 mm. long. Dorsal sepal narrowly oblong, obtuse, 10–12 mm. long, 4 mm. wide; lateral sepals obliquely oblong, rounded, 13–14 mm. long, 4 mm. wide. Petals narrowly elliptic-oblong, obtuse, 12–13 mm. long, 3–4 mm. wide. Lip concave, obscurely 3-lobed, ovate, obtuse, 15–17 mm. long, 7 mm. wide; spur S-shaped, dilated at the mouth and towards the apex, 13–17 mm. long. Column 7–8 mm. long; stipes bifid in apical half; viscidium very long, ligulate. Pl. 8. Fig.146/1–12.

TANZANIA. Lushoto District: W. Usambara Mts., Shagayu Forest, May 1953, *Procter* 245! & Mazumbai, 10 May 1976, *L. Tanner* 66! & E. Usambara Mts., Amani, 20 Feb. 1942, *Moreau* 38! & 2 km. from Amani on Kilosi road, 18 Jan. 1976, *Cribb & Grey-Wilson* 10278!
DISTR. **T** 3; not known elsewhere
HAB. In montane evergreen forest; 900–2100 m.

SYN. *Listrostachys refracta* Kraenzl. in E.J. 33: 74 (1902)
Cyrtorchis refracta (Kraenzl.) Schltr. in Beih. Bot. Centr. 36: 132 (1918)

8. **A. laxiflorus** *Mansf.* in N.B.G.B. 12: 705 (1935). Type: Tanzania, Uluguru Mts., Mindu Mt., *Schlieben* 3731 (B, holo., K, record of holo.!)

An erect to spreading epiphyte with a short stem 2–4 cm. long; roots very long, little branching, 1.5–3 mm. in diameter. Leaves 3–10, coriaceous, linear, unequally roundly bilobed at the apex, 13.5–35 cm. long, 0.6–0.8 cm. wide, articulated at the base to acutely angled imbricate leaf-bases 1.5–2 cm. long. Inflorescences densely 3–6-flowered, up to 3.5 cm. long; bracts papery, elliptic, obtuse, 8–14 mm. long, 6–7 mm. wide. Flowers white with a green mark on the lip; pedicel and ovary 10–15 mm. long, slightly furfuraceous. Dorsal sepal lanceolate, acute, 15–17 mm. long, 5–6 mm. wide; lateral sepals lanceolate to elliptic-lanceolate, acute or obtuse, 12.5–16 mm. long, 4.5–6 mm. wide. Petals oblong, obtuse, 12–16 mm. long, 4–5 mm. wide. Lip ovate, subacute to obtuse, 13–16 mm. long, 8–9 mm. wide; spur inflated in the basal half from a broad mouth, geniculate towards the apex, recurved and swollen at the apex, 9 mm. long. Column 6–7 mm. long; stipes divided in upper part; viscidium lanceolate, very large, 8 mm. long. Fig. 146/13–23.

TANZANIA. Morogoro District: Uluguru Mts., Mindu Mt., 4 Apr. 1933, *Schlieben* 3731!; Ulanga District: Sali, Muhulu Forest, 24 Jan. 1979, *Cribb, Grey-Wilson & Mwasumbi* 11178!; Iringa District: E. Muhanga, Uluti, 2 Nov. 1980, *Leedal* 6147!
DISTR. **T** 6, 7; not known elsewhere
HAB. In montane forest; 1200–1600 m.

9. **A. clandestinus** (*Lindl.*) *Schltr.* in Beih. Bot. Centr. 36: 138 (1918); Summerh. in Bot. Mus. Leafl. Harv. Univ. 11: 206 (1944) & in F.W.T.A., ed. 2, 3: 272 (1968). Type: Sierra Leone, without exact locality, cult. *Loddiges* (K, holo.!)

An erect or pendent epiphyte with a stem up to 7 cm. long; roots occasionally branching, 1.5–3 mm. in diameter. Leaves 5–10 in a fan, linear or tapering, acuminate at the unequally bilobed apex, 15–100 cm. long, 0.7–0.9 cm. wide, articulated to elongated

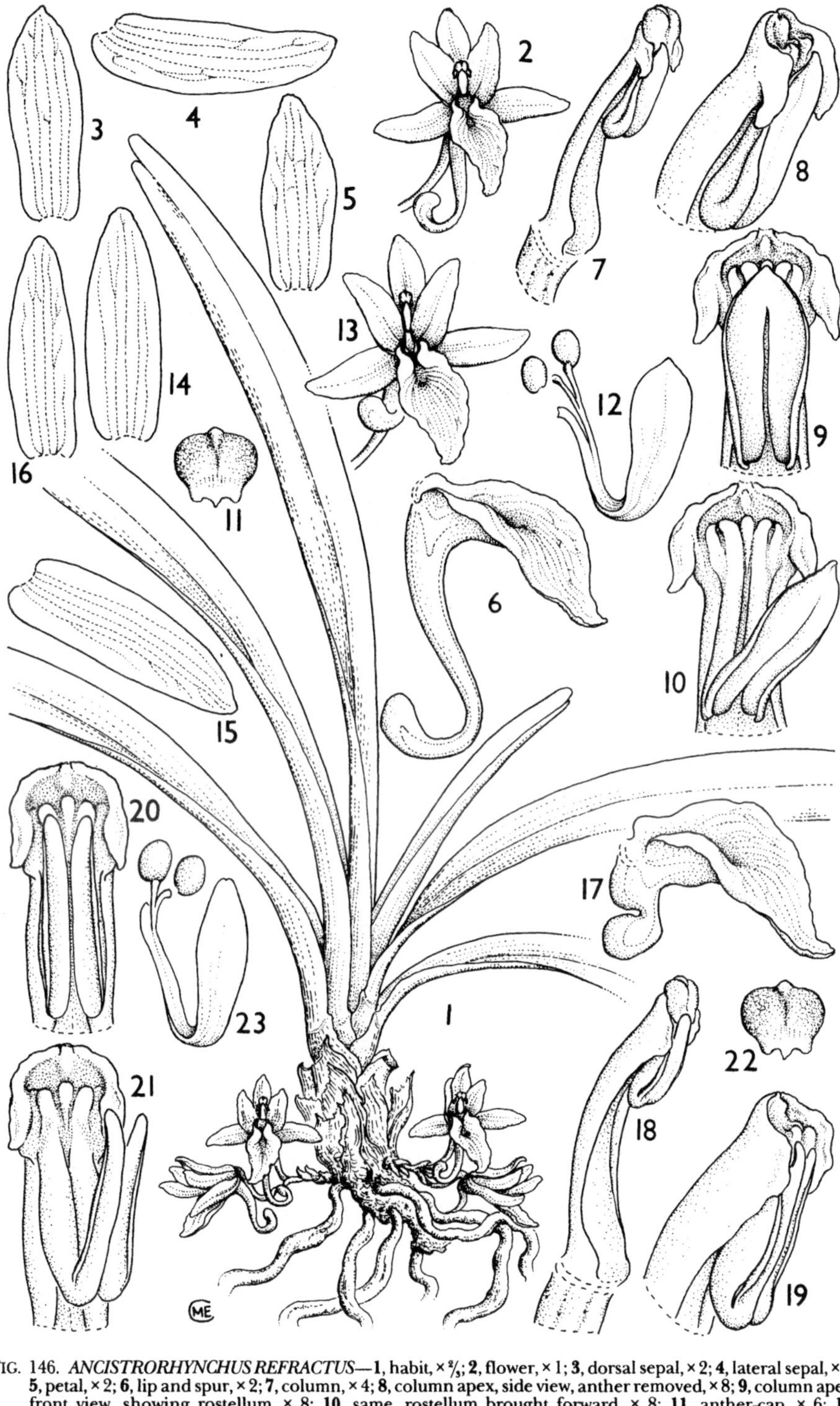

FIG. 146. *ANCISTRORHYNCHUS REFRACTUS*—**1**, habit, × ⅔; **2**, flower, × 1; **3**, dorsal sepal, × 2; **4**, lateral sepal, × 2; **5**, petal, × 2; **6**, lip and spur, × 2; **7**, column, × 4; **8**, column apex, side view, anther removed, × 8; **9**, column apex, front view, showing rostellum, × 8; **10**, same, rostellum brought forward, × 8; **11**, anther-cap, × 6; **12**, pollinarium, × 6. *A. LAXIFLORUS*—**13**, flower, × 1; **14**, dorsal sepal, × 2; **15**, lateral sepal, × 2; **16**, petal, × 2; **17**, lip and spur, × 2; **18**, column, × 4; **19**, column apex, side view, viscidium removed, × 8; **20**, column apex, front view, × 8; **21**, same, rostellum brought forward, × 8; **22**, anther-cap, × 6; **23**, pollinarium, × 6. 1, from *Procter* 245; 2–12, from *Moreau* 38; 13–23, from *Leedal* 6147. Drawn by Mrs M.E. Church.

imbricate leaf-bases 2–2.5 cm. long. Inflorescences capitate, densely many-flowered, up to 2 cm. long; bracts elliptic-lanceolate, acute, 2–3 mm. long. Flowers white; pedicel and ovary 7–9 mm. long. Dorsal sepal elliptic or oblong-elliptic, obtuse, 2–3.8 mm. long, 1.2–2 mm. wide; lateral sepals obliquely elliptic, obtuse, 2.5–4.7 mm. long, 1.3–2.6 mm. wide. Petals obliquely oblong-elliptic, obtuse, 2–4.1 mm. long, 0.6–1.4 mm. wide. Lip obscurely concave, 3-lobed, 4.5–6.2 mm. long, 3.4 mm. wide; side lobes obscure, elliptic, rounded; mid-lobe ovate, obtuse, with undulate margins; spur S-shaped, swollen at the apex and base, 3.2–6.2 mm. long. Column 1.5–2.4 mm. long; stipes 1, oblanceolate, bifid at apex; viscidium 1, lanceolate.

UGANDA. Ankole District: Lake Lutoto, fl. cult. Mar. 1948, *A.S. Thomas* 5773!
DISTR. **U** 2; Sierra Leone to Cameroon, Gabon and Zaire
HAB. In evergreen forest; 900–1100 m.

SYN. *Angraecum clandestinum* Lindl. in Hook., Comp. Bot. Mag. 2: 206 (1837)
Angorchis clandestina (Lindl.) O. Kuntze, Rev. Gen. Pl. 2: 651 (1891)
Listrostachys durandianus Kraenzl. in B.S.B.B. 38: 65 (1899). Type: Zaire, Pioka, *Laurent* (BR, holo.)
Angraecum clandestinum Lindl. var. *stenophyllum* Schltr., Westafr.-Kautsch.-Exped.: 283 (1900). Type: Cameroon, near Buea, *Schlechter* 12843 (B, holo.)
A. brunneo-maculatum Rendle, Cat. Talb. Nig. Pl.: 105, 146 (1913). Types: Nigeria, Oban, *Talbot* 943, 3744 & s.n. (BM, syn.!)
Ancistrorhynchus brunneo-maculatus (Rendle) Schltr. in Beih. Bot. Centr. 36: 138 (1918)
A. durandianus (Kraenzle.) Schltr. in Beih. Bot. Centr. 36: 138 (1918)
A. stenophyllus (Schltr.) Schltr. in Beih. Bot. Centr. 36: 139 (1918)

65. SPHYRARHYNCHUS

Mansf. in N.B.G.B. 12: 706 (1935)

Dwarf epiphytic monopodial herbs. Stems very short, covered by sheathing leaf-bases. Leaves distichous, fleshy-coriaceous, linear-elliptic. Inflorescences 1–many, shorter than to longer than the leaves, densely few-flowered. Flowers large for size of plant, white with a green mark on the lip. Sepals and petals free, subsimilar. Lip deflexed, entire, spurred at the base; spur pendent, clavate. Column short, fleshy; rostellum porrect, hammer-shaped, 2-lobed; pollinia 2, globose; stipites 2, small, very short, filiform; viscidium obovate, large.

A monotypic genus known only from Tanzania.

S. schliebenii *Mansf.* in N.B.G.B. 12: 705 (1935). Type: Tanzania, Morogoro District, Uluguru Mts., *Schlieben* 2892 (B, holo.)

A dwarf spreading to pendent epiphyte with a very short stem 0.3–0.6 cm. long; roots conspicuous, dorso-ventrally flattened, 2 mm. wide. Leaves 3–5, fleshy, linear, tapering to an unequally obtusely bilobed apex, 2–3.3 mm. long, 0.3–0.35 mm. wide, articulated at the base. Inflorescences 1–many, 1–3 cm. long, 6–10-flowered; peduncle 8–10 mm. long; bracts amplexicaul, triangular, subacute, 1–1.5 mm. long. Flowers rather variable in size, glistening white with a green mark on the lip; pedicel and ovary 5–6 mm. long. Dorsal sepal elliptic or lanceolate, acute or apiculate, (2.9–)5.5–14 mm. long, 1.5–4 mm. wide; lateral sepals obliquely oblong to lanceolate, acute or obtuse, (3–)5.5–14 mm. long, (1.3–)2.5–4 mm. wide. Petals oblong, obtuse, (2.7–)6–14 mm. long, (1.35–)3–3.5 mm. wide. Lip pendent, oblong or oblanceolate, subacute to obtuse, (2.6–)4–6 mm. long, 1.4–4 mm. wide; spur clavate, (1.5–)5.5–8 mm. long. Column 0.6 mm. long. Pl. 8. Fig. 147.

TANZANIA. Arusha District: Ngurdoto Crater, E. of Leopard Point, 27 Oct. 1965, *Greenway & Kanuri* 12212!; Lushoto District: E. Usambara Mts., Amani, 9 Oct. 1942, *Moreau* 96!; Morogoro District: Ukaguru Mts., 1 km. from Mandege Forest Station on Kilosa road, 10 Aug. 1972, *Mabberley* 1442!; Iringa District: Mufindi, 12 Nov. 1979, *Leedal* 5741!
DISTR. **T** 2, 3, 6, 7; not known elsewhere
HAB. In montane evergreen forest; 900–1600 m.

NOTE. The specimen from Mufindi (*Leedal* 5471) has very much smaller flowers than the others. However, because of the variable flower size found in *S. schliebenii*, this specimen is included here. Further collections are needed before a decision can be made on the status of the Mufindi specimen.

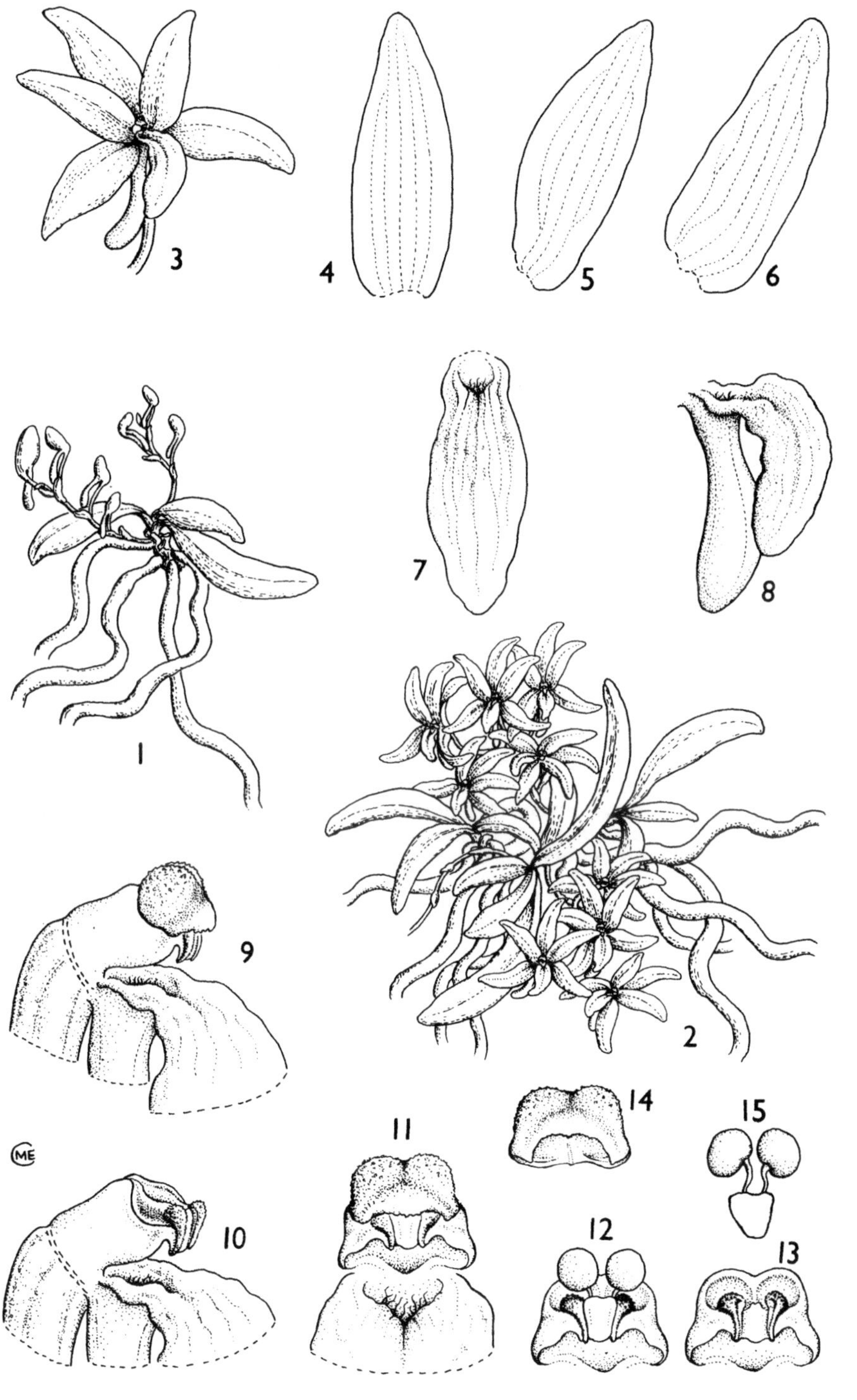

FIG. 147. *SPHYRARHYNCHUS SCHLIEBENII*—**1**, **2**, habit, × 1; **3**, flower, × 2; **4**, dorsal sepal, × 4; **5**, lateral sepal, × 4; **6**, petal, × 4; **7**, lip, front view, × 4; **8**, lip and spur, side view, × 4; **9**, column with anther-cap, side view, × 12; **10**, same, anther-cap removed, × 12; **11**, column and anther-cap, front view, × 12; **12**, same, anther-cap removed to show pollinarium in position, × 12; **13**, column and rostellum, front view, × 12, **14**, anther-cap, × 12; **15**, pollinarium, × 12. 1, 2, from living material cult. R.B.G. Kew; 3–15, from *Danahy* s.n. Drawn by Mrs M.E. Church.

66. MYSTACIDIUM

Lindl. in Hook., Comp. Bot. Mag. 2: 206(1836); Rolfe in Fl. Cap. 5(3): 75 (1912); Senghas in Schltr., Die Orchideen, ed. 3: 1120 (1986)

Dwarf or small monopodial epiphytic herbs. Stems short, covered by imbricate sheathing leaf-bases. Leaves distichous, fleshy or coriaceous, articulated to sheathing leaf-bases. Inflorescences axillary, 2–several-flowered; peduncles slender, wiry. Flowers small to medium-sized, white, pale green or pale yellow. Sepals and petals free, subequal, spreading or subconnivent. Lip entire or 3-lobed, spurred at the base, usually ecallose; spur tapering from a broad mouth. Column short; rostellum pendent, 3-lobed, with outer lobes much longer than the mid-lobe and papillate; pollinia 2, globose, sulcate; stipites 2, linear, slender; viscidia 2, rarely adherent, oblong or squamiform.

A small genus of about 13 species mainly in southern and south-central Africa, but with three or possibly four species in Tanzania.

1. Flowers yellow- or whitish green; lip obscurely 3-lobed, lanceolate, with a narrower mouth to the spur; sepals and petals linear-lanceolate . . . . 2. *M. tanganyikense*
   Flowers white; sepals and petals elliptic to ovate 2
2. Lip ecallose, broadly cordate with a broad mouth to the spur; spur 15–17 mm. long . . . . 1. *M. pulchellum*
   Lip with a longitudinal raised callus, ovate, obscurely 3-lobed, longer than broad, shortly apiculate; spur 20 mm. long 3. *M. nguruense*

1. **M. pulchellum** (*Kraenzl.*) *Schltr.* in Beih. Bot. Centr. 36: 126 (1918). Type: Tanzania, Iringa District, Uhehe, *Goetze* 573 (B, holo.)

A dwarf epiphyte with a very short stem 10 mm. long; roots many, clustered, fine, less than 1 mm. in diameter. Leaves 3–5, rather fleshy, obovate, unequally roundly bilobed at the apex, 1.3–5 cm. long, 0.4–1 cm. wide, articulated at the base to short leaf-bases. Inflorescences pendent, 1–several, 4–7-flowered; peduncle very fine, wiry, terete; rhachis zigzag, very fine; bracts ± 6 mm. apart, amplexicaul, ovate, subacute, 1 mm. long. Flowers white with a green anther and glaucous ovary; pedicel and ovary slender, straight, 12–18 mm. long. Dorsal sepal elliptic, obtuse, 6–7 mm. long, 2.5–3.7 mm. wide; lateral sepals spreading, elliptic, obtuse, 6–7.5 mm. long, 3–3.5 mm. wide. Petals subporrect, obliquely ovate, subacute, 5.5–7 mm. long, 3.5–5.1 mm. wide. Lip deeply concave, transversely cordate, obtuse, 4.5–5.5 mm. long, 6–8 mm. wide; spur tapering from a broad mouth, slightly dilated at the apex, 12–15 mm. long. Column 0.7–1 mm. long; rostellum trifid, papillose, pendent, mid-lobe shorter than the outer lobes; stipites 2, linear; viscidia 2, elliptic, adherent.

TANZANIA. Iringa District: Dabaga Highlands, Kibengu, 13 Feb. 1962, *Polhill & Paulo* 1450! & Mufindi, fl. Kew 21 Aug. 1986, *Keeley* & Uzungwa Mts., Udekwa, Dec. 1981, *Rodgers & Hall* 1455!
DISTR. **T** 7; not known elsewhere
HAB. In upland rain-forest; 1700–2400 m.

SYN. *Listrostachys pulchella* Kraenzl. in E.J. 28: 374 (1900)

2. **M. tanganyikense** *Summerh.* in Bot. Mus. Leafl. Harv. Univ. 12: 113 (1945) & in Hook., Ic. Pl. 35, t. 3489 (1950). Type: Tanzania, Arusha District, Lake Duluti, *Moreau* 317 (K, holo.)

A dwarf epiphyte with a very short stem 1–1.5 mm. long; roots numerous, flexuous, slender, 0.5–1 mm. in diameter. Leaves 4–5, ± falcate, ligulate or elliptic-oblanceolate, unequally roundly or obtusely bilobed, 1–6 cm. long, 0.3–0.9 cm. wide, articulated at the base. Inflorescences 1–several, pendent, 1.5–10 cm. long, 1–12-flowered; peduncle and rhachis very slender, wiry; bracts 3–7 mm. apart, triangular, obtuse, less than 1 mm. long. Flowers whitish to yellowish green, nocturnally scented; pedicel and ovary very slender, 5–12 mm. long. Dorsal sepal recurved, lanceolate, acuminate, 5–7.8 mm. long, 1.5–2 mm. wide; lateral sepals recurved, obliquely lanceolate-ligulate, acute, 6–8.5 mm. long, 1.2–1.8 mm. wide. Petals obliquely lanceolate, acute, 4.7–7.3 mm. long, 1.5–2.5 mm. wide. Lip

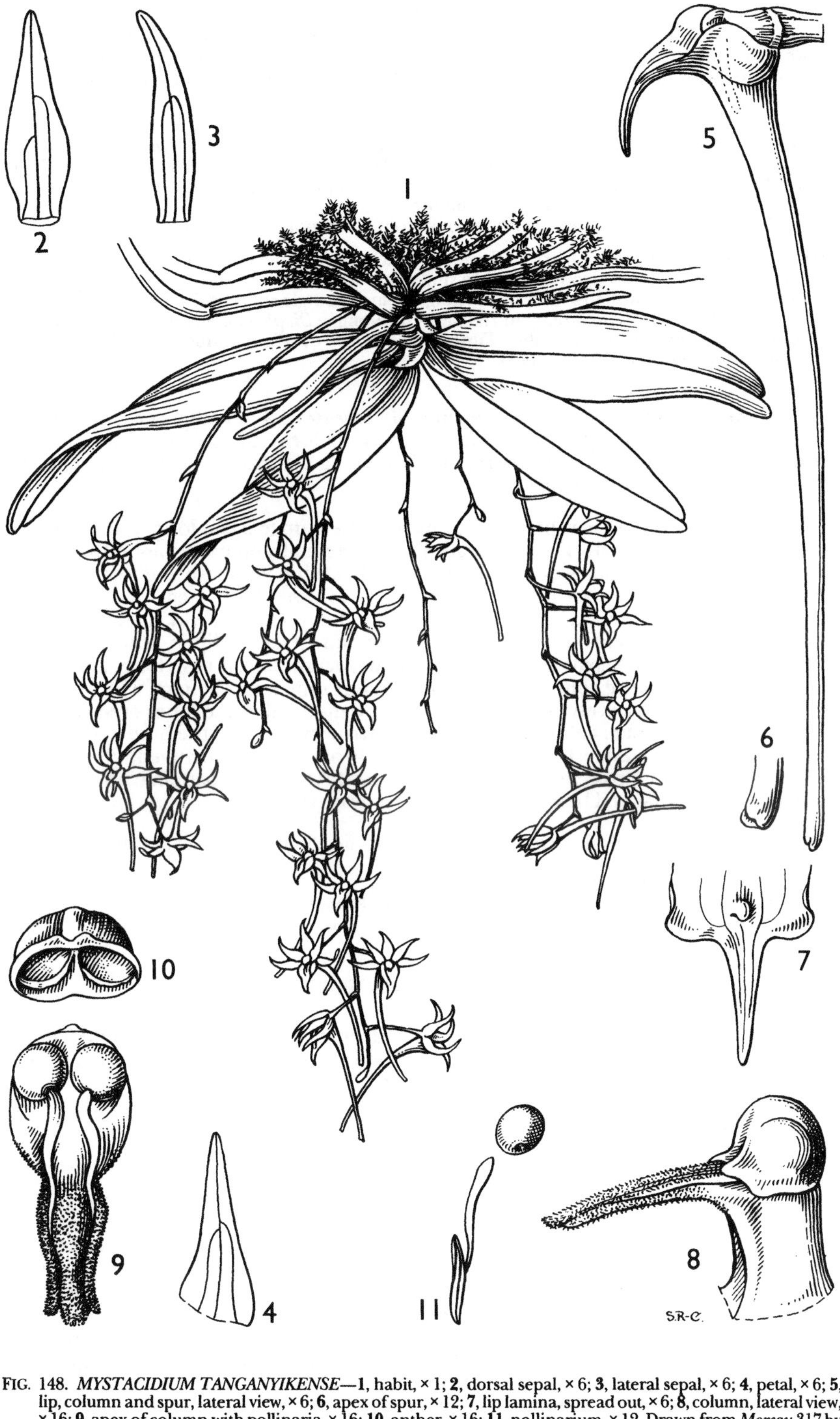

FIG. 148. *MYSTACIDIUM TANGANYIKENSE*—**1**, habit, × 1; **2**, dorsal sepal, × 6; **3**, lateral sepal, × 6; **4**, petal, × 6; **5**, lip, column and spur, lateral view, × 6; **6**, apex of spur, × 12; **7**, lip lamina, spread out, × 6; **8**, column, lateral view, × 16; **9**, apex of column with pollinaria, × 16; **10**, anther, × 16; **11**, pollinarium, × 12. Drawn from *Moreau* 317 by Miss Stella Ross-Craig. Reproduced from Hooker's Icones Plantarum.

obscurely 3-lobed, 4.7-6.6 mm. long, 2.7–3 mm. wide; side lobes upcurved, rounded; mid-lobe lanceolate, acute, recurved towards the apex; spur slender, pendent, tapering from a broad mouth, 12–17 mm. long. Column 1 mm. long; rostellum pendent, 3-lobed, the mid-lobe longer than the side lobes, papillose; stipites 2, linear; viscidia 2, linear. Fig. 148.

TANZANIA. Morogoro District: W. side of Lukwangule Plateau, 19 Jan. 1943, *Moreau* 474! & 27 Jan. 1976, *Cribb & Grey-Wilson* 10464!; Mbeya Range, World's End, 6 Feb. 1976, *Cribb & Grey-Wilson* 10571!; Njombe District: Kipengere Mts., Nyarere R., 10 Jan. 1957, *Richards* 7644!
DISTR. **T** 2, 6, 7; Malawi, Zambia and Zimbabwe
HAB. In upland evergreen forest, also in coniferous plantations; 1300–2500 m.

NOTE. *Schlieben* 3821 (from **T** 6 in the Uluguru Mts. at 1600 m.) is not unlike this species but differs in having broader leaves, up to 14 mm. wide, and much smaller flowers with 3–5 mm. long segments and a short distally swollen spur 5–6 mm. long. From this single collection, I cannot determine if it is correctly placed in *Mystacidium*, and it may well belong in *Diaphananthe*, near *D. stellata* which its flowers also superficially resemble. Further collections of this diminutive orchid are needed before this question can be resolved.

3. **M. nguruense** *Cribb*, sp. nov affine *M. capense* (L.f.) Schltr. sed foliis falcatis linearibus, pedunculo graciliore, labello ovato latiore calloso et calcare breviore distinguendo. Typus: Tanzania, Morogoro District, Nguru Mts., Divue R. headwaters, *Pócs & Knox* 89052/C (K, holotypus)

An epiphytic herb with a short spreading leafy stem up to 5 cm. long; roots slender, 1 mm. in diameter, elongate. Leaves falcate, linear, unequally obtusely bilobed at the apex, 12–18 cm. long, 1.3–2 cm. wide, articulated to a sheathing leaf-base and twisted to lie in one plane. Inflorescences pendent, densely 5–8-flowered, 6–9 cm. long; peduncle slender, wiry; bracts ovate, amplexicaul, 1.5 mm. long. Flowers white, stellate but not opening widely; pedicel and ovary 8–10 mm. long. Dorsal sepal elliptic, acute, 7 mm. long, 3.5 mm. wide; lateral sepals obliquely lanceolate, acuminate, 8 mm. long, 2 mm. wide. Petals decurrent on the column-foot, ovate, acute, 7 mm. long, 3.5 mm. wide. Lip ovate, shortly apiculate, obscurely 3-lobed, 7 mm. long, 6 mm. wide; side lobes erect, rounded; mid-lobe broadly ovate; callus a longitudinal narrow ridge from the base to the base of the midlobe; spur slenderly cylindrical from a broad mouth, incurved, 20 mm. long. Column 2 mm. long; rostellum 3-lobed, 2 mm. long, with lobes papillate and subequal; pollinia 2, each attached by a linear stipe to an ovate viscidium.

TANZANIA. Morogoro District: Nguru Mts., Divue R. headwaters, 4–6 Feb. 1989, *Pócs & Knox* 89052/C!
DISTR. **T** 6; known only from the type collection
HAB. Epiphyte in mossy montane forest in valley; 2000 m.

67. ANGRAECOPSIS

Kraenzl. in E.J. 28: 171 (1900); Schltr., Die Orchideen: 600 (1914) & in Beih. Bot. Centr. 36: 139 (1918); Perrier in Fl. Madag., Orch. 2: 80 (1941); Summerh. in Bot. Mus. Leafl. Harv. Univ. 14: 242 (1951)

Holmesia Cribb in K.B. 32: 175 (1977) *non* Agardh (1890)
Microholmesia Cribb in Mabb., The Plant Book: 371 (1987)

Small epiphytes or rarely lithophytes with short or very short monopodial stems. Roots long, flexuous, terete or flattened, simple or branching. Leaves distichous, ligulate to oblanceolate, often falcate, unequally bilobed at the apex, sometimes twisted at the base and articulated to imbricate sheathing bases. Inflorescences axillary, 1–several, racemose, few–many-flowered; peduncle wiry, often longer than the rhachis; bracts small. Flowers small, white, pale green or yellow-green. Dorsal and lateral sepals similar or dissimilar; lateral sepals often oblanceolate and produced at one side at the base. Petals often adnate at the base to the lateral sepals and obliquely triangular. Lip 3-lobed or rarely entire, ecallose, spurred at the base; spur sometimes inflated at the apex. Column fleshy, short; rostellum short or rarely elongate, 3-lobed; pollinia 2, globose; stipites 2, linear or rarely oblanceolate; viscidia 2 or rarely 1, elongate.

The genus has been most recently revised by Summerhayes (1951). He recognised 3 sections, all of which are in East Africa:

Sect. **Angraecopsis**. Plants small to medium-sized; peduncle usually much longer than the rhachis; lateral sepals much longer than the dorsal; petals triangular, acute; lip 3-lobed, with linear or ligulate side lobes; pollinia each attached by long stipites to a separate viscidium. Species 1–4.

Sect. **Coenodenium** *Summerh.* Plants small; peduncle equalling or shorter than the rhachis; sepals ± similar; petals ovate or lanceolate; lip entire or 3-lobed; pollinia attached by 2 short stipites to a common viscidium. Species 5–7.

Sect. **Cardiochilos** *Summerh.* Plants minute; peduncle shorter than rhachis; sepals unequal; petals elliptic-ovate, rounded; lip entire, cordate or ovate; viscidia 2. Species 8, 9 and *A. pusilla* Summerh., known only from Zaire.

Material recently collected in spirit has enabled a better understanding of the relationships of *Microholmesia parva* (Cribb) Cribb. I have therefore transferred it to *Angraecopsis* because of its similarity to *A. lovettii* and *A. pusilla* Summerh.

Spur more than 2.5 cm. long:
  Spur 5–6.5 cm. long; lip side lobes ± equalling the mid-lobe in length, slightly dilated at the apex; petals 3–3.6 mm. long, 3.8–4.5 mm. wide; leaves 1–2.2 cm. wide; roots verrucose . . . . 1. *A. tenerrima*
  Spur 2.5–4.5 cm. long; lip side lobes about half the length of the mid-lobe, tapering to the apex; petals 2–2.7 mm. long, 2.5–2.7 mm. wide; leaves mostly less than 1 cm. wide; roots tomentose . . . . 2. *A. gracillima*
Spur less than 2.3 cm. long, mostly less than 1.5 cm. long:
  Lip entire or auriculate at the base; lip side lobes if present much less than a third the length of the mid-lobe; leaves 9 cm. or less long; inflorescences secund:
    Lip entire; roots ± terete, less prominent than the leaves:
      Spur filiform, 2 cm. or more long . . . . 7. *A. holochila*
      Spur clavate, 1–2 mm. long:
        Sepals and petals 2–3 mm. long; spur clavate, 2 mm. long, notched at the apex . . . . 8. *A. lovettii*
        Sepals and petals 1–2 mm. long; spur clavate, 1 mm. long, not notched at the apex . . . . . . 9. *A. parva*
    Lip auriculate at the base; roots dorso-ventrally flattened on the substrate, much more prominent than the leaves:
      Spur filiform, 7–13 mm. long . . . . 5. *A. amaniensis*
      Spur clavate, 4–5 mm. long . . . . 6. *A. breviloba*
  Lip 3-lobed in the middle, with side lobes prominent, half to almost as long as the mid-lobe; leaves mostly more than 10 cm. long; inflorescences not secund:
    Leaves over 1.9 cm. wide; roots verrucose; rhachis 4–7 cm. long; spur over 1 cm. long . . . . 3. *A. macrophylla*
    Leaves 0.5–1.4 cm. wide; roots smooth; rhachis 1–3 cm. long; spur less than 1 cm. long . . . . 4. *A. parviflora*

1. **A. tenerrima** *Kraenzl.* in E.J. 28: 171 (1900); Schltr. in Beih. Bot. Centr. 36: 141 (1918); Summerh. in Bot. Mus. Leafl. Harv. Univ. 14: 247 (1951). Type: Tanzania, Usambara Mts., Nquelo, *Scheffler* (B, holo., K, drawing of holo.!)

An epiphyte or rarely a lithophyte with a short pendent stem 1–4(–7) cm. long. Leaves falcate, oblanceolate to linear, unequally roundly bilobed at the apex, 6–20 cm. long, 1.4–2.1 cm. wide, twisted at the base to lie in one plane and articulated to imbricate leaf-sheaths, very dark green. Inflorescences 1–several, 7.5–20 cm. long, laxly 3–7-flowered; peduncle slender, 6–13 cm. long; bracts amplexicaul, conical, obtuse, 2–4 mm. long. Flowers white with the tip of the spur green; pedicel and ovary 2.2–3.1 cm. long. Dorsal sepal elliptic, obtuse, 2.7–3.3 mm. long, 1.6–1.8 mm. wide; lateral sepals spathulate, obtuse, 8.5–11.7 mm. long, 2–2.5 mm. wide. Petals obliquely triangular, acute, 3–3.3 mm.

long, 3.8–4.6 mm. wide, joined to the lateral sepals for the basal 3–5 mm. Lip 3-lobed, 9–9.5 mm. long, 6.5–8 mm. wide; side lobes oblanceolate, somewhat bilobed or entire and rounded at the apex, 4–5 mm. long; mid-lobe lanceolate, acute, equal in length to the side lobes; spur cylindrical, pendent, 5–6 mm. long from a broad mouth. Column 1 mm. long; rostellum 3-lobed; stipites 2; viscidia 2. Pl. 8.

TANZANIA. Moshi District: Weru Weru Gorge, 5 Feb. 1942, *Harris* in *Moreau* 141!; Lushoto District: E. Usambara Mts., Amani, 8 May 1938, *Moreau* 14!; Ulanga District: Muhulu Mts., 26 Feb. 1932, *Schlieben* 1844! & 23 Jan 1979, *Cribb, Grey-Wilson & Mwasumbi* 11139!
DISTR. T 2, 3, 6; not known elsewhere
HAB. In lower montane forest; 400–1600 m.

SYN. *Angraecum tenerrimum* (Kraenzl.) Schltr. in E.J. 38: 164 (1906)
A. amaniense Kraenzl. in E.J. 51: 396 (1914). Type: Tanzania, E. Usambara Mts., Amani, *Braun* in *Herb. Amani* 3211 (B, holo.)

2. **A. gracillima** (*Rolfe*) *Summerh.* in K.B. 1937: 465 (1937) & in Bot. Mus. Leafl. Harv. Univ. 14: 248 (1951). Type: Uganda, Mengo District, without exact locality, cult. Glasnevin, *E. Brown* (K, holo.!)

An epiphyte with a short pendent stem, 1–7 cm. long. Leaves falcate, linear, obtusely bilobed at the oblique apex, twisted to lie in one plane at the base and articulated to imbricate leaf-sheaths. Inflorescences 4–16.5 cm. long, 4–14-flowered; peduncle (2–)7–11 cm. long; bracts sheathing, conical, acute, 2–3 mm. long. Flowers white; pedicel and ovary 1.7–2.7 cm. long. Dorsal sepal ovate, obtuse, 2–3.1 mm. long, 1.3–1.5 mm. wide; lateral sepals dependent, spathulate, rounded, 7.5–9 mm. long, 1.6–1.7 mm. wide. Petals obliquely ovate-triangular, acute, 2.2–5 mm. long, 2.7–5.5 mm. wide, adnate to the lateral sepals for the basal 2–4 mm. Lip 3-lobed, 5–5.5 mm. long, 2–2.5 mm. wide; side lobes linear, 1.3–1.5 mm. long, less than half the length of the mid-lobe; mid-lobe oblanceolate, 4.5–4.7 mm. long, 1–1.3 mm. wide, rounded at the apex; spur cylindrical, 3.8–4 cm. long. Column 1 mm. long; stipites 2, linear; viscidia 2, subcircular.

UGANDA. Toro, fl. cult. May 1939, *Chandler* 2792!; Mbale District: Elgon, Tracey Falls, 27 July 1917, *Snowden* 505!; Masaka District: S. Buddu, *Fyffe* 184!
KENYA. Kericho District: Chepalungu Forest, 19 May 1964, *Polhill & Moscoff* 2418!; N. Kavirondo District: Kakamega Forest, Aug. 1942, *Tweedie* 590!; S. Kavirondo District: Kisii, *Napier* 6355!
DISTR. U 3, 4; **K** 3, 5; Zambia
HAB. In evergreen forest; 1500–1850 m.

SYN. *Mystacidium gracillimum* Rolfe in K.B. 1913: 144 (1913)

3. **A. macrophylla** *Summerh.* in Bot. Mus. Leafl. Harv. Univ. 14: 249 (1951); F. Rasmussen in Norw. Journ. Bot. 25: 141 (1978). Type: Uganda, Bunyoro District, Budongo Forest, *Eggeling* 5386 (K, holo.!)

A pendent epiphyte with a short stem 3–6 cm. long; roots verrucose. Leaves 2–4, falcate, linear to oblanceolate, unequally acutely or subacutely bilobed at the apex, 13–33.5 cm. long, 1.4–2.2 cm. wide, twisted at the base and articulated to imbricate leaf-sheaths, 2–3 cm. long. Inflorescences 10–20 cm. long, many-flowered; peduncle 6.5–8.5 cm. long; bracts triangular, 1 mm. long. Flowers white; pedicel and ovary 0.8–1.3 cm. long. Dorsal sepal elliptic, obtuse, 2.4–2.6 mm. long, 2–2.2 mm. wide; lateral sepals spathulate, rounded, 5.5–5.7 mm. long, 1.7–1.8 mm. wide. Petals obliquely ovate-triangular, acute, 2.5 mm. long, 2.7–3 mm. wide. Lip 3-lobed in the middle, 5.5–6.5 mm. long, 4 mm. wide, subauriculate at the base; side lobes linear-lanceolate, subacute, 1–1.5 mm. long; mid-lobe narrowly elliptic, obtuse, longer than the side lobes, 2.5 mm. long; spur incurved, cylindrical, swollen at the apex, 10–11 mm. long. Column 1.5–2 mm. long; foot 1.5 mm. long; rostellum deeply 3-lobed; stipites 2, oblanceolate; viscidia 2, linear.

UGANDA. Bunyoro District: Budongo Forest, June 1944, *Eggeling* 5386!
DISTR. U 2; Ethiopia and Ivory Coast
HAB. In forest on the bole of a tree on the bank of a stream; ± 1000 m.

4. **A. parviflora** (*Thou.*) *Schltr.*, Die Orchideen: 601 (1914), in Beih. Bot. Centr. 33: 428 (1915) & 36: 140 (1918); Perrier in Fl. Madag., Orch. 2: 85 (1941); Summerh. in Bot. Mus. Leafl. Harv. Univ. 14: 255 (1951); la Croix et al., Malawi Orch. 1: 28 (1983). Type: Mauritius, without exact locality, *Thouars* 30 (P, holo.!)

A spreading or pendent epiphyte with stems 1–3 cm. long. Leaves 4–6, falcate, linear or oblanceolate, unequally acutely or subacutely bilobed at the apex, 6–16(–22) cm. long, 0.5–0.6(–1.1) cm. wide, twisted at the base, dark green. Inflorescences 1-several, 5–10 cm. long, densely 2–8-flowered; peduncle slender, wiry, 4–9 cm. long; bracts ovate-triangular, acute, 1–1.5 mm. long. Flowers pale green or greenish white; pedicel and ovary 5–6 mm. long. Dorsal sepal ovate, obtuse, 2 mm. long, 1 mm. wide; lateral sepals oblanceolate, obtuse, 3.5–4 mm. long, 1 mm. wide. Petals obliquely triangular, acute, 1.5–1.8 mm. long, 1.3–1.5 mm. wide. Lip 3-lobed in the middle, 3–3.6 mm. long and wide; side lobes linear, acute, slightly shorter than the mid-lobe; mid-lobe linear, acute; spur incurved, cylindrical, slightly inflated at the apex, 5–9(–17) mm. long. Column 1 mm. long; rostellum short, bifid; stipites 2, linear; viscidium apparently 1. Fig. 149/1–10, p. 600.

TANZANIA. Rungwe District: Kyimbila, sterile, without date, *Stolz* K51! & Poroto Mts., ridge W. of Lufirio R., sterile, Feb. 1976, *Cribb & Grey-Wilson* 10624!
DISTR. **T** 7; Cameroon, Malawi, Mozambique, Zimbabwe, Madagascar and Mascarene Is.
HAB. In lower montane forest; 600–1600 m.

SYN. *Angraecum parvifolium* Thou., Orch. Iles Austr. Afr., t. 60 (1822)
Aerobion parviflorum (Thou.) Sprengel, Syst. Veg. ed. 16, 3: 718 (1826)
Listrostachys parviflora (Thou.) S. Moore in Baker, Fl. Maurit. & Sech.: 355 (1877)
Epidorchis parviflora (Thou.) O. Kuntze, Rev. Gen. Pl. 2: 660 (1891)
Saccolobium parviflorum (Thou.) Jacob Cordemoy, Fl. Ile Réunion: 197 (1895)
Mystacidium pedunculatum Rolfe in F.T.A. 7: 175 (1897). Types: Mozambique, Namuli, *Last* (K, syn.!); Malawi, Chiradzulu, *Miller* 3 (K, syn.!) & Mlanje, *McClounie* 29 (K, syn.!)

5. **A. amaniensis** *Summerh.* in Bot. Mus. Leafl. Harv. Univ. 11: 259 (1945) & 14: 260 (1951); Williamson, Orch. SC. Afr.: 201 (1977); la Croix et al., Malawi Orch. 1: 25 (1983). Type: Tanzania, E Usambaras, Amani, *Moreau* 68 (K, holo.!)

A dwarf epiphyte with very short stems less than 5 mm. long, sometimes appearing leafless; roots very long, flattened, 1–2 mm. across, much more prominent than the leaves. Leaves 1–3, deciduous, linear, obtuse, 1–3.5 cm. long, 0.3–0.5 cm. wide, twisted and articulated at the base. Inflorescences pendent, secund, 3–13 cm. long, densely 10–20-flowered; peduncle 1–2.5 cm. long, wiry, very slender; bracts sheathing, ovate, obtuse, 1–1.5 mm. long. Flowers cream or pale green; pedicel and ovary 6–7 mm. long. Dorsal sepal elliptic-oblanceolate, obtuse, 4–4.2 mm. long, 1–1.7 mm. wide; lateral sepals linear, acute, 6.5–7 mm. long, 1 mm. wide. Petals obliquely lanceolate, acute, 3.5–4.5 mm. long, 1.5–1.6 mm. wide. Lip 3-lobed, 4 mm. long, 1.5 mm. wide; side lobes obscurely auriculate; mid-lobe linear; spur cylindrical, straight or incurved, 7–13 mm. long. Column 0.5–0.6 mm. long; rostellum 3-lobed, with truncate side lobes; stipites 2, short; viscidium 1, oblong. Fig. 149/11–21, p. 600.

KENYA. N. Kavirondo District: S. Elgon, Apr. 1943, *Tweedie* 603!
TANZANIA. Lushoto District: W. Usambara Mts., Mkuzi to Kifungilo road, Mkussu [Mkusi] valley, 23 Apr. 1953, *Drummond & Hemsley* 2219! & E. Usambara Mts., near Amani, 1 May 1941, *Moreau* 68! & 23 July 1944, *Moreau* 443!
DISTR. **K** 3/5; **T** 3; Mozambique, Zambia, Malawi and Zimbabwe
HAB. In montane forest, on *Podocarpus* and other trees, also in conifer plantations; 900–2100 m.

SYN. *Angraecopsis tenuicalcar* Summerh. in Bot. Mus. Leafl. Harv. Univ. 11: 258 (1945). Type: Tanzania, W. Usambara Mts., Shume, *Moreau* 325 (K, holo.!)

6. **A. breviloba** *Summerh.* in Bot. Mus. Leafl. Harv. Univ. 11: 256 (1945) & 14: 261 (1951). Type: Tanzania, Masai District, Monduli [Mondul] Mt., *Moreau* 304 (K, holo.!)

A dwarf epiphyte with a very short stem less than 5 mm. long, often appearing leafless; roots much more prominent than the leaves, dense, flattened on the substrate, 1.5–3 mm. wide. Leaves linear-elliptic to oblanceolate, acute, 1.5–4 cm. long, 0.3–0.4 cm. wide, dull green. Inflorescences 7–9 cm. long, secund, densely up to 15-flowered; peduncle 1 cm. long, bearing 3–4 bracts; bracts distichous, funnel-like, 1–1.5 mm. long, blackish. Flowers pale green or pale yellow-green with a brown tip to the spur, slightly scented; pedicel and ovary 5–6 mm. long. Dorsal sepal oblong-elliptic, obtuse, 3–5 mm. long, 1.5–1.6 mm. wide; lateral sepals similar. Petals lanceolate, obtuse, 2.7–4 mm. long, 1–1.4 mm. wide. Lip obscurely 3-lobed at the base, lanceolate, acute, 3.7–4.6 mm. long, 1.5–1.8 mm. wide; side lobes auriculate; spur clavate, 4.5–5 mm. long. Column 1 mm. long; rostellum 3-lobed;

FIG. 149. *ANGRAECOPSIS PARVIFLORA*—**1**, habit, × ⅔; **2**, flower, × 6; **3**, dorsal sepal, × 8; **4**, lateral sepal and petal, × 8; **5**, column and lip, side view, × 12; **6**, lip, front view, × 12; **7**, column with pollinaria, front view, × 16; **8**, same, anther removed, × 16; **9**, anther-cap, × 16; **10**, pollinaria, × 20. *A. AMANIENSIS*—**11**, habit, × ⅔; **12**, flower, × 6; **13**, dorsal sepal, × 6; **14**, lateral sepal and petal, × 6; **15**, column and lip, side view, × 8; **16**, lip, front view, × 8; **17**, column, side view, × 10; **18**, column and pollinaria, front view, × 12; **19**, same, pollinaria removed, × 12; **20**, anther-cap, × 16; **21**, pollinia, stipites and viscidium and back and side views of stipites and viscidium, × 20. *A. BREVILOBA*—**22**, flower, × 6. 1–10, from *la Croix* 49; 11–21, from *la Croix* 168; 22, details not given. Drawn by Mrs M.E. Church.

stipites 2, short; viscidium 1, linear-elliptic. Fig. 149/22.

KENYA. Fort Hall District: Tana R. basin, Thika(-thika), 16 July 1893, *Gregory*!; Kisumu-Londiani District: Londiani, 20 July 1957, *Smart* E9!; Masai District: Ngong Hills, 7 Nov. 1934, *C. van Someren* 3653! & 25 Jan. 1985, *Khayota* 120!
TANZANIA. Masai District: Ngorongoro, May 1983, *Rodgers & Chuwa* 2414!; Moshi District: Kilimanjaro, Marangu, 27 Nov. 1941, *Moreau* 107!; Morogoro District: Uluguru Mts., Bunduki, 19 Mar. 1953, *Drummond & Hemsley* 1689!
DISTR. **K** 4–6; **T** 2, 3, 6, 7; not known elsewhere
HAB. In montane forest, often by streams or on ridges, also in citrus plantations; 1300–2400 m.

7. **A. holochila** *Summerh.* in Bot. Mus. Leafl. Harv. Univ. 12: 115 (1945) & in 14: 260 (1951); F. Rasmussen in Norw. Journ. Bot. 25: 143 (1978). Type: Uganda, Karamoja District, Napak, *A.S. Thomas* 3646 (K, holo.!)

A dwarf epiphyte or rarely a lithophyte with a very short stem less than 5 mm. long; roots flattened, 2 mm. wide. Leaves linear, obtuse, 4–9 cm. long, 0.4–0.6 cm. wide, twisted and articulated at the base. Inflorescences 4–5 cm. long, laxly few–8-flowered; peduncle 1–2 cm. long, very slender, wiry; bracts ovate, acute, 1–1.5 mm. long. Flowers pale green turning yellow with age; pedicel and ovary 5–10 mm. long. Dorsal sepal elliptic, obtuse, 3–3.3 mm. long, 1.5–1.8 mm. wide; lateral sepals oblanceolate, rounded, 4.5 mm. long, 1.6–1.7 mm. wide. Petals obliquely ovate, obtuse, 3 mm. long, 1.8–2 mm. wide. Lip entire, ovate, obtuse, 3 mm. long, 1.6–1.8 mm. wide; spur narrowly cylindrical, slightly incurved, 14–22.5 mm. long. Column 0.7 mm. long; rostellum 3-lobed, short; stipites 2, slender; viscidia 2.

UGANDA. Karamoja District: Napak, May 1940, *A.S. Thomas* 3646! & June 1950, *Eggeling* 5971!
DISTR. **U** 1; Ethiopia
HAB. In montane forest in ravine, on trees and on rocks; 2150–2300 m.

8. **A. lovettii** *Cribb*, sp. nov. affinis *A. holochila* Summerh. et *A. pusilla* Summerh. sed planta minore, inflorescentia breviore, calcare 1–2 mm. longo clavato ad apicem bilobato satis differt. Typus: Tanzania, Iringa District, Image Mt., *Lovett* 642 (K, holo.!)

A dwarf epiphyte with a very short stem up to 0.5 cm. long; roots wiry, 0.5 mm. in diameter. Leaves 3–5, linear-ligulate to oblanceolate, unequally roundly bilobed at the apex, 2–2.75 cm. long, 3–4 mm. wide, twisted to lie in one plane at the base. Inflorescences secund, less than 2.5 cm. long, 4–8-flowered; peduncle 4–9 mm. long; bracts amplexicaul, ovate-triangular, acute, 0.5–0.8 mm. long. Flowers tiny, pale greenish white, all segments somewhat recurved; pedicel and ovary 3–5 mm. long. Dorsal sepal elliptic-oblong, acute to obtuse, 3 mm. long, 1 mm. wide; lateral sepals obliquely narrowly elliptic, acute to obtuse, 3 mm. long, 1 mm. wide. Petals oblong-elliptic, acute, 2.5 mm. long, 1 mm. wide. Lip slightly recurved, ovate, acute and fleshy at the apex, 2 mm. long, 1 mm. wide; spur slightly incurved, clavate, 1–2 mm. long, notched at the apex. Column 0.5 mm. long; rostellum 3-lobed, with papillate side lobes produced above and below into acute tips and mid-lobe pendent and acute; pollinia 2; stipites 2, linear; viscidia 2, narrowly elliptic.

TANZANIA. Iringa District: Image Mt., 13 Apr. 1986, *J. & J. Lovett* 642! & 4 Mar. 1962, *Polhill & Paulo*!
DISTR. **T** 7; known only from the above collections
HAB. Epiphyte in montane forest; 2000–2300 m.

NOTE. This tiny orchid is superficially like *Angraecum decipiens* and the relationship of *Angraecopsis* sect. *Cardiochilus* to that species needs further investigation.

9. **A. parva** (*Cribb*) *Cribb*, comb. nov. Type: Malawi, Nyika Plateau, *Holmes* 0230 (K, holo.!)

A dwarf epiphyte with a very short stem, 2–4 mm. long; roots slender, terete, less than 1 mm. in diameter. Leaves 2–4, linear-oblanceolate, obtuse, 1.5–2 cm. long, 2–4 mm. wide, twisted at the base to lie in one plane. Inflorescences 1.5–2.5 cm. long, 2–5-flowered; peduncle very finely wiry, 5–10 mm. long; bracts amplexicaul, ovate-triangular, acute, 0.5 mm. long. Flowers minute, greenish white; pedicel and ovary 2–3 mm. long. Dorsal sepal elliptic, obtuse or subacute, 1.5 mm. long, 0.7–0.8 mm. wide; lateral sepals oblanceolate, acute, 2 mm. long, 0.5 mm. wide. Petals obliquely elliptic, obtuse, 1.2–1.3 mm. long, 0.5 mm. wide. Lip concave, fleshy especially at apex, very obscurely 3-lobed at apex, elliptic-ovate, shortly apiculate, 1.7 mm. long, 1.3 mm. wide; spur clavate, somewhat sigmoid, 1 mm. long. Column 0.4 mm. long; rostellum 3-lobed, with side lobes shortly produced above and below and mid-lobe deflexed and blunt.

TANZANIA. Iringa District: Mufindi, Caldarara, 13 Apr. 1971. *Paget-Wilkes* 872!
DISTR. **T** 7; Malawi
HAB. Epiphyte in montane forest on branches of forest under-shrubs; 1800–2000 m.

SYN. *Holmesia parva* Cribb in K.B. 32: 175 (1977)
Microholmesia parva (Cribb) Cribb in Mabb., The Plant Book: 371 (1987)

68. DISTYLODON

Summerh. in K.B. 20: 197 (1966)

Dwarf epiphytic herbs with a short stem; roots many, slender, elongate. Leaves distichous, conduplicate. Inflorescences as long as the leaves, 1-flowered. Flowers small, insignificant. Sepals and petals subsimilar, free. Lip obscurely 3-lobed, ecallose, spurred; spur slender, cylindrical, with a narrow mouth. Column short, fleshy; anther convex, scarcely hemisphaerical; pollinia 2, ovoid; stipites 2, linear; viscidia 2, linear, longer than the stipes; rostellum erect, geniculate at base, bifid, each lobe linear, porrect, parallel.

A monotypic genus known only from Uganda.

The single species superficially resembles some of the small-flowered species of *Angraecum* but is probably most closely allied to *Triceratorhynchus*.

D. comptum *Summerh.* in K.B. 20: 197 (1966). Type: Uganda, Bunyoro District, Budongo Forest, *Eggeling* 5381 (K, holo.!)

A dwarf epiphyte with a short bilaterally flattened stem, 1.5 cm. long, 1 mm. diameter. Leaves ovate or elliptic-lanceolate, acute and shortly unequally bilobed at apex, 0.7–1.7 cm. long, 0.4–0.8 cm. wide, twisted at base to lie in one plane, articulated to a short sheath at base. Inflorescence 1-flowered, up to 2.2 cm. long; peduncle slender, 1–1.8 cm. long; bract ovate, acute, 3 mm. long. Flower not opening widely, pale green; pedicel and ovary ± 9 mm. long. Dorsal sepal lanceolate, acute, 6.5 mm. long, 2.25 mm. wide; lateral sepals obliquely lanceolate, acute or acuminate, 7–8 mm. long, 2 mm. wide. Petals lanceolate, acute or acuminate, 6–6.5 mm. long, 2–2.5 mm. wide. Lip 3-lobed in basal half, 6.5 mm. long, 3–3.5 mm. wide, concave at base; side lobes narrowly oblong, erect, subacute at apex; mid-lobe much longer, lanceolate, acute, fleshy, 4–4.5 mm. long, 1 mm. wide; spur cylindrical, straight, 6.5–7 mm. long. Column 0.7–1 mm. long; rostellar arms 0.8 mm. long. Fig. 150.

UGANDA. Bunyoro District: Budongo Forest, fl. cult. Apr. 1944, *Eggeling* 5381!
DISTR. **U** 2; known only from the type collection
HAB. Epiphyte in forest; ± 1000 m.

69. TRICERATORHYNCHUS

Summerh. in Bot. Mus. Leafl. Harv. Univ. 14: 232 (1951)

Epiphytic herbs with a short many-leaved stems. Leaves distichous, linear, unequally bilobed at apex. Inflorescence spreading, elongate, laxly few–many-flowered. Flowers small, pedicellate. Sepals and petals free, subsimilar. Lip obscurely 3-lobed, concave at base, ovate, acute, spurred, ecallose; spur filiform, elongate, longer than or equalling lip in length. Column short, fleshy; anther hemisphaerical, erose on apical margin; pollinia 2, ovoid; stipites 2, very short, slender; viscidia 2, oblong-elliptic, twice as long as each stipes; rostellum complex, porrect, 3-pronged in front and behind, the central prong in front shorter than the outer ones.

A monotypic genus. The solitary species superficially resembles some of the smaller flowered species of *Angraecum* but is readily distinguished by its bizarre rostellum.

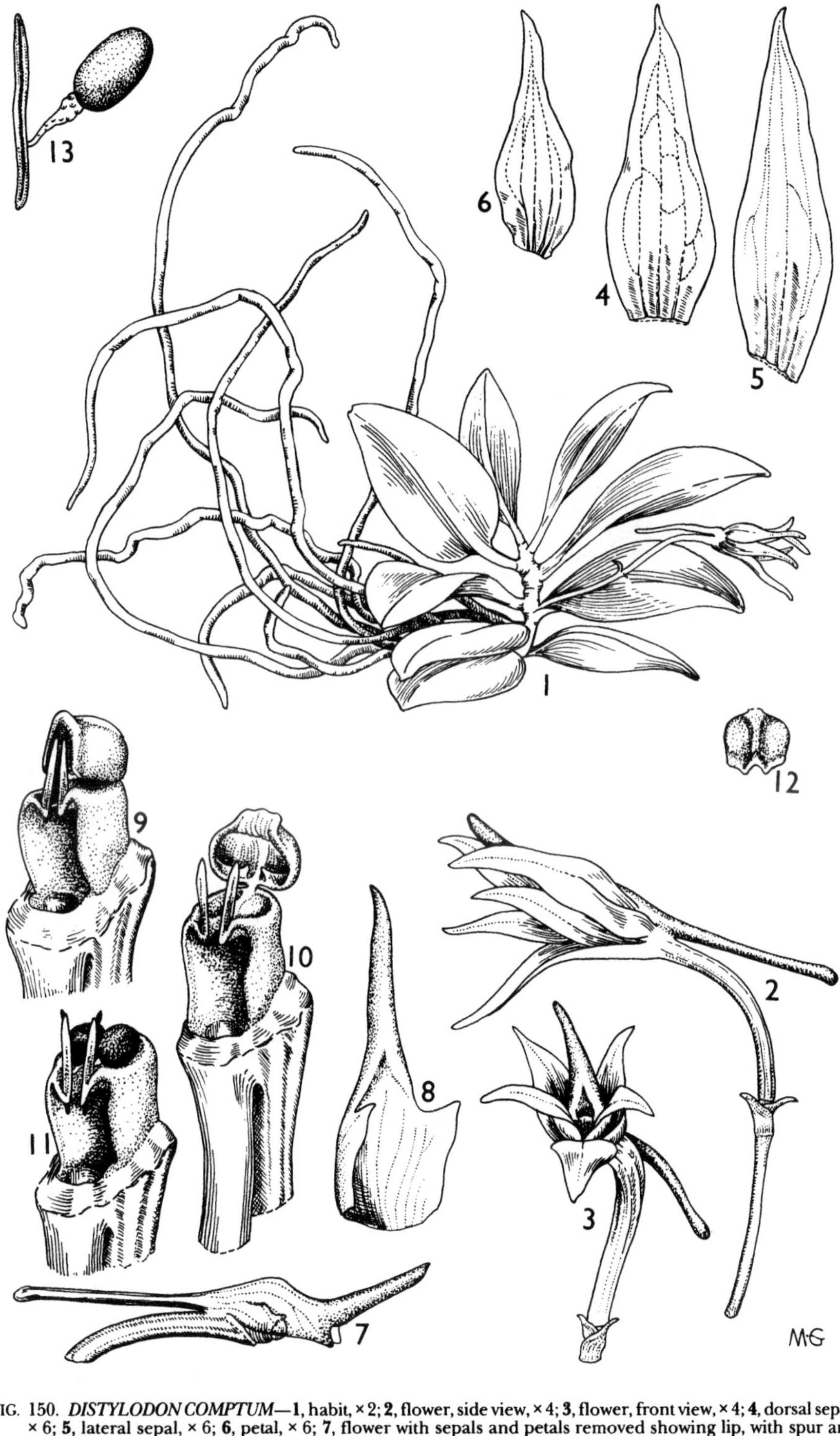

FIG. 150. *DISTYLODON COMPTUM*—**1**, habit, × 2; **2**, flower, side view, × 4; **3**, flower, front view, × 4; **4**, dorsal sepal, × 6; **5**, lateral sepal, × 6; **6**, petal, × 6; **7**, flower with sepals and petals removed showing lip, with spur and column, × 4; **8**, lip, one side flattened, × 6; **9**, column, three-quarter view, × 8; **10**, column with anther-cap raised and pollinaria removed, × 8; **11**, column with anther-cap removed and pollinia in position, × 8; **12**, anther-cap, × 8; **13**, pollinarium, × 12. Drawn from *Eggeling* 5381 by Miss Mary Grierson.

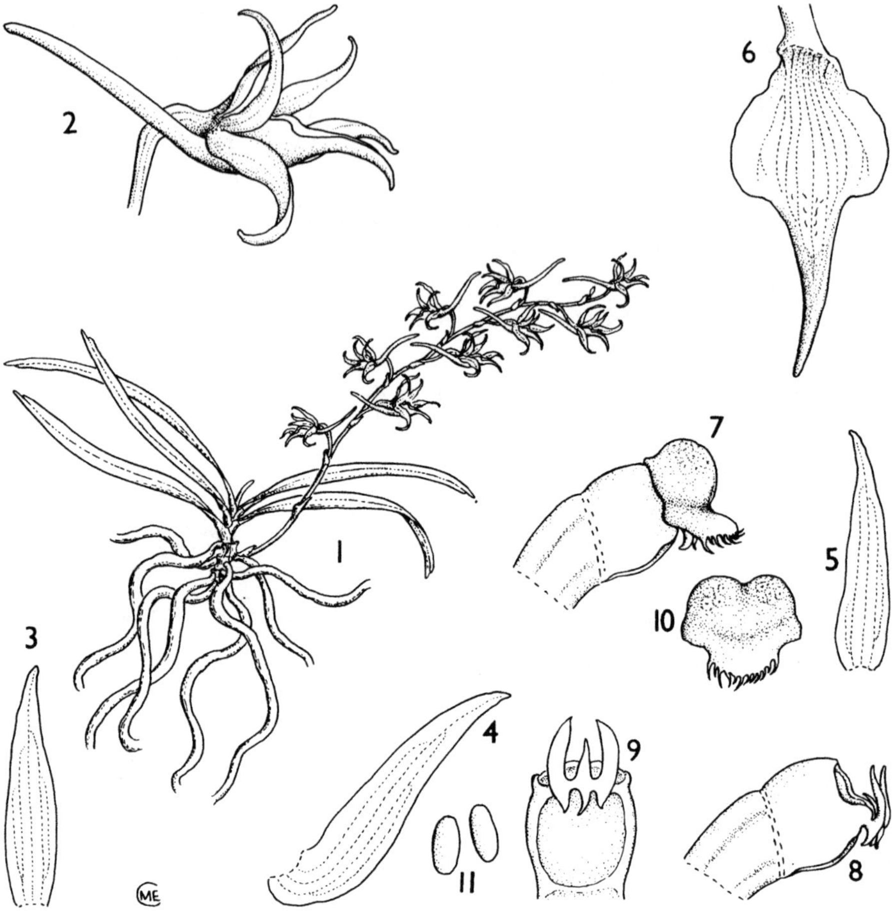

FIG. 151. *TRICERATORHYNCHUS VIRIDIFLORUS*—**1**, habit, × 1; **2**, flower, × 6; **3**, dorsal sepal, × 8; **4**, lateral sepal, × 8; **5**, petal, × 8; **6**, lip, × 8; **7**, column with anther-cap, side view, × 20; **8**, same, anther removed, × 20; **9**, column, front view, showing rostellum, × 20; **10**, anther-cap, × 20; **11**, pollinia, × 20. 1–6, from *Eggeling* 2058; 7–11, from *Eggeling* 5492. Drawn by Mrs M.E. Church.

T. viridiflorus *Summerh.* in Bot. Mus. Leafl. Harv. Univ. 14: 234 (1951). Type: Uganda, Bunyoro District, Budongo Forest, *Eggeling* 5492 (K, holo.!)

A dwarf epiphytic plant with a short stem ± 5 mm. long; roots numerous, elongate, ± 0.5 mm. diameter. Leaves suberect or reflexed, often falcate, linear, obscurely unequally bilobed or entire at the apex, 1–5 cm. long, 1–3.5 mm. wide, somewhat concave on upper surface. Inflorescences spreading or horizontal, longer than the leaves, up to 6 cm. long, laxly few–9-flowered; peduncle wiry, up to 2 cm. long; bracts broadly ovate, acute, 1–1.5 mm. long. Flowers green or greenish yellow; pedicel and ovary 4–5 mm. long, often geniculate in the middle. Dorsal sepal linear-lanceolate, acute, 4.2–5.5 mm. long, 0.7–1.25 mm. wide; lateral sepals falcate, obliquely linear-lanceolate, subacute, 5.2–6.3 mm. long, 0.9–1.2 mm. wide. Petals lanceolate, acute or acuminate, 4.2–5.3 mm. long, 1 mm. wide. Lip obscurely 3-lobed, lanceolate, acute, 4.7–5.5 mm. long, 3.5 mm. wide, slightly recurved and fleshy towards apex; spur curving up above ovary, slender, cylindrical, 5.5–8 mm. long. Column 0.7–0.8 mm. long; anther 1 mm. long; rostellar arms ± 0.5 mm. long. Fig. 151.

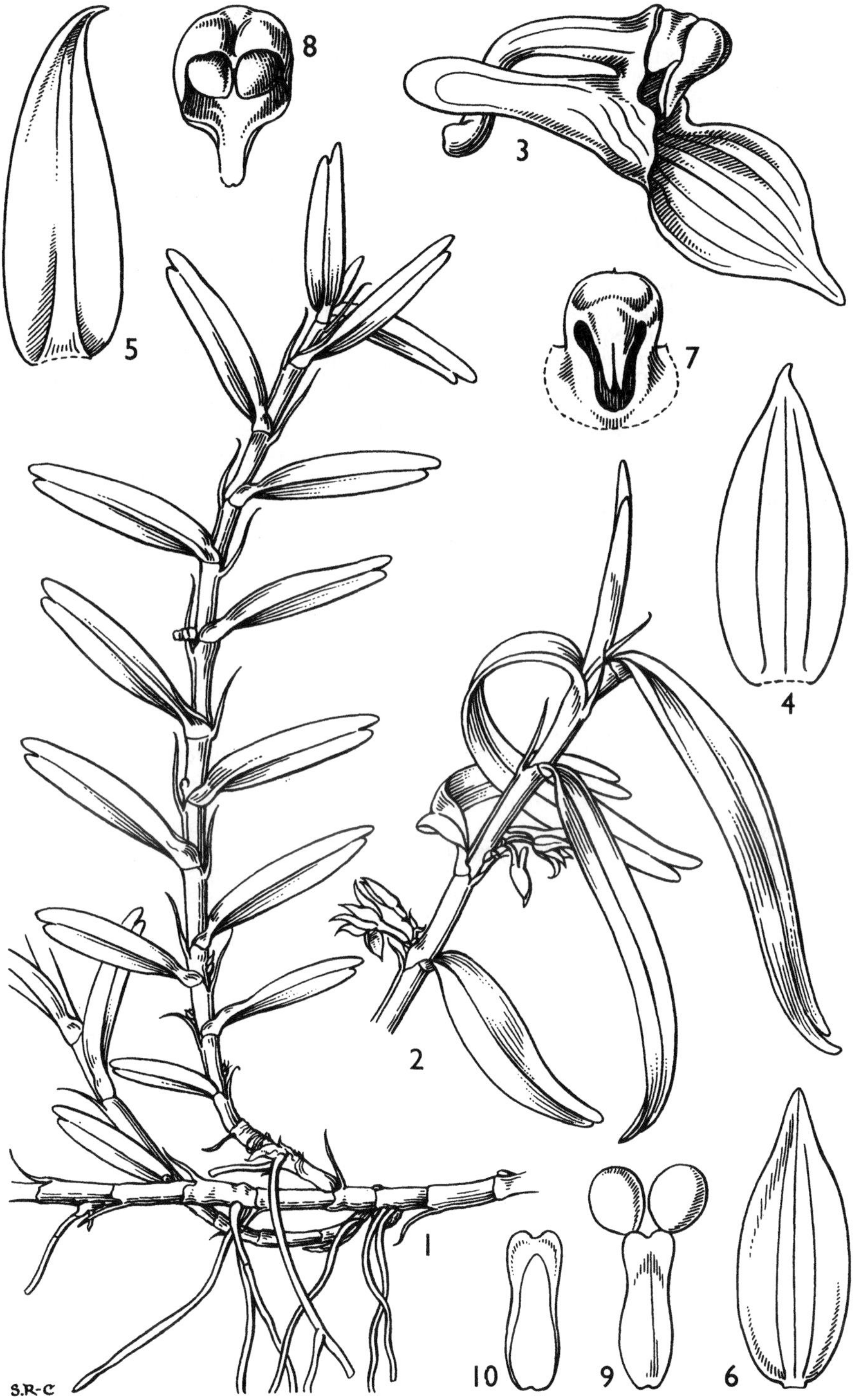

FIG. 152. *EGGELINGIA LIGULIFOLIA*—**1**, habit, × 1; **2**, inflorescence, × 1; **3**, flower, sepals and petals removed, × 6; **4**, dorsal sepal, × 6; **5**, lateral sepal, × 6; **6**, petal, × 6; **7**, column, showing rostellum, × 12; **8**, anther-cap with pollinia in position, × 12; **9**, pollinarium and stipes, × 12; **10**, stipes, back view, showing viscidium, × 12. Drawn from *Eggeling* 5372 and *Eggeling* 5845 by Miss Stella Ross-Craig. Reproduced from Hooker's Icones Plantarum.

UGANDA. Bunyoro District: Budongo Forest, June 1935, *Eggeling* 2058! & fl. cult. June–July 1944, *Eggeling* 5492!
KENYA. N. Kavirondo District: Kakamega Forest, June 1944, *Tweedie* 648!
DISTR. U 2; **K** 5; not known elsewhere
HAB. Epiphytic in evergreen forest; 1500–1600 m.

70. EGGELINGIA

Summerh. in Bot. Mus. Leafl. Harv. Univ. 14: 235 (1951)

Epiphytic or rarely lithophytic herbs with elongate monopodial stems, with few branches, leafy along length; roots in basal part, elongate, aerial. Leaves distichous, spreading, ligulate, unequally bilobed at apex, articulated to persistent leaf-sheaths; leaf-sheaths bearing a hair-like ligule at apex opposite leaf. Inflorescences axillary, very short, few-flowered. Flowers small, pedicellate. Sepals and petals free, subsimilar. Lip entire or obscurely 3-lobed at base, ovate, acute, ecallose, spurred; spur cylindric or subclavate, as long as the lip. Column short, fleshy; anther-cap hemispherical, produced in front into an acute apex; pollinia 2, ovoid or subpyriform-ovoid; stipes ligulate, dilated at apex, sometimes Y-shaped; viscidium rectangular-elliptic, almost as large as the stipes; rostellum deflexed, triangular, deeply bifid.

A small genus of three species only in tropical Africa from Ghana across to Uganda and south to Malawi.

E. ligulifolia *Summerh.* in Bot. Mus. Leafl. Harv. Univ. 14: 236 (1951). Type: Uganda, Toro District, Itwara Forest, *Eggeling* 5372 (K, holo.!)

An epiphytic or rarely lithophytic herb with pendent or arcuate occasionally branching stems, 7–45 cm. long, 1.5–3 mm. diameter, leafy in apical part. Leaves fleshy-coriaceous, narrowly oblong-lanceolate, unequally roundly or obtusely bilobed at apex, 1.7–7 cm. long, 0.4–0.8 cm. wide, twisted to lie in one plane at base, articulated to a 6–8 mm. long slightly compressed sheath at base, with a setose 4–5 mm. long ligule at apex of sheath. Inflorescences subsessile, 4–6 mm. long, 2–3-flowered; peduncle very short; rhachis 3 mm. long, fractiflex; bracts triangular, acute, 1–1.5 mm. long. Flowers white, scented; pedicel and ovary 5–5.5 mm. long, slightly furfuraceous. Dorsal sepal lanceolate, acute, 5.3–6.8 mm. long, 2.2–2.6 mm. wide; lateral sepals falcate, obliquely lanceolate, acute, 5.3–6.8 mm. long, 2–2.3 mm. wide. Petals oblong, obtuse, 4.6–6 mm. long, 1.7–2.25 mm. wide. Lip entire, auriculate at base, concave, oblong-ovate, acute, 4.4–5.7 mm. long, 1.7–3.5 mm. wide; spur cylindrical, straight, 3.8–5 mm. long. Column 1.2–1.75 mm. long. Fig. 152, p. 605.

UGANDA. Toro District: Itwara Forest, Aug. 1943, *Eggeling* 5372!; Kigezi District: Kayonza Forest, Ishasha Gorge, fl. cult. Sept. 1944, *Eggeling* 5379! & Impenetrable Forest, Nov. 1948, *Eggeling* 5845!
DISTR. U 2; Zaire
HAB. Epiphyte in rain-forest and rarely on lava flows; 1300–1950 m.

71. TRIDACTYLE

Schltr., Die Orchideen: 601 (1914) & in Beih. Bot. Centr. 36: 142 (1918); Summerh. in K.B. 3: 282 (1948)

Epiphytic monopodial herbs with short to long unbranched or little branched, erect to pendent stems, covered by sheathing leaf-bases. Leaves thin-textured or fleshy, all along stem, distichous, usually twisted at base to lie in one plane, linear to elliptic or narrowly lanceolate, unequally bilobed at apex, articulated at base to a sheathing leaf-base. Inflorescences emerging through the leaf-sheaths opposite the leaves, 2–many-flowered; peduncle usually short; rhachis often zig-zag; bracts amplexicaul. Flowers usually rather small and seldom showy, white, yellow, orange or green, usually scented; pedicel and ovary short, often lepidote. Sepals and petals subsimilar, free, spreading. Lip entire to 3-lobed, auriculate at base, with a short to long spur at the base. Column short, fleshy, lacking a foot; anther-cap thin-textured, smooth or papillate; pollinia 2; stipes 1, entire or

Y-shaped; viscidium circular or elliptic, usually rather small; rostellum elongate, tapering to apex, slender.

About 42 species in tropical and in South Africa.

The genus *Tridactyle* has been revised by Schlechter (1918) and most recently by Summerhayes (1948). Summerhayes recognised two sections: the monotypic section *Scandentes* Summerh. based on *T. wakefieldii* (Rolfe) Summerh. and section *Tridactyle* (*Eutridactyle*) (Schltr.) Summerh. comprising the remaining species. He subdivided the latter into eight groups based on inflorescence length, lip dissection and stipes structure but gave none of these formal taxonomic recognition. *T. wakefieldii* has been subsequently transferred to the genus *Solenangis* by Cribb & J. Stewart in K.B. 40: 413 (1985). Consequently all of the species now recognised in the genus are considered to belong to the typical section.

Spur more than 3 cm. long:
- Stem short, usually less than 6 cm. long, leaf-sheaths distichously overlapping, acute on each side when the leaf is shed . . . . 26. *T. citrina*
- Stem elongate, usually more than 15 cm. long; leaf-sheaths not overlapping:
  - Lip side lobes broadest across apex, flabellate, erose on front margins; bracts 6 mm. long . . . . . . 21. *T. unguiculata*
  - Lip side lobes slender, broadest across base, dissected at the apex; bracts 3 mm. long . . . . 20. *T. gentilii*

Spur less than 3 cm. long, often much shorter:
- Lip fan-shaped in outline when flattened, entire . . . 16. *T. flabellata*
- Lip not fan-shaped when flattened:
  - Leaves terete, sometimes grooved on their upper surface:
    - Leaves very slender, aciculate, 1–1.5 mm. in diameter, sharply pointed at the apex; spur usually less than 8 mm. long . . . . 3. *T. filifolia*
    - Leaves 1.5–2 mm. in diameter, rather blunt at the apex; spur usually more than 8 mm. long . . . . . 17. *T. tridentata*
  - Leaves dorso-ventrally flattened:
    - Leaf-bases distichously overlapping, leaving acute or acuminate points at each side when the leaves drop; leaves arranged in a somewhat fan-like manner on the stem:
      - Leaves spotted dark green on a paler background, each apical lobe strongly bidentate; side lobes of the lip obscure, shoulder-like . . . . . . 25. *T. tanneri*
      - Leaves unspotted, apical lobes entire or very rarely weakly bidentate; side lobes of lip, if present, spreading, acute, acuminate or dissected at the apex:
        - Spur inflated towards the apex, 2 mm. in diameter, leaves 15–19 cm. long . . . . 24. *T. sarcodantha*
        - Spur slender, not inflated; leaves less than 15 cm. long, unequally roundly bilobed at the apex:
          - Flowers small; tepals 5–7.5 mm. long; spur 12–14 mm. long . . . . 22. *T. tricuspis*
          - Flowers larger; tepals 8–12 mm. long; spur 14–17 mm. long . . . . 23. *T. furcistipes*
    - Leaf-bases not distichously overlapping, cylindrical, somewhat truncate at the apex when the leaves are shed:
      - Spur upcurved, strongly dilated at the apex . . . 27. *T. sp. A*
      - Spur straight or incurved:
        - Stems less than 10 cm. long . . . . 15. *T. brevifolia*
        - Stems elongate, over 20 cm. long, sometimes branching:
          - Lip entire or very obscurely 3-lobed with side lobes shoulder-like or dentate:
            - Leaves fleshy-coriaceous or coriaceous; lip entire; spur straight:

- Leaves 3–7.5 cm. long, fleshy, margins reflexed when dry; stem branching at right-angles regularly . . . . . 1. *T. anthomaniaca*
- Leaves 8–9.5 cm. long, leathery, margins incurved when dry; stems not branching . . . . 2. *T. oblongifolia*

Leaves relatively thin-textured; lip obscurely 3-lobed; spur decurved:

- Spur very short, 3 mm. long, inflated at the apex; tepals ± 2 mm. long . . . . 5. *T. minuta*
- Spur 10–12 mm. long, not inflated at the apex; tepals 2.8–4 mm. long . . . 4. *T. virgula*

Lip prominently 3-lobed; side lobes triangular to linear-elongate, often dissected at the apex:

Leaves somewhat fleshy and leathery:

- Leaves 7–10 mm. wide, obtusely bifid at the apex; plants erect; stems stout, ± 8 mm. in diameter; roots verrucose . . . 12. *T. verrucosa*
- Leaves 3–5 mm. wide, acutely bifid at the apex; plants pendulous; stems slender, less than 4 mm. in diameter; roots smooth . . . . 7. *T. nigrescens*

Leaves relatively thin-textured:

- Flowers glistening white with a red anther-cap . . . . 10. *T. virginea*
- Flowers yellow, green or ochre:
  - Inflorescences very short, less than 12 mm. long:
    - Spur inflated in the apical half; roots verrucose . . . . 9. *T. inflata*
    - Spur not inflated, slender; roots smooth:
      - Leaves elongate, 9–19 cm. long; spur 9.5–10 mm. long; flowers greenish yellow to apricot 6. *T. scottellii*
      - Leaves 4–6 cm. long; spur 6–7 mm. long; flowers whitish or green 8. *T. eggelingii*
  - Inflorescences more than 13 mm. long, often much longer:
    - Leaves slender, grass-like, less than 2.5 mm. wide; lip 3-lobed in the basal half, with recurved side lobes, erose on their front margins; roots markedly verrucose; stem very slender, 2 mm. or less in diameter . . . . 14. *T. cruciformis*
    - Leaves linear to ligulate, 4–19 mm. wide; lip 3-lobed in the middle or apical half; side lobes spreading, entire or dissected; roots smooth or verrucose; stem more than 3 mm. in diameter:
      - Roots verrucose; inflorescences 1.3–1.6 cm. long, 4–6-flowered; lip less than 3 mm. long . . . 11. *T. inaequilonga*
      - Roots smooth or rarely slightly rough; inflorescences more than 2 cm. long, usually 6- or more-flowered; lip more than 3 mm. long:

Leaves 4–5 mm. wide, with erosely subacute apical lobes; flowers green or pale yellowish green; spur slightly inflated in the apical half . . . . . . 13. *T. phaeocephala*
Leaves 5–19 mm. wide, with entire rounded apical lobes; flowers ochre, yellow or creamy white; spur slender, not inflated:
Spur 8.5–10 mm. long; side lobes of the lip ± equal in length to mid-lobe, entire or rarely slightly bifid at the apex; inflorescences 1.7–5.5 cm. long . . . . . 18. *T. tridactylites*
Spur 11–17 mm. long; side lobes of the lip much longer than the mid-lobe, divided at the apex; inflorescences usually 8–11 cm. long . . . . 19. *T. bicaudata*

1. **T. anthomaniaca** (*Reichb. f.*) *Summerh.* in K.B. 3: 284 (1948) & in F.W.T.A., ed. 2, 3: 274 (1968); J. Stewart in U.K.W.F.: 802 (1974); Williamson, Orch. SC. Afr.: 202 (1977); Ball, S. Afr. Epiph. Orch.: 220 (1978); la Croix et al., Malawi Orch.: 134 (1983). Type: Nigeria, Bonny, *Monteiro* (W, holo., K, iso.!)

A pendulous epiphyte with stems up to 45 cm. or more long, 5–6 mm. diameter, little branched, often producing sizeable festoons; roots 2 mm. diameter, verrucose. Leaves fleshy-coriaceous, linear to narrowly elliptic-oblong, unequally obtusely bilobed at apex, 3.7–8 cm. long, 0.6–1.9 cm. wide, pale green, twisted at base to lie in one plane; leaf-bases ridged and rough, 12–16 mm. long. Inflorescences very short, up to 1 cm. long, 2–4-flowered; peduncle 1–2 mm. long; bracts ovate, 1 mm. long. Flowers apricot, dirty orange or brown; pedicel and ovary 3–4.5 mm. long, lepidote. Dorsal sepal oblong or elliptic, acute to obtuse, 3.7–5.7 mm. long, 1.7–2.5 mm. wide; lateral sepals similar but oblique at base, 3.7–5.3 mm. long, 2.2–2.8 mm. wide. Petals lanceolate, acute, 4.6–5.2 mm. long, 1.1–1.7 mm. wide. Lip entire, lanceolate, acute, 3–6 mm. long, 1–1.6 mm. wide, with 2 basal fleshy auricles at mouth of spur; spur filiform, slightly sigmoid or straight, (6–)11–16 mm. long. Column 1.2–2.2 mm. long; stipes oblanceolate, emarginate at apex; viscidium circular.

UGANDA. Masaka District: NW. side of Lake Nabugabo, 9 Oct. 1953, *Drummond & Hemsley* 4720; Mengo District: Entebbe, June 1905, *E. Brown* 252! & Mar. 1936, *Chandler* 1617!
KENYA. N. Kavirondo District: Kakamega, Dec. 1943, *Tweedie* 616! & 9 Apr. 1944, *Carroll* in *Moreau* 814!
TANZANIA. Lushoto District: E. Usambara Mts., Amani, 1 Dec. 1940, *Moreau* 25! & 20 Jan. 1950, *Verdcourt* 50!; Rungwe District: S. of Tukuyu, near head of Lake Malawi [Nyasa], fl. cult. 17 Nov. 1943, *Moreau* 733!
DISTR. **U** 4; **K** 5; **T** 3, 7; Sierra Leone to Zaire and Central African Republic and south to Mozambique, Malawi, Zambia and Zimbabwe
HAB. In riverine, swamp and lower montane forest; 750–1550 m.

SYN. *Angraecum anthomaniaca* Reichb. f. in Linnaea 41: 75 (1877)
A. lepidotum Rolfe in F.T.A. 7: 146 (1897).* Types: Liberia, without exact locality, cult. *Christy* (K, syn.!); Nigeria, Bonny, *Monteiro* (K, syn.!) & Lagos, *Moloney* (K, syn.!)
A. wittmackii Kraenzl. in Mitt. Deutsch. Schutzgeb. 2: 158 (1889). Type: Cameroon, without exact locality, *Braun* Angraecum no. 2 (B, holo.)
Listrostachys wittmackii (Kraenzl.) Rolfe in F.T.A. 7: 158 (1897)
Angraecum trachyrhizum Schltr. in E.J. 26: 343 (1899). Type: Mozambique, near km. 25 in territory of Mozambique Co., *Schlechter* 12256 (B, holo., K, iso.!)
Tridactyle lepidota (Rolfe) Schltr. in Beih. Bot. Centr. 36: 146 (1918)
T. trachyrhiza (Schltr.) Schtr. in Beih. Bot. Centr. 36: 147 (1918)

* Mentioned as a name by Reichenbach in Gard. Chron., ser. 2, 13: 806 (1880).

2. **T. oblongifolia** *Summerh.* in K.B. 3: 286 (1948) & in F.W.T.A., ed. 2, 3: 274 (1968). Type: Central African Republic, Yadele, *Le Testu* 4314 (K, holo.!)

An epiphyte with few-branched stems, up to 8 cm. long, 4–5 mm. in diameter; roots verrucose. Leaves coriaceous, narrowly oblong or ligulate, unequally obtusely bilobed at apex, 5.8–10.5 cm. long, 0.8–1.6 cm. wide, twisted at base to lie in one plane, articulated to reticulately veined sheaths, 16–20 mm. long. Inflorescences very short, 2–3-flowered; peduncle up to 4 mm. long; bracts triangular. Flowers greenish white turning yellow with age, lepidote on outside of sepals; pedicel and ovary 3.5 mm. long, lepidote. Dorsal sepal oblong-ovate, acute, 4.8–5.2 mm. long, 1.8–2 mm. wide; lateral sepals similar but oblique, 4–4.4 mm. long, 2–2.5 mm. wide. Petals lanceolate, acuminate, 4.4–5 mm. long, 1–1.2 mm. wide. Lip entire, lanceolate, acute, 3.7–4.5 mm. long, 1–1.3 mm. wide, with a tooth on each side at the mouth of the spur; spur filiform, slightly geniculate in basal half, slightly dilated towards apex, 13–14.5 mm. long. Column 1.6–1.8 mm. long.

UGANDA. Masaka District: Buddu County, *Dawe* 311!
DISTR. U 4; Sierra Leone, Ivory Coast, Nigeria, Central African Republic and Zaire
HAB. In riverine forest; 900–1300 m.

3. **T. filifolia** (*Schltr.*) *Schltr.* in Beih. Bot. Centr. 36: 144 (1918); Williamson in Journ. S. Afr. Bot. 48: 15 (1982). Type: Gabon, by Sanga R., *Schlechter* 12791 (B, holo., K, holo.!)

A pendent epiphyte, often forming large masses. Stem with a few branches, 10–80 cm. or more long, 1.2–1.5 mm. in diameter; roots basal, smooth, 1–2 mm. in diameter. Leaves suberect to falcate, terete, 3.5–14 cm. long, 0.3–1 mm. wide, usually grooved on the upper surface, articulated to 0.8–1 cm. long bases. Inflorescences 2–8 mm. long, 2–3-flowered; pedicel 1–2 mm. long; bracts amplexicaul, triangular, 0.5 mm. long. Flowers dingy white; pedicel and ovary 2.2–4 mm. long, lepidote. Dorsal sepal lanceolate, acute, 2.3–3.3 mm. long, 1–1.8 mm. wide; lateral sepals obliquely ovate, acute, 2.4–3.5 mm. long, 1.2–2 mm. wide. Petals linear-lanceolate, acute, 2–2.9 mm. long, 0.6–1.1 mm. wide. Lip auriculate at the base, 3-lobed in the apical half, 1.7–3.4 mm. long, 0.8–2 mm. wide; side lobes spreading, acute, half the length of the mid-lobe; mid-lobe triangular, acute, ± 1 mm. long; spur filiform, incurved, 6–8.3 mm. long. Column 0.7–1.2 mm. long. Fig. 153.

UGANDA. Bunyoro District: Budongo Forest, Oct. 1935, *Eggeling* 2235! & Sept. 1943, *Eggeling* 5186!; Masaka District: Sese Is., *Dawe* 12! & 23 Feb. 1933, *A.S. Thomas* 858!; Mengo District: Entebbe, Dec. 1905, *E. Brown* 329! & Mar. 1943, *Eggeling* 5279!
KENYA. N. Kavirondo District: Kakamega Forest, fl. cult. 17 Oct. 1943, *Carroll* in *Moreau* 527A! & 15 Oct. 1953, *Drummond & Hemsley* 4794!
TANZANIA. Masai District: Samatwa, 3 May 1943, *Pullon* in *Moreau* 613!; Lushoto District: E. Usambara Mts., Amani, 6 Apr. 1941, *Moreau* 57!
DISTR. **U** 2, 4; **K** 5; **T** 2, 3; Sierra Leone to Ethiopia, and south through Zaire to N. Zambia
HAB. In rain-forest; 850–1600 m.

SYN. *Angraecum filifolium* Schltr., Westafr. Kautsch.-Exped.: 284 (1900)
Listrostachys linearifolium De Wild., Not. Pl. Utiles Congo 1: 149 (1903). Types: Zaire, Lomami, *E. Laurent* & near Eala, *M. Laurent* 25 (BR, syn.)
Tridactyle linearifolia (De Wild.) Schltr. in Beih. Bot. Centr. 36: 146 (1918)
T. tridentata (Harv.) Schltr. var. *subulifolia* Summerh. in K.B. 3: 287 (1948). Type: Uganda, Bunyoro District, Budongo Forest, *Eggeling* 5277 (K, holo.!)

NOTE. For a discussion of the relationship of this species to *T. tridentata* see the note under species 16 here.

Summerhayes distinguished as var. *subulifolia* those specimens with leaves lacking the groove on their upper surface. Now that more specimens are available for study, it seems that this distinction cannot be maintained as several specimens have been seen in which the leaves are partially grooved. Even Summerhayes suggested that his variety was difficult to distinguish on floral features alone.

4. **T. virgula** (*Kraenzl.*) *Schltr.* in Beih. Bot. Centr. 36: 148 (1918). Type: Zaire, Ruwenzori, *Stuhlmann* 2336 (B, holo.)

A pendent tufted epiphyte with arcuate little-branched stems 10–65(–400) cm. long, 2.5 mm. diameter; roots smooth, 2–3 mm. in diameter. Leaves linear, unequally acutely bilobed at the apex, 9–16 cm. long, 0.3–0.5 cm. wide, twisted at the base and articulated to spotted sheaths, 1–2 cm. long. Inflorescences 4–5 mm. long, 2–4-flowered; rhachis fractiflex; bracts ovate, 1 mm. long. Flowers greenish white or whitish; pedicel and ovary

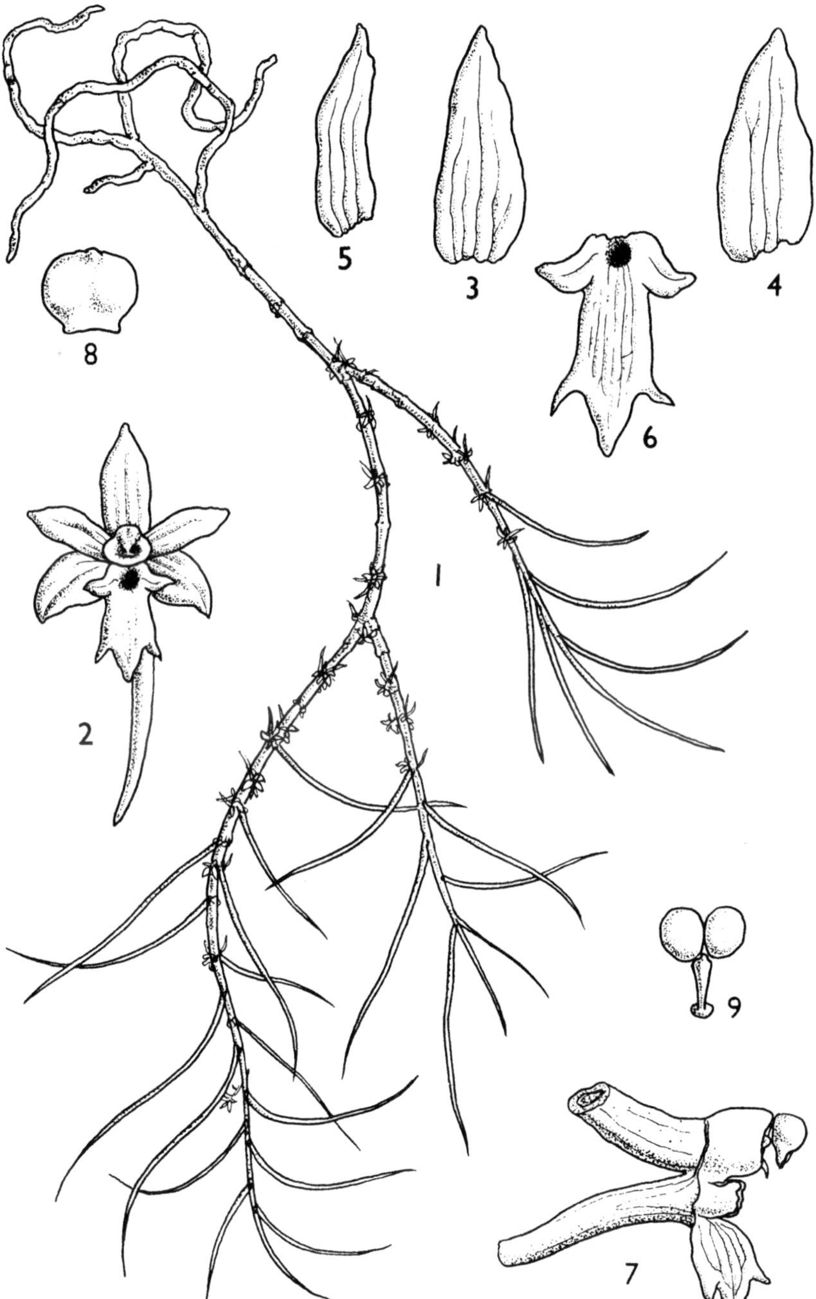

FIG. 153. *TRIDACTYLE FILIFOLIA*—**1**, habit, × ⅔; **2**, flower, × 2⅔; **3**, dorsal sepal, × 12⅔; **4**, lateral sepal, × 12⅔; **5**, petal, × 12⅔; **6**, lip, × 10⅔; **7**, ovary, column and lip, × 6⅔; **8**, anther-cap, × 13⅓; **9**, pollinarium, × 13⅓. Drawn from *Eggeling* 2235 by Cherry Ann Lavrih.

3–6 mm. long. Dorsal sepal ovate-elliptic, acute, 3–3.7 mm. long, 1.6–2.2 mm. wide; lateral sepals obliquely ovate, acute, 3.2–4 mm. long, 1.9–2.1 mm. wide. Petals oblong-linear, acute, 2.5–3.6 mm. long, 1–2 mm. wide. Lip cuneate below, 3-lobed in apical part, 2.5–3.3 mm. long, 1.5–2.5 mm. wide, with fleshy auricles at base; side lobes spreading, short, subacute; mid-lobe slightly longer, acute, 0.7 mm. long, 0.5 mm. wide; spur decurved, slightly apically inflated, 8–12 mm. long. Column 1–1.5 mm. long.

UGANDA. Toro District: May 1939, *Chandler* 2798!; Kigezi District: Impenetrable Forest, Sept. 1936, *Eggeling* 3306! & Rutenga, Apr. 1948, *Purseglove* 2621!
DISTR. U 2; Zaire, Rwanda
HAB. In undergrowth of montane forest; 1770–2500 m.

SYN. *Angraecum virgulum* Kraenzl. in E.J. 22: 27 (1895)

5. **T. minuta** *Cribb*, sp. nov. in habitu *T. virgulae* (Kraenzl.) Schltr. similis sed radicibus verrucosis, inflorescentia breviore, floribus minutis luteo-viridis, calcare clavato incurvato 3 mm. longo satis differt; e *T. cruciforme* Summerh. foliis latioribus, floribus minoribus, labello angustiore et calcare breviore clavato distinguenda. Type: Tanzania, Iringa District, Mwanihana Forest, *D.W. Thomas* 3764 (K, holo.!)

An epiphyte with pendent straggling little-branched arcuate stems, 2–2.5 mm. in diameter, leafy in apical part; roots 2 mm. in diameter, verrucose. Leaves grass-like, suberect, linear, very unequally acutely bilobed at the apex, 7.3–11.5 cm. long, 1.5–3 mm. wide, twisted and articulated at base to a spotted sheath 15–22 mm. long. Inflorescences 5–7 mm. long, 3–5-flowered; peduncle 1–2 mm. long; rhachis fractiflex; bracts ovate, 0.5 mm. long. Flowers greenish yellow with an orange column; pedicel and ovary 2.5 mm. long. Dorsal sepal elliptic-ovate, obtuse, 1.9–2 mm. long, 1.2–1.3 mm. wide; lateral sepals oblong-ovate or oblong, obtuse, 1.8–2 mm. long, 1–1.2 mm. wide. Petals elliptic, acute, 1.6–1.7 mm. long, 0.7–0.8 mm. wide. Lip 3-lobed in apical half, deflexed, 1.5 mm. long, 1.8 mm. wide; side lobes spreading, triangular, ± as large as the mid-lobe; mid-lobe triangular, acute; spur slightly incurved, clavate, 3 mm. long. Column 0.5–0.8 mm. long. Fig. 154.

TANZANIA. Iringa District: Uzungwa, Muhanga, 1 Nov. 1980, *Leedal* 6138! & Mwanihana Forest Reserve, above Sanje, 10 Oct. 1984, *D.W. Thomas* 3764!
DISTR. **T** 7; known only from the above collections
HAB. Epiphyte in dwarf montane mist-forest on steep slopes; 1400–1700 m.

6. **T. scottellii** (*Rendle*) *Schltr.* in Beih. Bot. Centr. 36: 147 (1918); J. Stewart in U.K.W.F.: 802 (1974). Type: Zaire, Ruwenzori, *Scott Elliott* 8100 (BM, holo., K, iso.!)

An epiphyte with straggling compressed pendent stems, 8–50 cm. long, 2–3 mm. in diameter; roots 2–3 mm. in diameter, smooth. Leaves grass-like, linear, unequally subacutely bilobed at the apex, 7–19 cm. long, 3.5–6 mm. wide, twisted and articulated at the base to slightly ancipitous sheaths, 1.4–2 cm. long. Inflorescences 3–4 mm. long, 1–3-flowered; peduncle very short; bracts ovate, 1–2 mm. long. Flowers greenish yellow becoming orange-brown with age; pedicel and ovary 6 mm. long. Dorsal sepal ovate-elliptic, acute, 4.3–6 mm. long, 1.7–2.6 mm. wide; lateral sepals oblong-ovate, acute, 4.8–5.9 mm. long, 1.9–2.6 mm. wide. Petals lanceolate, acuminate, 4.4–5.5 mm. long, 0.9–1.4 mm. wide. Lip strongly 3-lobed in apical half, auriculate at the base, 4–5.3 mm. long, 6.5–7 mm. wide; side lobes longer than the mid-lobe, filiform, occasionally bilobed at the apex, 3–3.8 mm. long; mid-lobe triangular, acute, 2.5 mm. long, 1 mm. wide; spur straight, 9–12 mm. long. Column 1.7–1.9 mm. long.

UGANDA. Mbale District: Elgon, 29 June 1924, *Snowden* 903!
KENYA. Northern Frontier Province: Karisia Hills, SE. of Maralal, 16 Feb. 1969, *J. Stewart* 971!; Elgeyo District: Kamorin, fl. cult. July 1931, *Tweedie* 20!; Fort Hall District: Thika, Chania, Sept. 1945, *Copley* 67A!
DISTR. **U** 3; **K** 1, 3–5; Zaire and Rwanda
HAB. In upland evergreen forest and riverine forest; 2000–2800 m.

7. **T. nigrescens** *Summerh.* in K.B. 3: 288 (1948). Type: Uganda, Ankole, Kyamahunga, near Kalinzu Forest, *Eggeling* 3722 (K, holo.!)

An epiphyte or lithophyte, with stems 13–30 cm. long, 1–1.5 mm. in diameter; roots 1–1.5 mm. in diameter, smooth. Leaves drying black, fleshy-coriaceous, conduplicate,

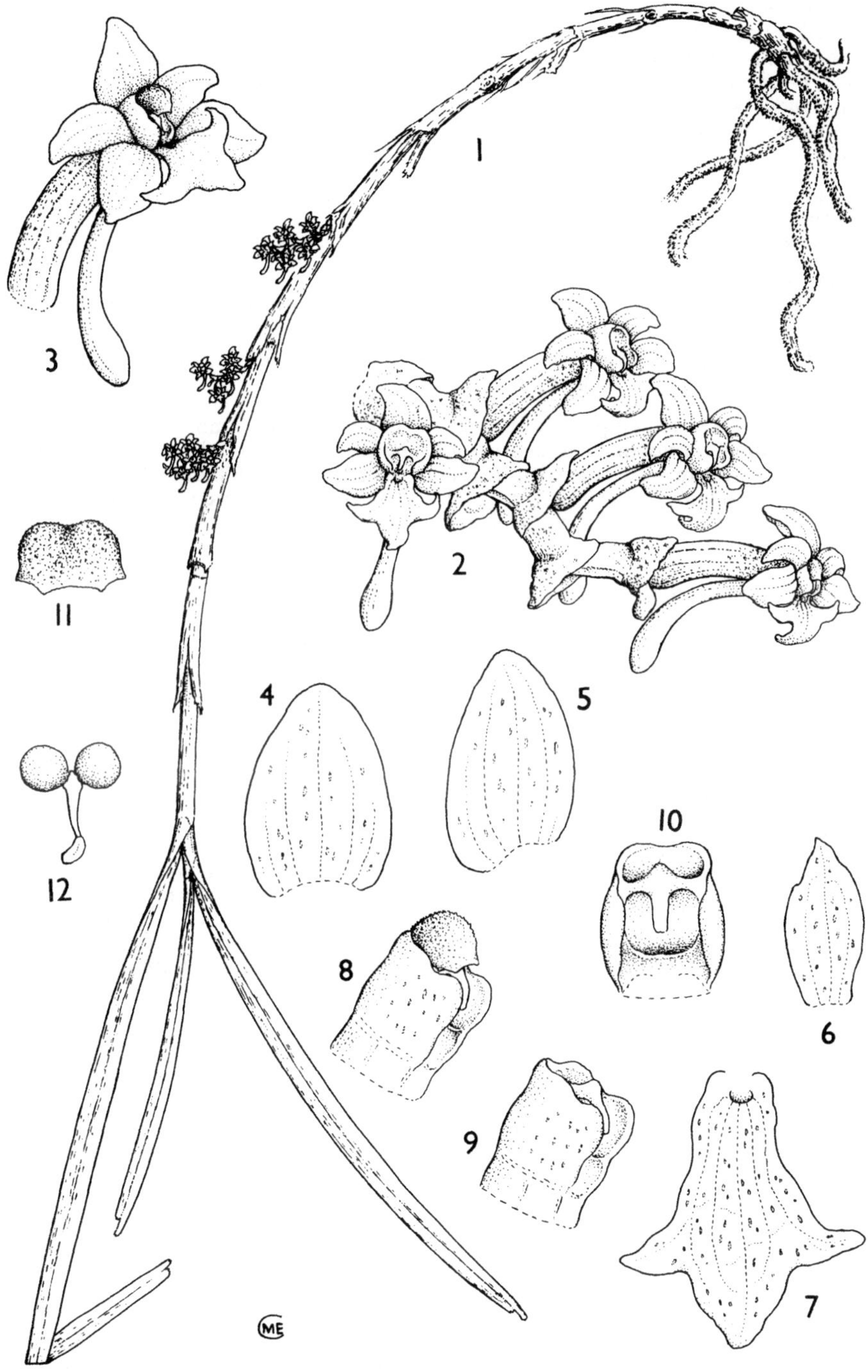

FIG. 154. *TRIDACTYLE MINUTA*—**1**, habit, × 1; **2**, inflorescence, × 8; **3**, flower, × 10; **4**, dorsal sepal, × 14; **5**, lateral sepal, × 14; **6**, petal, × 14; **7**, lip, × 20; **8**, column, side view, × 16; **9**, same, without anther-cap, × 16; **10**, column, front view, × 20; **11**, anther-cap, × 20; **12**, pollinarium, × 20. Drawn from *Leedal* 6138 by Mrs M.E. Church.

reflexed or spreading, linear, unequally acutely bilobed at the apex, 3–6.5 cm. long, 1–2.5 mm. wide, articulated to 1 cm. long sheaths. Inflorescences 5–6 mm. long, 2–3-flowered; peduncle 1 mm. long; bracts ovate, amplexicaul, 0.5 mm. long. Flowers pale lime-green or dull yellow-orange, sweetly scented; pedicel and ovary 2.5–3 mm. long, sparsely lepidote. Dorsal sepal elliptic, obtuse, 3–3.5 mm. long, 1.6–2.1 mm. wide; lateral sepals oblong-ovate, obtuse, 3–3.5 mm. long, 1.7–2.3 mm. wide. Petals lanceolate, acute, 2.7–3.8 mm. long, 1–1.3 mm. wide. Lip 3-lobed at the apex, auriculate at the base, 2.6–4 mm. long, 2–2.2 mm. wide; side lobes longer than the mid-lobe, 3-lacerate at the apex; mid-lobe triangular; spur pendent, slightly inflated in middle, 6–7.5 mm. long. Column 1–1.3 mm. long.

UGANDA. Masaka District: Katera, 3 May 1972, *Lye* 6808! & Kyanamulnalna, Nov. 1942, *A.S. Thomas* 4088!
KENYA. Trans-Nzoia District: Cherangani Hills, Kapolet [Kabolet], *Jackson* in *Tweedie* 3492!
TANZANIA. Bukoba, Dec. 1940, *Tweedie* 557!
DISTR. **U** 2, 4; **K** 3; **T** 1; Zaire
HAB. In dry sclerophyllous forest, woodland, riverine forest and on rocks above lake; 1150–2200 m.

8. **T. eggelingii** *Summerh.* in K.B. 3: 289 (1948). Type: Uganda, Kigezi District, Impenetrable Forest, *Eggeling* 4186 (K, holo.!)

A pendulous tufted epiphyte with stems up to 60 cm. long, 3 mm. in diameter; roots 2 mm. in diameter, smooth. Leaves spreading, linear, unequally bilobed at the apex, 2.7–6.5 cm. long, 2.5–5 mm. wide, twisted and articulated at the base to lepidote sheaths, 1–1.3 cm. long. Inflorescences very short, 8 mm. long, 2–3-flowered; bracts obtuse, scarcely 1 mm. long, ochreate. Flowers whitish; pedicel and ovary 3–4 mm. long, lepidote. Dorsal sepal ovate, obtuse, 4.2–4.5 mm. long, 2.2–2.5 mm. wide; lateral sepals oblong-ovate, obtuse, 4.3–4.6 mm. long, 2.5–2.7 mm. wide. Petals oblong, acute, 3.8–4.3 mm. long, 1.2–1.7 mm. wide. Lip 3-lobed in the apical half, basally auriculate, 4–4.3 mm. long, 4.3–6 mm. wide; side lobes spreading, longer than the mid-lobe, linear-tapering, 2–3-toothed or -fimbriate at the apex, ± 3 mm. long, papillose at the apex; mid-lobe triangular, 1.7–1.9 mm. long. 0.7–1.2 mm. wide; spur ± straight, filiform, 6–7 mm. long. Column 1.5–1.6 mm. long.

UGANDA. Kigezi District: near Karungu, Mar. 1962, *C. Leakey* OR 17!
DISTR. **U** 2; Zaire
HAB. On exposed trees; 1600–2200 m.

9. **T. inflata** *Summerh.* in K.B. 3: 291 (1948). Type: Tanzania, S. Masailand, Njoro Hill, *Elliott* in *Moreau* 326 (K, holo.!)

An epiphyte forming dense clumps with stems 12–20 cm. or more long, 5 mm. in diameter; roots 1.5–2 mm. long, verrucose. Leaves linear, unequally obtusely or roundly bilobed at the apex, 5–7.3 cm. long, 0.6–0.8 cm. wide, articulated to sheaths 8 mm. long. Inflorescences up to 11 mm. long, 3-flowered; bracts ovate-cupulate, amplexicaul, ± 1 mm. long. Flowers pale sea-green with an orange column, with a slight diurnal scent; pedicel and ovary 5–6 mm. long. Dorsal sepal oblong-elliptic, obtuse, 3.5–4 mm. long, 1.5–1.8 mm. wide; lateral sepals obliquely oblong, obtuse, 3.7–4 mm. long, 2–2.3 mm. wide. Petals oblong, obtuse, 3.5–3.8 mm. long, 1.3–1.5 mm. wide. Lip flabellate, obscurely 3-lobed at the apex, 3 mm. long, 3.1–3.3 mm. wide, auriculate at the base; side lobes triangular, several-toothed on the apical margin; mid-lobe shortly triangular, 0.7–0.8 mm. long, 1.2–1.3 mm. wide; spur straight, clavate, 5.5 mm. long. Column 1–1.2 mm. long.

TANZANIA. Masai District: S. Masailand, Njoro Hill, *Elliott* in *Moreau* 326!
DISTR. **T** 2; known only from the type collection
HAB. Epiphyte on thorn and *Combretum* trees in mist zone; 1600–1800 m.

10. **T. virginea** *Cribb & la Croix* sp.nov. affinis *T. inaequilonga* (De Wild.) Schltr. sed vaginis foliorum ligulatis, floribus albis, anthera lateritia et labelio obscure trilobato satis differt. Typus: Malawi, N. Viphya Plateau, *la Croix* 862 (K, holo.!)

An erect to trailing epiphyte with slender rather flattened stems, 8–40 cm. long, 4 mm. in diameter; roots verrucose, ± 3 mm. in diameter. Leaves 3–7, linear, unequally acutely bilobed at the apex, 6–13 cm. long, 0.7–1.1 cm. wide, twisted at the base and articulated to 6–13 mm. long leaf-sheaths, each with a slender setose ligule opposite the leaf. Inflorescences 0.7–2.7 cm. long, 3–11-flowered; peduncle very short; bracts amplexicaul, ovate-triangular, 1–1.5 mm. long. Flowers slightly campanulate, glistening white with a red anther-cap; pedicel and ovary 6–7 mm. long. Dorsal sepal ovate, acute to obtuse, 3.5–5 mm. long, 1.5–2.5 mm. wide; lateral sepals similar. Petals elliptic to ovate, acute or subacute, 3.5–4 mm. long, 1.3–1.5 mm. wide. Lip 3-lobed in the apical part, auriculate at the base, 3–4 mm. long, 2–3 mm. wide; side lobes spreading, triangular, acute, shorter than the mid-lobe; mid-lobe triangular, acute, 1 mm. long; spur ± parallel to the ovary, 5–7 mm. long, slightly inflated in the middle, acute or obtuse at the apex. Column 1 mm. long.

TANZANIA. ? Mpanda District: Kikongwa, Mipande Forest, 29 Oct. 1982, *Leedal* 7258!
DISTR. T 4; N. Malawi
HAB. In montane forest; ± 1600 m.

11. **T. inaequilonga** (*De Wild.*) *Schltr.* in Beih. Bot. Centr. 36: 145 (1918); Ball, S. Afr. Epiph. Orch.: 230 (1978); la Croix et al., Malawi Orch. 1: 137 (1983). Type: Zaire, Injolo, *Laurent* 2044 (BR, holo., K, record of holo.!)

An epiphyte or lithophyte with stems 20 cm. or more long, 3 mm. in diameter; roots 2–3 mm. in diameter, verrucose. Leaves linear, unequally acutely bilobed at the apex, 5–14 cm. long, 0.4–0.7 cm. wide, twisted at the base and articulated to spotted sheaths up to 13 mm. long. Inflorescences 13–17 mm. long, 4–6-flowered; bracts amplexicaul, ovate, 1 mm. long. Flowers greenish yellow to dull orange; pedicel and ovary 2.5 mm. long, lepidote. Dorsal sepal oblong, obtuse, 3 mm. long, 1.2–1.4 mm. wide; lateral sepals obliquely ovate, obtuse, 3 mm. long, 1.5 mm. wide. Petals oblong, acute, 3 mm. long, 1 mm. wide. Lip 3-lobed in the middle, auriculate at the base, 2.5 mm. long, 5.5 mm. wide; side lobes longer than the mid-lobe, 0–3-toothed on the front margins; mid-lobe narrowly triangular, sub-acute; spur straight, filiform, 7–8 mm. long. Column 1–1.3 mm. long.

TANZANIA. Lushoto District: W. Usambara Mts., Mazumbai, 18 Nov. 1974, *Baeslev* 301!; Songea District: Mbinga, 4 Mar. 1987, *Congdon* 169!
DISTR. T 3, 8; Zaire, Malawi and Zimbabwe
HAB. In montane forest; 1850–1900 m.

SYN. *Angraecum inaequilongum* De Wild. in B.J.B.B. 5: 187 (1916)
A. acutoemarginatum De Wild. in B.J.B.B. 5: 181 (1916). Type: Zaire, Eala, *Laurent* 1214 (BR, holo.)
Tridactyle acutoemarginatum (De Wild.) Schltr. in Beih. Bot. Centr. 36: 143 (1918)

12. **T. verrucosa** *Cribb* in K.B. 40: 415 (1985). Type: Tanzania, Songea District, Matengo Hills, *Milne-Redhead & Taylor* 8984 (K, holo.!)

A lithophyte with erect stems 8–20 cm. long, 5–8 mm. in diameter; roots basal, verrucose, 1–3 mm. in diameter. Leaves fleshy-coriaceous, suberect-spreading, narrowly oblong-elliptic, unequally roundly bilobed at the apex, 3–5.3 cm. long, 0.7–1 cm. wide. Inflorescences 1–1.7(–3) cm. long, densely 3–5-flowered, spreading-arcuate; peduncle and rhachis fleshy; bracts reflexed, ovate, 1–3 mm. long. Flowers green or yellow, sweetly scented, column orange; pedicel and ovary 3 mm. long. Dorsal sepal lanceolate, acuminate, 5 mm. long, 2 mm. wide; lateral sepals obliquely lanceolate, acuminate, 5 mm. long, 2 mm. wide. Petals lanceolate, acuminate, 5 mm. long, 1.5 mm. wide. Lip 3-lobed in the apical half, 4.5–5.5 mm. long, 6–7 mm. wide; side lobes longer than the mid-lobe, spreading, 3 mm. long, entire or bifid at the apex; midlobe narrowly triangular, acute; spur subclavate, decurved, 8–12 mm. long. Column 3–4 mm. long; stipes oblanceolate. Fig. 155, p. 616.

TANZANIA. Iringa District: Image Mt., Nov. 1953, *Carmichael* 315! & 15 Dec. 1961, *Richards* 15658! & 28 Dec. 1986, *Lovett & Congdon* 1221!; Songea District: Matengo Hills, Liwiri Kiteza [Luwiri Kitesa], 5 Mar. 1956, *Milne-Redhead & Taylor* 8984!
DISTR. T 7, 8; Malawi
HAB. On rocks in exposed montane situations and in *Brachystegia* woodland; 1500–2200 m.

13. **T. phaeocephala** *Summerh.* in K.B. 3: 292 (1948). Type: Tanzania, Uluguru Mts., Lukwangule Plateau, *E.M. Bruce* 729 (K, holo.!)

An epiphyte with erect or pendent stems, 9–30 cm. long, 2–3 mm. in diameter; roots 2 mm. in diameter, smooth. Leaves suberect-spreading, linear, unequally obtusely or roundly bilobed at the apex, 3–8(–14) cm. long, 0.4–0.5 mm. wide, sap-green, glossy;

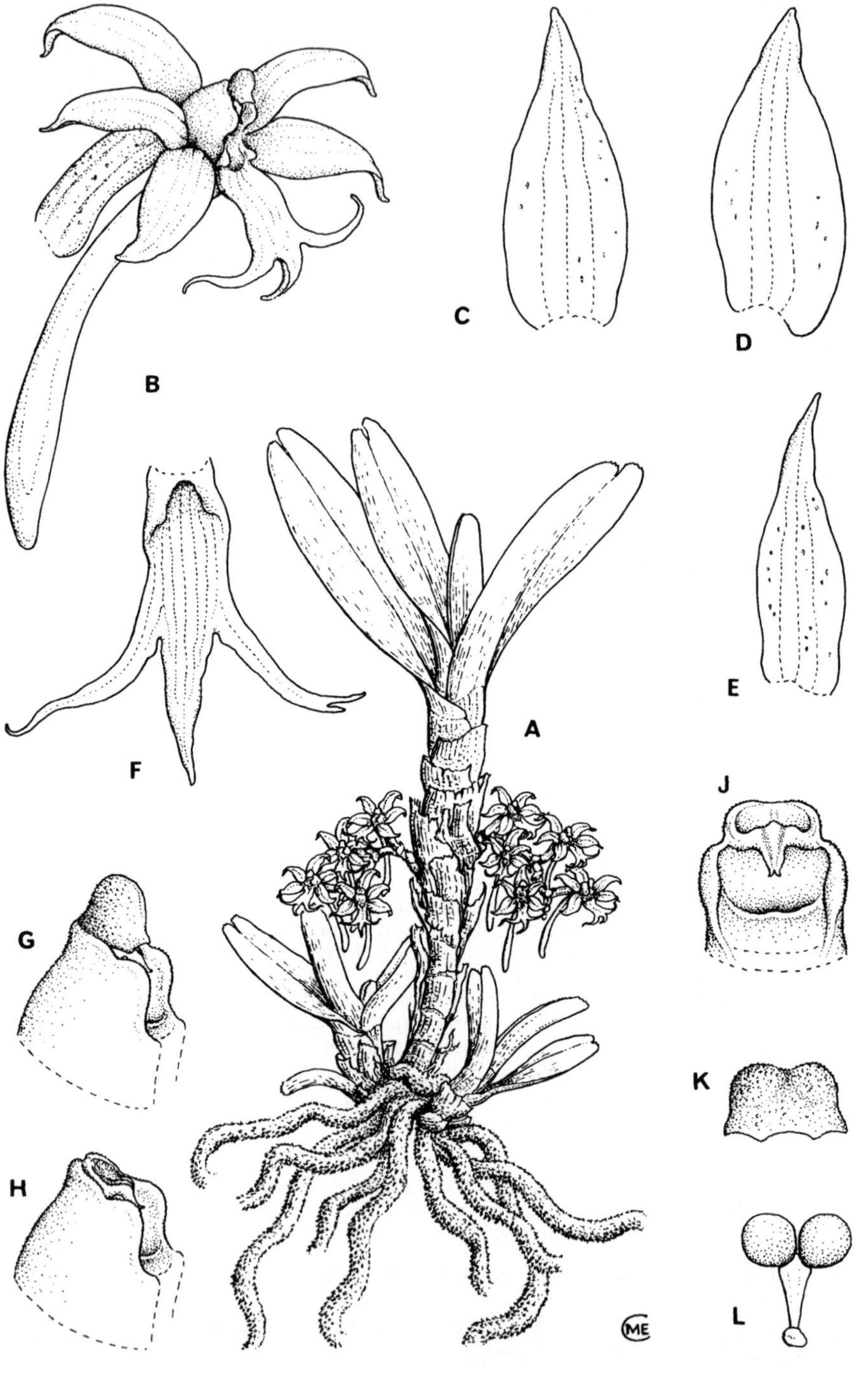

FIG. 155. *TRIDACTYLE VERRUCOSA*—**A**, habit, × 1; **B**, flower, × 6; **C**, dorsal sepal, × 8; **D**, lateral sepal, × 8; **E**, petal, × 8; **F**, lip, × 8; **G**, column, side view, × 12; **H**, same, anther removed, × 12; **J**, column, front view, showing rostellum, × 12; **K**, anther-cap, × 14; **L**, pollinarium, × 14. Drawn from *Milne-Redhead & Taylor* 8984 by Mrs M.E. Church.

twisted at base and articulated to sheaths, 9–14 mm. long. Inflorescences 1–4 cm. long, 3–7-flowered; bracts ovate, amplexicaul, 1.5–2 mm. long. Flowers pale green with a brown stigma; pedicel and ovary 3.5–4.5 mm. long, lepidote. Dorsal sepal oblong-elliptic, acute or obtuse, 4.8–6 mm. long, 2.1–2.4 mm. wide; lateral sepals obliquely ovate, acute or obtuse, 5–5.2 mm. long, 2.4–3 mm. wide. Petals oblong or lanceolate, acuminate, 4.5–5.7 mm. long, 1.7–1.8 mm. wide. Lip 3-lobed in the middle, auriculate at the base, 3.6–5.8 mm. long, 7 mm. wide; side lobes spreading, linear, 2.5–4 mm. long, 2–6-lacerate at apex; midlobe linear-tapering, acute, 2–2.5 mm. long; spur straight, inflated at apex, 8–9 mm. long. Column 1–2 mm. long.

TANZANIA. Pare Mts., fl. cult. 30 Aug. 1942, *Greenway* in *Moreau* 334!; Morogoro District: Uluguru Mts., W. side of Lukwangule Plateau, 19 Jan. 1943, *Moreau* 472! & 471! & Mar. 1955, *Semsei* 2014! & Jan. 1976, *Cribb & Grey-Wilson* 10462! & Bondwa Peak, by radio station, 22 Jan. 1976, *Cribb & Grey-Wilson* 10323!
DISTR. **T** 3, 6; not known elsewhere
HAB. In montane and dwarf montane forest; 2200–2400 m.

14. **T. cruciformis** *Summerh.* in K.B. 3: 293 (1948); J. Stewart in U.K.W.F.: 802 (1974). Type: Tanzania, E. Usambara Mts., Amani, *Moreau* 17 (K, holo.!)

An epiphyte with pendent stems, 9–60 cm. long, 2 mm. in diameter, ascending at the apex; roots 1.5–2.5 mm. in diameter, verrucose. Leaves grass-like, linear, unequally acutely bilobed, 7–23 cm. long, 1–2.5 mm. wide, glossy sap-green, twisted at base and articulated to spotted leaf-sheaths 15–17 mm. long. Inflorescences 2–5 cm. long 5–9-flowered; peduncle 2–8 mm. long; bracts amplexicaul, 1–2 mm. long. Flowers pale lime-green, green or yellow-green, with a sweet nocturnal scent; pedicel and ovary 3–3.6 mm. long, lepidote. Dorsal sepal elliptic-ovate, obtuse, 2.3–2.8 mm. long, 1.5–1.6 mm. wide; lateral sepals obliquely elliptic-oblong, obtuse, 2.5–3.3 mm. long, 1.5–1.6 mm. wide. Petals linear, acute or obtuse, 2–3 mm. long, 0.5–1 mm. wide. Lip 3-lobed in the middle, 1.7–2.3 mm. long, 1.8–3 mm. wide; side lobes spreading, narrower than the mid-lobe, acute; mid-lobe blunt; spur filiform, incurved, 11–13 mm. long. Column 0.7–0.8 mm. long. Fig. 156, p. 618.

KENYA. Masai District: Chyulu Hills, *Bally* 7856; Teita Hills, Mbololo Hill, Mraru Ridge, Jan. 1971, *Faden et al.* 71/46!
TANZANIA. Lushoto District: W. Usambara Mts., Mazumbai Forest. Jan. 1976, *Cribb & Grey-Wilson* 10051!, 10170! & 10239! & Dec. 1979, *L. Tanner* 225! & E. Usambara Mts., Amani, Jan. 1976, *Cribb & Grey-Wilson* 10277! & fl. cult. 28 Dec. 1941, *Moreau* 116!
DISTR. **K** 6, 7; **T** 3; not known elsewhere
HAB. In upland rain-forest; 600–1550 m.

15. **T. brevifolia** *Mansf.* in N.B.G.B. 11: 1063 (1934). Type: Tanzania, Morogoro District, Uluguru Mts., NW. side, *Schlieben* 2982 (B, holo., K, part of holo.!)

A small erect epiphyte with 6–10 cm. long stems, 2 mm. in diameter, leafy in apical part; roots 1–2 mm. in diameter, smooth. Leaves linear, unequally roundly bilobed at the apex, 2–4.5 cm. long, 2–4 mm. wide, articulated to 6–7 mm. long leaf-sheaths. Inflorescences 2–3 cm. long, 4–9-flowered; peduncle 5 mm. long; bracts amplexicaul, triangular, 0.5–1.5 mm. long. Flowers butterscotch-yellow with a reddish anther-cap; pedicel and ovary 4–7 mm. long. Dorsal sepal elliptic, obtuse, 3.5–3.7 mm. long, 2 mm. wide; lateral sepals obliquely ovate, obtuse, 3.5–4.2 mm. long, 2.7–2.8 mm. wide. Petals oblong-elliptic, obtuse, 3–3.5 mm. long, 1.5–1.6 mm. wide. Lip 3-lobed at the apex, auriculate at the base, 3–3.5 mm. long, 2–2.2 mm. wide; side lobes obscure; mid-lobe oblong, 3 mm. long; spur clavate, pendent, 6–7 mm. long. Column 1 mm. long.

TANZANIA. Morogoro District: Uluguru Mts., Bondwa Peak, below ridge to the summit, Jan. 1976, *Cribb & Grey-Wilson* 10311 & 10381! & Jan. 1975, *Mwasumbi & Kinunda* 11312!; Nguru Mts., N. source of Chazi R., 4 Feb. 1989, *Pócs* 89053B!
DISTR. **T** 6; known only from the Uluguru and Nguru Mts.
HAB. In montane mist-forest, on small trees; 1650–2100 m.

16. **T. flabellata** *Cribb*, sp. nov. affinis *T. inflatae* Summerh. sed habitu minore, caulibus angustioribus brevioribus, foliis angustioribus brevioribus, inflorescentiis suberectis longioribus, floribus candidis, petalis obovatis ad apicem rotundatis et labello flabellato

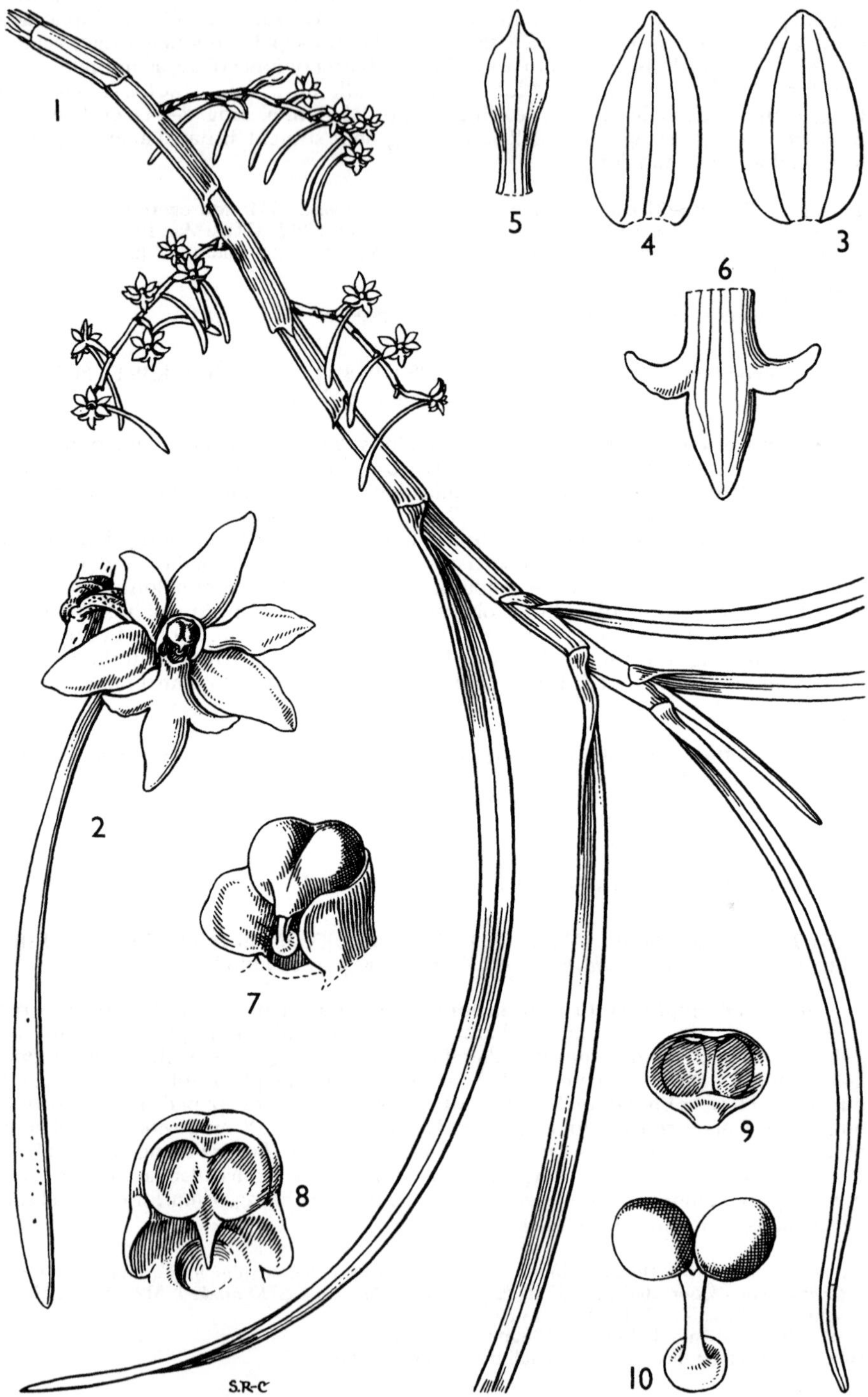

FIG. 156. *TRIDACTYLE CRUCIFORMIS*—**1**, habit, × 1; **2**, flower, × 6; **3**, dorsal sepal, × 8; **4**, lateral sepal, × 8; **5**, petal, × 8; **6**, lip, × 8; **7**, column, × 16; **8**, same, front view, anther removed, × 16; **9**, anther-cap, × 16; **10**, pollinarium, × 32. Drawn from *Moreau* 17 by Miss Stella Ross-Craig. Reproduced from Hooker's Icones Plantarum.

ad apicum truncato leviter eroso satis distinguenda. Type: Tanzania, Iringa District, Mwanihana Forest Reserve, *D.W. Thomas* 3763 (K, holo.!, MO, iso.)

An erect to arcuate epiphyte; stems 6–16 cm. long, 2–2.5 mm. in diameter; roots slender, smooth, 1–2 mm. in diameter. Leaves suberect-arcuate, linear, unequally acutely bilobed at the apex, 2.5–5 cm. long, 2–4 mm. wide, twisted at the base and articulated to black-spotted leaf-sheaths 5–6 mm. long. Inflorescences 1-several, 1–3.7 cm. long, 3–12-flowered; peduncle 2–6 mm. long; bracts amplexicaul, ovate, acute or obtuse, 1–2.5 mm. long. Flowers white with a brown anther-cap, cupulate; pedicel and ovary 2–3 mm. long. Dorsal sepal elliptic, obtuse, 2.5–4 mm. long, 1.5–2 mm. wide; lateral sepals obliquely oblong-elliptic, obtuse, 4–5 mm. long, 2 mm. wide. Petals obovate, obtuse, 3 mm. long, 2 mm. wide. Lip flabellate when flattened, 2.5–3 mm. long, 3–4 mm. wide, erose on the apical margin; spur longer than the ovary, slightly decurved, 4–5 mm. long. Column 1 mm. long.

TANZANIA. Iringa District: Mwanihana Forest Reserve, above Sanje, 10 Oct. 1984, *D.W. Thomas* 3763! & 3763A!
DISTR. **T** 7; known only from the Uzungwa Mts.
HAB. In montane forest on steep slopes and in patches of elfin forest on ridge-tops; 1400–1600 m.

NOTE. *Tridactyle flabellata* is a distinctive dwarf species which resembles in its lip shape *T. inflata* Summerh., another narrow endemic from nearby Masailand in Tanzania. It differs, however, in being a much smaller plant with shorter narrower leaves, and in having longer inflorescences with white flowers with a brown anther-cap, obovate rounded petals, a shorter more clavate spur and a less markedly erose lip.

17. **T. tridentata** (*Harv.*) *Schltr.* in E.J. 53: 603 (1915) in obs.; Summerh. in K.B. 3: 287 (1948), excl. syn.; J. Stewart et al., Wild Orch. S. Afr.: 273 (1982). Type: South Africa, Natal, *Sanderson* 562 (TCD, holo.)

A lithophyte or epiphyte with erect, arcuate or pendent few-branched stems, 10–50 cm. or more long, 3–5 mm. in diameter; roots basal, smooth, 2–3 mm. in diameter. Leaves cylindrical, grooved on the upper surface, 6–10 cm. long, 1.5–3 mm. wide, articulated to 7–10 mm. long leaf-sheaths. Inflorescences 1.8–2.7 cm. long, 4–5-flowered; peduncle 5 mm. long; bracts amplexicaul, obconical, 1 mm. long. Flowers green to yellow or dull orange; pedicel and ovary 5 mm. long, lepidote. Dorsal sepal oblong-elliptic, obtuse, 4.5–5 mm. long, 2–2.3 mm. wide; lateral sepals obliquely ovate, acute, 4–5.3 mm. long, 2.5–3 mm. wide. Petals falcate, lanceolate, acute, 4.3–5.4 mm. long, 1.2–1.3 mm. wide. Lip auriculate at the base, 3-lobed in the apical half, 4.5–5 mm. long, 3.5–4 mm. wide; side lobes acute, shorter than or as long as the mid-lobe; mid-lobe triangular, acute; spur filiform or dilated in apical half, incurved, 6–18 mm. long. Column 1–2 mm. long.

UGANDA. Ankole District: near Nsongezi, beside Kafu R., fl. cult. July 1945, *Eggeling* 5497!
TANZANIA. Ufipa District: Mbisi Forest, fl. cult. 2 Jan. 1943, *Bell* in *Moreau* 651!; Iringa District: Mt. Selegu, 28 Dec. 1986, *Lovett & Congdon* 1219!; Songea District: Matagoro Hills, 27 Feb. 1956, *Milne-Redhead & Taylor* 8891!
DISTR. **U** 2; **T** 1, 4, 7, 8; S. Zaire, Zambia, Mozambique, Malawi, Zimbabwe and South Africa
HAB. In *Brachystegia* and in open *Euphorbia candelabra-Combretum-Terminalia* woodlands; 1300–2250 m.

SYN. *Angraecum tridentatum* Harv., Thes. Cap. 2: 6 (1863)
A. goetzeanum Kraenzl. in E.J. 30: 289 (1901). Type: Tanzania, Rungwe District, Kiwira [Kivira] valley, *Goetze* 1491 (B, holo., K, record of holo.!)
A. bolusii Rolfe in Fl. Cap. 5 (3): 73 (1912). Type: South Africa, Natal, *Maxwell* in *BOL* 6319 (BOL, holo.)
Tridactyle goetzeana (Kraenzl.) Schltr. in E.J. 53: 602 (1915)
T. bolusii (Rolfe) Schltr. in E.J. 53: 603 (1915), in obs.
T. teretifolia Schltr. in R.E. Fries, Wiss. Ergebn. Schwed. Rhod.-Kongo-Exped. 1911–1912, 1: 251 (1916). Type: Zambia, Mbala [Abercorn], *R.E. Fries* 1273 (B, holo.)

NOTE. G. Williamson (Journ. S. Afr. Bot. 48: 12–19 (1982)) has suggested that *T. teretifolia* is a synonym of *T. tridentata* and that specimens previously ascribed to the latter from West Africa across to Uganda and Ethiopia, and also some from Zaire and N. Zambia, should be referred to *T. filifolia*. I have followed this treatment here. However, It seems that the remaining specimens from Tanzania southwards to South Africa are extremely variable in habit and in vegetative and floral morphology. I am not at all convinced that they can all satisfactorily be referred to a single taxon. In the absence of a critical study, I have followed with some trepidation Williamson's treatment in placing these specimens in *T. tridentata*.

T. tridentata, as treated here, is an extremely variable species. Those specimens with narrow leaves, short inflorescences and small flowers in which the spur reaches only ± 7 mm. long agree well with the type of the species. Those with fleshier leaves, longer inflorescences and larger flowers with the spur up to 18 mm. long agree well with the type of *T. teretifolia.* Unfortunately the extremes are linked with specimens showing the complete range of intermediate sizes of vegetative and floral characters.

18. **T. tridactylites** (*Rolfe*) *Schltr.* in Beih. Bot. Centr. 36: 148 (1918); Summerh. in F.W.T.A., ed. 2, 3: 274 (1968); Williamson, Orch. SC. Afr.: 203 (1977); Ball, S. Afr. Epiphytic Orch.: 226 (1978); la Croix et al., Malawi Orch. 1: 140 (1983). Type: Sierra Leone, without exact locality, cult. Kew 19 June 1888, *T.R. Griffith* (K, holo.!)

An epiphyte with a straggling pendent habit. Stems 40–160 cm. or more long, 4–5 mm. in diameter; roots smooth, 3 mm. in diameter. Leaves linear, unequally subacutely bilobed at the apex, 6–21 cm. long, 0.6–1.2 cm. wide, twisted at the base and articulated to leaf-sheaths up to 2.5 cm. long. Inflorescences 3–6 cm. long, up to 11-flowered; bracts reflexed, ovate, 1–2 mm. long. Flowers yellow-green; pedicel and ovary 3 mm. long. Dorsal sepal elliptic-ovate, acute, 3.5–5.3 mm. long, 1.5–2.6 mm. wide; lateral sepals obliquely ovate, acute, 3.7–5.6 mm. long, 1.7–2.7 mm. wide. Petals lanceolate, acuminate, 3.6–4.5 mm. long, 0.7–1.3 mm. wide. Lip auriculate at the base, 3-lobed in the apical part, 4–5 mm. long, 5–6.5 mm. wide; side lobes spreading, filiform, longer than the mid-lobe, 2.3–3 mm. long; mid-lobe triangular, 1.9–2.8 mm. long, 1.1–1.2 mm. wide; spur incurved, filiform, 6–8.5 mm. long. Column 1.4–1.9 mm. long; stipes oblanceolate.

UGANDA. Toro District: Kibale Forest, Sept. 1938, *A.S. Thomas* 2526!; Mbale District: Elgon, near Namatala R., 23 Oct. 1917, *Snowden* 525!; Masaka District: Buganda, Nov. 1970, *Griesbach* in *Tweedie* 3862!
TANZANIA. Morogoro District: Uluguru Mts., NW. side, 18 Oct. 1932, *Schlieben* 2832!
DISTR. **U** 2–4; **T** 6; Sierra Leone to Zaire, Mozambique, Malawi, Zambia, Angola and Zimbabwe
HAB. In rain-forest; 1000–1800 m.

SYN. *Angraecum tridactylites* Rolfe in Gard. Chron., ser. 3, 4: 34 (1888)
Aeranthes deistelianus Kraenzl. in E.J. 33: 75 (1902). Type: Cameroon, Buea, *Deistel* 593 (B, holo., K, iso.!)
Mystacidium ledermannianum Kraenzl. in E.J. 51: 393 (1914). Type: Cameroon, Mfongu, *Ledermann* 5950 (B, holo., K, record of holo.!)
Tridactyle ledermanniana (Kraenzl.) Schltr. in Beih. Bot. Centr. 36: 146 (1918)

19. **T. bicaudata** (*Lindl.*) *Schltr.* in Die Orchideen: 602 (1914); Summerh. in K.B. 3: 296 (1948) & in F.W.T.A., ed. 2, 3: 276 (1968); J. Stewart in U.K.W.F.: 802 (1974); Williamson, Orch. SC. Afr.: 203 (1977); Ball, S. Afr. Epiphytic Orch.: 222 (1978); J. Stewart et al., Wild Orch. S. Afr.: 274 (1982); la Croix et al., Malawi Orch. 1: 135 (1983). Type: South Africa, Zwartkopsrivier, *Drège* (K, holo.!)

A pendent epiphyte often forming large hanging masses on trees; stems 12–80 cm. long, 3.5–6 mm. in diameter, few-branched; roots smooth, 2–3 mm. in diameter. Leaves linear, unequally roundly bilobed, 9–14.5 cm. long, 0.5–1.3 cm. wide, usually twisted at base to lie in one plane or less commonly conduplicate, articulated to 1–1.5 cm. long leaf-sheaths. Inflorescences suberect-spreading, 3.5–13 cm. long, 8–25-flowered; bracts amplexicaul, 1–2 mm. long. Flowers white, parchment-yellow or greenish yellow, fragrant; pedicel and ovary 2–3 mm. long, glabrous. dorsal sepal oblong-ovate, acute or apiculate, 4.4–5.9 mm. long, 2.3–3.2 mm. wide; lateral sepals obliquely ovate, acute or apiculate, 4.7–6 mm. long, 2.6–3.6 mm. wide. Petals oblong, acute or obtuse, 3.7–6.2 mm. long, 1.2–12.2 mm. wide. Lip 3-lobed in the middle, auriculate at the base, 3–5.7 mm. long, 8–11.5 mm. wide; side lobes spreading, linear, laciniate at the apex, 3.6–5.5 mm. long; mid-lobe narrowly triangular, fleshily apiculate at the apex, 2–2.5 mm. long; spur straight, (7.5–)10–16 mm. long. Column 1–1.6 mm. long; stipes oblanceolate, notched at the apex. Fig. 157.

UGANDA. Toro Distrist: Nyinabitaba, 9 Sept. 1951, *Osmaston* 1187!; Kigezi District: Impenetrable Forest, Sept. 1936, *Eggeling* 3309! & Mafuga, Oct. 1947, *Purseglove* P2541!; Mengo District: Entebbe, Oct. 1905, *E. Brown* 339!
KENYA. Kisumu-Londiani District: Koru, Mar. 1937, *C. van Someren* 11!; Masai District: Mara R., Ngebai Langata Endul, 3 Apr. 1961, *Glover, Gwynne & Samuel* 358!; Teita District: Kasigau, Sept.–Oct. 1938, *Bally* 8863!
TANZANIA. Arusha District: Ngurdoto National Park, Little Momela Lake, Nov. 1965, *Greenway & Kanuri* 12277!; Lushoto District: W. Usambara Mts., Magamba Forest, Oct. 1943, *Eggeling* in *Moreau* 69A!; Ufipa District: Lake Sundu, Dec. 1958, *Richards* 10293!; Zanzibar I., near Ndagaa, Sept. 1935, *Vaughan* 2283!

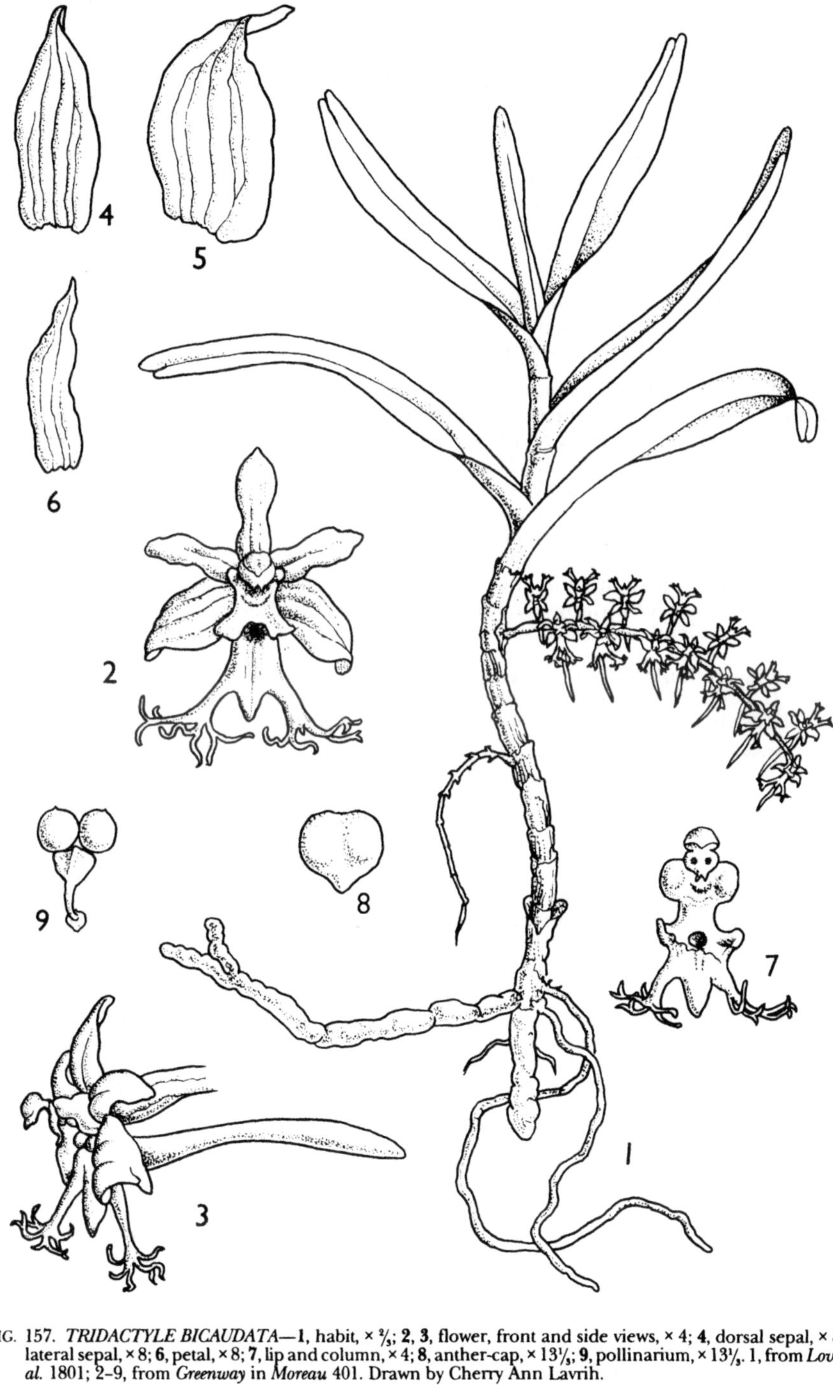

FIG. 157. *TRIDACTYLE BICAUDATA*—**1**, habit, × ⅔; **2**, **3**, flower, front and side views, × 4; **4**, dorsal sepal, × 8; **5**, lateral sepal, × 8; **6**, petal, × 8; **7**, lip and column, × 4; **8**, anther-cap, × 13⅓; **9**, pollinarium, × 13⅓. 1, from *Lovett et al.* 1801; 2–9, from *Greenway* in *Moreau* 401. Drawn by Cherry Ann Lavrih.

DISTR. **U** 2–4; **K** 3–7; **T** 2–4, 6, 7; **Z**; Sierra Leone to Ethiopia and south to Zimbabwe and South Africa
HAB. In lowland to montane and riverine forest, also in secondary forest and in plantations; sea-level–2500 m.

SYN. *Angraecum bicaudatum* Lindl. in Hook., Comp. Bot. Mag. 2: 205 (1837)
A. fimbriatum Rendle in J.L.S. 30: 387, t. 32, figs. 11–13 (1895). Type: Kenya, Kilifi District, Giriama [Giryama] and Kwale District, Shimba Hills, *W.E. Taylor* (BM, holo.)
Listrostachys cirrhosa Kraenzl. in E.J. 33: 73 (1902). Type: Tanzania, Usambara Mts., Derema, *Scheffler* 125 (B, holo.)
L. fimbriata (Rendle) Kraenzl. in E.J. 33: 74 (1902), in obs.
L. bicaudata (Lindl.) Finet in Mém. Soc. Bot. Fr. 9: 51 (1907), excl. tab.
Tridactyle pulchella Schltr. in E.J. 53: 602 (1915). Type: Tanzania, Rungwe District, near Kyimbila, *Stolz* 716 (B, holo., K, iso.!)
T. fimbriata (Rendle) Schltr. in E.J. 53: 602 (1915), in obs.
Angraecum laciniatum Kraenzl. in Viert. Nat. Ges. Zürich 74: 101 (1929). Type: Uganda, Mengo District, Kipayo, *Dummer* (Z, holo.!, K, record of holo.!)
Tridactyle polyschista Mansf. in F.R., Beih. 68: 11, t. 104, fig. 415 (1932). Type: Cameroon, without exact locality, *Mildbraed* 10125 (B, holo., K, iso.!)

20. **T. gentilii** (*De Wild.*) *Schltr.* in Beih. Bot. Centr. 36: 145 (1918); J. Stewart in U.K.W.F.: 802 (1974); J. Stewart et al., Wild Orch. S. Afr.: 275 (1982). Type: Zaire, Kinumbi, Haut Lomami, *Gentil* (BR, holo., K, iso.!)

An epiphyte with long stems 30–85 cm. long, 4–6 mm. in diameter; roots 3–3.5 mm. in diameter, smooth. Leaves linear, unequally bilobed at the apex, 10–21 cm. long, 0.8–2.1 cm. wide, twisted at base and articulated to 1–1.8 cm. long sheaths. Inflorescences 3.5–15 cm. long, 7–15-flowered; peduncle and rhachis stout, 1–1.5 mm. in diameter; bracts amplexicaul, 1.5–2 mm. long. Flowers white or cream, with pinkish tips to the segments, fragrant; pedicel and ovary 6–7 mm. long. Dorsal sepal oblong, obtuse, 7–9 mm. long, 3–4.7 mm. wide; lateral sepals obliquely ovate, obtuse, 8.5–9 mm. long, 3–4.3 mm. wide. Petals linear-lanceolate, acute, 7–9 mm. long, 1.5–2 mm. wide. Lip 3-lobed in the middle, 9–10 mm. long, 13–16 mm. wide; side lobes spreading, oblong, longly lacerate at the apex, 6–8 mm. long; mid-lobe triangular, with a fleshy apiculum, 2.6–4 mm. long, 2.5 mm. wide; spur incurved, slender cylindrical, (3.3–)4.3–6.5 cm. long. Column 2–2.75 mm. long. Pl. 8.

UGANDA. Kigezi District: Impenetrable Forest, Dec. 1974, *C. Leakey* 50!
DISTR. **U** 2; Ghana, Nigeria, Cameroon, Zaire, Zambia and South Africa
HAB. In swampy forest and montane rain-forest; 1700–1800 m.

SYN. *Angraecum gentilii* De Wild., Not. Pl. Utiles Congo 1: 140 (1903)

21. **T. unguiculata** *Mansf.* in N.B.G.B. 9: 1064 (1924). Type: Tanzania, Njombe District, Lupembe, *Schlieben* 1352 (B, holo., K, drawing of holo.!)

An epiphyte with stems up to 42 cm. long, 3.5 mm. in diameter. Leaves linear-ligulate, unequally bilobed at the apex, up to 15 cm. long, 7 mm. wide, articulated to leaf-sheaths up to 2 cm. long. Inflorescences several-flowered, up to 8 cm. long; bracts cucullate, broad, 6 mm. long. Flower colour not known; pedicel and ovary 5–9 mm. long. Dorsal sepal oblong, obtuse, 8.3–12 mm. long, 4–4.5 mm. wide; lateral sepals elliptic-oblong, obtuse, 9.2–12 mm. long, 3–5.3 mm. wide. Petals oblong-spathulate, subacute or obtuse, 8.7–12 mm. long, 4–4.5 mm. wide. Lip broadly flabellate-cucullate, 3-lobed in the apical half, auriculate at the base, 9 mm. long, 15 mm. wide; side lobes oblong-flabellate, erose-dentate on apical margin, 4.5 mm. long and wide; mid-lobe triangular, acute, 2.5 mm. long, 3 mm. wide; spur filiform, straight, 4–5 cm. long. Column 2 mm. long.

TANZANIA. Njombe District: Lupembe, N. of Upper Ruhudje R., *Schlieben* 1352!
DISTR. **T** 7; known only from the type collection
HAB. In mist-forest; 1800 m.

22. **T. tricuspis** (*Bolus*) *Schltr.* in E.J. 53: 601 (1915), in obs.; Williamson, Orch. SC. Afr.: 205 (1977); Ball, S. Afr. Epiphytic Orch.: 228 (1978); la Croix et al., Malawi Orch. 1: 140 (1983). Types: South Africa, Natal, *McKen* 14 (BOL, syn., K, isosyn.!), *Cooper* 1398 (BOL, syn.) & *Sanderson* (BOL, syn.)

An erect, spreading or pendent epiphyte with a stem 1–10 cm. long, 5–8 mm. in diameter; roots smooth, 2–3 mm. in diameter. Leaves spreading or arcuate, linear,

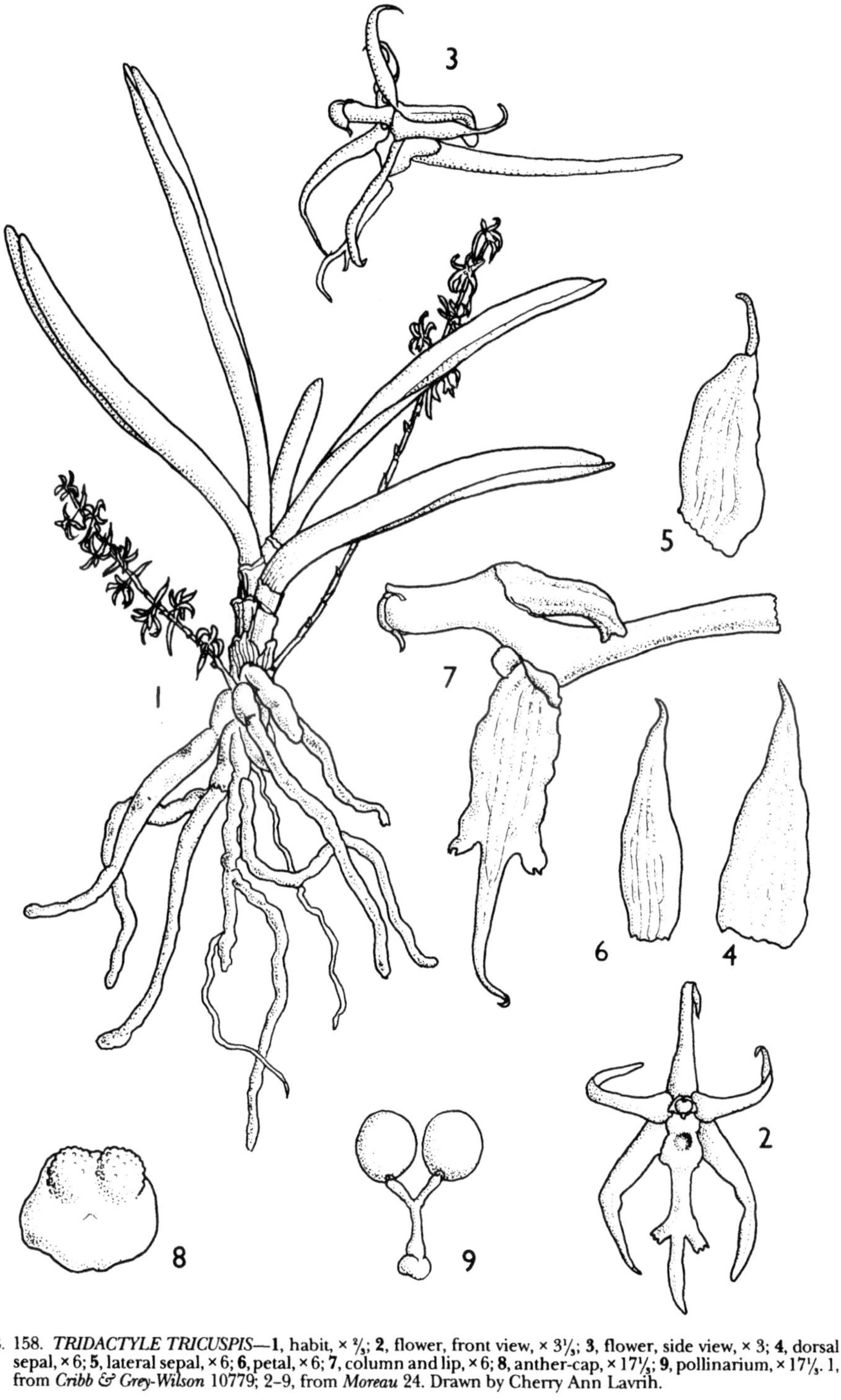

FIG. 158. *TRIDACTYLE TRICUSPIS*—**1**, habit, × ⅔; **2**, flower, front view, × 3⅓; **3**, flower, side view, × 3; **4**, dorsal sepal, × 6; **5**, lateral sepal, × 6; **6**, petal, × 6; **7**, column and lip, × 6; **8**, anther-cap, × 17⅓; **9**, pollinarium, × 17⅓. 1, from *Cribb & Grey-Wilson* 10779; 2–9, from *Moreau* 24. Drawn by Cherry Ann Lavrih.

unequally roundly bilobed, 4–15 cm. long, 0.6–1.2 cm. wide, conduplicate or twisted at the base to lie in one plane. Inflorescences suberect, almost as long as the leaves, 7–30-flowered, 4–15 cm. long; peduncle 6–10 mm. long; bracts amplexicaul, ovate, apiculate, reflexed, 1–2 mm. long, dark brown. Flowers distichous, pale green to pale butterscotch-yellow, scented of mignonette (*Reseda lutea*); pedicel and ovary 2–3 mm. long, lepidote. Dorsal sepal lanceolate, acuminate, 6–7.5 mm. long, 2–2.4 mm. wide; lateral sepals obliquely lanceolate, acuminate, 6–7.5 mm. long, 1.8–2.2 mm. wide. Petals lanceolate, acuminate, 5–6.5 mm. long, 1.5–2 mm. wide. Lip 3-lobed in the apical half, cuneate at base and auriculate, 5.7–8 mm. long, 3.5–5 mm. wide; side lobes short, spreading, erose or bifid at the apex, 0.7–2.5 mm. long; mid-lobe narrowly triangular, acuminate, 2.5–4 mm. long, 0.5–0.6 mm. wide; spur filiform, 12–14 mm. long. Column 2–2.3 mm. long; stipes Y-shaped. Fig. 158.

TANZANIA. Moshi District: Kilimanjaro, Rongai, Apr. 1934, *Schlieben* 5086!; Mbulu District: Mt. Hanang, above Katesh, May 1962, *Polhill & Paulo* 2310!; Iringa District: Image Mt., Mar. 1962, *Polhill & Paulo* 1662!; Njombe District: Kitulo Plateau, top edge of Ndumbi [Ndombe] Forest, Feb. 1976, *Cribb & Grey-Wilson* 10779!

DISTR. **T** 2–4, 7, 8; Zaire, Malawi, Mozambique, Zambia, Zimbabwe and South Africa

HAB. In montane forest and associated bushland also in *Brachystegia* woodland; 900–2600 m.

SYN. *Angraecum tricuspe* Bolus in J.L.S. 25: 163 (1889)

A. rhodesiacum Rendle in J.L.S. 40: 208 (1911). Type: Zimbabwe, Melsetter, *Swynnerton* 755 (BM, holo.!)

A. frommianum Kraenzl. in E.J. 51: 398 (1914). Type: Tanzania, Ufipa District, Mfimbwa Mt., *Fromm* 226 (B, holo.)

Tridactyle nyassana Schltr. in E.J. 53: 601 (1915). Type: Tanzania, Rungwe District, Kyimbila, *Stolz* 658 (B, holo., K, iso.!)

T. frommiana (Kraenzl.) Schltr. in Beih. Bot. Centr. 36: 144 (1918)

T. rhodesiaca (Rendle) Schltr. in Beih. Bot. Centr. 36: 147 (1918)

T. fragrans Williamson in Pl. Syst. & Evol. 134: 75 (1980). Type: Malawi, Nyika Plateau, *Williamson* 227 (SRGH, holo., K, iso.!)

23. **T. furcistipes** *Summerh.* in K.B. 3: 300 (1948); J. Stewart in U.K.W.F.: 802 (1974). Type: Kenya, Aberdare Mts., Kinangop, *Chandler* 2401 (K, holo.!)

An epiphyte with 7–18 cm. long stems, 6–8 mm. in diameter; roots smooth, 2 mm. in diameter. Leaves falcate-spreading, linear, unequally roundly bilobed at the apex, 11–18 cm. long, 0.7–1.3 cm. wide, conduplicate or twisted at the base to lie in one plane. Inflorescences 4.5–11 cm. long; peduncle and rhachis stout; bracts amplexicaul, broadly ovate, obconical, 2–4.5 mm. long. Flowers greenish white or cream fading to orange; pedicel and ovary 3 mm. long, lepidote. Dorsal sepal lanceolate, acuminate, 9.5–11.5 mm. long, 2.5–3.7 mm. wide; lateral sepals falcate, lanceolate, acuminate, 10.5–12 mm. long, 2–3.5 mm. wide. Petals lanceolate, acuminate, 8.5–10 mm. long, 2–3 mm. wide. Lip 3-lobed in the apical half, auriculate at the base, 9–12.5 mm. long, 3.5–6.5 mm. wide; side lobes spreading, shorter than the mid-lobe, 1.4–2 mm. long, toothed at the apex; mid-lobe ligulate, acuminate, 4.5–6 mm. long, 0.8–1.2 mm. wide; spur filiform, straight, 14–27 mm. long. Column 2–3 mm. long; stipes Y-shaped.

UGANDA. Karamoja District: Mt. Debasien, May 1939, *A.S. Thomas* 2957A! & May 1948, *Eggeling* 5813!

KENYA. Elgeyo District: Cherangani Hills, Kaisungur [Kaisangor], July 1961, *Tweedie* 2173! & Elgeyo, Apr. 1943, *Bally* 2538!

TANZANIA. Masai District: Lolkisale, Mar. 1967, *Beesley* 261!

DISTR. **U** 1; **K** 3; **T** 2; not known elsewhere

HAB. In montane forest of *Juniperus* and in montane scrub; 2500–2700 m.

NOTE. *T. furcistipes* resembles a large-flowered *T. tricuspis* and its relationship to that taxon needs further investigation.

24. **T. sarcodantha** *Mansf.* in N.B.G.B. 9: 1063 (1924). Type: Tanzania, Morogoro District, NW. Uluguru Mts., *Schlieben* 3705 (B, holo., BR, iso.)

An epiphyte with a short suberect stem ± 5 cm. long. Leaves linear, unequally bilobed at the apex, each lobe with 2 blunt teeth, 9–16.5 cm. long, 1.2–1.3 cm. wide, twisted at the base. Inflorescences up to 9 cm. long; rhachis fleshy; bracts ovate, cucullate, 4 mm. long. Flowers fleshy, yellowish; pedicel and ovary 5 mm. long, lepidote. Dorsal sepal ovate-lanceolate, shortly acuminate, 9–10 mm. long, 3.5 mm. wide, lepidote on outer surface; lateral sepals similar. Petals triangular-lanceolate, shortly acuminate, 8 mm. long, 2.5 mm.

wide. Lip cuneate-rectangular, auriculate at the base, 3-lobed in front, 11 mm. long, 3.5 mm. wide; side lobes spreading obliquely, short, 2 mm. long, 1 mm. wide; mid-lobe narrowly triangular, acuminate, 6 mm. long, 2.5 mm. wide; spur incurved, slightly inflated apically, 2.2 cm. long. Column 4 mm. long.

TANZANIA. Morogoro District: Uluguru Mts., NW. side, 31 Mar. 1933, *Schlieben* 3705!
DISTR. **T** 6; known only from the type collection
HAB. In montane forest; 1800 m.

25. **T. tanneri** *Cribb* in K.B. 34: 338 (1979). Type: Tanzania, W. Usambara Mts., Mazumbai, *Cribb & Grey-Wilson* 10038 (K, holo.!)

An erect to spreading epiphyte with a short stem 1–3.5 cm. long; roots smooth, 2 mm. in diameter. Leaves linear, unequally obtusely bilobed, 6–11 cm. long, 0.6–1.1 cm. wide, each lobe 2-toothed at apex, twisted at apex to lie in one plane, spotted with dark green on a paler green background. Inflorescences 2–5 cm. long, 2–8-flowered; bracts reflexed, ovate, 3.5–4.5 mm. long. Flowers green or dull green; pedicel and ovary 6–8 mm. long. Dorsal sepal oblong, acute, 7–7.5 mm. long, 2.5–3 mm. wide; lateral sepals obliquely oblong-lanceolate, acuminate, 8.5–9.5 mm. long, 2.5 mm. wide. Petals lanceolate, acuminate, 7–7.5 mm. long, 2 mm. wide. Lip obscurely 3-lobed in the apical part, 8–9.5 mm. long, 3–3.5 mm. wide, auriculate at the base; side lobes tooth-like, blunt, 1 mm. long; mid-lobe narrowly triangular, fleshy at the apex, 3.5 mm. long, 1.5 mm. wide; spur pendent, slightly upcurved at the apex, slightly dilated in apical part, 18.5–19.5 mm. long. Column 3 mm. long; stipes Y-shaped, rugulose. Pl. 8.

KENYA. Teita Hills, Mbololo Hill, 29 Dec. 1971, *R. & A. Faden, Sneenk & McNee* 71/971!
TANZANIA. Lushoto District: W. Usambara Mts., Mazumbai, Jan. 1976, *Cribb & Grey-Wilson* 10038!
DISTR. **K** 7; **T** 3; not known elsewhere
HAB. In montane evergreen forest and dwarf forest, on tree trunks in deep shade; 1450–1600 m.

26. **T. citrina** *Cribb* in K.B. 32: 183 (1977); la Croix et al., Malawi Orch. 1: 136 (1983). Type: Zambia/Malawi, Nyika Plateau, *Williamson* 230 (K, holo.!)

An erect or spreading epiphyte forming small clumps; stems 1–9 cm. long, 7–8 mm. in diameter; roots smooth or slightly rough, 1.5–3.5 mm. in diameter. Leaves spreading-falcate, conduplicate, linear, unequally roundly bilobed at the apex, 8–13 cm. long, 0.7–0.8 cm. wide. Inflorescences 7.5–11 cm. long, 9–11-flowered; rhachis 1 mm. in diameter; bracts amplexicaul, ovate, 3 mm. long. Flowers pale green to creamy yellow; pedicel and ovary 4 mm. long. Dorsal sepal lanceolate, acuminate, 8–11 mm. long, 3.5–4 mm. wide; lateral sepals similar but oblique. Petals lanceolate, acuminate, 7–8 mm. long, 2.5–3 mm. wide. Lip 3-lobed in the apical half, 10–16 mm. long, 3–5 mm. wide, auriculate at the base; side lobes obliquely spreading, acute, half the length of the mid-lobe; mid-lobe narrowly triangular, 3.5–5 mm. long; spur filiform, decurved-pendent, 3.6–4.3 cm. long. Column 4 mm. long.

TANZANIA. Ufipa District: Malonje Farm, Dec. 1966, *Richards* 7272! & Namwele Escarpment, Dec. 1961, *Richards* 15780! & Kito Mt., Dec. 1959, *Richards* 11857!
DISTR. **T** 4; Malawi and Zambia
HAB. In *Brachystegia* woodland, on the branches of trees in dappled shade; 1800–2100 m.

27. **T. sp. A**

An epiphyte with stems up to 30 cm. long; roots grey, densely squamose. Leaves 1.5 cm. apart, ± 6 cm. long, 2 cm. wide. Inflorescences not seen. Flowers salmon-coloured with green margins to the segments, pleasantly fragrant; anther-cap brown; pedicel and ovary 4–4.5 mm. long, lepidote. Dorsal sepal oblong, obtuse, convex, 8 mm. long, 3.5–4 mm. wide; lateral sepals obliquely oblong, obtuse, convex, 8–9 mm. long, 4 mm. wide. Petals oblong-lanceolate, subacute, convex, 7 mm. long, 2 mm. wide. Lip auriculate at the base, 3-lobed in apical half, 5 mm. long, 3.5 mm. wide; side lobes obliquely spreading, linear, 2 mm. long; mid-lobe triangular, acute, 1.5 mm. long; spur recurved, clavate, 15 mm. long. Column 2 mm. long.

UGANDA. Kigezi District: Impenetrable Forest, fl. cult. Dec. 1978, *H. Meyer* 162!
DISTR. **U** 2; known only from the above collection
HAB. In montane forest; 2000–2500 m.

72. **CARDIOCHILOS**

Cribb in K.B. 32: 183 (1977); la Croix et al., Malawi Orch. 1: 61 (1983); Senghas in Schltr, Die Orchideen, ed. 3, 1: 1084 (1986)

Small epiphytic herbs with short stems; roots branching, clustered in lower part of the stem. Leaves distichous, arcuate-spreading, linear, unequally roundly or obtusely bilobed at the apex, conduplicate or twisted at the base to lie in one plane, articulated at base to persistent imbricate leaf-bases. Inflorescences one–several, shorter than the leaves, 1(–2)-flowered; peduncle slender, terete, elongate, bearing 2 sterile sheathing bracts along length. Flowers green to ochre-orange, non-resupinate, scabrid on outer surface. Sepals and petals subsimilar; lateral sepals fused at their oblique bases to enclose base of spur. Lip obscurely 3-lobed, concave, ecallose, spurred at the base. Column short, fleshy; rostellum narrow, bifid, dilated and scutelliform at apex; pollinia 2, subglobose; stipes solitary, linear; viscidium small, reniform.

A monotypic genus known only from Malawi and adjacent Tanzania.

C. williamsonii *Cribb* in K.B. 32: 184 (1977); la Croix et al., Malawi Orch. 1: 61 (1983). Type: Malawi, Nyika Plateau, Kasaramba, *Williamson* 388 (K, holo.!, SRGH, iso.)

An epiphyte up to 12 cm. tall with a short stem up to 8 cm. long; roots clustered in lower part of the stem, 1.5–2 mm. in diameter. Leaves spreading-falcate, linear, unequally roundly or obtusely bilobed at apex, 4–8 cm. long, 3–6 mm. wide; leaf-bases truncate, up to 8 mm. long. Inflorescences 3–4 cm. long, 1(–2)-flowered; peduncle 1–2.5 cm. long; bracts ovate to obovate, acute to obtuse, 2–3.5 mm. long. Flowers green to ochreous orange; pedicel and ovary 7–13 mm. long, scabrid. Dorsal sepal elliptic, obtuse, 4 mm. long, 2.2 mm. wide; lateral sepals obliquely falcate-elliptic, obtuse, 4.5 mm. long, 2.5 mm. wide, fused at the base. Petals oblong-lanceolate, acute, 4 mm. long, 2 mm. wide. Lip obscurely 3-lobed, cordate, shortly apiculate, 4 mm. long and wide; spur straight or undulate, narrowly cylindrical from a broad mouth, 13–16 mm. long. Column 1–1.5 mm. long.

TANZANIA. Rungwe District: Ibaba area, Kulenyeli, *Leedal* 273A! (sterile); Njombe District: Rudewa, Mdando Forest, 3 Apr. 1989, *Spurrier* 76!
DISTR. **T** 7; Malawi
HAB. Montane forest; 1800–2500 m.

73. **YPSILOPUS**

Summerh. in K.B. 4: 439 (1949); Senghas in Schltr, Die Orchideen, ed. 3, 1: 1086 (1986)

Erect to pendent epiphytic monopodial herbs with short stems, less than 6 cm. long; roots from basal part of stem, not branching. Leaves arranged in a fan, linear, conduplicate or twisted at the base to lie in one plane, articulated to short persistent leaf-bases, leaving sharp acuminate points when the leaves fall. Inflorescences usually shorter than the leaves, racemose, laxly few–many-flowered; peduncle and rhachis slender, terete; bracts amplexicaul, rather small. Flowers white or pale green, with spreading segments, scabrid on outer surface and on ovary. Sepals and petals free, subsimilar. Lip entire or obscurely 3-lobed, ecallose, long-spurred at the base. Column short, terete; rostellum geniculate and recurved at base, 2-lobed, usually dilated and scutelliform at the apex; pollinia 2, subglobose; stipes solitary, slender, Y-shaped; viscidium large, reniform or oblong.

A small genus of four species from East and South-central Africa. All four species are found in the Flora region, three being endemic.

The genus is related to *Tridactyle* and its relationship to that genus needs further investigation.

Inflorescences 1-flowered; flowers green; lip linear-lanceolate, acuminate; rostellar lobes not dilated at the apex ... 4. *Y. viridiflorus*
Inflorescences 2–many-flowered; flowers white; lip subrhombic to oblong-flabellate, acute to obtuse; rostellar arms dilated at the apex:

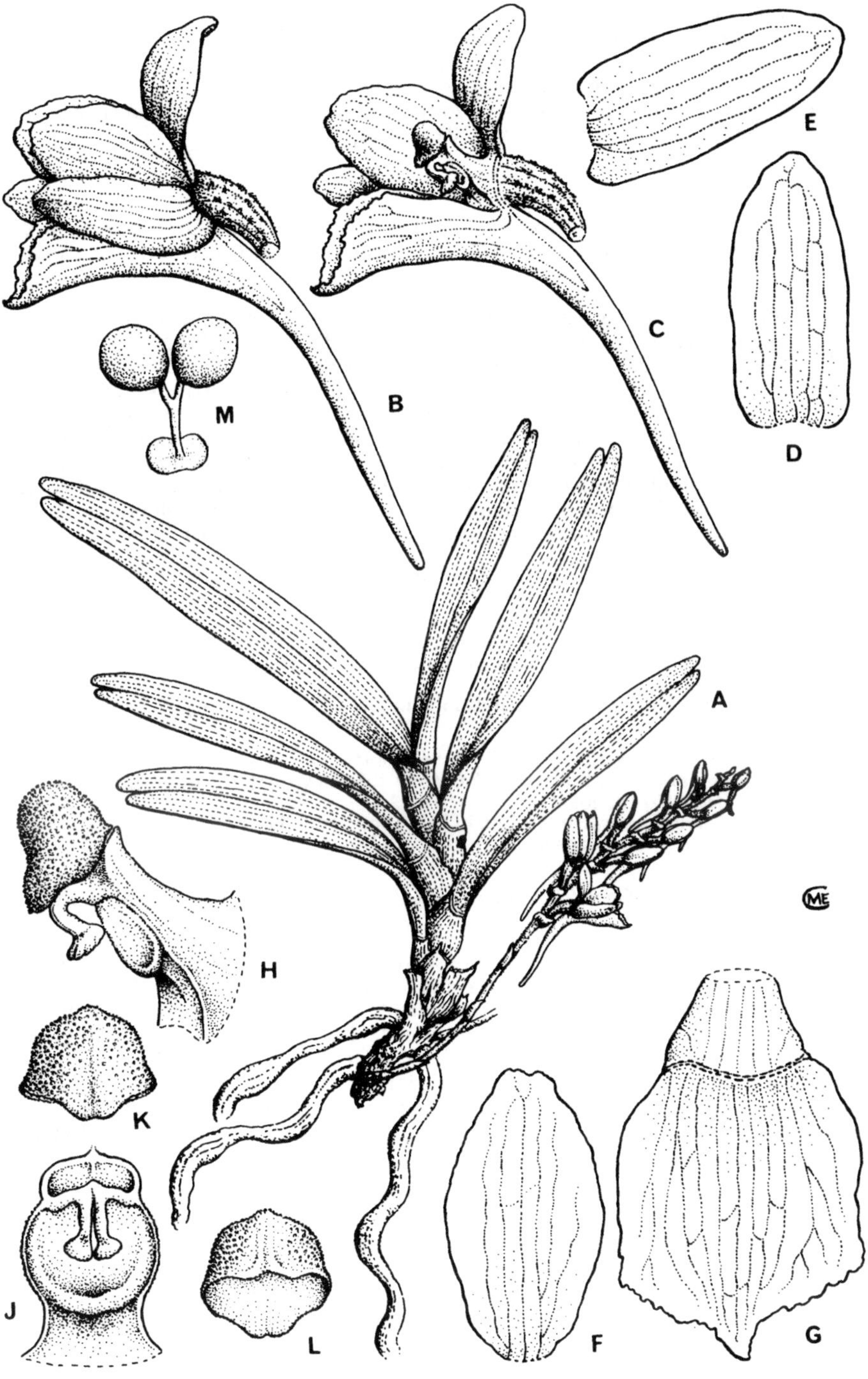

FIG. 159. *YPSILOPUS LEEDALII*—**A**, habit, × 1; **B**, flower, × 4; **C**, flower with lateral sepal and petal removed, × 4; **D**, dorsal sepal, × 6; **E**, lateral sepal, × 6; **F**, petal, × 6; **G**, lip, × 6; **H**, column and anther, side view, × 12; **J**, column, front view, showing rostellum, × 12; **K**, anther-cap, front view, × 12; **L**, anther-cap, back view, × 12; **M**, pollinarium, × 12. Drawn from *Leedal* 5135 by Mrs M.E. Church.

Lip oblong-flabellate, obtuse, 5 mm. wide, erose on apical margins; spur 12–14 mm. long; leaves 8 mm. or more wide; petals elliptic, obtuse . . . . 3. *Y. leedalii*
Lip subrhombic to ovate, acute, 2–3.5 mm. wide, with entire margins; spur 3 cm. or more long; leaves 1.5–6 mm. wide; petals lanceolate, acute or acuminate:
Plant pendent; leaves up to 50 cm. long, 1.5–3 mm. wide 1. *Y. longifolius*
Plant erect to spreading; leaves 5–10 cm. long, 3.5–6 mm. wide . . . . 2. *Y. erectus*

1. **Y. longifolius** (*Kraenzl.*) *Summerh.* in K.B. 4: 440 (1949); Piers, Orch. E. Afr., ed. 2: 298 (1968); J. Stewart in U.K.W.F.: 796 (1974). Type: Kenya, Mt. Kenya, Ndoro, *von Hoehnel* 31 (B, holo.†)

A pendent epiphyte with a short stem up to 4 cm. long; roots 2 mm. in diameter, white. Leaves conduplicate, linear, acute or subacute, 5–25(–50) cm. long, 1.5–3 mm. wide. Inflorescences pendent, 3–8 cm. long, 2–9-flowered; bracts broadly ovate-triangular, acute or obtuse, 2–2.5 mm. long. Flowers white, with greenish tips to the segments; pedicel and ovary scabrid, 9–12 mm. long. Dorsal sepal ovate, acute, 6–7.3 mm. long, 2.5–3.8 mm. wide; lateral sepals obliquely ovate-lanceolate or oblong-lanceolate, acute to acuminate, 6–9.2 mm. long, 2.1–3.1 mm. wide. Petals lanceolate, acuminate, 5.5–6.7 mm. long, 1.9–2.8 mm. wide. Lip slightly recurved, obscurely 3-lobed in the middle, ovate to subrhombic, acute and fleshy at the apex, 6.5–7.5 mm. long, 2–3.7 mm. wide; spur slender, cylindrical, pendent-decurved, 3.5–4.2 cm. long. Column 1.7–2.3 mm. long; viscidium reniform. Pl. 8.

KENYA. Laikipia District: Rumuruti, Mar. 1943, *Copley* 115! & Laikipia, 28 Mar. 1935, *D. Leakey* 8540!; Nairobi District: near Nairobi, 26 Apr. 1942, *Moreau* 158A!; Masai District: Ngong Hills, 20 Mar. 1942, *C. van Someren* in *Moreau* 158!
TANZANIA. Lushoto District: W. Usambara Mts., Sunga to Manolo road, 26 May 1953, *Drummond & Hemsley* 2780! & Mazumbai Forest Reserve, June 1975, *L. Tanner* 34!
DISTR. **K** 3, 4, 6; **T** 3; not known elsewhere
HAB. Epiphyte in montane rain-forest and montane juniper forest; also in plantations on *Grevillea*; 1450–2400 m.

SYN. *Mystacidium longifolium* Kraenzl. in E.J. 17: 57 (1893)
Listrostachys graminifolia Kraenzl. in P.O.A. C: 158 (1895). Type: Tanzania, Usambara Mts., Mbalu, *Holst* 2607 (B, holo.†)
Aerangis graminifolia (Kraenzl.) Schltr. in Beih. Bot. Centr. 36: 122 (1918)
Ypsilopus graminifolius (Kraenzl.) Summerh. in K.B. 4: 440 (1949)

2. **Y. erectus** (*Cribb*) *Cribb & J. Stewart* in K.B. 40: 419 (1985). Type: Zambia, Kundalila Falls, *Williamson* 1545 (K, holo.!)

An erect or spreading epiphyte with a short stem up to 6 cm. long; roots 2 mm. in diameter. Leaves conduplicate, often recurved, linear, obtuse, 3–15 cm. long, 3.5–6 mm. wide. Inflorescences spreading to suberect, 5–20 cm. long, laxly 4–12-flowered; rhachis somewhat fractiflex; bracts ovate-triangular, acute, 1.5–2 mm. long. Flowers white; pedicel and ovary scabrid, 9–12 mm. long. Dorsal sepal ovate, acute or acuminate, 5–6.5 mm. long, 3 mm. wide; lateral sepals obliquely oblong-lanceolate, acute, 5–7 mm. long, 2.3–3 mm. wide, keeled on the reverse. Petals oblong-lanceolate, acuminate, 5–6.5 mm. long, 1.5–2 mm. wide. Lip obscurely 3-lobed, subrhombic to lanceolate, acute, 5–6.5 mm. long, 2.3–3 mm. wide; spur slender, cylindrical, 3.5–5.3 cm. long. Column ± 2 mm. long.

TANZANIA. Mpanda District: Mahali Mts., Musenabantu [Muhensaboutu], 9 Jan. 1976, *Nishida* 187!; Rungwe District: 3 km. N. of Rungwe mission, 13 May 1944, *Moreau* 871! & Igogwe, 20 May 1976, *Leedal* 3564!; Songea District: Matagoro Mts., *McLoughlin* 83!
DISTR. **T** 4, 7, 8; Zambia, Malawi, Zimbabwe and Swaziland
HAB. Epiphyte in woodland on *Brachystegia*, *Uapaca* and *Albizia*; 1300–2100 m.

SYN. *Y. longifolius* (Kraenzl.) Summerh. var. *erectus* Cribb in K.B. 32: 177 (1977)

3. **Y. leedalii** *Cribb* in K.B. 40: 417 (1985). Type: Tanzania, Rungwe District, Umalila, Ibaba, *Leedal* 2625 (K, holo.!)

A small epiphyte with a short stem ± 4 cm. long; roots elongate, 2–2.5 mm. in diameter. Leaves ligulate, unequally roundly bilobed at the apex, 5–7 cm. long, 8–9 mm. wide,

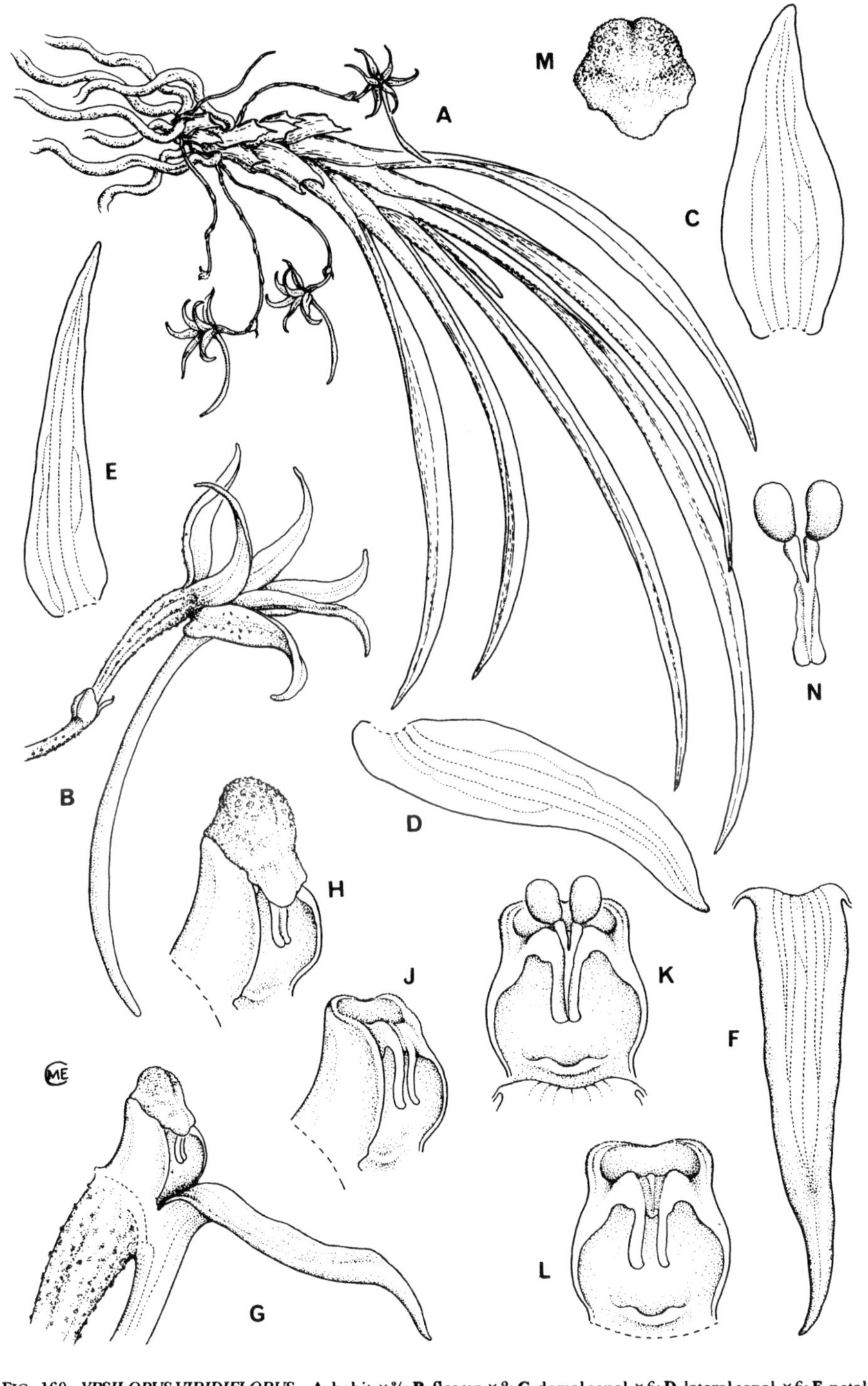

FIG. 160. *YPSILOPUS VIRIDIFLORUS*—**A**, habit, × ⅔; **B**, flower, × 3; **C**, dorsal sepal, × 6; **D**, lateral sepal, × 6; **E**, petal, × 6; **F**, lip, front view, × 8; **G**, lip, side view and column, × 6; **H**, column, side view, × 12; **J**, same, anther-cap removed, × 12; **K**, column and pollinarium, front view, × 12; **L**, same, pollinarium removed, showing rostellum, × 12; **M**, anther-cap, × 12; **N**, pollinarium, × 14. Drawn from *Greenway et al.* 13621 by Mrs M.E. Church.

twisted at the base to lie in one plane. Inflorescences 4–5.5 cm. long, 6–10-flowered; bracts broadly triangular-ovate, obtuse, 1.5 mm. long. Flowers white, subcampanulate; pedicel and ovary scabrid, 3–4 mm. long. Dorsal sepal oblong, rounded or obtuse, 5.5 mm. long, 2.5 mm. wide; lateral sepals oblique, oblong-elliptic, obtuse, 5.5–6 mm. long, 2.5–3 mm. wide. Petals elliptic, obtuse, 5.5–6 mm. long, 3 mm. wide, slightly erose on the margins. Lip concave, oblong-flabellate, obtuse, 8.5–9 mm. long, 5–6 mm. wide, with erose apical margins; spur cylindrical, tapering to the apex from a broader mouth, 13–14 mm. long. Column 2–2.4 mm. long. Fig. 159, p. 627.

TANZANIA. Rungwe District: Umalila, Ibaba, 19 Mar. 1978, *Leedal* 5135! & Saswago, 31 Mar. 1975, *Leedal* 2625!
DISTR. **T** 7; not known elsewhere
HAB. Epiphyte in montane rain-forest; 2200–2300 m.

4. **Y. viridiflorus** *Cribb & J. Stewart* in K.B. 40: 417 (1985). Type: Tanzania, Mt. Meru, *Greenway & Kanuri* 13493 (K, holo.!, EA, iso.)

A pendent epiphyte 6–27 cm. long with a short stem, 0.5–3 cm. long, entirely hidden by the leaf-bases. Leaves bilaterally compressed, iridiform, linear-tapering, acuminate, 4–25 cm. long, 2–5 mm. wide, articulated obliquely to prominent leaf-bases up to 2 cm. long. Inflorescences 1–several, 1(–2)-flowered, 2–6 cm. long; peduncle slender, wiry, up to 5 cm. long; bracts elliptic-ovate, acute, 2–3 mm. long. Flowers pale yellow-green or whitish green, scabrid on the outer surface; pedicel and ovary 5–7 mm. long. Dorsal sepal lanceolate, acuminate, 7–9 mm. long, 1.5–2 mm. wide; lateral sepals similar but oblique at the base. Petals linear-lanceolate, acuminate, 6–8 mm. long, 1.5–1.8 mm. wide. Lip entire, lanceolate, acuminate, deflexed, 7–8 mm. long, 2 mm. wide; spur decurved-pendent, slenderly cylindrical, 15–18 mm. long. Column 1 mm. long. Fig. 160, p. 629.

TANZANIA. Arusha District: Mt. Meru, 14 July 1968, *V.C. Gilbert* 2928! & Mt. Meru, Jekukumia [Jekekumia] Rangers' Post, 27 Apr. 1969, *Greenway, Fitzgerald & Kanuri* 13621! & Jekukumia [Jekekumia] River Camp, 22 Apr. 1968, *Greenway & Kanuri* 13492!; Lushoto District: W. Usambara Mts., W. Shagai Forest, 15 May 1953, *Drummond & Hemsley* 2555!
DISTR. **T** 2, 3; not known elsewhere
HAB. Epiphyte on the branches of *Halleria lucida* and *Agauria salicifolia* in mixed montane forest of *Podocarpus gracilior*, *Juniperus procera* and *Nuxia congesta*; 2100–2350 m.

74. NEPHRANGIS*

Summerh. in K.B. 2: 301 (1948)

Epiphytic, monopodial herb. Stems elongated, simple or branching, leafy. Leaves distichous, aciculiform, terete. Inflorescences abbreviated, solitary, paired or fasciculate, 1–3(–4)-flowered, emerging from the base of the leaf-sheaths, much shorter than leaves; bracts minute. Flowers small, translucent, pale brown to brownish green. Sepals and petals spreading to reflexed. Sepals ovate to almost orbicular, obtuse, laterals connate at base. Petals narrowly elliptic to linear, acute. Lip spurred, abruptly dilated from a narrow, clawed base into a broad, bilobed lamina; lobes reniform; spur narrowly conical, obtuse. Column short, without a foot or stelidia; rostellum elongated, porrect, apex shortly recurved; anther ovate, cucullate, broadly rostrate, obtuse; pollinia 2, ellipsoid, each bearing a short, spathulate-liguliform stipes; viscidium solitary, triangular, rounded, truncate.

A monotypic genus widely distributed in tropical Africa.

N. filiformis (*Kraenzl.*) *Summerh.* in K.B. 2: 302 (1948); Robyns & Tournay, F.P.N.A. 3: 537 (1955); Piers, Orch. E. Afr., ed. 2: 279 (1968); Summerh. in F.W.T.A., ed. 2, 3: 276 (1968); Williamson, Orch. SC. Afr.: 207 (1977). Type: Zaire, Njavekesi-Berge, *Stuhlmann* 2229 (B, holo.)

An epiphytic herb. Stems slender, usually branched, covered in persistent leaf-sheaths, leafy distally, pendulous, rooting at base only, 3–35(–50) cm. long. Leaves aciculiform,

* By J.J. Wood

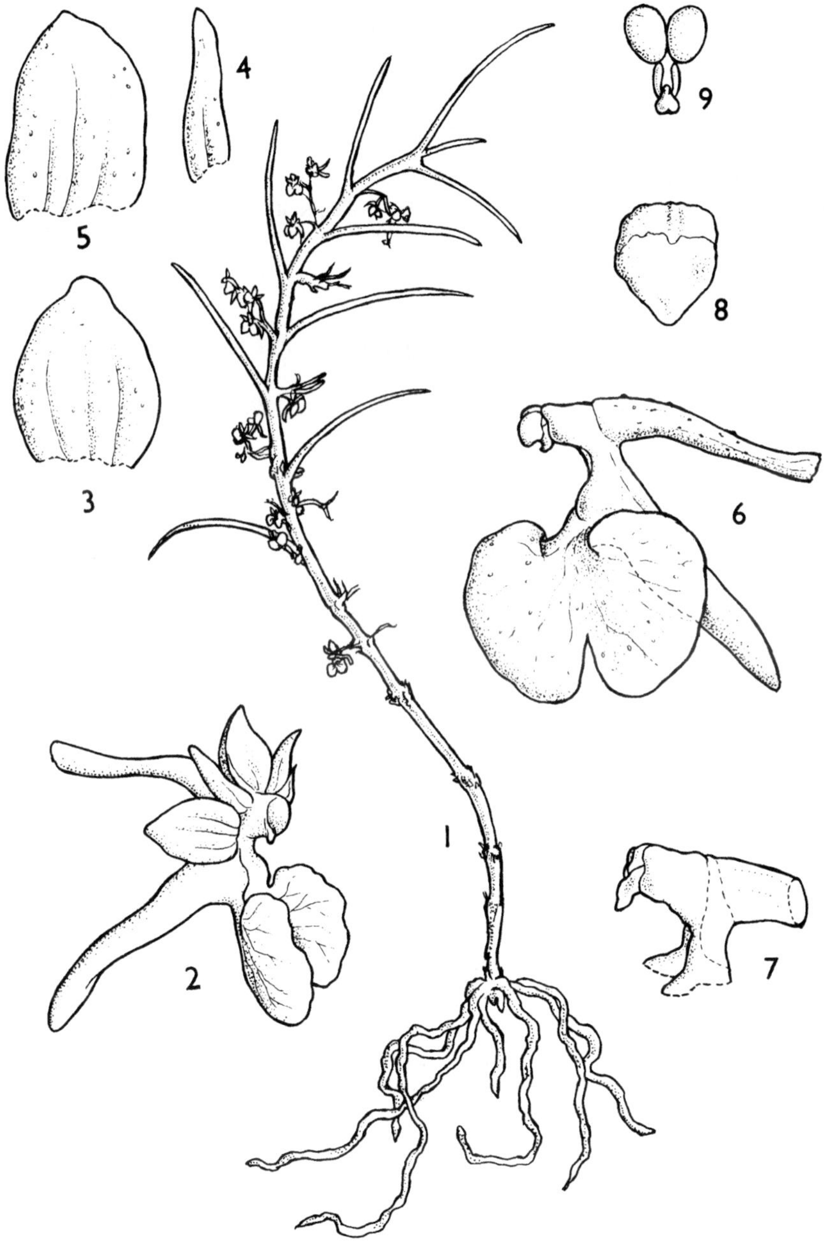

FIG. 161. *NEPHRANGIS FILIFORMIS*—**1**, habit, × ⅔; **2**, flower, × 5 ⅓; **3**, dorsal sepal, × 10; **4**, lateral sepal, × 10; **5**, petal, × 10; **6**, column and lip, × 6⅔; **7**, column, side view, × 8; **8**, anther-cap, × 14; **9**, pollinarium, × 14. 1, from *Eggeling* 5842; 2–9, from *Pfennig* 1287. Drawn by Cherry Ann Lavrih.

terete, somewhat flattened, usually arcuate, acute, glabrous or minutely pubescent at base, 1.5–9 cm. long, 0.1–0.2 cm. broad, sheaths obscurely rugulose, glabrous or minutely pubescent. Inflorescences abbreviated, 0.5–1(–1.5) cm. long, 1–3(–4)-flowered; bracts ovate, acute, amplexicaul, scarious, 0.5–1 mm. long. Flowers translucent, pale brown or brownish green, the lip creamy-white; pedicel and ovary slender, 3–4 mm. long. Sepals ovate to almost orbicular, obtuse, 1.5–2.7 mm. long, 1.5–2 mm. broad; laterals connate at base and slightly oblique. Petals narrowly elliptic to linear, acute, 1.7–2.3 mm. long, 0.4–0.6 mm. broad. Lip unguiculate and concave at base, abruptly dilated into 2 reniform, rounded lobes, total length 4–5 mm., bilobed apical lamina 3.5–5 mm. broad; spur narrowly conical, obtuse, 4–9 mm. long. Column 0.6–1.2 mm. long. Fig. 161.

UGANDA. Bunyoro District: Budongo Forest, Aug. 1935, *Eggeling* 2143!; Kigezi District: Bufumbira, Rubuguli, Oct. 1947, *Purseglove* 2540! & 8 km. behind Kanungu, 3 Sept. 1977, *Pfennig* 1287!
KENYA. N. Kavirondo District: Kakamega Forest, *Piers* (in lit.)
TANZANIA. Tanga District: near Muheza, *Piers* (in lit.)
DISTR. **U** 2; **K** 5; **T** 3; Liberia, Zaire, Zambia
HAB. On forest and woodland trees; 600–2000 m.

SYN. *Listrostachys filiformis* Kraenzl. in P.O.A.C: 158 (1895); Rolfe in F.T.A. 7: 169 (1897)
Tridactyle filiformis (Kraenzl.) Schltr. in Beih. Bot. Centr. 36: 148 (1918); Moreau in Journ. E. Afr. Nat. Hist. Soc. 17: 25 (1943)

Addenda

The following additions relate to Part 1 (Summerhayes, 1968).

THULINIA

Cribb in K.B. 40: 401 (1985)

Terrestrial orchids with underground root-stem tuberoids. Leaves basal, lanceolate, acuminate. Inflorescence erect, few-flowered. Flowers relatively showy. Sepals and petals subsimilar. Lip much longer than other segments, flabellate, bilobed, with a basal complex callus and spur united to the ovary. Column short; anther erect, bilocular stigmas very short, sessile.

A monotypic genus known only from the Nguru Mts. of E. Tanzania.

Thulinia keys out with *Brachycorythis* in Summerhayes' (1968) key to the tribe *Orchideae*. It can be readily distinguished, however, by its complex callus and spur which is fused to the ovary for its entire length.

T. albo-lutea *Cribb* in K.B. 40: 401 (1985). Type: Tanzania, Morogoro District, Nguru Mts., *Thulin & Mhoro* 3060 (K, holo.!, UPS, iso.)

An herbaceous plant up to 20 cm. tall arising from a basal tuber. Stem erect, slender, about 8 cm. long, ± 7-leaved. Leaves suberect to arcuate, narrowly lanceolate, acuminate, somewhat sheathing at the base, to 6.5 cm. long, 0.7 cm. wide. Inflorescence erect, ± 9 cm. long, few-flowered; peduncle and rhachis slender, wiry; bracts lanceolate, acuminate, up to 9 mm. long. Flowers white with a yellow mark on the lip, ascending. Dorsal sepal elliptic, rounded at apex, 6 mm. long, 3.5 mm. wide; lateral sepals oblong-elliptic, obtuse, 6.5 mm. long, 3.5 mm. wide. Petals ovate-elliptic, rounded at apex, 5.5 mm. long, 3.5 mm. wide. Lip flabellate, deeply emarginate at apex, mucronate in apical sinus, 1.2 cm. long, 0.8 cm. wide, with erose margins; spur at base of lip cylindrical, united for entire length to ovary, 4.5 mm. long; callus tapering from base to apex of lip, verruculose. Column erect, 3 mm. high.

TANZANIA. Morogoro District: Nguru Mts., Mabega Mt., near Maskati Mission, 8 June 1978, *Thulin & Mhoro* 3060! & 16 Mar. 1988, *Bidgood, Mwasumbi, Pócs & Vollesen* 520!
DISTR. **T** 6; known only from the cited collections
HAB. Upland grassland, on steep rocky slopes near summit of mountain; 1900–2100 m.

Other taxa to be added to the account in Part 1 include:

Neobolusia stolzii *Schltr.* var. **bomlylifora** *Cribb* in K.B. 34: 321 (1979)
Holothrix hydra *Cribb*, loc. cit. (1979)
Habenaria tortilis *Cribb* loc. cit. 323 (1979)
Disa ukingensis *Schltr.* in E.J. 53: 539 (1915); Cribb in K.B. 34: 325 (1979).

INDEX TO ORCHIDACEAE

GEOGRAPHICAL DIVISIONS OF THE FLORA

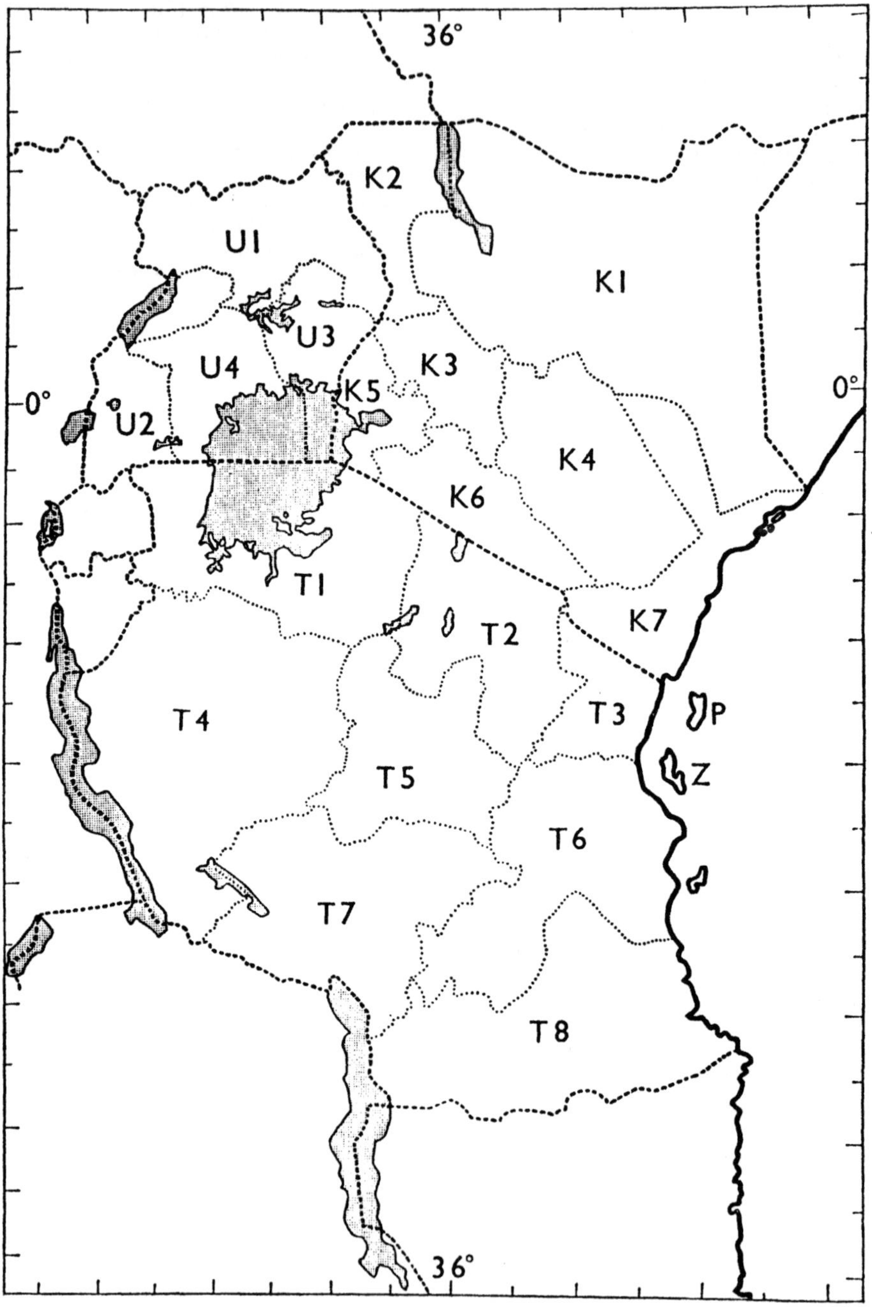

For Product Safety Concerns and Information please contact our EU representative GPSR@taylorandfrancis.com Taylor & Francis Verlag GmbH, Kaufingerstraße 24, 80331 München, Germany

Batch number: 10725802

Printed by Printforce, the Netherlands